The
Molecular Biology
of Viruses

The Molecular Biology of Viruses

Proceedings of the Symposium of the Molecular Biology of Viruses held at the University of Alberta, Canada, June 27th to 30th, 1966 in conjunction with the Faculty of Medicine of the University of Alberta.

Edited by

John S. Colter and William Paranchych
DEPARTMENT OF BIOCHEMISTRY
UNIVERSITY OF ALBERTA
EDMONTON, ALBERTA, CANADA

1967

ACADEMIC PRESS New York and London

ACADEMIC PRESS INC.
111 Fifth Avenue, New York, New York 10003

United Kingdom Edition published by
ACADEMIC PRESS INC. (LONDON) LTD.
Berkeley Square House, London W.1

LIBRARY OF CONGRESS CATALOG CARD NUMBER: 67–19847

PRINTED IN THE UNITED STATES OF AMERICA

List of Contributors

Numbers in parentheses indicate the pages on which the authors' contributions begin.

AMMANN, J., Max-Planck-Institut for Biochemistry, Munich, Germany (321)

AUGUST, J. THOMAS, Department of Molecular Biology, Albert Einstein College of Medicine, Bronx, New York (343)

BADER, J. P., Chemistry Branch, National Cancer Institute, National Institutes of Health, Bethesda, Maryland (697)

*BALTIMORE, D., Departments of Biochemistry and Cell Biology, Albert Einstein College of Medicine, Bronx, New York (375)

†BEER, HERMAN, Department of Microbial and Molecular Biology, University of Pittsburgh, Pittsburgh, Pennsylvania (251)

**BELLO, LEONARD J., Department of Microbiology, University of Pennsylvania School of Medicine, Philadephia, Pennsylvania (547)

BEN-PORAT, TAMAR, Department of Microbiology, Research Laboratories, Albert Einstein Medical Center, Philadelphia, Pennsylvania (527)

BISHOP, J. MICHAEL, Laboratory of Biology of Viruses, National Institute of Allergy and Infectious Diseases, National Institutes of Health, Bethesda, Maryland, (355)

BLACK, LINDSAY W., Department of Biochemistry, Stanford University School of Medicine, Palo Alto, California (91)

BRINTON, CHARLES C., JR., Department of Microbial and Molecular Biology, University of Pittsburgh, Pittsburgh, Pennsylvania (251)

BUCHANAN, JOHN M., Division of Biochemistry, Department of Biology, Massachusetts Institute of Technology, Cambridge, Massachusetts (31)

BURGE, B. W., Department of Bacteriology and Immunology, Harvard Medical School, Boston, Massachusetts (403)

---

* Present address: The Salk Institute for Biological Studies, San Diego, California.
† Present address: Department of Microbiology, University of Florida School of Medicine, Gainesville, Florida.
** Present address: Laboratory of Microbiology, School of Veterinary Medicine, University of Pennsylvania, Philadelphia, Pennsylvania.

BUTLER, B., Department of Biochemistry, University of Wisconsin, Madison, Wisconsin (125)

CAMPBELL, JAMES B., The Wistar Institute of Anatomy and Biology, Philadelphia, Pennsylvania (449)

CHAMPE, SEWELL P., Department of Biological Sciences, Purdue University, Lafayette, Indiana (55)

COHEN, SEYMOUR S., Department of Therapeutic Research, University of Pennsylvania, Philadelphia, Pennsylvania (3)

COHEN, STANLEY N., Department of Developmental Biology and Cancer, Albert Einstein College of Medicine, Bronx, New York (159)

*COTO, CELIA, Department of Microbiology, Research Laboratories, Albert Einstein Medical Center, Philadelphia, Pennsylvania (527)

DARNELL, J. E., Departments of Biochemistry and Cell Biology, Albert Einstein College of Medicine, Bronx, New York (375)

DEFENDI, V., The Wistar Institute of Anatomy and Biology, and the Department of Pathology, University of Pennsylvania, Philadelphia, Pennsylvania (645)

†DIGGELMANN, HEIDI, Department of Virology, Swiss Institute for Experimental Cancer Research, Lausanne, Switzerland (593)

**DOERFLER, WALTER, Department of Biochemistry, Stanford University School of Medicine, Palo Alto, California (91)

DOVE, WILLIAM F., McArdle Memorial Laboratory for Cancer Research, University of Wisconsin, Madison, Wisconsin (111)

ECHOLS, H., Department of Biochemistry, University of Wisconsin, Madison, Wisconsin (125)

††EDDLEMAN, HAROLD L., Department of Biological Sciences, Purdue University, Lafayette, Indiana (55)

EGAN, J. BARRY, Department of Biochemistry, Stanford University School of Medicine, Palo Alto, California (91)

ERDALH, WILLIAM S., Department of Bacteriology, University of Wisconsin, Madison, Wisconsin (219)

FRANCKE, B., Max-Planck-Institut for Biochemistry, Munich, Germany (321)

GINSBERG, HAROLD S., Department of Microbiology, University of Pennsylvania School of Medicine, Philadelphia, Pennsylvania (247)

---

* Present address: Faculty of Medicine, Department of Microbiology, University of Buenos Aires, Argentina.
† Present address: Department of Biophysics, University of Chicago, Chicago, Illinois.
** Present address: The Rockefeller University, New York, New York.
†† Present address: Division of Biology, California Institute of Technology, Pasadena, California.

*GIRARD, M., Departments of Biochemistry and Cell Biology, Albert Einstein College of Medicine, Bronx, New York (375)

GREEN, HOWARD, Department of Pathology, New York University School of Medicine, New York, New York (667)

GREEN, MELVIN H., Department of Biology, University of California, San Diego, La Jolla, California (139)

GRESLAND, LUCE, Research Institute for the Normal and Cancerous Cell, Villejuif, France (463)

HAREL, LOUISE, Research Institute for the Normal and Cancerous Cell, Villejuif, France (463)

HENRY, TIMOTHY J., Department of Bacteriology, University of Wisconsin, Madison, Wisconsin (219)

HOFSCHNEIDER, P. H., Max-Planck-Institut for Biochemistry, Munich, Germany (321)

HOGNESS, DAVID S., Department of Biochemistry, Stanford University School of Medicine, Palo Alto, California (91)

†HOSODA, JUNKO, Department of Biology, Massachusetts Institute of Technology, Cambridge, Massachusetts (71)

HUPPERT, JOSEPH, Gustave Roussy Institute, Department of Molecular Biology, Villejuif (Val de Marne), France (463)

HURWITZ, JERARD, Department of Developmental Biology and Cancer, Albert Einstein College of Medicine, Bronx, New York (159)

HUTCHISON, CLYDE A., III, Division of Biology, California Institute of Technology, Pasadena, California (175)

JENSEN, F., The Wistar Institute of Anatomy and Biology, Philadelphia, Pennsylvania (645)

JOKLIK, WOLFGANG K., Department of Cell Biology, Albert Einstein College of Medicine, Bronx, New York (473)

JOYNER, A., Department of Biochemistry, University of Wisconsin, Madison, Wisconsin (125)

**JUNGWIRTH, C., Department of Cell Biology, Albert Einstein College of Medicine, Bronx, New York (473)

KAESBERG, PAUL, Laboratory of Biophysics and the Biochemistry Department, University of Wisconsin, Madison, Wisconsin (241)

KAPLAN, ALBERT S., Department of Microbiology, Research Laboratories, Albert Einstein Medical Center, Philadelphia, Pennsylvania (527)

---

* Present address: Department of Microbe Physiology, Pasteur Institute, Paris, France.
† Present address: Space Sciences Laboratory, University of California, Berkeley, California.
** Present address: Department of Virology, University of Würzburg, Würzburg, Germany.

*KAPLAN, MARTIN M., The Wistar Institute of Anatomy and Biology, Philadelphia, Pennsylvania (449)

†KÁRA, JINDŘICH, Department of Virology, Swiss Institute for Experimental Cancer Research, Lausanne, Switzerland (593)

KIT, SAUL, Division of Biochemical Virology, Baylor University College of Medicine, Houston, Texas (495)

**KOCH, GEBHARD, Laboratory of Biology of Viruses, National Institute of Allergy and Infectious Diseases, National Institutes of Health, Bethesda, Maryland (355)

KOPROWSKI, HILARY, The Wistar Institute of Anatomy and Biology, Philadelphia, Pennsylvania (449)

††LEVINE, ARNOLD J., Department of Microbiology, University of Pennsylvania School of Medicine, Philadelphia, Pennsylvania (547)

LEVINTHAL, CYRUS, Department of Biology, Massachusetts Institute of Technology, Cambridge, Massachusetts (71)

LEVINTOW, LEON, Department of Microbiology, University of California School of Medicine, San Francisco, California (355)

LINDQVIST, BJÖRN H., Division of Biology, California Institute of Technology, Pasadena, California (175)

MACHATTIE, L. A., Department of Biophysics, The Johns Hopkins University, Baltimore, Maryland (9)

MAES, ROLAND F., The Wistar Institute of Anatomy and Biology, Philadelphia, Pennsylvania (449)

MAITRA, UMADAS, Department of Developmental Biology and Cancer, Albert Einstein College of Medicine, Bronx, New York (159)

MAIZEL, J. V., Department of Cell Biology, Albert Einstein College of Medicine, Bronx, New York (375)

***McCORQUODALE, D. JAMES, Division of Biochemistry, Department of Biology, Massachusetts Institute of Technology, Cambridge, Massachusetts (31)

ODA, K., Department of Cell Biology, Albert Einstein College of Medicine, Bronx, New York (473)

---

* Present address: World Health Organization, Geneva, Switzerland.

† Present address: Institute of Experimental Biology and Genetics, Czechoslovak Academy of Sciences, Prague, Czechoslovakia.

** Present address: Heinrich Pette-Institut, Hamburg, Germany.

†† Present address: Division of Biology, California Institute of Technology, Pasadena, California.

*** Present address: Department of Biochemistry, Division of Basic Health Sciences, Emory University, Atlanta, Georgia (31).

OLESON, ARLAND E., Division of Biochemistry, Department of Biology, Massachusetts Institute of Technology, Cambridge, Massachusetts (31)

PETURSSON, GUDMUNDUR, Department of Virology, Swiss Institute for Experimental Cancer Research, Lausanne, Switzerland (593)

PFEFFERKORN, E. R., Department of Bacteriology and Immunology, Harvard Medical School, Boston, Massachusetts (403)

PILARSKI, L., Department of Biochemistry, University of Wisconsin, Madison, Wisconsin (125)

PRATT, DAVID, Department of Bacteriology, University of Wisconsin, Madison, Wisconsin (219)

*RADA, B., National Institute of Allergy and Infectious Diseases National Institutes of Health, Bethesda, Maryland (427)

†RITCHIE, D. A., Department of Biophysics, The Johns Hopkins University, Baltimore, Maryland (9)

ROBINSON, W. S., Department of Molecular Biology and Virus Laboratory, University of California, Berkeley, California (681)

ROSENBERGOVA, MARTA, Institute of Virology, Czechoslovak Academy of Sciences, Mlynska Dolina, Bratislava, Czechoslovakia (463)

SAUER, G., The Wistar Institute of Anatomy and Biology, Philadelphia, Pennsylvania (645)

SCOTT, JUNE ROTHMAN, The Rockefeller University, New York, New York (211)

SHATKIN, A. J., National Institute of Allergy and Infectious Diseases, National Institutes of Health, Bethesda, Maryland (427)

SHEININ, ROSE, Ontario Cancer Institute, and the Department of Microbiology, University of Toronto, Toronto, Ontario (627)

SHUB, DAVID, Department of Biology, Massachusetts Institute of Technology, Cambridge, Massachusetts (71)

SINSHEIMER, ROBERT L., Division of Biology, California Institute of Technology, Pasadena, California (175)

SUMMERS, D. F., Department of Microbiology and Immunology, Albert Einstein College of Medicine, Bronx, New York (375)

TEMIN, H. M., McArdle Memorial Laboratory for Cancer Research, University of Wisconsin, Madison, Wisconsin (709)

---

* Present address: Institute of Virology, Czechoslovak Academy of Sciences, Bratislava, Czechoslovakia.

† Present address: Institute of Virology, University of Glasgow, Glasgow, Scotland.

TESSMAN, ETHEL S., Department of Biological Sciences, Purdue University, Lafayette, Indiana (193)

THOMAS, C. A., JR., Department of Biophysics, The Johns Hopkins University, Baltimore, Maryland (9)

TODARO, GEORGE J., Department of Pathology, New York University School of Medicine, New York, New York (667)

TZAGOLOFF, HELEN, Department of Bacteriology, University of Wisconsin, Madison, Wisconsin (219)

*WATANABE, MAMORU, Department of Molecular Biology, Albert Einstein College of Medicine, Bronx, New York (343)

WEIL, ROGER, Department of Virology, Swiss Institute for Experimental Cancer Research, Lausanne, Switzerland (593)

WEISSMANN, CHARLES, Department of Biochemistry, New York University School of Medicine, New York, New York (291)

WIKTOR, TADEUSZ J., The Wistar Institute of Anatomy and Biology, Philadelphia, Pennsylvania (449)

WILLARD, M., Department of Biochemistry, University of Wisconsin, Madison, Wisconsin (125)

WINOCOUR, ERNEST, Section of Genetics, The Weizmann Institute of Science, Rehovoth, Israel (577)

**Woodson, B., Department of Cell Biology, Albert Einstein College of Medicine, Bronx, New York (473)

ZINDER, NORTON D., The Rockefeller University, New York, New York (211)

---

* Present address: Department of Microbiology, San Francisco Medical Center, University of California, San Francisco, California.

** Present address: Departments of Medicine and Biochemistry, University of Alberta School of Medicine, Edmonton, Alberta.

Preface

This volume contains the proceedings of the Third Annual International Symposium sponsored by the Faculty of Medicine of the University of Alberta. The choice of the symposium topic, "The Molecular Biology of Viruses," was dictated, quite simply, by our own personal interests; yet it seemed to us to be one of very broad interest and significance. It has been suggested that molecular biology had its origins in investigations of virus (or more specifically, bacteriophage) systems. Although this suggestion may be regarded by some people as being altogether too sweeping, it is unquestionably true that virus systems have played a crucial role in the development of modern biology and in the evolution of the discipline of molecular biology.

The meeting was, by design, divided evenly between contributions from investigators working with bacterial and mammalian virus systems. It was hoped, thereby, to create an opportunity for an extensive and meaningful dialogue between investigators in these two areas. In this respect, we feel that the symposium was a success. It is our further hope—and expectation—that the published works will prove to be of considerable interest to the biochemist, biophysicist, geneticist, microbiologist, and virologist (molecular biologists all) working in either area.

We would like to express our gratitude to the Medical Research Council of Canada, Hoffmann-La Roche Ltd., Smith, Kline & French Inter-America Corp., The Upjohn Co., American Cyanamid Co., and Merck, Sharpe & Dohme of Canada, Ltd., whose financial support helped to make the symposium possible, and to the distinguished speakers and session chairmen, whose contributions and splendid cooperation spelled its success. Finally, a special vote of thanks must go to Laura Randall for her unselfish and tireless efforts in every phase of the planning of the symposium and the publication of its proceedings.

Edmonton, Alberta, Canada
June, 1967

JOHN S. COLTER

WILLIAM PARANCHYCH

Contents

PART III

Single-Stranded DNA Bacteriophages

PART IV

RNA Bacteriophages

PART V

Mammalian RNA Viruses

Contents

PART VI

Mammalian DNA Viruses

PART VII

Oncogenic Viruses I

PART VIII

Oncogenic Viruses II

PART I

VIRULENT BACTERIOPHAGES

Introductory Remarks

Seymour S. Cohen

DEPARTMENT OF THERAPEUTIC RESEARCH,
UNIVERSITY OF PENNSYLVANIA, PHILADELPHIA, PENNSYLVANIA

I note that the participants in this symposium have been in the field some fifteen years or less, and, therefore, are scarcely as qualified to present some sort of historical perspective as I, who can remember the good old days (defined as when the field was less crowded). However, since a symposium is best remembered by its new contributions rather than new summaries of the past, I would like to start by presenting some new observations which I hope will be relevant to virology in general. Nevertheless, I do have a few remarks as to where we have come in the last twenty years.

In 1951, I participated in a symposium on viruses and spoke on biochemical studies on the multiplication of bacterial viruses. As a very few of you may remember, by that time we had been engaged in this type of study for about five years; but, it was popular in some circles to minimize the possibilities inherent in biochemical work. Indeed, the chairman of the session had been particularly dubious of the potentialities of biochemistry in virology, and I added the following footnote to my paper:

"Some recent essays on virus multiplication have implied a vigorous separation of the levels of organization being examined by chemical and biological investigations, as well as of the deductions possible from the two lines of work. The inacceptability of this viewpoint in general is especially manifest in the field of virology where it is the task of the biochemist to study genetic duplication at molecular and intermolecular levels."

Some fifteen years later, almost everyone engaged in this field, even that chairman, now finds it necessary to become a biochemist of sorts. It is now clear to all that the biology of the viruses is peculiarly amenable to exploration at molecular and intermolecular levels; hence, the title of our present Symposium and the nature of its papers. The theoretical bases of the particular utility of biochemistry in virology derive from the biological fact, evident long before the Hershey-Chase experiment, that a virus particle loses its integrity as such when it is engaged in multiplication. In contrast to other parasitic organisms such as bacteria, which multiply when bounded by their membranes and contain much of their own metabolic machinery, viruses can complete a cycle of multiplication in a cell even after divesting their viral chromosome of a considerable assortment of coatings which facilitate initial functions, such as attachment and penetration. To meet the exigencies of a world in which molecular interactions must operate at 3 to 5 Å or less, viral genomes have learned how to leave their coats in order to interact with the enzymatic systems of a host called upon to supply both metabolites and most of the metabolic machinery. The cycle requires the reproduction of new genomes; most of the viruses we meet have even compelled the host to make new metabolic machinery, i.e., early virus-induced enzymes to help in the duplication of genomes. Finally, the cycle is completed with the repackaging of the new genomes and the release from the cell of intact virus particles.

Most of the work of the last twenty years has been concerned with the problems involved in reproducing the genomes rather than the beginning or final steps of the cycle of virus multiplication. We showed in 1946 and 1947 that T-even phage multiplication involved extensive net synthesis of DNA; in 1952, Hershey and Chase showed that viral DNA was, in fact, the bulk of the viral chromosome, which, when freed of its wrappings, controlled both DNA and protein synthesis within the cell. In 1956, it was learned that the isolated nucleic acid of certain viruses could be infectious and that the chromosome might be freed before placing it within the cell. In that same year, Volkin and Astrachan discovered the intermediate between viral DNA and protein, i.e., messenger RNA; but, they made the disastrous error of not coining the name. In 1957, we showed that a phage not only makes use of the enzymes of the cellular host to make virus, but also, that it finds it helpful to compel the infected cell to make new types of enzymes for the specialized needs of individual viruses. Following the discovery of a new DNA component by Wyatt and myself, work in our laboratory showed the biosynthetic requirements of the new pyrimidine, demonstrated the existence of an enzyme that generated the compound, and revealed the requirements for the production of the enzyme, which we then showed to be made completely *de novo*.

As the Table of Contents of this book reveals, most of the chapters are concerned with the study of the central biosynthetic steps involved in polymer duplication. In a sense, it appears that virology has in the past two decades been taken over by the more theoretically oriented biologists, who have tended to neglect those specialized processes of the cycle of virus multiplication which will possibly be more amenable to control by an enlightened medical interest. Thus, although biochemical areas, such as lipid and mucoprotein metabolism, are not covered here, we can anticipate that they will be important in such cellular functions as entry and release.

However, I wish to call your attention to the fact that we have also neglected some areas which may well be of central importance, even within the narrow limits of our research interests in viral chromosomes. I have long been bothered by references to the nucleic acids which are never encountered in nature as other than nucleates. It has occurred to me that we rarely describe salts in the literature without defining the cations. But, we do this all the time in referring to the nucleates. What is even more disturbing is that we do not appear concerned by this deficiency. Since 1957, when Hershey described the polyamines in phage (in everything but name—a deficiency later made up by Ames), it has been evident that much of the cations associated with the phage nucleates are organic bases derived from amino acids. It is known that an organism, as *E. coli,* is rich in these polyamines, putrescine and spermidine, which account, as cations, for about a fifth of all the nucleic acid phosphate in the cell. However, the most elementary physiological data concerning these substances in such a cell have been unavailable. It has not been clear, for instance, whether T-even virus multiplication involves net synthesis. This year, my colleague, Dr. Aarne Raina, and I have been exploring the physiology of the polyamines. We have begun with uninfected *E. coli* and have moved on to phage infection; I will briefly indicate some of our recent observations.

Working with the polyauxotrophic *E. coli* strain 15 TAU, we have shown that RNA synthesis is stringent in the absence of the amino acid, arginine. RNA synthesis under these conditions is relaxed by chloramphenicol and by streptomycin. In the latter instance, we have shown that such a relaxation produces 16 S ribosomal RNA, whose synthesis parallels the lethality of the aminoglycoside antibiotics. In the chloramphenicol-relaxed system, spermidine synthesis is sharply stimulated without such an effect on putrescine synthesis. In the streptomycin system, putrescine is pushed from the cell while spermidine production is stimulated. Thus, in both instances, relaxation of synthesis of ribosomal RNA by the antibiotics coincides with an increased level of spermidine to total polyamine. Which comes first, RNA or spermidine? Can spermidine be controlling synthesis of ribosomal RNA?

Adding spermidine alone to the amino-acid-deficient system, we produce a far more complete relaxation with this natural polyamine. Addition of putrescine to the relaxed system reverses the stimulation. Thus it appears that not only is the absolute level of spermidine critical but its level relative to putrescine is also important. These spermidine effects, demonstrated in *in vivo* systems, appear to be similar to *in vitro* systems with RNA polymerase, in which stalled RNA synthesis can be begun again by addition of spermidine.

In growing bacteria the increments in these polyamines seem to relate in most instances to RNA synthesis. What happens in phage infection in which RNA content is essentially constant and DNA is accumulated? As can be seen in Fig. 1, DNA synthesis is paralleled by net increments in the organic cations, putrescine and spermidine. We can say that the rates of production of both cations approach that of synthesis in the uninfected cell and are at similar ratios; however, now the production of the cations is tied to DNA synthesis instead of RNA synthesis. Preliminary experiments suggest that the absolute level of spermidine may determine the absolute level of DNA made, as well as the rapidity of lysis in T-even phage systems.

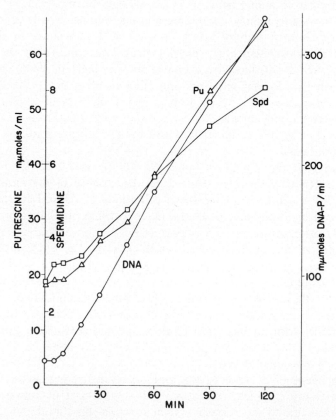

Fig. 1. The synthesis of putrescine and spermidine in T6r+-infected *E. coli* strain B (Raina and Cohen, in press).

Finally, I would note that animal cells contain spermine, as well as spermidine, and that putrescine, although a precursor of both compounds, is usually present at very low levels as compared to *E. coli*. We can anticipate, then, that the polyamine relations of growing and virus-infected animal cells will be significantly different from these relations in growing and infected *E. coli*.

The Natural History of Viruses as Suggested by the Structure of Their DNA Molecules

C. A. Thomas, Jr., D. A. Ritchie, and L. A. MacHattie*

DEPARTMENT OF BIOPHYSICS, THE JOHNS HOPKINS UNIVERSITY,
BALTIMORE, MARYLAND

History, natural or otherwise, is not essentially scientific. History may be truly beyond science; it is too difficult to separate the real events from the imagination of the historian. On the other hand, natural history is something that concerns biologists very much. They realize that their subject is more than structure as adapted to function: Every object of their study—even viruses—is the present result of a long chain of evolutionary events. Since this is so, part of this evolutionary record must be represented in the present species. Our present preoccupation with viral structure-function relationships, which are susceptible to scientific inquiry, does not absolve us from trying to generate a unified picture of their Natural History, even though this is not now susceptible to scientific test. It is unfortunate that paleontological records of microorganisms and viruses have not yet provided us with enough information to be useful in this connection—however, there is some hope for this kind of study.

With this apology let us turn to the subject at hand. I would like to tell you what little we know about the anatomy and sequences of viral DNA molecules. Then, I would like to suggest that there is a unified way of looking at the four known forms of viral DNA molecules—there is a simple rule by which one form is converted into another. Finally, I would like to ask you to make a simple assumption, for or against which there is no good evidence. This assumption leads to some unusual predictions. Among them it provides a simple scheme to account for the origin of DNA viruses.

* Present address: Department of Virology, University of Glasgow, Glasgow, W.1, Scotland.

The Anatomy of Viral DNA Molecules

WEIGHT AND LENGTH

The invention of the protein-monolayer technique for electron microscopy visualization of DNA molecules has provided a direct method by which the molecular weight of DNA may be obtained, provided the linear density of the molecules is known (Kleinschmidt *et al.*, 1961). To determine this linear density, we have measured the contour length of T2, T7, and λ DNA molecules using a "standard" protein-monolayer spreading procedure. The molecular weight of the first two of these phage DNA's has been determined by ^{32}P-radioautography, while the last was known from an empirical sedimentation-molecular weight relationship based on T2 DNA (see Thomas, 1966a). A compilation of our present results for these and other DNA's is shown in Fig. 1. As can be seen, length is a good measure of molecular weight, and there can be little doubt that these molecules are linear duplexes of polynucleotide chains over the majority of their length. The same holds for some mammalian viruses—adeno viruses, pseudo rabies, fowl pox virus and others.

CHAIN CONTINUITY AND DISCONTINUITY

This subject has a long history, and will undoubtedly reappear in unusual ways as we learn more about the anatomy of DNA. Our sedimentation and EM studies indicate that for most phage types so far studied, the majority of the DNA molecules (indeed, all but a small undetermined fraction) are composed of two continuous polynucleotide chains. This applies to T2, T4, P1, P22, λ, T3, T7, and T1 (Abelson and Thomas, 1966; Rhoades, MacHattie, and Thomas, 1966). In contrast, T5, PB, and perhaps certain of the *subtilus* phages contain specifically-located discontinuities in the component polynucleotide chains. These molecules containing specific gaps are of great interest, but, in what follows, we will be concerned only with the phages containing uninterrupted chains.

PERMUTATION AND TERMINAL REPETITION

Streisinger and his collaborators (1964) and Séchaud *et al.* (1965) have suggested that the linear chromosomes of phage T4 are circular permutations of a common sequence. This is one model which has been advanced to explain the observation that a linear DNA molecule produces a circular genetic map. Physical evidence for the permuted character of the chromosomes of the related phage T2 was provided by Thomas and Rubenstein (1964). If T2 or T4 DNA molecules are denatured by alkali, reneutralized and annealed, a large fraction of the resulting duplexes are in the form of circles (Thomas and MacHattie, 1964). This finding is in exact accord with the notion of circular permutation as depicted in Fig. 2.

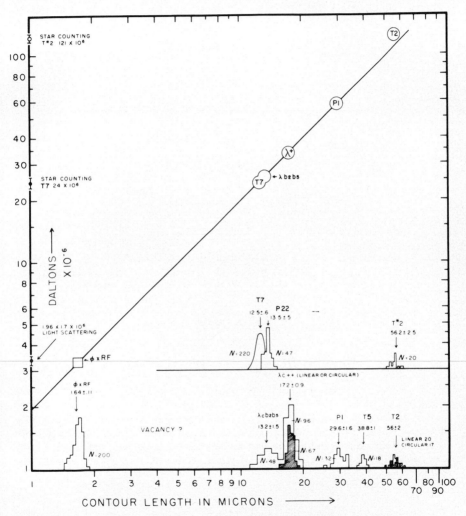

Fig. 1. The molecular weight and length of phage DNA molecules. The measured molecular lengths are plotted as small histograms on the abscissa against molecular weight derived directly or indirectly from ^{32}P-autoradiography. The straight line is drawn for a linear density of 192 daltons/Å—the value obtained from the X-ray crystallography of the B-form of DNA. This same graph is appearing elsewhere with the sources of information documented (Thomas, 1966a).

As we shall see later, these molecules begin and end with the same sequence of nucleotides—they are terminally repetitious. In the experiment depicted in Fig. 2, it can be seen that one repetitious terminal of each chain has no complement and will be left out of the circular duplex, forming a short single-chain tail. In our protein film technique these single chains, which mark the respective ends of the two component chains in

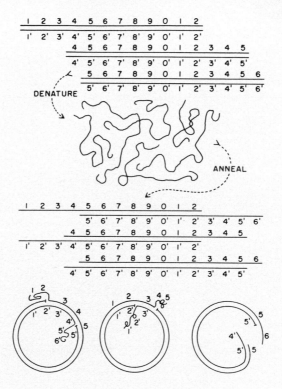

Fɪɢ. 2. Circle formation by denaturing and annealing a permuted collection of duplexes. Notice that each permutation is also terminally repetitious. One repetitious terminal from each strand cannot find a complementary partner and is left out of the circular duplex. Their separation depends on the relative permutation of the partner chains.

the circle, usually show up as ill-defined "bushes." Figure 3 shows two circles formed by denaturing and annealing T2 DNA. As can be seen, the distance along the duplex between the two bushes is variable. Since this distance depends on the relative permutation of the new partner chains, the distribution of "interbush" distances should reveal whether different permutations are equally abundant, or whether there are preferred groups of permutations. The results of such an analysis, shown in Fig. 4, indicate that there is no preferred "interbush" distance in T2. We take this to mean that all permutations are equally likely.

P22 DNA also cyclizes after denaturation. We have tentatively concluded that these molecules are a permuted collection (Rhoades, Mac-Hattie, and Thomas, 1967).

When this experiment is repeated with T3 or T7 DNA molecules, no circles are found. About 85% of the chains reform full-length linear duplexes; the remainder are fragments of various sizes (Ritchie *et al.,* 1966). This is true even though these molecules are terminally repetitious, as we will presently see. Thus, terminal repetition does not play a decisive role in circle formation by annealing denatured molecules—circle formation requires the presence of permutations.

In summary, T2, T4, and P22 are permuted, while T3, T7, λ, and T5 are nonpermuted collections of duplexes.

REPETITION AT THE ENDS

If T*2, T3, T7, or P22 DNA molecules are partially degraded by exonuclease III, a large fraction of them will form circles on subsequent annealing. We take this as evidence that these molecules are terminally repetitious (MacHattie *et al.,* 1966; Ritchie *et al.,* 1966; Rhoades *et al.,* 1967). The plan for this experiment is shown in Fig. 5.

Exonuclease III releases nucleotides in a stepwise manner from the 3' end of each of the component chains in a linear duplex molecule (see Richardson, 1966). The extent of degradation is conveniently assayed by acid-soluble radiolabel. No circles can be formed unless some degradation has occurred. The only known exceptions to this rule are the lambdoid phage DNA molecules (Hershey *et al.,* 1963; Yamagishi *et al.,* 1965).

Figure 6 shows a circular T*2 DNA molecule derived from a linear molecule by the scheme outlined in Fig. 5A. As indicated in Fig. 5, if exonuclease III degradation has proceeded beyond the length of the terminal repetition, a "gap," presumed to be single chain, would be expected. In our experiments, we found about a dozen T*2 circles that were unambiguously continuous over their entire length; the rest contained gaps of variable length. In order to obtain an objective way of judging circles, all molecules were treated as linear, and the straight line distance between the beginning and end points was measured. In case a putative circle contained 2 or more gaps, the largest one was considered for this analysis. The results can be seen in the histogram in Fig. 7A. Here it can be seen that the distribution of end-to-end distances is that expected for a

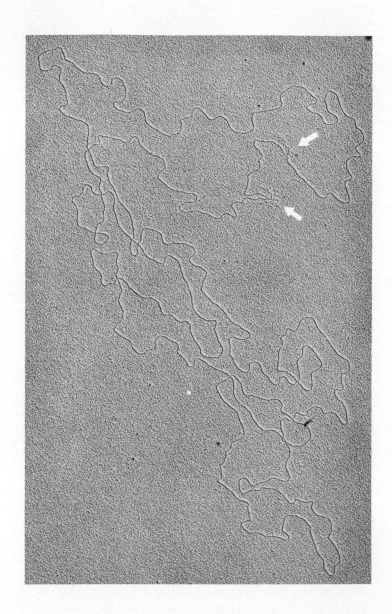

FIG. 3. Examples of denatured and annealed artificial circular T2 duplexes bearing two single-chain "bushes." These bushes (indicated by arrows) should be the repetitious terminal segments of the two component chains. Notice that the separation of the two bushes is significantly different in A and B.

A

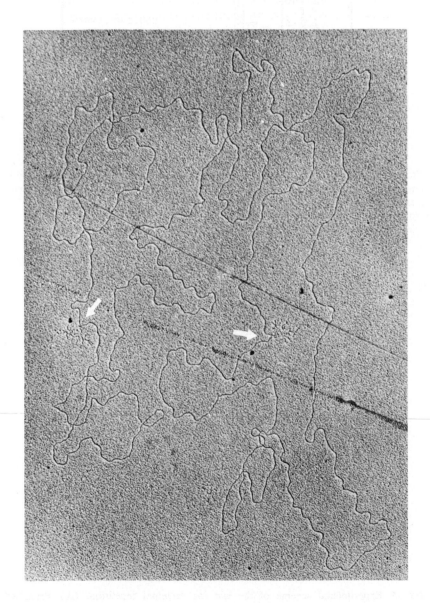

FIG. 3B.

B

FIG. 6. A circular T*2 DNA molecule produced by annealing a molecule degraded by exo III to the extent of 3.5%. Contour length 54.9 microns.

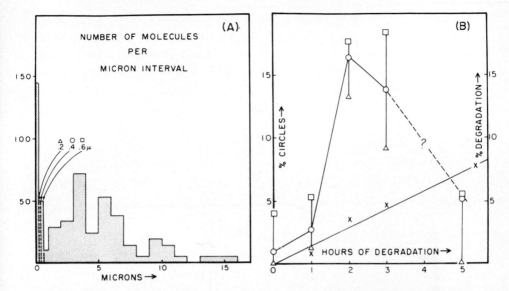

FIG. 7. The frequency of circles and their definition. (A) Histogram of end-to-end separation of all molecules seen in experiment III. This includes the undegraded control molecules. Since most of the molecules that were visually classified as circles actually contained short gaps, the longest such gap in each molecule has been treated as an end-to-end separation. This distribution is seen to depart from the theoretical expectation for linear molecules in that the frequency of near-zero values (0 to 0.6 μ) is disproportionately high. Clearly, we have the overlapping of two distributions: truly linear and truly circular molecules. For purposes of this analysis, molecules with apparent ends no more than 0.2, 0.4, or 0.6 μ apart (see arrows) are classed as circles. The consequence of these limits on the frequency of circles is shown in B. By the 0.6 μ criterion, more than 50 circles have been identified and measured. (B) Frequency of circular molecules as related to extent of exonuclease III degradation prior to annealing. (X)—fraction of nucleotides rendered TCA-soluble by exonuclease III. $\triangle$, $\bigcirc$, and $\square$ = the frequency of circles defined as molecules having apparent end separations of less than 0.2, 0.4, and 0.6 μ respectively. (See A.) The low "% circles" value at 5 hours is probably spurious.

length of the terminal repetition, as estimated by circle frequency and by direct measurement are in agreement. An estimate of the length of the terminal repetition can be derived from the genetic data of Séchaud *et al.* (1965). Agreeably enough, this is found to be about 1%.

CYCLIZATION OF T3 AND T7

Turning now to T3 and T7 DNA molecules, we see the same kind of thing: Molecules do not cyclize before exonuclease III degradation, but do so very efficiently after less than 1% of the nucleotides have been released (Fig. 8). Since these molecules are nonpermuted, all the ends within the collection are the same. Thus, not only can the two ends of a single molecule unite, but also the ends of different molecules, leading to the frequent production of dimers, trimers and higher concatemers (Fig. 9).

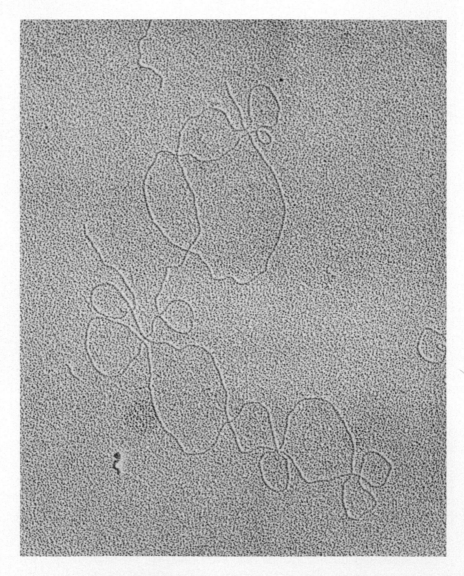

FIG. 8. T7 DNA molecules. (A) Annealed before partial exonuclease III degradation. (B) Annealed after partial exonuclease III degradation.

A

LENGTH OF OVERLAP IN T3 AND T7

As expected from Fig. 5, overdigestion reveals the terminal overlap. When as much as 6% of the nucleotides have been removed by exonuclease III, about 20–45% of the circles show pairs of gaps separated by a duplex segment 0.10 μ ± .03 SD (±.01 SEM) μ long. We interpret this as the terminal repetition: This length corresponds to 0.7% of the genome, or 260 nucleotide pairs. Figure 10 shows an example of a

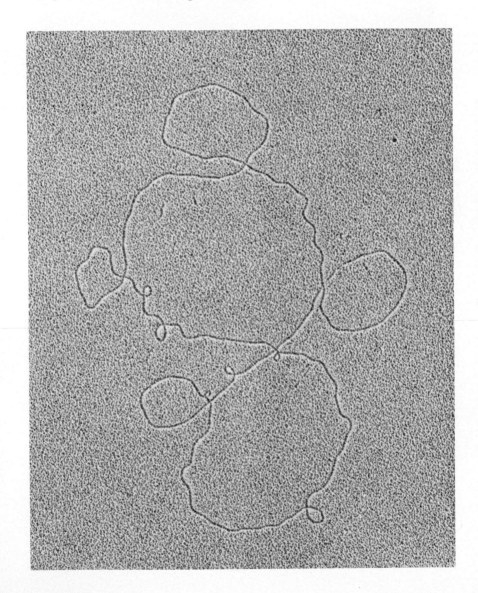

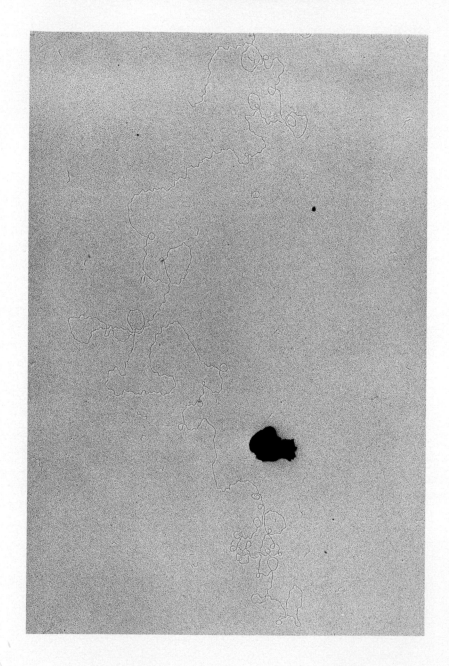

A

suspected overlap region in T7. Circles and concatemers of similarly over-degraded T3 DNA have not so far shown as high an incidence of measurable overlaps as was obtained with T7. However, the few that we have been able to measure suggest that the length of the repetition in T3 may be the same as, or slightly smaller than that in T7. Our interpretation of these overlap structures as the terminal repetition

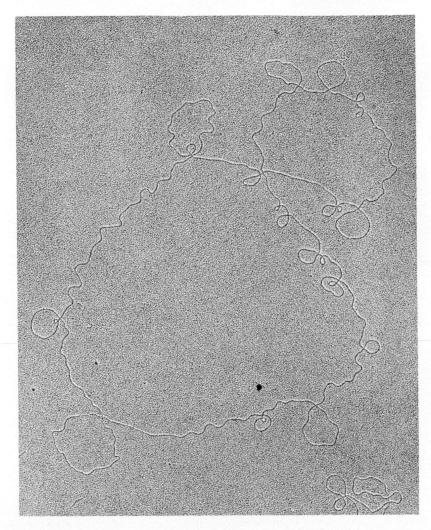

FIG. 9. Concatenates of T7 DNA molecules. (A) A fourfold linear concatenate produced by annealing following partial exonuclease III degradation. Length 49.4 μ. (B) A circular dimer produced in the same way. Length 27.4 μ. The length of T7 monomers is 12.5 $\pm$ .6 μ.

B

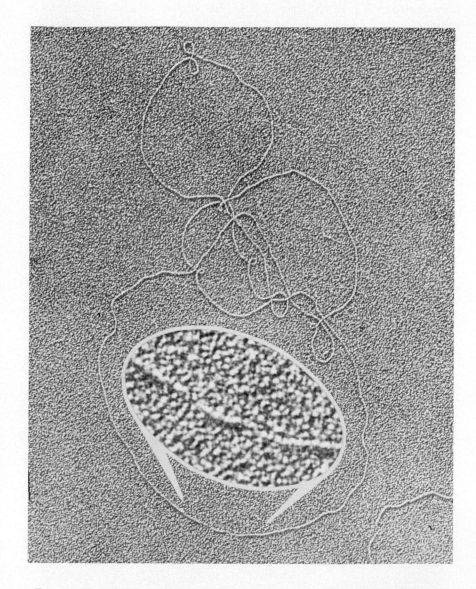

Fig. 10. The duplex terminal repetition bracketed by single chains—the putative terminal repetition. A T7 circle formed by annealing after 6.1% degradation by exonuclease III. The extent of overlap in base sequence between the two ends of the molecule should be visible as a short ($< .15 \mu$) duplex segment between two single-chain regions. The duplex segment in this case is 0.10μ long, or 0.7% of the circle's total contour length of 14.9μ.

has been strengthened by the finding of two dimers in T3 in which such an overlap occurs within 0.2 μ of the center of the molecule.

We are trying to extend these observations to other virus DNA molecules. Our present information indicates that P22 DNA can be cyclized efficiently after treatment with exonuclease III (Rhoades *et al.,* 1967). Some preliminary experiments with adeno virus DNA (in collaboration with Drs. H. Ginsberg and M. Green) indicate that these molecules are cyclizable, but with much lower efficiency.

CONTROLS

It might be thought that any exposed single chains would complex to form circles, and that cyclization is therefore not evidence for terminal repetition. To test this idea, shear-broken fragments were mixed with intact molecules, and the mixture degraded with exonuclease III and annealed. The results shown in Fig. 11 reveal that the intact T3 and T7 molecules cyclize readily after degradation, but the fragments do not. It

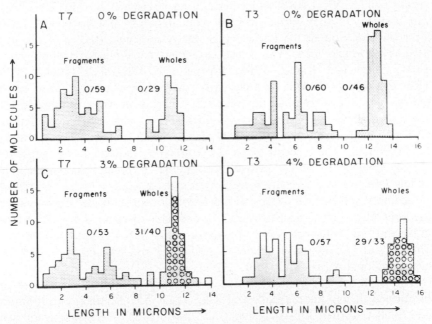

FIG. 11. Whole molecules and fragments of T3 and T7, partially degraded by exonuclease III and annealed. Length distributions measured by electron microscopy on a mixture of intact molecules with shear-produced fragments. A and B: controls, annealed only. C and D: the same mixtures, exonuclease III degraded and annealed. The circles (○) in the histogram represent circular molecules. Note that only full-length molecules form circles, and then only after partial degradation.

might be possible to reject this experiment by supposing that exonuclease III will not operate at shear-broken ends. This has been ruled out by showing that when T3 or T7 molecules are sheared into 2% fragments, as much as 35–45% of the DNA is hydrolyzed by exonuclease III. If only the original molecular ends were susceptible to this enzyme, a maximum of 2% hydrolysis would be expected. This means that cyclization requires a stronger interaction than that between noncomplementary single chains, and that only at the ends of intact molecules do we find sequences which interact this strongly. It then appears that circle formation is a true indication of terminal repetition.

In summary, we see that both permuted and nonpermuted molecules are terminally repetitious, and it seems possible that this may be a general rule for all DNA viruses.

The Basic Forms of Viral DNA Molecules

So far we have been considering only the DNA molecules from mature virus particles. Viral DNA molecules can also exist within cells. Here they can assume 3 other forms—circles, concatenates, and prophage. Soon after the injection of the linear phage DNA molecule, the molecule can assume a circular form. This is the case with λ (Young and Sinsheimer, 1964; Bode and Kaiser, 1965; Lipton and Weissbach, 1966), P22 (Rhoades et al., 1967), and perhaps other phages as well. It is most reasonable to assume that the first step in this in vivo process is the same as that which occurs in vitro during exonuclease III degradation. Cyclization after injection may result from the annealing of complementary terminals. There seems to be one difference, however—the gaps are rapidly sealed, resulting in two continuous (circular) polynucleotide chains that are topologically linked. When these molecules are examined in ordinary solvents, this leads to the formation of a twisted form, called the superhelix. This superhelical form was first discovered by Dr. Vinograd and his collaborators in polyoma virus DNA (Vinograd et al., 1965), and has also been found in φXRF (Jansz and Pouwels, 1965), and papilloma virus (Crawford, 1965).

The third form, the concatenate, is least well identified. It appears to be formed during the replication of T2, λ, and T5 (Frankel, 1966; Smith and Skalka, 1966). This structure could be formed by the joining of two or more molecules by the same process that unites the ends of the same molecule to form a circle. Alternatively, it may be a direct product of replication.

The final form, the prophage, is shown (Fig. 12) as a linear insertion in the bacterial DNA molecule. While there is no physical evidence to this point, the genetic evidence is very strong. The prophage seems to be incorporated by a single crossing-over event. If this is accomplished by the joining of complementary polynucleotide chains, then this crossing event will produce a sequence at the beginning of the prophage that is identical to a sequence at the end of it.

There is an amusing observation about these four forms: They all begin and end with the same nucleotide sequence. They are all terminally repe-

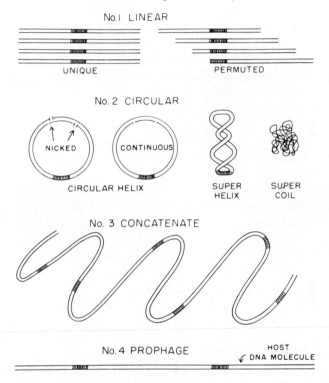

FIG. 12. The basic forms of viral DNA molecules. Notice that in all the basic forms, the viral genome begins and ends with the same sequence of nucleotides.

titious. This is true of the DNA from virus particles (1); it is true of circular molecules (2); it is true of the concatenate (3); it is true of the prophage (4). Thus, a viral DNA molecule is a parenthetical genetic sentence, bracketed by repetitions.

NONREPETITION

Having reviewed the reasons for believing that viral DNA molecules are repetitious at their terminals, it is interesting to turn this conclusion upside down and suppose that DNA molecules are nonrepetitious everywhere—except at their terminals. This has been called the "Principle of Nonrepetition" to underscore the fact that there is as yet not a shred of evidence to support it (Thomas, 1966b). The known cases of repeating sequences are associated with some degree of genetic instability. It may be that this genetic instability could provide enough selective pressure to eliminate most of the repetitions in DNA.

The "principle" of nonrepetition will always be true if one considers as the repeating unit a very long segment of DNA containing several hundreds or thousands of nucleotides; it will never be true if one considers segments only 3, 4, or 5 nucleotides long. Indeed, it is a mathematical impossibility to construct the larger phage DNA molecules nonrepetitively unless one is willing to consider blocks 9 nucleotides long. For *E. coli* one must consider at least 12 nucleotides, since there are more nucleotide pairs in *E. coli* than there are different sequences of 11 nucleotides.*

For simplicity, we suppose that no sequence 12 nucleotides long is found more than once in a genome like *E. coli*. This has many implications that need not concern us here. One thing is clear. In a long, nonrepetitive sequence, there will be frequent examples of sequences that are nearly repetitive. If one nucleotide is replaced by another by a point mutation, a formerly nonrepeated sequence is converted into one that is repeated elsewhere in the genome. Such an event produces a segment that is bracketed by repetitions. This is the basic form of the prophage (basic form 4, Fig. 12). It is suggested that this point mutation could, in a sense, "produce" a virus by violating the hypothetical rule of syntax, namely, nonreptition. Once a repetition is established, the parent genome becomes unstable, and the "normal" processes of genetic recombination (whatever these may be) will (a) delete the text between the repetitions, and (b) produce a circular DNA molecule. These events are suggested in Fig. 13. According to this suggestion, the basic form of a virus DNA molecule is the product (usually lost) of that event that produces deletion mutations.

It must be understood that the genetic text of the viral DNA molecule so created may not contain the necessary information for autonomous growth and development, but the opportunity to include useful functions is there. Meanwhile, such an episome would be highly dependent upon the functions provided by the host. Indeed, all presently known viruses retain this dependence upon their hosts to a greater or lesser degree.

---

* This calculation refers to a given sequence of 11 nucleotides and its inverted complement. Both "ordinary" and "inverted" repetitions are excluded by the "principle" of nonrepetition (see Thomas, 1966b).

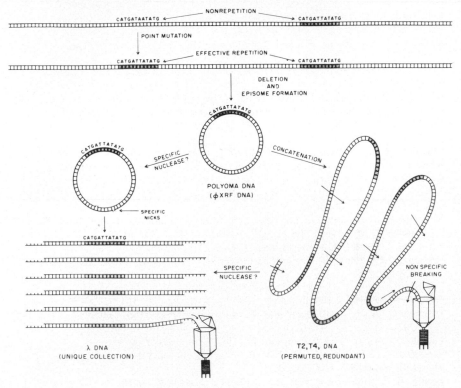

Fig. 13. Possible events leading to the production of different forms of viral DNA molecules. One pictures a DNA molecule containing two similar yet nonrepetitious 12-mers CAT GAT AAT ATG and CAT GAT TAT ATG (reading from the 5′ to 3′ on one chain). A point mutation leads to the substitution of a single base pair giving two sequences of CAT GAT AAT ATG (or two of CAT GAT TAT ATG). These repetitious 12-mers produce an instability to recombination and lead to a deletion and the possibility of episome formation. This is essentially the scheme of Campbell (1962) for prophage release and insertion, restated in chemical terms for a minimum recognition length of 12 nucleotides.

Either the circular helix can replicate as such, or multiple copies of it may concatenate for exactly the same reason that circles are formed. If these circles or concatenates are cut by a sequence-specific nuclease, a nonpermuted collection of linear duplexes can be formed. Molecules of this type are found in λ, T5, T3, and T7 phage. If the concatenate is broken on the basis of the amount of DNA only, then permuted and terminally repetitious molecules could be formed. Such molecules are found in T2, T4, and perhaps P22.

REFERENCES

Abelson, J. N., and Thomas, C. A., Jr. (1966). *J. Mol. Biol.* **18,** 262.
Bode, V. C., and Kaiser, A. D. (1965). *J. Mol Biol.* **14,** 399.
Campbell, A. (1962). *Advan. Genet.* **11,** 101–145.

Crawford, L. V. (1965). *J. Mol. Biol.* **13,** 362.

Frankel, F. R. (1966). *J. Mol. Biol.* **18,** 109, 127, and 144.

Hershey, A. D., Burgi, E., and Ingraham. (1963). *Proc. Natl. Acad. Sci. U.S.* **49,** 748.

Jansz, H. S., and Pouwels, P. H. (1965). *Biochem. Biophys. Res. Commun.* **18,** 4 and 589.

Kleinschmidt, A., Lang, D., and Zahn, R. K. (1961). *Z. Naturforsch.* **16b,** 730.

Lipton, A., and Weissbach, A. (1966). *Biochem. Biophys. Res. Commun.* **23,** 436.

MacHattie, L. A., Ritchie, D. A., Richardson, C. C., and Thomas, C. A., Jr. (1967). *J. Mol. Biol.* (in press).

Rhoades, M., and Thomas, C. A., Jr. (1966). *J. Mol. Biol.* **18,** 288.

Rhoades, M., MacHattie, L. A., and Thomas, C. A., Jr. (1967). In preparation.

Richardson, C. C. (1966). *J. Biol. Chem.* **239,** 242.

Ritchie, D. A., Thomas, C. A., Jr., MacHattie, L. A., and Wensink, P. C. (1967). *J. Mol. Biol.* (in press).

Séchaud, J., Streisinger, G., Emrich, J., Newton, J., Lanford, H., Reinhold, H., and Stahl, M. M. (1965). *Proc. Natl. Acad. Sci. U.S.* **54,** 1333.

Smith, M. G., and Skalka, A. (1966). *J. Gen. Physiol.* **49,** 127.

Streisinger, G., Edgar, R. S., and Denhardt, G. H. (1964). *Proc. Natl. Acad. Sci. U.S.* **51,** 775.

Thomas, C. A., Jr. (1966a). *J. Gen. Physiol.* **49,** 6 and 143.

Thomas, C. A., Jr., (1966b). *Prog. Nucleic Acid Res.* **5,** 315.

Thomas, C. A., Jr., and MacHattie, L. A. (1964). *Proc. Natl. Acad. Sci. U.S.* **52,** 1297

Thomas, C. A., Jr., and Rubenstein, I. (1964). *Biophys. J.* **4,** 93.

Vinograd, J., Lebowitz, J., Radloff, R., Watson, R., and Laipis, P. (1965). *Proc. Natl. Acad. Sci. U.S.* **53,** 1104.

Young, E. T., and Sinsheimer, R. L. (1964). *J. Mol. Biol.* **10,** 3.

Yamagishi, H., Nakamura, K., and Ozeki, H. (1965). *Biochem. Biophys. Res. Commun.* **20,** 727.

Control of Virus-Induced Enzyme Synthesis in Bacteria[*]

D. James McCorquodale,[†] Arland E. Oleson,[**] and John M. Buchanan

DIVISION OF BIOCHEMISTRY, DEPARTMENT OF BIOLOGY,
MASSACHUSETTS INSTITUTE OF TECHNOLOGY, CAMBRIDGE, MASSACHUSETTS

Since the initial biochemical studies by Cohen (1963) and his collaborators, investigators have come to recognize that the production of bacteriophage occurs by a sequence of very ordered processes requiring a highly controlled metabolic environment. This realization was reinforced by the finding from the same laboratory (Flaks and Cohen, 1959; Flaks *et al.,* 1959) that new enzymes, concerned with the synthesis of phage DNA and in particular with the unique pyrimidine nucleotide, 5-hydroxymethyl deoxycytidylic acid, are induced after phage infection. A study of the kinetics of the synthesis of dCMP hydroxymethylase in *E. coli* cells infected with T-even phage at $37°C$ revealed that the synthesis is initiated approximately 3 to 4 min after infection, and continues until approximately 10 to 12 min post-infection, at which time any further increase in activity abruptly ceases. At 8 to 9 min after infection, the synthesis of phage coat proteins and lysozyme begins, and continues more or less steadily until lysis occurs. On the basis of the time of synthesis of these two classes of proteins, they have been designated the "early" and "late" functions, and it has been generally accepted that these syntheses are

[*] This work was supported by grants from the National Cancer Institute (CA 02015) to J. M. Buchanan and the Institute of General Medical Sciences (GM-06298) to D. J. McCorquodale.

[†] Career Development Awardee of the National Institute of General Medical Sciences (5-K3-GM15421) on leave of absence from Department of Biochemistry, Division of Basic Health Sciences, Emory University, Atlanta, Georgia.

[**] Postdoctoral Fellow of the Institute of Allergy and Infectious Diseases (5-F2-A1-21,560-02) 1964–1966.

probably regulated by different metabolic controls. Recently, Hosoda and Levinthal (personal communication) have recognized that, very early after infection with T4 bacteriophage, a third class of proteins is formed. This new group of proteins of unknown function has been designated as Class I, the early enzymes as Class II, and the late functions as Class III proteins.

It has been possible to disrupt the organized pattern of enzyme induction and phage production by a number of methods. One of the first to be studied was irradiation of the phage with ultraviolet light prior to infection (Flaks *et al.*, 1959; Vidaver and Kozloff, 1957). As shown in Fig. 1, irradiation of the bacteriophage with an appropriate dose of ultraviolet light results in the loss of the control system responsible for the shut-off of the synthesis of the early enzymes; under certain experimental conditions, though, the initial slope of the curve showing the rate of enzyme

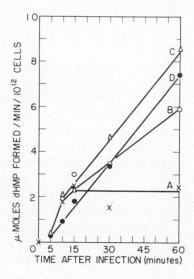

FIG. 1. The rate of synthesis of deoxycytidylate hydroxymethylase in *E. coli* B cells infected with irradiated (ultraviolet light) T2 phage (from Dirksen *et al.*, 1960). *E. coli* input, 8×10^8 cells per ml. Phage input, 4.7×10^9 particles per ml.

Designation on graph	Time of irradiation (min)	Fraction of phage particles able to form plaques after irradiation
X	0	1
○	0.5	1.2×10^{-3}
△	1	7.2×10^{-6}
●	1.5	2.8×10^{-7}

synthesis may be the same as that obtained with cells infected with un-irradiated phage (Dirksen *et al.,* 1960; Delihas, 1961).

Irradiation of the infecting phage results in other metabolic changes. The synthesis of phage DNA is not initiated at 7 min post-infection, and the production of phage coat proteins and of lysozyme (which normally begins a few minutes thereafter) is blocked (Flaks *et al.,* 1959; Vidaver and Kozloff, 1957). On the presumption that the failure to initiate the synthesis of phage DNA might be related to the extended synthesis of early enzymes, other means of blocking DNA synthesis were sought—the objective was to compare the kinetics of early enzyme synthesis under these circumstances to those seen in the case of infection with irradiated phage. Towards this end, several amber mutants of phage T4 were ob-tained from Drs. R. Edgar and R. Epstein (Epstein *et al.,* 1963), among which was a group unable to initiate DNA synthesis in the nonpermissive *E. coli* host, but able to induce the synthesis of most of the early enzymes (Wiberg *et al.,* 1962). The kinetics of the synthesis of these enzymes in cells infected with one such mutant, T4 *am* N82 (Fig. 2), resemble those

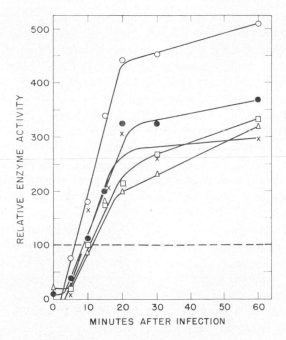

Fig. 2. Formation of enzymes after infection with T4 *am* N82. ○ = dCMP hydroxymethylase. ● = dTMP synthetase. □ = dHMP kinase. △ = dTMP kinase. × = dCTPase. The horizontal dotted line represents the level of enzyme activities 15. min after infection with T4 *am*⁺ (from Wiberg *et al.,* 1962).

observed in cells infected with irradiated T4—that is, the enzymes are synthesized in larger amounts and for a longer period of time than is the case with infection with wild type T4. However, with some of the enzymes induced by the amber mutant, cut-off of synthesis does occur at 20 min post-infection (or possibly even earlier), an observation not usually found with the irradiated phage system.

In animal cell systems, an effect similar to that observed with irradiated T4 phage has been described in the case of the induction of thymidine kinase in HeLa cells infected with UV irradiated pox virus (McAuslan, 1963). The same effect may be achieved with unirradiated pox virus if the infection is carried out in the presence of aminopterin, an inhibitor of thymidylate production and hence of DNA synthesis. These experiments taken together provide circumstantial evidence, albeit incomplete, that DNA synthesis may in some way be concerned with the shut-off of production of the "early" enzymes.

Messenger RNA—A Focal Point of Control of Enzyme Synthesis

The development of the concept of mRNA, which gives a central role in protein synthesis to a form of RNA complementary in base composition to the DNA from which it is copied, has made possible another approach to the study of the control of enzyme synthesis in E. coli after phage infection. According to the theory of Jacob and Monod (1961), control of enzyme or protein synthesis is achieved at the level of the transcription of DNA by the action of the enzyme RNA polymerase. DNA is normally kept in an inactive or repressed state through the action of a regulator gene, which controls the production of a repressor of DNA transcription. An operon may consist of several structural genes, each of which contains the information for a particular protein product. The selective derepression of the operator gene could thus result in the transcription of the structural genes of a given operon.

It might be assumed that the synthesis of proteins involved in phage multiplication is under the direction of three operons, each concerned with one of the three classes of proteins cited above. The problem, then, is how each of these operator genes is selectively switched on and off during phage replication.

Since, at present, we cannot study the control of enzyme synthesis at such a precise level (at least in this system), we have undertaken an investigation of phage-induced factors that control the general activity of DNA-directed RNA polymerase of E. coli. The experiments were undertaken in the hope that these factors will prove to be of physiological significance.

Inhibition of RNA Polymerase Activity in *E. coli* after Phage Infection

The experiments of Astrachan and Volkin (1958), in which pulses of phosphate-^{32}P were administered to phage-infected *E. coli* during 5-min intervals throughout the course of the infection, indicated that the turnover of "labeled" (messenger) RNA decreased considerably in the later stages. These experiments led to a study of the levels of RNA polymerase following bacteriophage infection.

A decrease in the activity of RNA polymerase in phage-infected *E. coli* was first observed by Khesin and co-workers (1962) in the T2 system, and by Sköld and Buchanan (1964) in the T4 system. Subsequently, it was reported that uninfected cells of *E. coli* contain a factor that exerts an inhibitory effect on the assay of purified RNA polymerase from *E. coli* (Furth and Pizer, 1966; Oleson, 1966). Recently we have found that the inhibitory activity of this material can be abolished if an ATP-generating system is included in the assay mixture. This effect is illustrated by the data summarized in Fig. 3. In all experiments to be reported here we included creatine phosphate and creatine kinase in the incubation mixture to minimize the effects of ATPases present in the crude extracts. In addi-

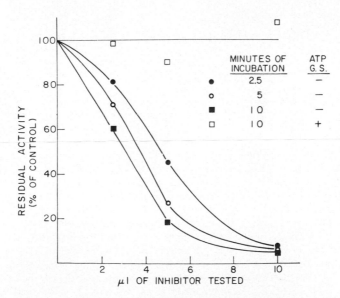

	MINUTES OF INCUBATION	ATP G.S.
●	2.5	−
○	5	−
■	10	−
□	10	+

Fig. 3. Effect of an ATP-generating system and of varying times of incubation upon the inhibition of purified RNA polymerase by an inhibitor preparation from uninfected *E. coli* B. The assay was performed as described in the legend of Fig. 4, except that the incubation was carried out at 37°C and the creatine phosphate and creatine kinase were not added to the vessels so indicated. The RNA polymerase was purified by the procedure of Chamberlin and Berg (1962).

tion, the crude extracts were assayed at 25°C, and at three levels of enzyme, to detect any possible interference by other degradative enzymes. The specific activity of RNA polymerase in crude extracts was calculated from the slope of a straight line obtained by plotting enzyme activity as a function of amount of enzyme added. An example of this assay procedure is shown in Fig. 4(a) and (b). Preincubation of the assay mixture and of the crude extracts (separately) with RNase-free pancreatic DNase was found to almost completely prevent the incorporation of radioactivity into acid-insoluble material. This result, shown in Fig. 4(c), indicates that the enzyme activity studied in crude extracts is almost entirely DNA-dependent.

The effect of infection with bacteriophages T4E, T4 *am* N82, and T4 *am* N130, on the RNA polymerase activity of *E. coli* B cells is shown in Fig. 5. In all three systems, the activity of RNA polymerase was found to decrease rapidly after infection, dropping to a level of only 10% of the activity in extracts of uninfected cells by 10 to 15 min after infec-

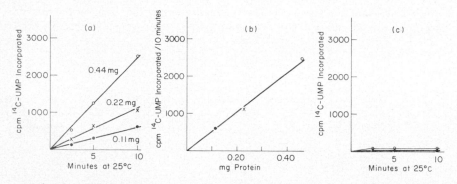

Fig. 4. (a) RNA polymerase activity as a function of time of incubation. Aliquots of a crude extract of *E. coli* B (containing 0.11, 0.22, and 0.44 mg of protein respectively) were added to an assay mixture containing 80 mµmoles each of ATP, CTP, GTP, and ^{14}C-UTP (specific activity of 1000 cpm/mµmole), 200 mµmoles of T4 DNA, 16 µmoles of Tris-Cl buffer (pH 7.9), 4 µmoles of mercaptoethanol, 1.6 µmoles of MgCl$_2$, 0.4 µmoles of MnCl$_2$, 4 µmoles creatine phosphate, 4 µg of creatine kinase, and 0.4 mg of bovine serum albumin in a final volume of 0.4 ml. Vessels were incubated at 25°C for 0, 2.5, 5 and 10 min. The reaction was terminated by chilling in ice and adding 1 ml of a cold solution containing 1 mg of bovine serum albumin and 10 µg of RNase-free DNase. After 2 min at 0°C, 1.5 ml of cold 10% TCA were added. The acid-insoluble material was sedimented, washed two times with 5% TCA, and counted in a liquid scintillation counter. (b) RNA polymerase activity as a function of amount of crude extract tested. (c) DNA dependence of the observed activity. The assay was carried out as in (a), except that the crude extract was preincubated for 1 hour at 0°C with RNase-free DNase (final concentration, 1 µg/ml), and the assay mixture was preincubated for 30 min at 25°C with 1 µg of RNase-free DNase before adding the enzyme preparation.

tion. The decrease in the RNA polymerase activity in cells infected with T4 *am* N82, a phage mutant unable to synthesize any phage DNA (Wiberg *et al.,* 1962), indicates that the inhibition is not dependent upon binding of the RNA polymerase to newly synthesized phage DNA. The prolonged synthesis of the phage-induced early enzymes that one sees with this mutant, even though the kinetics of decrease of RNA polymerase are normal, suggests that the shut-off of synthesis of the early enzymes is not a direct result of the inhibition of the RNA polymerase. The decrease in the level of RNA polymerase in *E. coli* infected with T4 *am* N130, a phage mutant unable to induce the degradation of the host DNA (Wiberg, 1966), indicates that the inhibition does not result from a deleterious effect on the enzyme of oligonucleotides derived from the host DNA.

The addition of chloramphenicol to a culture of *E. coli,* shortly after

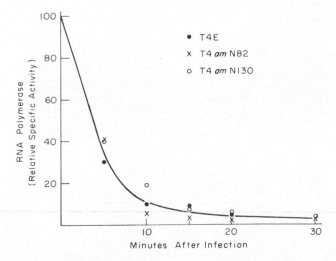

FIG. 5. Effect of bacteriophage infection on the level of RNA polymerase in crude cell extracts. Cells of *E. coli* B were grown with vigorous aeration to a concentration of 1.3×10^9 cells/ml in a modified M9 medium at 37°C. L-Tryptophan was added to the culture at a final concentration of 25 µg/ml; then, a portion of the culture was removed and chilled. The remainder of the culture was infected at an input multiplicity of 4 at zero time; and, it was superinfected at the same multiplicity 4 min later. The points on the graph indicate the times at which portions of the infected culture were removed and chilled. The chilled cells were sedimented and, then, resuspended in a small volume of Buffer A (10 m*M* MgCl₂, 10m*M* Tris-Cl (pH 7.9), $+ 10^{-4}$ *M* EDTA). The suspension was frozen in liquid nitrogen and then passed through a Hughes press. The suspension of ruptured cells was thawed, mercaptoethanol was added at a final concentration of 10 m*M,* and the mixture was centrifuged at 100,000 *g* for 2 hours to sediment cell debris and ribosomes. The extracts were assayed for RNA polymerase as described in the legend to Fig. 4.

infection with T4 *am* N82, results in an immediate cessation of the phage-induced decrease in RNA polymerase activity (Fig. 6). This effect indicates that the complete expression of the inhibition in the infected cell requires protein synthesis; it does not reflect merely a general change in the intracellular environment of the cell to one unfavorable for the stability of RNA polymerase. The abrupt change in the kinetics of decrease of RNA polymerase activity, after the addition of chloramphenicol, also suggests that the inhibition is not caused by a phage-induced enzyme acting in a catalytic manner to produce an inhibitory substance which binds to the RNA polymerase molecule. The result observed, however, is consistent with the hypothesis that the inhibitor is itself a phage-induced protein, which binds in a stoichiometric manner to the RNA polymerase molecule.

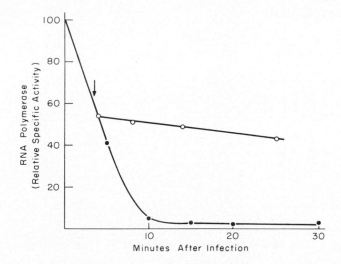

FIG. 6. Effect of chloramphenicol on the level of RNA polymerase after T4 *am* N82 infection. Solid circles: The experiment was performed as described in the legend of Fig. 5. Open circles: The experiment was performed as described in the legend of Fig. 5, except that chloramphenicol was added to the culture at a final concentration of 110 μg/ml at 3½ min after infection, as indicated by the arrow.

A further important piece of information has come from an extension of our studies to the T5 system. As shown in Table I, infection of *E. coli* F (a derivative of *E. coli* FCb) with T5 does not produce a decrease in the activity of RNA polymerase by 20 min post-infection. We have not yet had an opportunity to determine whether other conditions of assay (e.g., use of other DNA samples as primer, etc.), would have any influence on this result. However, we know that the control systems required for the production of T5 are of a complexity equivalent to those involved in the synthesis of T2 and T4.

At the moment we are unable to correlate the development of an inhibition of RNA polymerase with any one of the specific control systems concerned with the production of T2, T4, or T5. In the T5 system, where normal control is operating, there is no observed inhibition of RNA polymerase, whereas in cells infected with T4 *am* N82, the activity of RNA polymerase is inhibited under circumstances in which the control of the synthesis of early enzymes, at least, is abnormal.

The role of the inhibition of RNA polymerase in the metabolism of the phage system remains to be clarified.

TABLE I

RNA POLYMERASE ACTIVITY IN CELL EXTRACTS OF UNINFECTED AND
T5-INFECTED *E. coli* F[a]

Sources of crude extract	Relative specific activity of RNA polymerase
Uninfected cells	100
Cells infected for 20 min	98

[a] Cells of *E. coli* F were grown to a concentration of 5×10^8 cells/ml in maleate-glucose medium (MGM) at 37°C. A 100-ml sample was centrifuged and the pellet of bacterial cells was resuspended in 1.4 ml of Buffer A. The remainder of the culture was also centrifuged but then the pellet was resuspended at 0°C in buffer·Ca at 5×10^9 cells/ml. After 5 min at 37°C for temperature equilibration, sufficient T5st was added to give a multiplicity of 5, and a 10 min period was allowed for adsorption. At zero time the phage-bacterium complexes were diluted tenfold into MGM·Ca medium at 37°C, and aeration was begun. Twenty min after dilution, a 100-ml sample of the infected culture was poured over 50 gm of cracked ice, centrifuged, and resuspended in 1.4 ml of Buffer A. The uninfected and infected samples were then frozen in liquid nitrogen, passed through a Hughes press at −20°C, and assayed for RNA polymerase activity as described in the legend of Fig. 4.

The Production of T5 Bacteriophage

Studies of metabolic control systems in *E. coli* infected with bacteriophage T5 have many advantages, in spite of the fact that, hitherto, most investigations have centered around the T-even systems. One advantage of considerable importance is that the process of infection by bacteriophage T5 can be divided into a number of discrete steps (Lanni, 1960). After adsorption of the phage, 8% of its DNA is initially transferred into the host in the absence of protein synthesis, and in a temperature-dependent manner (McCorquodale and Lanni, 1964; Lanni *et al.*, 1964). If protein synthesis is blocked at this stage, the phage-bacterium complexes remain in a state of partial DNA transfer. When protein synthesis is allowed to continue, the remaining 92% of the phage DNA is transferred, but only after a 3 to 4 min period of protein synthesis. One of the proteins synthesized during this period is required for the completion of the transfer of phage DNA (Lanni, 1965). After the transfer is complete, a calcium-dependent event occurs; this is followed by the synthesis of additional phage-specific proteins, which are required for productive infection.

The DNA which is initially transferred is termed FST-DNA (for "first-step transfer" DNA). The specificity of the FST-DNA has been demonstrated in two ways. First, some phage functions [e.g., degradation of host-cell DNA (Lanni and McCorquodale, 1963), and complete transfer of phage DNA], but not others (e.g., synthesis of phage DNA) are induced by the FST-DNA. A further division of coded phage functions between the FST-DNA and the rest of the T5 DNA molecule will be demonstrated in what follows. Second, the FST-DNA carries some, but not all, genetic markers. For example, wild-type FST-DNA will complement only one of six widely spaced cistrons in the genome of a superinfecting phage (Lanni *et al.,* 1966). These combined data strongly support the conclusion of Thomas and Rubenstein (1964)—a population of T5 DNA molecules all have the same unique nucleotide sequence, and therefore do not represent circular permutations of a nucleotide sequence as is the case in phage T2.

Further Delineation of Proteins Formed during the First Step Transfer of Viral DNA

During the last year, a somewhat different approach to the study of the function of the FST-DNA was taken—namely the identification and kinetics of formation of the proteins induced by this DNA, and an investigation of the role that these proteins may play in control mechanisms of enzyme synthesis (McCorquodale, 1966).

As previously mentioned, two classes of phage-induced proteins have long been recognized, and distinguished from each other. To one class, the "early" proteins, belong the enzymes concerned with nucleotide interconversions, while the second class, which arises later in the infection, includes lysozyme and the phage structural proteins. Among the early proteins formed after infection with phage T5 (that is, after complete transfer of the viral DNA) are thymidylate kinase (Kornberg *et al.,* 1959), DNase (Paul and Lehman, 1965), thymidylate synthetase, dihydrofolate reductase (Mathews and Cohen, 1963) and DNA polymerase (Orr *et al.,* 1965). The kinetics of the synthesis of these enzymes after T5 infection are quite similar to those found after infection with the T-even phages, except that enzyme synthesis begins later (6 min), and continues for a longer period (20 min) after infection than is the case with the latter agents.

The synthesis of these two classes of proteins (early and late) in *E. coli* cells infected with phage T5 can be demonstrated readily by employing a combination of the techniques of electrophoresis in polyacrylamide gel, and radioautography (Fairbanks *et al.,* 1965). In this investigation, the synthesis of various protein components was followed by exposing the infected cells to 2 min pulses of a radioactive amino acid. After separating the individual protein components by gel electrophoresis, the concentration of radioactivity in the various bands was measured by radioautography.

When such an analysis is made of extracts of *E. coli* F cells infected with T5, it is possible to distinguish a class of proteins (Class II) whose synthesis takes place during the period between 5 to 20 min after infection, and which undoubtedly represent the early enzymes. Class III proteins (the late proteins) are formed between 12 min post-infection and the time of lysis, which comes at 45 to 60 min after infection. In addition to these two groups, still another group of proteins (Class I) can be recognized, and these are synthesized between 1 and 8 min post-infection (Fig. 7). Thus, there are 3 classes of proteins formed in *E. coli* after T5 infection, in analogy to the corresponding classes observed in the cases of infection with the T4 and SPO1 phages (Hosoda and Levinthal, and Shub and Levinthal, personal communications). The experiments described in the following sections provide information regarding some of the functions of the Class I proteins.

Proteins Formed after Infection with FST-DNA

When cells were infected by the first step transfer technique (only the FST-DNA of the phage was transferred to the host), and the subsequent synthesis of proteins was analyzed by polyacrylamide gel electrophoresis, only the proteins of Class I were found to be formed (Fig. 8). When extracts of cells that had undergone FST-DNA infection were assayed for four of the enzymatic activities mentioned above, the results indicated clearly that none of these early enzymes of Class II are formed in response to the presence of the FST-DNA.

To establish that Class I proteins are not precursors of Class II proteins, a typical "chase" experiment was performed. Fully infected cells were incubated during the first 4 min with a radioactive amino acid, at which time the radioactive substrate was diluted out with a 1000-fold

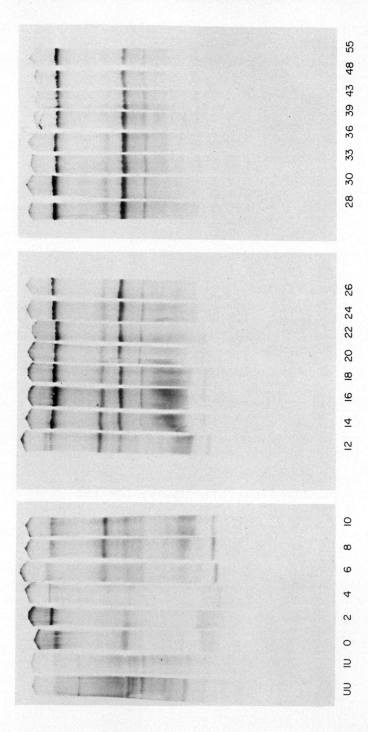

excess of nonradioactive amino acid, and the incubation continued for 6, 16, and 36 min. At these times, aliquots of cells were disrupted, and the extracts were subjected to analysis by gel electrophoresis and radioautography. Radioactivity was found only in the Class I proteins at all three times, from which one can rule out the possibility that Class II proteins arise by modification or reorganization of Class I proteins.

In another experiment, the kinetics of synthesis of total proteins, including Class I proteins, were followed by measuring the radioactivity of ^{14}C-labeled amino acids found in all proteins synthesized after infection of irradiated *E. coli* F cells with the FST-DNA of T5 phage. Irradiation of the *E. coli* F cells reduces substantially the endogenous synthesis of bac-

---

FIG. 7. Autoradiographs of sliced, dried gels obtained from gel electrophoresis of extracts of T5-infected *E. coli* cells. Cells of *E. coli* were grown to a concentration of 2×10^8/ml in a maleate-glucose medium (MGM) at 37°C, centrifuged, and resuspended at 0°C in M9 medium at 4×10^8 cells/ml. A sample was taken for the uninfected, unirradiated control. The remainder of the suspension was given a dose of UV light sufficient to reduce the fraction of surviving colony formers to 10^{-4}. This treatment reduces, but does not eliminate, host-specific protein synthesis. The irradiated cells were sedimented and resuspended at 0°C in a buffer·Ca medium at a concentration of 5×10^9 cells/ml. A sample was removed for the uninfected but irradiated control. The remainder of the cells were infected with T5st exactly as described in the legend to Table I. The phage-bacterium complexes (at 5×10^9/ml) were diluted tenfold into a MGM·Ca medium at 37°C (zero time); two-min labeling periods of the proteins with leucine-^{14}C were begun by withdrawing samples from the main culture at the times (in min) indicated by the numbers below each gel pattern and adding them to tubes containing the leucine-^{14}C for the two-min labeling period. The labeled samples were sonically disrupted and subjected to disc electrophoresis. Autoradiography was then performed on the sliced, dried gels. The gel patterns designated UU and IU represent the unirradiated-uninfected, and the irradiated-uninfected controls, respectively. Characteristic bands, which represent radioactive proteins of three classes as determined by their period of synthesis, can be identified by their R_f, sharpness, and intensity as follows.

Class	Period of synthesis	R_f	Sharpness	Intensity
I	1 to 8 min	0.075	Sharp	Heavy
		0.60	Diffuse	Light–medium
II	5 to 20 min	0.26	Sharp	Medium–heavy
		0.61	Sharp	Medium
III	12 min to lysis	0.092	Ragged	Heavy
		0.32	Sharp–ragged	Heavy
		0.40	Sharp	Light

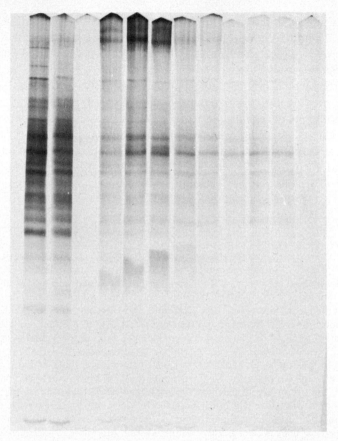

UU IU 0 2 4 6 8 10 15 25 40 60

FIG. 8. Autoradiographic pattern of proteins synthesized in *E. coli* cells after FST-DNA (T5) infection. Cells were grown, irradiated, and infected exactly as described in the legend to Fig. 7. The phage-bacterium complexes (in buffer·Ca at 5×10^9/ml) were diluted twofold into ice cold buffer·Ca, and blended for 6 min in an Omnimixer. The blended phage-bacterium complexes (at 2.5×10^9/ml) were diluted fivefold into MGM·Ca medium at $37°C$, and two-min labeling periods with leucine-^{14}C were begun at the times (in min) indicated below each gel pattern by the method described in the legend to Fig. 7. The gel patterns designated UU and IU have the same connotation as in Fig. 7.

terial proteins, and makes it possible to obtain a clearer picture of the synthesis of the phage-induced components. From the data summarized in Fig. 9, it may be seen that the incorporation of ^{14}C-amino acids into protein occurs during the first 8 min after infection, and ceases thereafter. It is also clear that there is a drastic reduction of bacterial protein synthesis during this time.

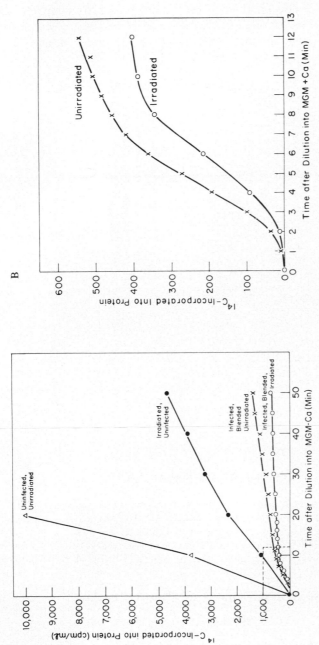

FIG. 9A. The incorporation of leucine-^{14}C into the proteins of *E. coli* F, and *E. coli* F infected with first step transfer-DNA of phage T5st. Cells were grown and, where indicated, were irradiated, infected, and blended as described in the legend to Fig. 8. After dilution into MGM·Ca medium containing leucine-^{14}C, one-ml samples were removed at intervals, and the radioactivity in the hot acid-insoluble material was determined. FIG. 9B. This figure is an enlargement of part of the lower two curves shown in Fig. 9A. Hence, the upper curve represents the unirradiated cells and the lower curve the irradiated cells; both types of cells have been infected only with FST-DNA.

In the experiments with irradiated bacteria infected with FST-DNA, radioautographs were also obtained (Fig. 10). In the nearly complete absence of synthesis of background host proteins, it is possible to see clearly three bands, representing at least three, and probably more, induced proteins of Class I.

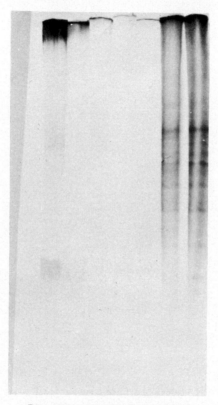

B3   B8 B13 B18 B33 UU UI

FIG. 10. Autoradiographic pattern of the proteins produced after infection of heavily irradiated host cells with first step transfer-DNA. Experimental conditions were identical to those described in the legend to Fig. 8, except that the host cells were irradiated until the fraction of surviving colony-formers was 10^{-7}. Two-min labeling periods were begun at the times (in min) indicated below each gel pattern.

Control Systems Related to Class I Proteins

CESSATION OF SYNTHESIS OF BACTERIAL HOST PROTEINS

The drastic reduction of total protein synthesis which is observed when cells are infected with FST-DNA (see Fig. 9a), suggests that one role of this DNA is the production of a substance responsible for the shut-off of synthesis of bacterial proteins.

CESSATION OF PRODUCTION OF CLASS I PROTEINS

Experiments with Chloramphenicol

The FST-DNA also carries the genetic information for the synthesis of a protein which shuts off the production of Class I proteins. Evidence which supports this statement was obtained from experiments involving the use of chloramphenicol. Cells infected with FST-DNA were incubated with complete medium for 10 min at 37°C in the presence of chloramphenicol (Fig. 11). The cells were then centrifuged, resuspended in complete medium (in the absence of chloramphenicol), and were incubated for an additional period at 37°C. Protein synthesis was examined during both incubation periods by the previously described techniques (2 min pulses with leucine-^{14}C, gel electrophoresis, radioautography). As expected, the synthesis of all proteins was prevented throughout the initial 10 min incubation period (that is, in the presence of chloramphenicol); but, synthesis of Class I proteins could still be initiated and then shut off after 8 min of the second incubation period. These experiments support the hypothesis that chloramphenicol prevents the formation of a protein that shuts off the synthesis of Class I proteins.

Superinfection of FST-DNA Infected Cells

The finding that the FST-DNA contains the genetic information for regulating the shut-off of synthesis of Class I proteins, prompted us to carry out additional experiments designed to clarify this point. If the FST-DNA codes for a protein that irreversibly cuts off the synthesis of Class I proteins, the introduction of the complete T5 genome by superinfection of the FST-DNA-infected cells should lead to no further synthesis of Class I proteins during the second infection period.

E. coli F cells were infected with FST-DNA, and were incubated for 10 min at 37°C in complete medium. The cells were then permitted to undergo full infection (transfer of the total viral genome) by the addition of T5, and measurements were made of the appearance of enzyme activity (Fig. 12). Radioautography, after gel electrophoresis (Fig. 13A and B), was also carried out on extracts of cells that had been exposed to radioactive leucine for two-min intervals during the period of superinfection. These data were compared to data obtained from cells that were infected with FST-DNA, but kept at 0°C for a 10-min period before

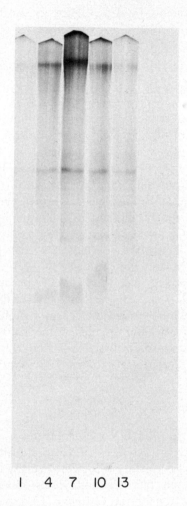

FIG. 11. Effect of chloramphenicol on the "shut off" of proteins induced by FST-DNA in *E. coli* F. Cells were grown, infected with FST-DNA, and blended as described in the legend to Fig. 8, except that the cells were not irradiated. The blended complexes (at 5×10^8/ml) were incubated at $37°C$ with aeration for 10 min in an MGM·Ca medium containing chloramphenicol (50 μg/ml). The complexes were quickly cooled, sedimented, and resuspended at 5×10^9/ml in chilled MGM·Ca medium lacking chloramphenicol. Two-min labeling periods were then begun after a tenfold dilution into MGM·Ca medium at $37°C$ as described in Fig. 8. The numbers under each gel pattern have the same meaning as described in the legend to Fig. 7.

being superinfected and shifted to a temperature of 37°C (at 0°C the cells are unable to synthesize the Class I proteins that are under the control of the FST-DNA). The kinetics of formation of one early enzyme of the Class II proteins, thymidylate synthetase, are shown in Fig. 12. The initiation of synthesis of this enzyme in cells, that had been infected with FST-DNA and incubated for 10 min at 37°C, was shown to take place almost immediately after the introduction of the complete T5 DNA molecule by superinfection. In contrast, the initiation of enzyme synthesis in those cells that had been held for a period of 10 min at 0°C, after FST-DNA infection, was found to be delayed for 12 min after superinfection.

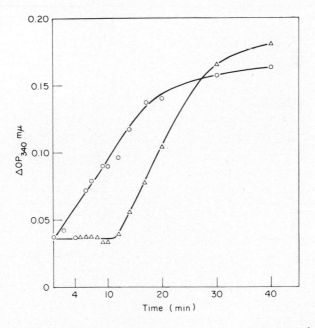

FIG. 12. Effect of a previous infection by first-step transfer-DNA on the formation of thymidylate synthetase after infection with T5 phage. Cells were grown, infected with FST-DNA and blended as described in Fig. 11. One portion of the blended phage-bacterium complexes was incubated for 10 min in MGM·Ca medium at 37°C with aeration at $5 \times 10^{\text{x}}$/ml, sedimented, and resuspended in buffer·Ca. Another portion was sedimented immediately, resuspended in buffer·Ca, and kept at 0°C. Each portion of the blended complexes (one with and one without presynthesized proteins of class I) was then superinfected with T5st phage, diluted into MGM·Ca medium at 37°C, and two-min labeling periods were begun as described in Fig. 8. Thymidylate synthetase activity was assayed by the procedure of Friedkin (1963). ○, cells kept at 37°C during infection with FST-DNA; △ cells kept at 0°C during infection with FST-DNA.

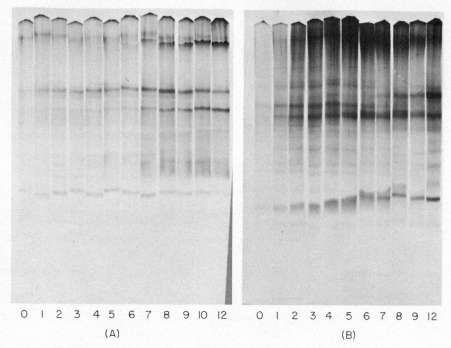

0 1 2 3 4 5 6 7 8 9 10 12   0 1 2 3 4 5 6 7 8 9 12

(A)   (B)

Fig. 13. Effect of previous infection by first-step transfer-DNA on the autoradio-graphic pattern of proteins formed after infection with T5st phage. The labeled proteins in samples of the sonically disrupted cells from both superinfected cultures described in the legend of Fig. 12 (see conditions A and B) were subjected to disc electrophoresis and autoradiography. The numbers under each gel pattern have the same meaning as in Fig. 7. (A) Cells kept at 37°C during infection with FST-DNA. (B) Cells kept at 0°C during infection with FST-DNA.

The extraordinarily long period required for the appearance of the first enzyme activity is consistently seen when a full infection is made of un-infected cells that have undergone treatment in a Waring blender.

The conclusions drawn on the basis of these enzyme assays were con-firmed by the radioautograms shown in Figs. 13A and B. Class II pro-teins were produced immediately upon superinfection of cells that had been incubated with FST-DNA for 10 min at 37°C. In contrast, when cells were infected with FST-DNA, but kept at 0°C until superinfection, there was a delay of several minutes before the synthesis of Class II pro-teins began.

The virtual absence of radioactivity in Class I proteins in those cells, which had been incubated with FST-DNA at 37°C prior to superinfection and pulse labeling (see Fig. 13A), leads to the conclusion that the incu-bation at 37°C with FST-DNA resulted in the production of a substance responsible for the permanent cut-off of synthesis of Class I proteins. The results of the previous experiment with chloramphenicol make it seem likely that this material is a protein. A second, but tentative, suggestion arose from the superinfection experiments: A substance is produced during the infection with FST-DNA that sets the stage for the immediate synthesis of Class II proteins, when the genetic material carrying the code for proteins of this class finally finds its way into the bacterial cell.

Discussion

Several hypotheses have been advanced to explain the control of the synthesis of phage-induced proteins after phage infection. These hypotheses include control at the level of (1) transcription of specific segments of the DNA molecule, (2) translation of specific messenger RNA's, (3) regulation of total mRNA synthesis, and (4) regulation of messenger RNA destruction. Some investigators have felt that one or another of these hypotheses is adequate to explain the degree of control observed; others have postulated that the overall control systems may represent the interplay of controls operating at more than one of the levels indicated above.

The theories of Jacob and Monod (1961) would suggest that control mechanisms operate principally at the level of transcription of specific DNA segments. On the other hand, Stent (1964), Ames and Hartman (1963), and Bautz *et al.* (1966) have postulated that the regulation of protein synthesis may occur at the level of translation. Each has presented experimental evidence supporting his particular variation of this hypothesis. A specific and attractive suggestion bearing on this hypothesis has been made by Sueoka and Kano-Sueoka (1964), who detected modifications of Leucyl tRNA after T2 and T4 infection, but not after T3, T5, T7, and λ infection (Kano-Sueoka and Sueoka, 1966). As yet, however, no connection has been established between the change in transfer RNA patterns and the regulation of the synthesis of a specific protein.

The possible role of a general inhibitor of RNA polymerase activity, as a regulator of the total amount of messenger RNA produced after T2 and T4 infection, has also been discussed (Sköld and Buchanan, 1964; Guthrie and Buchanan, 1966). At present, it is difficult to reconcile the strong inhibition of RNA polymerase, which is observed at 10 to 15 min after infection both *in vivo* and *in vitro*, with the observation that early and late messenger RNA's are formed at a relatively significant rate during the later stages of the infection period as measured by the hybridization technique (Hall *et al.*, 1964; Bautz *et al.*, 1966). Bautz *et al.* (1966), expressed their findings of "deletion specific RNA" in terms of "percent of total hybridizable RNA." In view of the low rate of RNA turnover and synthesis *in vivo,* noted by Astrachan and Volkin (1958) in the later stages of infection, it would have been of interest to determine the relative quantities of each messenger RNA at the different times at which pulse labeling was carried out. An analysis of the kinetics of early enzyme synthesis in the systems used by Bautz *et al.* (1966) and by Hall *et al.* (1964) would have been helpful in interpreting the data obtained from their experiments concerning the hybridization of messenger RNA, especially since the exact cut-off time of early enzyme synthesis may vary with the different experimental conditions used in various laboratories.

The experiments herein reported on the effect of T5 infection on the metabolism of *E. coli* support the following conclusions: (1) A new class of proteins (Class I), distinct from early enzymes and late proteins, is produced after T5 infection of *E. coli* F. Proteins of this class are synthesized from 1 to 8 min after infection and, thus, represent the earliest phage-specific proteins synthesized in the infected cell; (2) the proteins synthesized during this time contain a component that is responsible for the shut-off of both bacterial protein synthesis and of synthesis of Class I proteins themselves; (3) it is possible that a protein required for the initiation of synthesis of Class II proteins is present in the Class I group; (4) the 8% section of T5 DNA that initially penetrates the *E. coli* cell (FST-DNA) is responsible for the synthesis of Class I proteins.

In summary, it is believed that the eventual explanation of the operation of metabolic control systems of phage-infected cells will have to take into account, or at least consider further, the following points: (1) There are now at least three classes of phage-induced proteins in addition to the bacterial proteins whose regulation must be examined. It may be an oversimplification to assume that one principle or factor will account for all situations. It is possible that the selective synthesis and utilization of messenger RNA will be involved at one stage or another of phage production. A further examination of factors concerned with DNA transcription and messenger RNA translations is required—in particular the role of variations in transfer RNA patterns after phage infection; (2) the quantitative aspects of RNA production, and hence of inhibitors of RNA synthesis, may have importance at least in the T2 and T4 systems, if not in the T5 system. The limitation of the activity of RNA polymerase during the course of phage infection may possibly vary with respect to the nutritional state, or the environment of the microorganism, or to other, presently unknown factors. Hybridization experiments with messenger RNA, although yielding valuable information, should be coupled with other experimental approaches that correlate the synthesis of RNA with the appearance of enzyme activities or specific protein components; (3) DNA synthesis may play a role in the cut-off of Class II protein synthesis and in the initiation of the synthesis of Class III proteins. Sufficient account has not been taken of the reactions that permit the net accumulation of ribonucleotides for messenger RNA synthesis. Thus, the huge drain on the ribonucleotide pool, resulting from the initiation of phage DNA synthesis, may be responsible for the restriction of messenger RNA synthesis, particularly under circumstances where the production of ribonucleotides might be limited.

In view of the uncertainties and complexities of the general problem of control of the synthesis of the phage-induced products, it is probably not wise at this time to make too rigid an evaluation of any of the approaches taken or of the factors involved. However, there is a conviction that an understanding of the problems of regulation of protein synthesis in this system will have ramifications in other, still more complex, areas of biology.

REFERENCES

Ames, B. N., and Hartman, P. (1963). *Cold Spring Harbor Symp. Quant. Biol.* **28**, 349.

Astrachan, L., and Volkin, E. (1958). *Biochim. Biophys. Acta* **29**, 536.

Bautz, E. K. F., Kasai, T., Reilly, E., and Bautz, F. A. (1966). *Proc. Natl. Acad. Sci. U.S.* **55**, 1081.

Chamberlin, M., and Berg, P. (1962). *Proc. Natl. Acad. Sci. U.S.* **48**, 81.

Cohen, S. S. (1963). *Federation Proc.* **20**, 641.

Delihas, N. (1961). *Virology* **13**, 242.

Dirksen, M. L., Wiberg, J. S., Koerner, J. F., and Buchanan, J. M. (1960). *Proc. Natl. Acad. Sci. U.S.* **46**, 1425.

Epstein, R. H., Bolle, A., Steinberg, C. M., Kellenberger, E., Boy De La Tour, E., Chevalley, R., Edgar, R. S., Susman, M., Denhardt, G. H., and Lielausis, A. (1963). *Cold Spring Harbor Symp. Quant. Biol.* **28**, 375.

Fairbanks, G., Jr., Levinthal, C., and Reeder, R. H. (1965). *Biochem. Biophys. Res. Commun.* **20**, 393.

Flaks, J. G., and Cohen, S. S. (1959). *J. Biol. Chem.* **234**, 1501.

Flaks, J. G., Lichtenstein, J., and Cohen, S. S. (1959). *J. Biol. Chem.* **234**, 1507.

Friedkin, M. (1963). *Methods Enzymol.* **6**, 128–129.

Furth, J. J., and Pizer, L. I. (1966). *J. Mol. Biol.* **15**, 124.

Guthrie, G. D., and Buchanan, J. M. (1966). *Federation Proc.* **25**, 864.

Hall, B. D., Nygaard, A. P., and Green, M. H. (1964). *J. Mol. Biol.* **9**, 143.

Hosoda, J., and Levinthal, C. (1965). Personal communication.

Jacob, F., and Monod, J. (1961). *J. Mol. Biol.* **3**, 318.

Kano-Sueoka, T., and Sueoka, N. (1966). *J. Mol. Biol.* **20**, 183.

Khesin, R. B., Shemyakin, M. F., Gorlenko, Zh. M., Bogdanova, S. L., and Afanas'eva, T. P. (1962). *Biokhimiya* **27**, 1092.

Kornberg, A., Zimmerman, S. B., Kornberg, S. R., and Josse, J. (1959). *Proc. Natl. Acad. Sci. U.S.* **45**, 772.

Lanni, Y. T. (1960). *Virology* **10**, 514.

Lanni, Y. T. (1965). *Proc. Natl. Acad. Sci. U.S.* **53,** 969.

Lanni, Y. T., and McCorquodale, D. J. (1963). *Virology* **19,** 72.

Lanni, Y. T., McCorquodale, D. J., and Wilson, C. M. (1964). *J. Mol. Biol.* **10,** 19.

Lanni, Y. T., Lanni, F., and Tevethia, M. J. (1966). *Science* **152,** 208.

McAuslan, B. R. (1963). *Virology* **20,** 162.

McCorquodale, D. J. (1966). *Federation Proc.* **25,** 651.

McCorquodale, D. J., and Lanni, Y. T. (1964). *J. Mol. Biol.* **10,** 10.

Mathews, C. K., and Cohen, S. S. (1963). *J. Biol. Chem.* **238,** PC853.

Oleson, A. E. (1966). *Federation Proc.* **25,** 275.

Orr, C. W. M., Herriott, S. T., and Bessman, M. J. (1965). *J. Biol. Chem.* **240,** 4652.

Paul, A. V., and Lehman, I. R. (1965). *Federation Proc.* **24,** 287.

Shub, D., and Levinthal, C. (1965). Personal communication.

Sköld, O., and Buchanan, J. M. (1964). *Proc. Natl. Acad. Sci. U.S.* **51,** 553.

Stent, G. S. (1964). *Science* **144,** 816.

Sueoka, N., and Kano-Sueoka, T. (1964). *Proc. Natl. Acad. Sci. U.S.* **52,** 1535.

Thomas, C. A., Jr., and Rubenstein, I. (1964). *Biophys. J.* **4,** 93.

Vidaver, G. A., and Kozloff, L. M. (1957). *J. Biol. Chem.* **225,** 335.

Wiberg, J. S. (1966). *Proc. Natl. Acad. Sci. U.S.* **55,** 614.

Wiberg, J. S., Dirksen, M. L., Epstein, R. H., Luria, S. E., and Buchanan, J. M. (1962). *Proc. Natl. Acad. Sci. U.S.* **48,** 293.

Polypetides Associated with Morphogenic Defects in Bacteriophage T4

Sewell P. Champe and Harold L. Eddleman

DEPARTMENT OF BIOLOGICAL SCIENCES, PURDUE UNIVERSITY, LAFAYETTE, INDIANA

Introduction

The presence of low molecular weight substances in T2 bacteriophage particles was first reported by Hershey (1957). He found two compounds derived from arginine, which were later identified by Ames *et al.* (1958) as spermidine and putrescine, and another acid-soluble component which yielded amino acids upon hydrolysis and was concluded to be a peptide. In the course of searching for components in T4-infected cells which might be involved in phage maturation we encountered a set of components which corresponded, in part, to the peptide fraction detected by Hershey in phage particles. The details of this work have been reported elsewhere (Eddleman and Champe, 1966). In this chapter, we will summarize the evidence indicating the involvement of these polypeptides in the maturation process, and we will include some more recent information on their composition and properties.

Infection of *E. coli* by T4 Induces the Appearance of Several Acid-Soluble Components

Among the conditionally lethal (amber) mutants described by Epstein *et al.* (1963), there exists one which appears to be blocked in the synthesis of all of the late appearing proteins which compose the phage particle, but which is unaffected in the synthesis of the early appearing enzymes involved in DNA synthesis. The possibility that this mutant, *am*N134, might also fail to induce the synthesis of components involved in phage assembly processes (perhaps small molecules not necessarily incorporated into the phage particle) suggested a comparison of cell extracts of the nonpermissive host BB infected with this mutant and with the wild-type *am*+. Figure 1 shows a chromatographic comparison of the TCA-soluble extracts of infected cells which had been labeled with lysine-^{14}C, beginning at 10 min after infection. The chromatogram shows two strongly

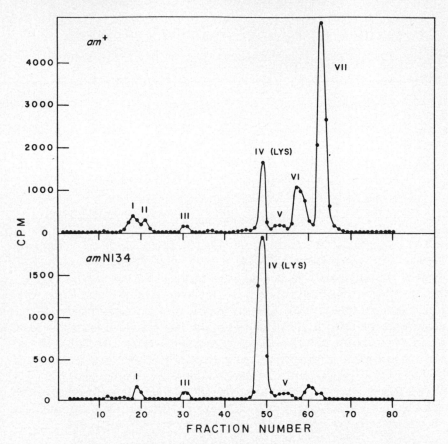

FIG. 1. Chromatographic comparison of the TCA-soluble fraction of *am*⁺-infected cells and *am*N134-infected cells on Dowex 50-X2. Five ml cultures of *E. coli* BB at 37°C were infected at a multiplicity of 5 phage per bacterium and labeled with 2 μCi of lysine-¹⁴C at 100 μCi/μmole beginning at 10 min after infection. At 30 min after infection the infected cells were chilled, sedimented, extracted with TCA, and the acid-soluble material applied to a 0.6 × 120 cm column of Dowex 50-X2. Elution was performed with a gradient from 0.1 *M* pyridinium acetate, pH 3.6, to 2.0 *M* pyridinium acetate, pH 4.8, as described in detail by Eddleman and Champe (1966).

labeled components, VI and VII, which are present in *am*⁺ infected cells but absent from *am*N134 infected cells. The weakly labeled component II is also absent from the mutant-infected extract. However, this component has been found to be highly variable in the wild-type and, therefore, it cannot be assayed with confidence. The components present in both chromatograms (I, III, and V) are also found in uninfected cell extracts. Component IV is free lysine; the component which elutes between VI and VII (seen in the analysis of the mutant extract) is found as well in uninfected cells. Analysis of the acid-soluble fraction of infected cells by chromatography on Dowex 50-X2, as shown in Fig. 1, has proved the most reproducible and convenient of several methods tried. It has been used for all of the quantitative assays of components VI and VII described below.

The fact that components VI and VII fail to appear in cells infected with *am*N134 suggests that these are substances which appear in wild-type infected cells sometime after 8 min, which is the time at which phage-precursor protein synthesis begins. To confirm this and to define more closely the time of appearance, a culture of *am*+ infected cells was continuously labeled with an excess of lysine-^{14}C from the time of infection; thereafter, it was sampled at various times for TCA extraction and analysis. As shown in Fig. 2, both of the phage-induced components first

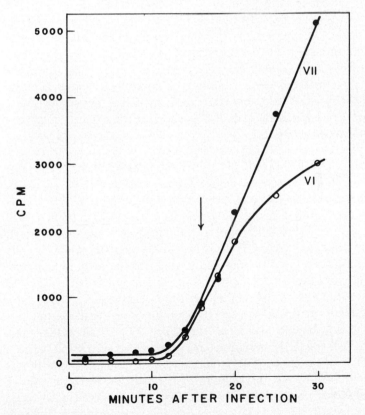

Fig. 2. The time of appearance of components VI and VII. A 60 ml culture of *E. coli* BB at 37°C was infected with T4D and labeled with 25 μCi of lysine-^{14}C at 16 μCi/μmole beginning at the time of infection. Five ml samples were taken at various times for TCA extraction and analysis. The arrow indicates the time (16 min) at which the cells contained an average of one progeny particle.

appear at 11 min after infection. Their kinetics differ, however; while VII accumulates at a linear rate until at least 30 min, the rate of accumulation of VI decreases after 20 min.

Attempts to demonstrate the components in cells opened by sonication prior to 10 min and in cells labeled prior to infection were unsuccessful. These components are not derived from material of bacterial origin, but they have, rather, the characteristics of products controlled by "late" phage genes.

Mutations Which Block Head Formation also Block the Appearance of the Acid-Soluble Components

The conditionally lethal mutations (temperature-sensitive and amber) of T4D described by Epstein *et al.* (1963) and Edgar *et al.* (1964) include a large number, distributed in over thirty different genes, which block either the appearance of specific morphological tail structures or the assembly of these with the phage head. Another phenotypic group of mutations, affecting six different genes, prevent phage head formation. The major protein of the head membrane is controlled by one of these genes (Sarabhai *et al.*, 1964), but the functions of the remaining five are unknown. A recent genetic map of T4, showing these genes, is given by Edgar and Wood (1966).

In an attempt to identify the gene or genes which control the appearance of the phage-induced acid soluble components detected in Fig. 1, we examined TCA-extracts of cells infected with a variety of morphogenically defective amber mutants. Figure 3 shows some of these analyses. It is seen that *am*B8 (gene 20) and *am*H36 (gene 23), both of which are defective in head formation, fail to produce both components VI and VII; whereas, *am*N66 (gene 16) and *am*N112 (gene 4), which are defective in tail assembly, produce the components in an amount comparable to the wild-type. Results for amber mutants affected in all six of the genes which control head formation and in eight genes which control tail assembly functions are given in Table I. The data show that all mutations that block head formation also block the appearance of the acid-soluble components VI and VII; mutations resulting in other morphogenic defects have little or no effect. The minor and variable component II was positively observed for six out of eight of the mutants for which VI and VII were observed, but the minor and variable component II was never observed when VI and VII were absent. The apparent differences in yield of components VI and VII among those mutants which produce them is not a reproducible property of the mutant; it appears to be due to a variation from one infected culture to another. In other experiments in which comparisons were made using the same cell culture infected at lower multiplicities, no significant differences in yield were observed.

Several of the head-defective mutants were tested using the permissive host CR63; and, all showed components VI and VII roughly in proportion to the yield of infective particles.

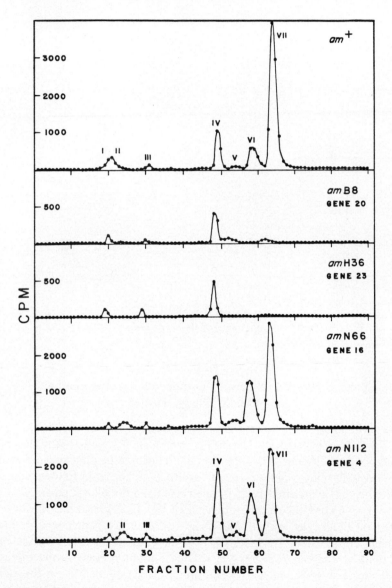

FIG. 3. Effect of amber mutations on the composition of the TCA-soluble fraction of phage-infected cells. *E. coli* BB infected with wild-type T4D (*am*⁺) or amber mutants were labeled, extracted, and analyzed as described in Fig. 1.

TABLE I

AMBER MUTATIONS AFFECTING THE APPEARANCE OF COMPONENTS VI AND VII

| | | Component | |
Gene	Mutant[a]	VI (cpm)	VII (cpm)
	T4 Dam+	1800	9400
Ambers which fail to make heads:			
20	*am* B8	110	240
21	*am* N90	130	470[b]
22	*am* B270	85	140
23	*am* H36	80	220
24	*am* B26	150	260
31	*am* N54	60	200
Ambers which make heads:			
2	*am* N51	1400	2200
4	*am* N112	3400	6600
11	*am* N93	5000	11000
16	*am* N66	3800	7000
25	*am* N67	3500	8700
48	*am* N85	4900	3700
50	*am* A458	1400	2200
51	*am* S29	4100	11400

[a] All mutants have been described either by Epstein *et al.* (1963), Edgar and Wood (1966), or Sarabhai *et al.* (1964).

[b] Mutants defective in gene 21 show small but definite amounts of component VII consistent with the fact that some (abnormal) heads are produced by mutants in this gene (Epstein *et al.*, 1963).

Two of the Acid-Soluble Components are Incorporated into Mature Phage Particles

The acid-soluble fraction of T4-infected cells should include the peptidic substance detected by Hershey (1957) in T2 phage particles if a corresponding component is made by T4. Thus, we examined purified T4 phage to determine whether one or another of the acid-soluble components detected in infected cells is incorporated into phage particles. Figure 4 shows a comparison on Dowex 50 of the TCA extracts of lysine-^{14}C labeled infected cells and of the phage purified by centrifugation from the same culture after lysis. It is seen that component VII is associated with purified phage particles; 74% of the amount present in the infected cell extract was sedimented with the phage. Component VI, on the other hand, cannot be detected in the phage fraction, the upper limit being 6% in this experiment. Nearly all of component VI was recovered, however, from the supernatant (nonphage) fraction. Component II, which, as mentioned before, is not a consistently appearing component, did appear in this preparation and, like VII, is phage associated.

We have never found less than about 25% of VII in the nonphage fraction of lysed cells, even for experiments in which the period of labeling is followed by a long period of incubation with a large excess of lysine-^{12}C. This is not, however, inconsistent with the observation of Koch and Hershey (1959) who found that about 17% of the protein labeled from 20 to 25 min after infection, and which was precipitable by antiphage serum, could not be chased into phage particles.

Hershey (1957) showed that the peptide associated with T2 is released from the phage by osmotic shock. This is also true for components II and VII—72% of II and 80% of VII, relative to the amount found in the TCA extract of the same phage preparation, were rendered nonsedimentable by rapid dilution from 4 M NaCl. In a control experiment, slow

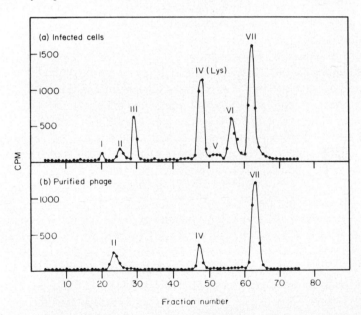

Fig. 4. The presence of components II and VII in phage particles. A 2.5 ml culture of T4D-infected *E. coli* BB was labeled with lysine-^{14}C as described in Fig. 1. One-half of the sedimented and resuspended cells were extracted directly with TCA. The remaining cells were lysed by shaking with one drop of chloroform; the phage were purified by centrifugation and extracted with TCA. To minimize damage and loss of phage during purification, high speed pellets were resuspended slowly in cold and low speed pellets were washed to recover any phage sedimented with bacterial debris. The figure shows analyses of the two TCA extracts an Dowex 50-X2.

dilution from 4 *M* NaCl resulted in the release of only 13% of VII, in agreement with an 11% loss of infective particles. This indicates that release of this component is in proportion to the amount of inadvertent shocking and furthermore it showed that release is not due merely to exposure of the phage to a concentrated salt solution. Since osmotic shock releases components from phage which are chromatographically identical to the TCA extractable components II and VII, it is concluded that these substances exist inside the phage head per se and do not arise as a consequence of chemical action of TCA.

The Phage-Associated Components are Rather Large Polypeptides

The fact that components II and VII can be released from phage particles provides a relatively easy means of obtaining these components in pure form for chemical characterization. Purification from a TCA extract of purified phage has been accomplished by the standard Dowex 50 chromatographic procedure described above after first desalting the extract on Sephadex G-25.

Figure 5 shows the fractionation on a G-25 Sephadex column of the TCA-soluble material obtained from 7.1 gm of purified phage. To facilitate assay of fractions and to assess recovery, a small amount of lysine-^{14}C labeled phage containing 2.8×10^6 cpm was added to the large amount of purified phage prior to TCA extraction. The extract contained 2.6% of the total ^{14}C added. Component A, which elutes from the Sephadex column just after the void volume includes II and VII, as will be shown below, while the material which emerges after fraction sixty includes the TCA, all of the salts from the phage suspension, and lysine and cadavarine (B and C).

Fractionation of component A on Dowex 50 is shown in Fig. 6, from which it is seen that the ^{14}C-components II and VII correspond precisely with ninhydrin-positive components. In addition there is revealed a strong ninhydrin-positive component (III-a) which is not labeled by lysine-^{14}C. Further analysis by paper electrophoresis at pH 6.4 of the isolated components II and VII is shown in Fig. 7 which is a tracing of the ninhydrin-developed spots. Each of the components yielded only one spot, each of which contained over 90% of the ^{14}C-label. Both components are seen to be acidic, component II strongly acidic.

The amino acid compositions of components II and VII given in Table II confirm that these substances are polypeptides; they also indicate minimum molecular weights of about 3900 and 2500, respectively—this is consistent with the behavior of the components on Sephadex G-25 and on larger pore gels.

Both components yield only six different amino acids including lysine (as expected) and large amounts of dicarboxylic amino acids. The number of residues of glutamic and aspartic for II and glutamic for VII can only be considered approximate, due to the large numbers obtained. Judging from the ammonia released on hydrolysis, the amide content of both components is estimated roughly at 10–20% of the released acidic residues. *N*-terminal determinations by Edman degradation indicate glycine

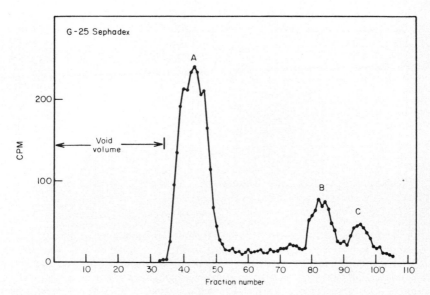

Fig. 5. Fractionation of the TCA-soluble components of purified T4 on Sephadex G-25. The TCA-soluble material resulting from extraction of 7.1 gm of purified phage (to which had been added a small amount of ^{14}C-labeled phage) was applied to a 5.0 × 71 cm column of Sephadex G-25 (medium grain size). The column had been equilibrated with water and was eluted with water at a flow rate of 3.5 ml/min. Fractions of 14 ml were collected of which 0.5 ml was assayed for ^{14}C. The void volume of the column was determined from a previous elution of a high molecular weight marker (Blue Dextran).

for II and lysine for VII. Other amino acids, including tryptophan (determined spectrophotometrically), were present in amounts less than 0.05 moles per mole of peptide. Assays for material absorbing maximally at 260 mμ and for polysaccharide were negative. The compositions of the two components II and VII taken together agree roughly with the isotopically determined composition of the material from T2 reported by Hershey (1957). The component IIIa has not yet been analyzed.

Estimates of the amounts of these components per phage based on the

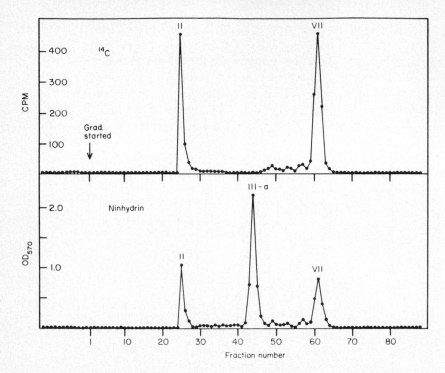

FIG. 6. Isolation of components II and VII by chromatography on Dowex 50-X2. Component A from the fractionation shown in Fig. 5 was adjusted to pH 1.0 with HCl and chromatographed on Dowex 50-X2 by the method described previously (Eddleman and Champe, 1966). The fraction enumeration has been adjusted to correspond to analytical runs by subtracting the volume of the applied sample. The fractions containing the components were pooled and lyophilized.

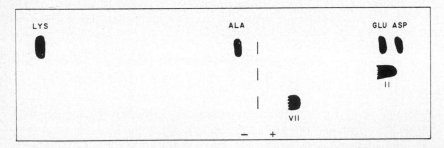

FIG. 7. Electrophoretic analysis of purified components II and VII. Samples of the purified components obtained from the fractionation shown in Fig. 6 were applied to Whatman 3MM paper and electrophoresed at pH. 6.4 by a gradient of 95 volts/cm for 15 min. The figure is a tracing of the ninhydrin-developed spots. Over 90% of the ^{14}C, assayed by means of a chromatogram scanner, was coincident with the ninhydrin spots.

dry weight of the recovered purified material and the recovery of ^{14}C-II and VII are given in Table III. The fact that the number of molecules of II and VII per phage turn out to be nearly equal may not be significant due to the variation in amount of II noted in small-scale labeled extracts. On the other hand, in large-scale preparations such as this one, the yield of ^{14}C-II is generally much more than ever observed in small-scale preparations. This suggests that the variability in yield arises in the extraction process rather than being a variation in the phage stock. It may be that

TABLE II

AMINO ACID COMPOSITION OF COMPONENTS II AND VII[a]

	Polypeptide			
	II		VII	
Amino acid	μMoles	Nearest integral residue	μMoles	Nearest integral residue
Aspartic acid	0.161	16	0.018	1
Glutamic acid	0.104	10	0.187	10
Lysine	0.031	3	0.105	6
Serine	0.020	2	—	—
Glycine	0.012	1	—	—
Proline	0.013	1	—	—
Alanine	—	—	0.075	4
Valine	—	—	0.019	1
Isoleucine	—	—	0.019	1

[a] Amino acid compositions were determined by automatic analysis following hydrolysis in 6 N HCl for 24 hours at 110°C. Free amino acids were absent in unhydrolyzed specimens.

component II is on the threshold of solubility in 5% TCA and that the yield is particularly high in large-scale preparations because the change in extraction conditions necessitated by handling a large quantity of phage (e.g., higher salt content, much greater ratio of protein to TCA) favor solubility.

In several preliminary attempts to purify these polypeptides, it was noted that both II and VII undergo changes in their chromatographic properties if the extract is lyophilized to dryness before the final purification on Dowex 50. The altered components elute from the column in the first fifteen fractions and are barely resolved. Electrophoresis of the altered components showed that they had both become more acidic.

Component VI, which is not associated with phage particles, also appears to be a polypeptide since it is sensitive to proteolytic enzymes and, like II and VII, behaves on gel filtration as a large molecule.

TABLE III

ESTIMATE OF THE AMOUNTS OF COMPONENTS II AND VII IN PHAGE

Phage	Estimation of amounts of components	
	II	VII
[14]C in unfractionated TCA extract	16,200 cpm	29,800 cpm
[14]C recovered from pooled Dowex 50 fractions	9,300 cpm	14,900 cpm
Dry weight of pooled Dowex 50 fractions	10.4 mg	5.1 mg
Estimated amount in 7.1gm[a] of phage	18.1 mg	10.3 mg
Percent of phage mass	0.26%	0.15%
Molecules per phage particle	140	120

[a] The infective phage content of the purified preparation accounted for 41% of the measured dry weight. The remaining 59% likely consists largely of damaged phage rather than contaminating nonphage material since 94% of the material was precipitable by anti-T4 serum.

The Acid-Soluble Polypetides are Formed from a Precursor

The experiments described above show that the appearance of two, and probably three polypeptides depends on the action of (at least) six different phage genes, all of which are involved in head formation. One way of explaining this result is to assume that the polypeptides arise from a precursor whose transformation requires successful head assembly.

To obtain evidence for the existence of such a precursor, an experiment was performed in which the kinetics of appearance of the polypeptides were measured after a short period of labeling. In this experiment lysine-^{14}C was added to the culture 20 min after infection and the incorporation of label was terminated two min later by addition of a 300-fold excess of lysine-[12]C. Figure 8 shows the subsequent appearance of label in the polypeptides, from which it is seen that both VI and VII continue to be labeled beyond the time (22 min) at which lysine-^{14}C incorporation is terminated. There was a large pool of lysine-^{14}C present in the cells at the time of termination of the pulse (7.2×10^4 cpm). However, most of this pool, upon addition of an excess of lysine-^{12}C, does not enter protein but is excreted into the medium. This can be seen from the fact that the

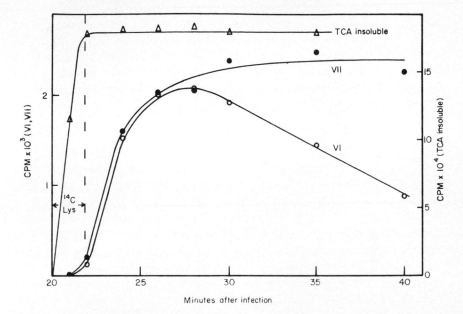

FIG. 8. Pulse labeling of components VI and VII. A 50 ml culture of *E. coli* BB at 37°C was infected with T4D and labeled with 7 μCi of lysine-¹⁴C at 240 μCi/μmole beginning at 20 min after infection. At 22 min, incorporation of ¹⁴C was terminated by the addition of a 300-fold excess of lysine-¹²C. Five ml samples, taken at the indicated times, were quickly chilled to 4°C, sedimented, extracted with TCA, and analyzed on Dowex 50-X2. The TCA-insoluble material was dissolved in 5 ml of 90% formic acid of which .01 ml was assayed.

TCA insoluble fraction shown in Fig. 8 increases little or none after 22 min; if the lysine pool were being incorporated into protein; the TCA insoluble fraction should have increased by some 40% after 22 min. A small fraction of the lysine-¹⁴C pool present at 22 min is probably used in peptide bond formation. If this pool is the origin of the ¹⁴C that appears in VI and VII, it must be concluded that the pool is being channeled preferentially into VI and VII. Although this possibility cannot be ruled out by the present data, the delay in the labeling of VI and VII relative to total protein can most reasonably be explained by assuming that these components are derived from a precursor which was synthesized during the pulse period.

The experiment in Fig. 8 also reveals a difference between components VI and VII. While component VII is a stable substance, component VI is a transient substance which disappears from the TCA-soluble fraction, decreasing to one-half of its maximum amount in about 8 min. The disappearance of VI from the TCA-soluble fraction is not due to its being excreted from the cells, since, when the medium is analyzed, component VI is not found. (None of the components II, VI, or VII is found in any significant amount in the medium.) Thus, component VI appears to be an intermediate which is transformed into a substance not detected by the methods used. This observation is consistent with the decrease in the rate of appearance of VI noted in Fig. 2.

Discussion

During the final stages in the reproduction of bacteriophage T4, the phage DNA becomes associated with several different kinds of protein subunits to form a structurally complex particle. In an electron microscopy analysis of thin sections of infected cells, Kellenberger *et al.* (1959) found that maturation begins with the condensation of the phage DNA into packets the size and shape of phage heads, a step which apparently precedes the encapsulation of the DNA within a protein membrane. This condensation step, however, is dependent on concomitant protein synthesis since the addition of chloramphenicol to infected cells, after DNA synthesis has begun, blocks packet formation and results in the accumulation of a large, diffusely distributed DNA pool. Koch and Hershey (1959) concluded that the phage head membrane, which forms on the condensed DNA, is initially an unstable structure which disintegrates into fragments upon lysis of the cells. They found that the membrane persists in this unstable state over a five-min waiting period and then stabilizes; the phage particle is then completed during a period of about one min.

The results of the experiments summarized here show that maturation of phage T4 is accompanied by the appearance of a set of TCA-soluble polypeptides. These polypeptides first appear at 11 min after infection, which is about 5 min prior to the intracellular appearance of mature phage particles. The polypeptides VI and VII (and II when possible) have been studied in some detail and have been shown to have the following properties in common: (1) Their appearance in the infected cell is blocked by mutations, in any of six different genes, that block head formation; (2) they originate from a precursor, since labeled polypeptides continue to appear for several min after incorporation of label into total protein has ceased. In addition, components II and VII, but not VI, are incorporated into phage inside the head together with the DNA.

If it is imagined that there exists a protein precursor which gives rise to the polypeptides by enzymatic fragmentation, and further that the fragmentation of the precursor protein occurs only inside the phage head, the properties of the polypeptides become consistent with one another: The polypeptides occur in the phage head together with the DNA because it is here that they are formed from the precursor; mutations that block head formation prevent the appearance of the polypeptides by preventing encapsulation of the protein precursor.

No proteolytic enzyme activity has ever been demonstrated in association with T4 particles but there are one or more acid-insoluble proteins contained inside the phage head (Hershey, 1955; Levine *et al.,* 1958; Minagawa, 1961). These might include additional fragments of the precursor which happen to be acid-insoluble; however, unlike the acid-soluble components we have described, at least one of the internal protein components appears in infected cells almost immediately after infection (Murakami *et al.,* 1959).

Although component VI is not found associated with phage particles, it could have the same origin as components II and VII. The fact that VI is unstable suggests that it may be an intermediate fragment associated with the phage particle at the time of maturation which, then becomes degraded into undetected components by the time the phage have been purified and extracted. Since component VI is detected only in the nonphage fraction of lysed infected cells, if it does originate inside the phage head it presumably enters the nonphage fraction from the immature unstable particles (Koch and Hershey, 1959) present at the time of lysis.

It is possible that the polypeptides associated with the phage are injected with the DNA and play some role in the early stages of phage reproduction. We prefer, however, the hypothesis that they are involved, directly or as a by-product, in phage assembly. Several possible reasons for the transformation of the precursor during maturation can be imagined: (1) The precursor might be a structural component of the phage which must be modified; for example, some of the head subunits may require alteration to allow attachment of the tail to the head. (2) The precursor might be a zymogen which is cleaved to yield one or more inactive fragments plus an active fragment which functions during phage assembly. (3) The precursor, itself, might be the active entity which must later be destroyed after its function has been performed. As an example of the last possibility the precursor might be the condensing factor for phage DNA postulated by Kellenberger *et al.* (1959). Such a condensing factor would be expected to become encapsulated with the DNA and might require elimination to allow the phage to inject its DNA.

ACKNOWLEDGMENT

This research was supported by Public Health Service Research Grant GM–10477 from the Division of General Medical Sciences. Harold L. Eddleman received support as a predoctoral trainee from Public Health Service Grant 5 TI GM 77907 BPS.

REFERENCES

Ames, B. N., Dubin, D. T., and Rosenthal, S. M. (1958). *Science* **127,** 814.
Eddleman, H. L., and Champe, S. P. (1966). *Virology* **30,** 471.
Edgar, R. S., and Wood, W. B. (1966). *Proc. Natl. Acad. Sci. U.S.* **55,** 498.
Edgar, R. S., Denhardt, G. H., and Epstein, R. H. (1964). *Genetics* **49,** 635.
Epstein, R. H., Bolle, A., Steinberg, C. M., Kellenberger, E., Boy de la Tour, E., Chevalley, R., Edgar, R. S., Susman, M., Denhardt, G. H., and Lielausis, A. (1963). *Cold Spring Harbor Symp. Quant. Biol.* **28,** 375.
Hershey, A. D. (1955). *Virology* **1,** 108.
Hershey, A. D. (1957). *Virology* **4,** 237.
Kellenberger, E., Séchaud, J., and Ryter, A. (1959). *Virology* **8,** 478.
Koch, G., and Hershey, A. D. (1959). *J. Mol. Biol.* **1,** 260.
Levine, L., Barlow, J. W., and Van Vunakis, H. (1958). *Virology* **6,** 702.
Minagawa, T. (1961). *Virology* **13,** 515.
Murakami, W. T., Van Vunakis, H., and Levine, L. (1959). *Virology* **9,** 624.
Sarabhai, A. S., Stretton, A. O. W., Brenner, S., and Bolle, A. (1964). *Nature* **201,** 13.

The Control of Protein Synthesis after Phage Infection

Cyrus Levinthal, Junko Hosoda,* and David Shub

DEPARTMENT OF BIOLOGY, MASSACHUSETTS INSTITUTE OF TECHNOLOGY,
CAMBRIDGE, MASSACHUSETTS

When a sensitive bacterium is infected by a virulent phage, drastic changes take place in the functioning of the synthetic machinery of the infected cell. However, the nature of these changes depends greatly on the bacterium, and the phage studied, and, in some ways, the more closely the process is examined, the more complex it appears to be. About 20 years ago, it was shown by S. S. Cohen (Cohen and Anderson, 1946) that the synthesis of certain enzymes normally made by the host cells is rapidly shut off after phage infection. It has been shown in Cohen's laboratory, as well as in a number of others, that the infected cell first makes a set of enzymes which are needed for the biosynthesis of phage DNA; at a time after phage DNA synthesis starts, a new set of proteins, which include the structural components of the new phage particles are synthesized. Most of the analyses of the proteins made soon after phage infection have depended on the demonstration of the appearance of particular enzymatic activities. The proteins, which were made later, have been identified by the appearance of new antigenic activities, as well as of structural components of the phage, and of the lysozyme which is made late in the infectious process (Cohen, 1963).

In general, the proteins made after phage infection have been classified as either "early" or "late." It has been assumed that it was necessary to have a control system containing only one switch—that is, from the early to the late synthesis. However, even if only two classes of protein were made after phage infection, the question of how the shift from one to another occurs, as well as the question of how the synthesis of host proteins is shut off, would still remain. One aspect of this question can

* Present address: Space Sciences Laboratory, University of California, Berkeley, California.

be formulated by asking whether the control is exercised at the level of the transcription of the DNA to make RNA, or at the level of the translation of the RNA to make protein. We will describe some experiments concerning the proteins made after phage infection; and, we discuss the data in relation to the question as to when various viral genes are expressed during the phage growth cycle.

In order to study the proteins synthesized after phage infection and the mechanisms which control these syntheses, we have developed an autoradiographic method for determining the radioactivity in proteins separated by high resolution gel electrophoresis (Fairbanks *et al.*, 1965). Three types of experiments have been carried out using this method, as illustrated diagramatically in Fig. 1. In the pulse labeling experiments (Fig. 1a), a short pulse of radioactive amino acids (or of ^{35}S) is given to

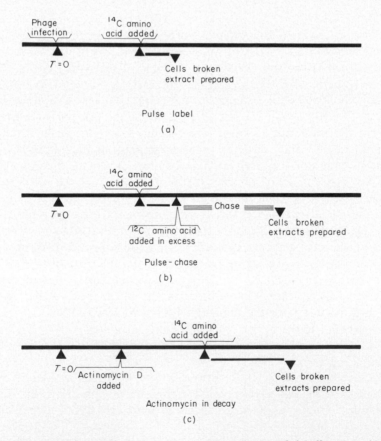

FIG. 1. Diagramatic representation of the three classes of labeling experiments (see text for full description).

the infected culture at various times after phage infection; immediately thereafter, the culture is harvested. Then, an extract is prepared. After separation by gel electrophoresis, the individual bands are examined for their radioactivity. In these experiments, the rate at which a particular protein is synthesized, during the time of the pulse, is reflected in the darkening of the photographic film at a single band in the gel.

In the second class of experiments (Fig. 1b), the culture is flooded with nonradioactive amino acid immediately after the pulse period. The specific activity of the labeled amino acid is thereby decreased sufficiently so that no further incorporation of label takes place; one can now determine whether the radioactivity incorporated into a band during the labeling period remains in that protein during the chase period. Label may be lost during the chase for two reasons: First, the protein may, itself, be unstable and become degraded at a later time during the infectious cycle, or, second, the protein may become incorporated into a larger structure which no longer moves with the same mobility in the gel electrophoresis. It is this latter possibility which makes it possible to establish that certain bands are the subunit protein of larger phage structures.

In the third class of experiments (Fig. 1c), the bacterium *B. subtilis,* which is sensitive to the antibiotic actinomycin D, is infected with the phage SPO1. One can determine the functional lifetime of a particular class of messenger RNA molecules by adding actinomycin to the culture in order to stop RNA synthesis and, subsequently, labeling the culture with radioactivity precursors of protein. Experiments of this kind cannot prove that the functional lifetime is the same as the physical lifetime of the RNA molecule. However, such experiments can be used to examine the validity of models of the phage control system, which are based on the hypothesis that the functional lifetimes of the various messengers for the early and the late proteins are different (for example, Edlin, 1965).

The Sequential Synthesis of Phage Proteins

Pulse labeling experiments have been carried out with the *B. subtilis*-SPO1 system, as well as with *E. coli* infected with phage T4. (See Figs. 2 and 3). In both of these systems, one observes that the simple description of the proteins as either early or late is not adequate. The biosyntheses of the late proteins all start at approximately the same time; and, they continue until the cells lyse. However, some of the early proteins start being synthesized immediately after infection and stop a short time later, while others start later and stop only when the late proteins begin to

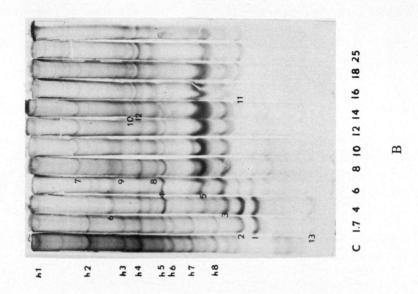

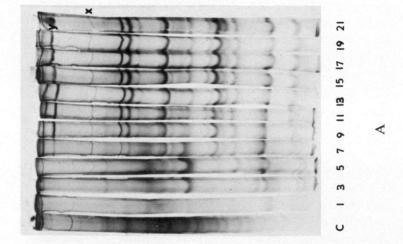

be synthesized. It is not clear from our observations of the labeling of the bands whether the pattern of synthesis can be classified into a small number of separate groups, or whether each protein must be considered as having its own control sequence. With the T4 proteins, it is clear that there are at least four different patterns, shown in Fig. 3b, as A, B, C, and D. We have, as yet, no information as to the function of any of the very early proteins; but, we can say from our experiments, as well as from the results of those performed in other laboratories, that the syntheses of different enzymes which are produced in the T4-infected cell are initiated at different times (Sekiguchi and Cohen, 1964). The phage functions responsible for at least some of the arrest of the synthesis of host cell macromolecules require protein synthesis for their expression. For example, it has been shown by Yarosh (personal communication) that T4 can stop the production of the RNA phage MS-2, even if the T4 infects the cell considerably later than the MS-2. The arrest of synthesis of MS-2 infectious RNA occurs immediately after T4 is added. This arrest requires that protein synthesis take place at the time of the T4 infection. There are numerous experiments which suggest that proteins encoded by the T4 are responsible for the early interruption of DNA, RNA, and protein synthesis of the host. Any phage protein acting in this manner is a likely candidate for one of the proteins made very early after phage infection.

FIG. 2. Pulse labeling pattern of phage infected cells. Each photograph shows several autoradiograms prepared from cells labeled at different times after infection. The origin of the electrophoresis is at the top of the photograph; the time after infection at which the pulse labeling started is indicated under each photoradiogram. The uninfected control is labeled c. For all of the gels, the pattern of the total protein, as revealed by staining, is approximately the same as that of the autoradiogram of the control.

(A) *E. coli* cells infected with phage T4 were given 2 min pulses of leucine-^{14}C at 37°C. A mutation in the gene producing lysozyme (*Am* 882) was used to reduce the loss of cells during centrifugation. The preparation of the radioactive extracts was done as described in the legend to Fig. 3a.

(B) A culture of *B. subtilis* 168 (1×10^8 cells/ml) was infected with phage SPO1 at a multiplicity of 10. At various times through the latent period, 10 ml samples were withdrawn and exposed to a mixture of uniformly labeled amino acids (0.67mg/mCi, 0.5 μCi/ml). After 2.2–2.3 min of labeling, cells were collected, resuspended in 1 ml of buffer containing chloramphenicol and lysozyme, incubated at 37°C for 5 min and then frozen. Two-tenths milliliter of the samples were analyzed by electrophoresis on a 10% acrylamide gel.

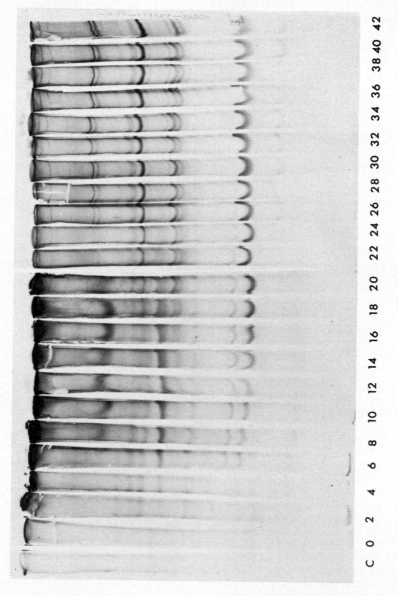

FIG. 3A.

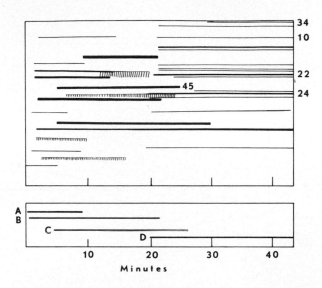

FIG. 3B.

FIG. 3. A suspension of *E. coli* B in M9 (4×10^8 per ml) was exposed for 90 sec to 3 ultraviolet lamps (Westinghouse Sterilamp G15T8) at a distance of 115 cm. The culture was then incubated with aeration at 25°C for 30 min. The UV treatment reduced the rate of leucine-^{14}C incorporation into the bacterial proteins to 5–10% that of the irradiated controls. After phage infection of the UV-treated cells, leucine uptake into phage proteins was 50% or more than that into phage infected cells without irradiation. The time of synthesis of the various phage proteins was not significantly altered by the irradiation.

(A) Tryptophan (50 µg/ml) and T4D (multiplicity of infection equal 5) were added at T = 0. Two min pulses of leucine-^{14}C (0.2 µCi/ml, 220 mCi/mmole) were given, starting at the times indicated. All incubations were at 25°C. The pulse was terminated by chilling, after which the cells were centrifuged in the cold, resuspended in buffer (Tris-HCl 0.01 *M;* MgCl$_2$ 0.01 *M;* pH 7.4), and sonicated for 1 min. Extracts corresponding to 8×10^8 infected cells were analyzed on 10% acrylamide gels according to the method described by Fairbanks *et al.* (1965).

(B) Upper panel: Schematic summary of the results of the pulse labeling experiments. The numbers on the figure refer to the genes which code for the proteins at the marked bands. Lower panel: Simplified classification of bands according to the time at which they are labeled.

(C) Phage specific enzyme synthesis in cells infected as described in (A). Deoxycytidilate (dCMP) hydroxymethylase assays were carried out according to the method of Dirksen *et al.* (1960) with slight modifications. Thymidilate kinase and thymidilate synthetase assays were performed according to the method of Wiberg *et al.* (1962). Curve 1, dCMP hydroxymethylase; curve 2, TMP kinase; curve 3, TMP synthetase.

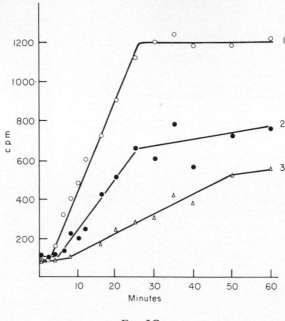

FIG. 3C.

The Shut-off of Host Protein Synthesis

There are two other points to be noted concerning the pulse labeling experiments with wild-type phage. One is that the shut-off of host protein synthesis occurs rapidly and seems to proceed at roughly the same rate for all of the host proteins. At least there is no evident change in the distribution of label in host cell proteins prior to the shut-off of their synthesis shortly after phage infection. The autoradiograms of *E. coli* cells, labeled earlier than 3 min post-infection, indicate that the protein made is a mixture of those that are characteristic of the *E. coli* host and those that are phage directed. The autoradiograms obtained with extracts of *E. coli* exhibit, in addition to discrete bands, a general darkening of the film due to the superposition of a very large number of weak bands. The individual protein bands of *E. coli*, as well as the continuous distribution of radioactivity, starts to decrease immediately after phage infection. The intensity of a sharp discrete band near the top of the autoradiogram, when pulse-labeled during the interval 1–3 min post-infection, is less than one-fourth as intense as is the same component pulse-labeled in uninfected bacteria; when the pulse is applied during the interval 3–5 min post-infection, no radiolabel is incorporated into this protein species.

The wild-type phage and all of the amber mutants so far examined turn off the synthesis of the host proteins. This is true of the DNA negative mutants (*DO*), with mutations in genes 1, 30, 32, 41, 42, 43, 44, or 45, of the maturation defective mutants (*MD*), with mutations in genes 33 and 35, and of those mutants with defects in genes 46 and 47, which Wiberg (1966) has shown are unable to degrade the host DNA. The fact, that the degradation of host DNA is not necessary for the shut-off of host protein synthesis, is in agreement with the earlier results of Nomura *et al.* (1962); he showed that the degradation of DNA, even with the wild-type phage, is not, per se, the reason for the shut-off.

The second point to be emphasized is that there are obvious differences in the extent to which each of the phage proteins, early as well as late, are labeled. This variation in labeling is not due to differences in the amino acid composition of the different proteins, since the commonly occurring amino acid leucine was used in most experiments, and since approximately the same pattern was observed when ^{35}S or a different labeled amino acid was used. Thus, we must ask not only how different genes are expressed at different times, but why these genes make varying amounts of their final products under conditions in which the variation does not seem to be taking place in response to a change in the environment. The expression of the phage genes seems to be preprogrammed, both with respect to the time when the genes function and to the extent to which they make their final products. In this respect, the development of the phage system seems to be more characteristic of a differentiating higher organism than of a bacterium, whose genes can change their functions rapidly in response to changes in the environment.

Identification of Bands

An amber mutation in a phage causes a termination in the polypeptide chain (Sarabhai *et al.,* 1964) at the point of mutation when the phage is grown in a restrictive host; but, it produces a complete protein in a host which suppresses the mutation. When pulse-labeling experiments are done using restrictive host cells infected with such a mutant phage, one would expect the protein to either appear at a different place on the radioautograph or not to appear at all. The results of such an experiment are shown in Fig. 4. The major head protein, made by gene 23 (Sarabhai *et al.,* 1964), is identified by the fact that mutants with defects in this gene produce partial peptides which either move more rapidly in the gel than does the complete protein, or fail to appear on the gel at all. There is a rough correspondence between the size of the peptide produced and the rate at which it migrates in the gel—the smaller the peptide, the

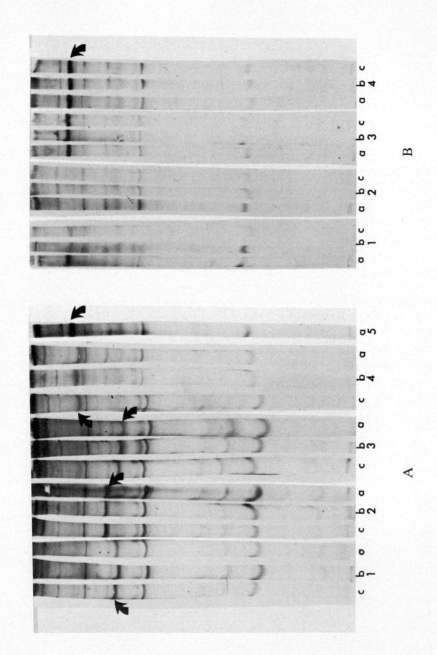

faster the band moves. This relationship is, of course, not followed in a strict sense, since it would assume that the charged groups on the protein were uniformly distributed along the polypeptide chain and that the only effect of the mutation would be to produce a shorter peptide with the same total charge per unit length. As can be seen in Fig. 4, all of the polypeptides produced by amber mutants with defects in gene 23 were found to migrate more rapidly than did the corresponding gene product of the wild-type phage. If the mutant gene produces a very short peptide, the material is likely to diffuse out of the gel prior to fixation and drying. It has been shown, by Sarabhai *et al.* (1964), that gene 23 produces approximately 60% of the total protein made late in infection; our results show that band 23 is much more heavily labeled than any other produced during the infectious cycle.

Loss of Label during a Chase

The product of gene 23 can also be identified as a protein which becomes incorporated into a large phage structure by the fact that its activity drops in a pulse chase experiment. Once the head protein has been incorporated into a completed head structure, urea treatment will no longer cause it to be solubilized; and, therefore, it will not run into the gel. This is demonstrated by the fact that in the case of infection with mutants with defects in any of the other genes producing head proteins, the label remains in band 23 in a pulse-chase experiment, whereas it is chased in a comparable experiment in which wild-type phage is used. The label in the band 23 protein is not chased when the infecting phage carries a mu-

---

Fig. 4. Cultures of *E. coli* B, infected with various mutants carrying defects in the genes controlling the synthesis of head proteins were labeled for two min with leucine-^{14}C. Aliquots of the cultures were then withdrawn and quickly chilled. Nonradioactive leucine (50 μg/ml) was added to the remaining portions of the cultures; after chase periods of 3 and 6 min, aliquots were withdrawn and chilled. Extracts were prepared as described in the legend to Fig. 3A. The extracts were treated overnight in the cold with 6 M urea in the presence of 0.02 M mercaptoethanol then analyzed in a 7.5% acrylamide gel containing 6 M urea. The position of the 23-protein or 23-peptides is indicated with an arrow (a) Labeled from 10 to 12 min. (b) Labeled from 10 to 12 min and chased from 12 to 15 min. (c) Labeled from 10 to 12 min and chased from 12 to 18 min. (A) 1. *am*E1056 (gene 23); 2. *am*E389 (gene 23); 3. *am*E1270 (gene 23); 4. *am*E509 (gene 23); and 5. *am*N76 (gene 21). (B) 1.*T4D* wild; 2. *am*B17 (gene 23); 3. *am*B8 (gene 20); and 4. *am*N76 (gene 21).

tation in genes 21, 22, 23, 24, or 31; but, it is chased if the phage contains a mutation in any of the tail genes (at least, with all those that we have tested). Thus, in order to stabilize the head protein, the formation of the intact phage is not necessary; but, formation of an intact head structure is necessary. Gene 22 makes a product containing about 15% as much label as does the product of gene 23; the behavior of this product in pulse-chase experiments is, in all respects, comparable to that of the gene 23 band. Therefore, it may be concluded that gene 22 also makes a protein which is a component of the head. It seems likely that it is one of these components which causes the head structure to assume its particular shape, since it is unlikely that the single structural protein could form a membrane having the angular morphology that characterizes the phage head. Radiolabel can also be chased, to a slight extent, from the product of gene 24; and it seems likely that the latter is also a component of the head structure itself. Although the other three head genes do not produce proteins which we have identified on the gels, there is no direct reason to assume that they do not do so and that the proteins are incorporated into the head structure. We can use the evidence that radiolabel can be chased from a particular band and that this property is lost in the case of infection with the appropriate mutants as evidence that a particular band is a structural protein. We cannot, however, use the fact that label cannot be chased from a particular protein as evidence that it is not a structural protein. The inability to chase radiolabel from a protein could be due to the fact that it is made in quantities far in excess of those required for its structural role in the assembly of the phage. If this were so, a large pool of the material would accumulate within the cell. There are a number of other interacting mutational effects which have enabled us to establish certain pathways in the early assembly of large units from small ones. For example, there is a particular band (labeled x, in Fig. 2a) from which label can normally be chased, but from which label is not chased if the infecting phage is carrying a mutation in gene 7 or 8. The same band disappears when the phage is carrying a mutation in either gene 10 or 11. Thus, band x must be an intermediate in the formation of a larger structure. It is formed by a combination of genes 10 and 11; its incorporation into the larger structure depends on the activity of genes 7 and 8. There is another band labeled y (see Fig. 2), which incorporates little radioactivity during pulse labeling; but, it becomes more heavily labeled during the chase period. Therefore, band y is presumed to be assembled from subunits; this assembly must take place at a time later than that at which the radioactive amino acids are incorporated into the individual polypeptide chains.

Protein Patterns in DNA Negative Mutants

As was expected from earlier work on this system (Wiberg *et al.*, 1962), it was found that mutants which fail to make DNA do not make any of the late proteins classified in Fig. 3b as D and that many of the earlier proteins (particularly those of classes B and C) continue to be synthesized for a longer time than is the case after infection with wild-type phage. However, all of the DNA negative mutants examined do allow the turn-off of synthesis of the very early proteins of class A and the initiation of synthesis of proteins of classes B and C. Enzyme analyses, carried out under the same conditions as those used for the pulse labeling experiments, indicate that dCMP hydroxymethylase and TMP kinase are of class B and that TMP synthetase is of class C. Mutants with defects in gene 30 have been classified as DNA negative. However, with at least one mutant with a mutation in this gene, the syntheses of the late proteins are turned on at the normal time. With another mutant in this class, the late proteins are synthesized, but their appearance is somewhat delayed. In the latter case, the class B and C proteins are generally turned off slowly, but at least one member of the class A group is turned off at the normal time. However, it is not clear that the behavior of gene 30 is an exception to the general rule concerning DNA negative mutants, since pulse labeling experiments with radioactive thymidine indicate that mutants of gene 30 do make small amounts of DNA (Geiduschek and Epstein, personal communication), most of which is broken down to acid-soluble form later in the infection (Hosoda and Levinthal, 1967). A recombinant between a mutant in gene 30 and one in gene 46 makes a small amount of DNA, but does not degrade it. Since genes 46 and 47 have been shown by Wiberg (1966) to be responsible for the breakdown of the host DNA, it seems likely that the small amount of DNA made when a mutation is present in gene 30 is not normal phage DNA. It is interesting to note that mutants with defects in genes 46 and 47, which do not cause the breakdown of host DNA, are still able to turn off host protein synthesis. In fact, all mutants examined by us are able to shut off host protein synthesis, even if they are not able to continue beyond the early stage of phage infection.

Mutants carrying mutations in genes 33 to 55 have been designated as maturation defectives (Epstein *et al.*, 1963), by virtue of the fact that they give rise to no late phage structures. Not only do they fail to make mature phage and large phage structures, but they fail to turn on the synthesis of any of the late proteins. In the case of leaky mutants, the initiation of synthesis of the late proteins may be delayed, with a corresponding delay in the shut-off of synthesis of proteins of classes B and C. However, in all cases, some of the very early proteins in class A are turned off at the normal time. Mutants with defects in genes 33 and 55 make DNA in normal amounts, from which it may be concluded that DNA synthesis itself is not sufficient, or responsible for the switch from early to late protein synthesis. Thus, all of the mtuants, which fail to make DNA, and the maturation defectives, which prevent synthesis of late proteins, allow the synthesis of proteins of classes B and C to continue longer than normal; but, they turn off the synthesis of the A proteins at the normal time. The differences in the times at which proteins of classes A, B, and C are synthesized persists even in the DNA negative and the maturation defective mutants. However, the results obtained with mutants having defects in genes 30, 33, and 55 show that the classification of the proteins into only four categories is probably insufficient. In addition to the gel data, measurements of enzyme activities show patterns of synthesis which are altered in different ways by the different mutations in these genes.

Decay of Phage Messengers

The work with the *B. subtilis*-SPO1 system was undertaken in order to examine the hypothesis put forward by Edlin (1965), which suggests that late phage proteins are made by messengers with long functional stability. There are many interesting aspects of this model. However, an essential feature concerns the switch from early protein synthesis to late protein synthesis that takes place because the different messenger RNA molecules have different inherent stabilities. Thus, late in infection, the more stable messenger components, which are assumed to code for the late proteins, would tend to dominate the protein synthesis machinery. More recently, Guthrie and Buchanan (1966) have reported experiments with T4 which seem to indicate the existence, early in infection, of a messenger which is considerably more stable than any found in uninfected cells. *B. subtilis* was used in our experiments since actinomycin D effectively blocks RNA synthesis in this organism, this made it possible to examine the rate of decay of preexisting messenger. Experiments of this kind in uninfected cells (Levinthal *et al.,* 1962) have shown that about 9% of the total RNA is unstable and that the decay time for this material is between 1 and 2 min at 37°C. The functional half-life of the messenger (that is, the rate at which the protein synthetic capacity decays) is longer than the average half-life of the pulse labeled RNA by about a factor of two. This difference may reflect the more efficient use of the other components of the protein synthetic machinery as the concentration of messenger RNA drops, or alternatively, it may reflect a heterogeneity in the half-life of various messenger molecules. These estimates of the amount and approximate half-life of the messenger RNA molecules have been confirmed by experiments in which the kinetics of labeling of the nucleotide triphosphate pool have been measured without the use of any metabolic inhibitor (Salser *et al.,* 1967). In uninfected cells then, we have reason to believe that an approximate measure of the half-life of the messenger molecules can be obtained by examining the decrease in the capacity of the cells to synthesize protein after actinomycin is added to the culture.

The phage SPO1 is a member of what Brodetsky and Romig (1964, 1965) have designated as the group one *B. subtilis* phages. The members of this group seem to be analogous to the T-even phages in morphology, and in the composition of their DNA in that one of the normal bases (in this case thymidine) is completely replaced by an unusual one, 5-hydroxymethyl uracil (Kallen *et al.,* 1962; Okubo *et al.,* 1964). When acrylamide gel electrophoresis of extracts of SPO1-infected *B. subtilis* cells is performed, one observes the labeling of a few bands very early in the infectious cycle. In fact, these very early proteins were first observed in this system, where they can be detected even while the host proteins are still being labeled. It was this observation that prompted us to carry out similar experiments with T4-infected *E. coli.* However, in the latter system, it is necessary to expose the host cells to UV light in order to reduce the amount of host proteins synthesized during the early period after phage infection. The synthesis of very early proteins has since been observed by McCorquodale and Buchanan in the phage T5 *E. coli* system (see Buchanan, Part I, this volume).

By measuring the rate at which labeled amino acids were incorporated into each of the various classes of proteins after the infected cells were treated with actinomycin D, it was possible to show that all of the different bands, and by implication, the corresponding messenger RNA's, have approximately the same functional half-life. This half-life, measured by the incorporation of amino acids into protein and by the decay of pulse labeled material, was approximately the same as that of the messenger RNA's in uninfected *B. subtilis* cells. Thus, in this system, we can find no evidence for any long-lived messenger. A similar conclusion was also reached on the basis of results of experiments in which the synthesis of the enzyme dCMP deaminase was measured. In the case of this enzyme, one finds that its rate of synthesis decays rapidly after the addition of actinomycin. If the actinomycin is added shortly before the enzyme would normally appear, no activity can be found. From this we conclude that the genes making the messenger RNA molecules must function immediately before the time at which the corresponding proteins appear.

These experiments do not imply that the control of protein synthesis in the phage-infected cell is entirely at the level of the expression of the genes. Translation control—that is, control at the level of messenger RNA's directing the synthesis of proteins—may still be involved. However, we can conclude that the early genes must be expressed early and the late genes must be expressed late; and, in all cases, the half-life of the corresponding messengers is short.

In the course of studying the decay of the RNA made after phage infection, a rather surprising observation was made. Although phage SPO1 shuts off host DNA synthesis almost completely and reduces host protein synthesis to a level of about 10% of that observed in the uninfected cell, the synthesis of ribosomes and of soluble RNA continues at about the same rate as in the uninfected cells. In this case, the synthesis of new ribosomal proteins and of ribosomal RNA continues after phage infection. This phenomenon has not been investigated; but, it is certainly an indication that a separate control mechanism, different from that which controls messenger RNA synthesis, operates on those genes which encode for ribosomes and soluble RNA. It is as yet too early to make any positive suggestions as to the functions of the very early phage-directed proteins, or the mechanism by which the phage shuts off the synthesis of host cell macromolecules, but it is possible that some of the very early phage proteins act as specific inhibitors of host functions. If there is more than one such inhibitor of the host function, then the failure to obtain mutants which are unable to shut off host synthesis may be explained by the fact that such a phenomenon would require the simultaneous appearance of several different mutational events. It seems likely that some of the very early proteins are responsible for the interruption of the normal biosynthetic activities of the host cell; therefore, a more detailed study of their function seems to be essential.

REFERENCES

Brodetsky, A. M., and Romig, W. R. (1964). *Bacteriol. Proc.* p. 118.

Brodetsky, A. M., and Romig, W. R. (1965). *J. Bacteriol.* **90**, 1655.

Cohen, S. S. (1963). *Ann. Rev. Biochem.* **32**, 83.

Cohen, S. S., and Anderson, T. F. (1946). *J. Exptl. Med.* **84**, 511.

Dirksen, M. L., Wiberg, J. S., Koerner, J. F., and Buchanan, J. (1960). *Proc. Natl. Acad. Sci. U.S.* **46**, 1425.

Edlin, G. (1965). *J. Mol. Biol.* **12**, 363.

Epstein, R. H., Bollé, A., Steinberg, C. M., Kellenberger, E., Boy de la Tour E., Chevalley, R., Edgar, R. S., Susman, M., Denhardt, G. H., and Leilausis, A. (1963). *Cold Spring Harbor Symp. Quant. Biol.* **28**, 375.

Fairbanks, G., Jr., Levinthal, C., and Reeder, R. H. (1965). *Biochem. Biophys. Res. Commun.* **20**, 393.

Geiduschek, E. P., and Epstein, R. H. (1966). Personal communication.

Guthrie, G. D., and Buchanan, J. M. (1966). *Proc. Natl. Acad. Sci. U.S.* **25,** 864.

Hosoda, J., and Levinthal, C. (1967). In preparation.

Kallen, R. G., Simon, M., and Marmur, J. (1962). *J. Mol. Biol.* **5,** 248.

Levinthal, C., Keynan, A., and Higa, A. (1962). *Proc. Natl. Acad. Sci. U.S.* **48,** 1631.

Nomura, M., Matsubara, K., Okamoto, K., and Fujimura, R. (1962). *J. Mol. Biol.* **5,** 535.

Okubo S., Strauss, B., and Stodtsky, M. (1964). *Virology* **24,** 552.

Sarabhai, A. S., Stretton, A. O. W., Brenner, S., and Bolle, A. (1964). *Nature* **201,** 13.

Salser, W. (1967). In preparation.

Sekiguchi, M., and Cohen, S. S. (1964). *J. Mol. Biol.* **8,** 638.

Wiberg, J. S. (1966). *Proc. Natl. Acad. Sci. U.S.* **55,** 614.

Wiberg, J. S., Dirksen, M., Epstein, R. H., Luria, S. E., and Buchanan, J. M. (1962). *Proc. Natl. Acad. Sci. U.S.* **48,** 293.

Yarosh, E. (1966). Personal communication.

PART II

TEMPERATE BACTERIOPHAGES

The Position and Orientation of Genes in λ and λ*dg* DNA*

David S. Hogness, Walter Doerfler,† J. Barry Egan, and Lindsay W. Black

DEPARTMENT OF BIOCHEMISTRY, STANFORD UNIVERSITY SCHOOL OF MEDICINE,
PALO ALTO, CALIFORNIA

Introduction

The molecule of DNA isolated from bacteriophage λ contains some 47,-000 base pairs, enough for 40 to 45 genes, each capable of specifying a polypeptide having a molecular weight of 40,000. In fact, some 25 genes have been identified and placed on a genetic map of the vegetative phage.

The ability of some of these λ genes to function in the transcription-translation process appears dependent upon this process having already occurred at other λ genes. This hierarchy of control may contain several levels. For example the product(s) of the immunity region (*i*λ; see Fig. 1) can be thought of as acting at a primary level, since it appears to restrict transcription-translation of the λ genome to the immunity region itself (Jacob and Wollman, 1961; Bode and Kaiser, 1965a; Isaacs *et al.,* 1965). That this repressive action of the immunity substance(s) may be indirect for all but one, or a few genes is indicated by the fact that independent inactivation by mutation of the "early" λ genes (*N, O,* or *P*; see Fig. 1) greatly reduces the transcription-translation of other, "late" genes. This is particularly clear in the case of the *N* gene, where such inactivation results in the loss of ability to synthesize any of the known λ-specific proteins, to transcribe most of the λ genome at significant rates,

* This paper was presented by the first author (D. S. H.) at Cold Spring Harbor, New York on June 3, 1966 as well as at this symposium about three weeks later. Consequently, it has also been published in the *Cold Spring Harbor Symp. Quant. Biol.* **31.** 1966 (in press).

† Present address: The Rockefeller University, New York, New York.

and to replicate λ DNA (Dove, 1966; Joyner *et al.,* 1966; Protass and Korn, 1966), but does not eliminate the synthesis of the immunity substance(s). Thus we may imagine that the immunity substance(s) exerts direct control only among the early genes, other genes being controlled secondarily by the products of the early genes.

In order to define this hierarchic control better and, more particularly, to facilitate its recreation *in vitro,* we are interested in determining the positions in λ DNA where transcription and replication are initiated as well as determining the directions along the DNA in which these processes advance after initiation. We have begun by attempting to describe λ DNA in terms of gene position and gene orientation. Thus we seek a determination of (1) the number of base pairs separating a given gene from some reference point on the DNA, and (2) the direction in which transcription of that gene proceeds along the DNA. It is this latter parameter which we term orientation.

Gene Position

GENERAL CONSIDERATIONS

Over the past few years we have performed a series of experiments which indicate that the position of several genes in the DNA isolated from bacteriophage λ (or λ*dg*) is that given in Fig. 1. The normal λ genes are ordered on the right half of the DNA as they are on the genetic map of the vegetative phage (Arber, 1958; Campbell, 1961; Amati and Meselson, 1965), not according to the genetic map of the prophage (Rothman, 1965). The genes of the galactose operon (*k, t,* and *e*), found in the left half of λ*dg* DNA, have not been ordered on the genetic map of λ*dg*. However, the order shown in Fig. 1 is consistent with their known order on the genetic map of *E. coli* (Buttin, 1963; Adler and

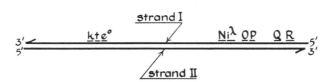

FIG. 1. The position of some genes in λ and λ*dg* DNA. The six genes on the right are found in both λ and λ*dg*. Some of their idiosyncrasies are defined later in the text. The three genes on the left (*k, t,* and *e*) are the structural genes for the three enzymes (galacto*kinase,* galactose-1-P uridyl *transferase,* UDP-galactose 4-*epimerase*) of the galactose operon in *E. coli.* The superscript *o* designates the controlling or operator end of this operon, which is found in λ*dg* but not λ.

Templeton, 1963), and with Campbell's model (Campbell, 1962) for the formation of λ*dg*. These points are illustrated in Fig. 2.

The solid lines at the top and bottom of the figure represent the partial genetic maps of λ prophage and vegetative phage, respectively. The top line also indicates the relative positions of the genes in the galactose operon and in the prophage on the genetic map of the lysogenic host.

In the formation of λ*dg* it is supposed that when a lysogenic bacterium is induced to yield phage there are rare occurrences of a looping out and crossover at regions of minor, accidental homology, such that genes of the host are included in the resulting closed form, thus displacing some λ genes. This is to be contrasted to the model for the release of normal

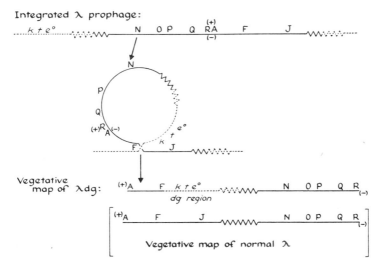

FIG. 2. Formation of λ*dg* from prophage according to Campbell (1962). The solid line indicates material derived from λ and the dashed line, material derived from *E. coli*. *A, F,* and *J* are additional λ genes not placed in Fig. 1.

λ, in which the loop is so placed that the crossover occurs between the two homologous regions indicated by the wavy lines, leaving only λ genes in the closed form.

If the closed form is now opened between genes *A* and *R,* the vegetative maps are formed. The model implies that at least three different forms of λ or λ*dg* DNA exist and that the gene orders in these forms are related by circular permutations. Furthermore, according to the model, the relative orientations of any two genes in one form must be the same as in any other form. Thus two genes on a single DNA molecule have identical or opposite orientations, and this relationship should not vary with the form of the λ or λ*dg* DNA.

The DNA we work with is the open linear molecule that is isolated from the mature phage. This is the DNA indicated in Fig. 1. The ends of this DNA, whether isolated from λ or λ*dg,* contain cohesive sites which can interact *in vitro* to close this open monomer (Hershey *et al.,* 1963). These sites appear to consist of single-strand protrusions; the base sequence in one is assumed to be complementary to that in the other so that these two single-strand regions can combine in a normal duplex DNA structure (Hershey and Burgi, 1965; Strack and Kaiser, 1965).

In addition to the open and closed monomers seen *in vitro,* there exist *in vivo* two different closed monomers (Young and Sinsheimer, 1964; Bode and Kaiser, 1965b), what appear to be concatenated forms (Smith and Skalka, 1966), and, of course, the prophage form. The point we wish to emphasize here is that knowledge of the position and orientation of the genes in the open monomer should be applicable to the various other forms of λDNA if the requisite circular permutation can be made.

SPECIFIC EXPERIMENTS

Our method for determining gene position depends, first, upon the formation of fragments of λDNA by subjecting the open monomers to hydrodynamic shear, and, second, upon an assay for the gene activity of these fragments. The basic assay was developed several years ago (Kaiser and Hogness, 1960) and consists of allowing the DNA to react with cells of *E. coli* which have been made competent by a previous infection with helper phage. To assay for the activity of a given gene, call it a^+, on a fragment of DNA, the bacteria are first infected with a^--helper phage that are defective in regard to the function of this gene. Recombination between the a^+-containing fragment and the a^--helper DNA can take place in the competent cell to yield whole λDNA containing a^+. Such DNA can either become prophage, or it can become mature phage that is released upon cell lysis.

The mature a^+-phage can be scored as plaques by plating with appropriate bacteria. It is this lytic response that is used in the assay of λ genes. When, however, a^+ represents a gene of the galactose operon, the assay largely depends upon the lysogenic response of whole λ*dg* DNA, created by recombination of a fragment with the non-*dg* helper DNA. In this case the competent bacteria must be galactose-negative so that their lysogenization by the λ*dg* can be easily scored as a conversion to the galactose-positive state.

In the early experiments the fragment size was at the level of half-molecules (Kaiser, 1962; Radding and Kaiser, 1963; Hogness and Simmons, 1964) and the resolution was therefore low. The experiments of Hogness and Simmons (1964) with λ*dg* DNA showed that the two types of half-molecules could be isolated from each other. One type (called left halves) contains the *A* gene (the m_6 site was identified; this resides in *A*, in *B*, or between these closely linked genes on the genetic map; Campbell, 1964) and the genes of the galactose operon, while the other type (right halves) contains *i*λ and *R*. This distribution of genes among the two halves tells little about their precise position, and, indeed, does not show whether the order of the genes in the isolated DNA is that given by the map of vegetative phage or that of prophage.

More recently, we (J. B. Egan) have increased the resolution by examining a wider distribution of fragment sizes. Equal weights of fragment populations, which we refer to loosely as halves, sixths and twelfths, were mixed; these designations indicate the length, as a fraction of the length of the open monomer, about which the populations center. The idea here was to achieve a very wide distribution of sizes. Since the distribution of lengths in populations of halves as observed in the electromicroscope is itself quite broad (Kaiser and Inman, 1965; Hogness, 1966), it was expected that the populations used in the mixture would overlap.

The mixture was sedimented as a zone in a sucrose gradient to distribute the fragments along the axis of the centrifuge tube according to their size. The fractions collected from the tube were then assayed for the activity of the following pairs of genes located on the right half of the vegetative map: *N-R*, *i*λ*-R*, *O-R*, *P-R*, and *Q-R*. The fractions were also assayed for *R* activity alone. Since the order of the genes on either map is *N-i*λ*-O-P-Q-R*, colinearity of map and DNA demands that the smallest fragments containing both genes of a given pair would have the following size relationship for the various pairs: $N\text{-}R > i\lambda\text{-}R > O\text{-}R > P\text{-}R > Q\text{-}R > R$ alone. The results of the experiment confirm this prediction and are given in Fig. 3.

The distribution of DNA mass (OD_{260}) in the relevant portion of the tube exhibits a broad peak consisting of fragments from the sets of sixths and twelfths, the peak of halves being off to the right and not shown in the figure. The important aspect of each activity distribution is the distance from the meniscus to the position at which the activity extrapolates to the background value at the left or small end of the distribution. This indicates the distance sedimented by the smallest molecules active for a given pair. Clearly, this distance increases in the order *R* (alone), *Q-R*, *P-R*, *O-R*, *N-R*. The *i*λ*-R* distribution was not included in Fig. 3 for the sake of simplicity. Its extrapolated position lies between that of *O-R* and

N-R, though its position is not significantly different from that for *N-R.*
Eisen *et al.* (1966) have shown that the *i*λ-region lies just to the right of
N and well to the left of *O* on the vegetative map of λ.

Actually three quantities are being measured in each case rather than
just the two activities of a gene pair. The third quantity is one of the two
cohesive sites. This results from the fact that only those fragments which

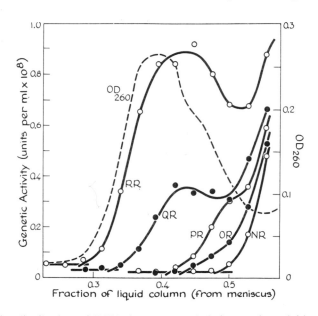

Fig. 3. The distribution of DNA fragments and their genetic activities after zone
centrifugation. The centrifuge tube initially contained 4.6 ml of a solution of
sucrose in 1 *M* NaCl, 0.01 *M* Tris-HCl buffer, pH 7.1, the sucrose concentration
varying linearly with volume from 5 to 20% (*w/v*). 0.1 ml of a mixture of
halves, sixths and twelfths (see text) in a 1 : 1 : 1 mass ratio containing a total of
19 *μ*g of DNA was layered on top and the tube then centrifuged for seven hours
at 38,000 rpm in a SW 39 rotor (Beckman-Spinco) at 2°C. Fractions were collected
and assayed for optical density at 260 m*μ* (OD$_{260}$) and for activity of the indicated gene pairs. The activity assay differs from that published (Kaiser and Inman,
1965) in respect to the helper phage and the recipient bacteria. The helper phage
used in assaying the gene pairs were the double mutants N_7R_{60}, $O_{29}R_{60}$, $P_{80}R_{60}$, and
$Q_{73}R_{60}$ where the subscript numbers refer to the *sus* mutants of Campbell (1961).
The RR curve in the figure designates the activity for the *R* gene independent of
other genes, the helper phage being $R_{54}R_{60}$. The recipient and indicator bacteria
were the nonpermissive W3350 except for the *N-R* assay in which a permissive
strain, C600, was used as recipient (Campbell, 1961). The solution layered on the
gradient also contained a small amount of purified *E. coli* β-galactosidase (gift of
M. Cohn) which was used as an internal standard for calculating the sedimentation
coefficient of the DNA. It sedimented as a single zone with a peak at 0.39 on the
abscissa.

contain a cohesive site are active in our assay system (Kaiser and Inman, 1965). Adding this fact to the above results, we can conclude that the order of the genes in the isolated DNA is: (*N, i*λ)*-O-P-Q-R*-end.

We can make a rough estimate of the distance of each of these genes from the right end of the DNA from the sedimentation coefficient of the smallest molecules active for a given pair. The sedimentation coefficients were calculated by comparing the distance sedimented by such molecules (after allowing for the shape of a sedimenting zone of homogeneous DNA) to the distance moved by a zone of molecules of known sedimentation coefficient—in this case, *E. coli* β-galactosidase, whose sedimentation coefficient is 16 S. From these admittedly rough values we can calculate, to a first approximation, the molecular weight of the respective DNA molecules (Studier, 1965), and consequently the number of included base pairs. The number of base pairs from the required end to the *R* gene is thus computed to be less than 3000, a maximum value being given because we run out of DNA at this end of the size spectrum. The *Q* gene appears between 4000 and 5000 base pairs from the end; the *P* gene, between 8000 and 9000; the *O* gene, a little farther along between 9000 and 10,000 base pairs; and finally, the *N* and *i*λ genes at about 13,000 base pairs from the end. The relative distances of these genes is indicated in the diagram of Fig. 1.

In another set of experiments employing a similar but simpler methodology, we have been able to order the genes of the galactose operon in the left half of λ*dg* DNA, the result being that shown in Fig. 1. As these experiments do not allow the positioning of the galactose operon in terms of the number of base pairs from the end, and furthermore have been reported elsewhere (Hogness, 1966), we can dispense with their further discussion here.

In the sense that the method represented in Fig. 3 is generally applicable to other genes, and can be developed to yield more accurate distances (e.g., the DNA in the critical fractions used to determine the extrapolated position could be resedimented), these experiments are preliminary. Their extension to other genes or more accurate values will depend upon the specific problem at hand. For the present we can say that the gene order in the DNA is the same as the gene order on the vegetative map, and that relative gene distances on the map and in the DNA are roughly comparable. (A strict comparison between map and DNA distances is beyond the scope of this article and is, in any case, an area where an increased precision of the above methods is needed). Further, our experiments indicate that it should be possible to obtain a catalogue of fragments, each member of which has the same cohesive site at one end, but which differ in length by as little as one or two genes.

Orientation

The Orientation of the Galactose Operon in λDG

We can specify the orientation of the genes of the galactose operon in
λ*dg* DNA from their order in the DNA, and the knowledge that the operator or controlling end of this operon is adjacent to the structural gene
for epimerase, *e.* The latter information comes largely from the experiments of Buttin (1963), who mapped o^c mutations at the *e* end. David
Wilson (1966), in our laboratory, has found that while the wild-type *E.
coli* produces the polypeptides derived from the *k* and *e* genes in a one-
to-one ratio, in certain t^- mutants this ratio (k/e) is as low as 0.2, indicating a polarity consistent with the o^c position. In a similar vein, Adler
and Kaiser (1963) have shown that certain mutations which drastically
reduce the formation of all three galactose enzymes reside in or very
close to *e* on the genetic map.

With the operator next to *e,* the orientation of the genes in the galactose operon shown in Fig. 1 must be from right to left. The argument is
as follows:

1. The galactose operon is assumed to be like the tryptophan operon of
E. coli, where it is known that the direction from the operator end to the
other end corresponds to the direction within a gene from the codon
specifying the amino-terminal residue to the codon specifying the carboxy-terminal residue (Yanofsky *et al.,* 1964; Imamoto *et al.,* 1966).
We call this the *N*-to-*C* direction, which for the galactose operon in Fig.
1 is therefore from right to left.

2. Translation proceeds along the mRNA in the 5'- to -3' direction as
the polypeptide is extended at its carboxy-terminus (Streisinger *et al.,*
1966; Salas *et at.,* 1965; Thach *et al.,* 1965). The 5'- to -3' direction of
the mRNA derived from the galactose operon is therefore from right-to-
left in Fig. 1.

3. Transcription proceeds by addition to the 3'-terminus of the mRNA
(Bremer *et al.,* 1965; Maitra *et al.,* 1966). The galactose operon is
therefore transcribed from right-to-left in Fig. 1.

The Orientation of *R*, the Structural Gene for λ-Lyosyme

Although there are obvious similarities between the regulation of λ
genes and of genes in an operon, no operon has been clearly defined in λ
DNA. (This excludes some very recent evidence of Eisen *et al.* (1966)
concerning a region just to the right of *i*λ and which will be discussed
later.) Lacking such information for the specification of gene orientation,
we are using two other techniques. One of these is the determination of

the *N*-to-*C* direction within the *R* gene. The *R* gene is the structural gene for the λ-lysozyme (Campillo-Campbell and Campbell, 1965), and represents a late function in the sense that the appearance of λ-lysozyme is dependent upon early functions, particularly the *N* function, and occurs after the appearance of the λ-nuclease, an early enzyme (Black and Hogness, 1966).

Because the approach is straightforward and our information incomplete, our present status on this problem will simply be summarized here.

L. W. Black has isolated the λ-lysozyme and shown that it consists of a single polypetide containing about 160 amino acid residues. There are three methionine residues, of which one is amino-terminal and the other two are internal. Reaction of the enzyme with cyanogen bromide should therefore yield one homoserine or its lactone, derived from the amino-terminal methionine, and three subpeptides, two of which should have homoserine at their carboxy-terminus, and one of which should have the same carboxy-terminal residue as the complete enzyme, i.e., valine. Determination of the amino-terminus of the subpeptides, and comparison of these to the residue in position two of the complete enzyme (determined by Edman degradation) should allow the ordering of the three subpeptides.

These determinations have been performed, as have the total amino acid analyses of the three subpeptides. The results, which will be reported in detail elsewhere, are summarized in the partial primary structure given in Fig. 4.

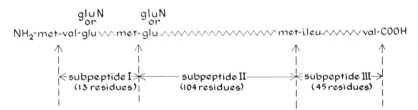

FIG. 4. A partial primary structure of λ-lysozyme.

The next step is to isolate the altered lysozymes from each of two *R* mutants, whose sites of mutation are sufficiently far apart that the respective positions of amino acid substitution will span an internal methionine residue. Direct amino acid analysis should indicate which subpeptide is altered in each case. This identification will yield the *N*-to-*C* direction for the two sites of mutation which, by the arguments given in the previous section, is the orientation of the *R* gene.

We are using *amber* mutants of the *R* gene, and isolating the lysozyme from lysates of bacteria lysogenic for the mutant phage and containing the *su-1* or *su-3* amber suppressor. These suppressors allow the *amber* nonsense codon of the mutant to be read as serine (*su-1*) or as tyrosine (*su-3*) with reasonable efficiency (Weigert *et al.,* 1966; Stretton *et al.,* 1966). The lysozyme isolated from at least one of the two suppressed sources should therefore differ from the wild-type enzyme by a single amino acid substitution. At present we are in the process of isolating these altered lysozymes. While further work is necessary to identify the orientation of the *R* gene, no significant technical problems remain to hinder obtaining the necessary results in the near future.

THE ORIENTATION OF THE *N* GENE

Clearly, the direct determination of the *N*-to-*C* direction is not sufficiently economical to be considered a generally applicable technique for determining gene orientation. It has, to date, been directly determined for only three genes: (1) the A gene of the *E. coli* tryptophan operon (Yanofsky *et al.,* 1964); (2) the structural gene for the head protein of phage T4 (Sarabhai *et al.,* 1964); and (3) the structural gene for the T4-lysozyme (Streisinger *et al.,* 1966). We are, therefore, developing a more general methodology based on our ability to isolate each of the intact single strands of λ DNA (Doerfler and Hogness, 1965, 1966). The goal is to determine which of the strands, I or II of Fig. 1, serves as template during transcription of a given gene. On the assumption that the transcribed mRNA and template strand of the DNA are antiparallel, the direction of transcription, and hence the gene orientation, will be opposite to the 5′- to -3′ direction of the template strand (see Hogness, 1966, for a discussion of this assumption and the possibility of its confirmation with the λ system). Thus the galactose operon, whose orientation is from right to left in Fig. 1, should utilize strand II for a template during transcription.

We have devised a technique that employs the above principle, but is limited in application to those genes whose function is necessary for replication of the infecting λ DNA (Doerfler and Hogness, 1966). It has been successfully applied to the N gene, and this case is summarized in the following paragraphs.

Consider the heteroduplex molecules indicated in Fig. 5. The first step in their construction is the isolation of the intact single strands from wild-type DNA, and from DNA which contains a point mutation in the *N* gene. This yields the four single strand preparations indicated in the middle of Fig. 5. The two strands from a single duplex DNA molecule are designated H and L because of the different buoyant densities they exhibit when centrifuged to equilibrium in alkaline (pH 12) CsCl gradi-

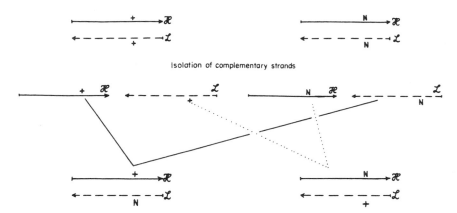

Isolation of complementary strands

FIG. 5. Scheme for the construction of heteroduplex molecules H^+/L^N and H^N/L^+.

ents. The H strand exhibits a density that is 0.004 gm cm^{-3} greater than that for the L strand at this pH, though at pH 7 no significant difference in density is observed.

The four strands can be paired in four ways if each is an H/L pair. Two pairs are the original homoduplex molecules shown at the top of Fig. 5 and two are the heteroduplex molecules shown at the bottom. Each of these four possible duplex DNA molecules was constructed by mixing equal amounts of the appropriate two strands and subjecting the mixture to the renaturation conditions of pH 10.5 and 37°C. While the single strands do not exhibit significant activity in our assay, the renatured duplex molecules exhibit activity for genes that have the wild-type homoduplex structure (see the bottom of the legend of Table I).

What is the N activity predicted for the two molecules which contain a heteroduplex structure in this gene? Since the helper phage must be an N mutant, the only source of the wild-type base sequence for this gene is the $+$ strand in each heteroduplex. However, as indicated earlier, the N function is necessary for replication of the infecting DNA—that is, synthesis of mRNA having the wild-type structure would appear to be a prerequisite for the formation of wild-type homoduplex DNA molecules necessary for observing a positive response in the assay. One would therefore predict that only the heteroduplex which contains the wild-type sequence in the template strand should be active. However the result, shown in Table I, is not the predicted one. Both heteroduplexes have about one-half the activity of the wild-type homoduplex.

We have constructed a working hypothesis which not only accounts for these results, but led to an experiment which allows a determination of the template strand for the N gene. We assume that both heteroduplexes

TABLE I

ACTIVITY OF GENE N IN HOMO- AND HETERODUPLEX MOLECULES[a]

Type of DNA	Relative activity of gene N
Homoduplexes:	
H^+/L^+	1.00
H^N/L^N	<0.01
Heteroduplexes:	
H^+/L^N	0.6
H^N/L^+	0.5

[a] The N mutant used to form the above DNA molecules is sus_7 obtained from Campbell (1961). After formation, the homo- and heteroduplexes were assayed for N and R activity according to the legend of Fig. 3. Recipient and indicator bacteria were the nonpermissive W3350. The helper phages were sus_7 and the double R mutant, $sus_{54} \ sus_{60}$, when assaying for N and R activity, respectively. The values given in the table represent the average ratio of N to R activity for each DNA relative to that for the H^+/L^+ DNA, the values from single assay sets lying within $\pm 25\%$ of these values. This ratio is used to take account of variations in the efficiency of renaturation during the formation of the different duplexes. The N activity of renatured H^+/L^+ DNA has varied between one- and two-thirds the activity of untreated wild-type DNA.

can be converted to the wild-type homoduplex, not by replication of the entire molecule, but rather by the processes of excision and repair which operate on lesions in DNA caused by ultraviolet irradiation (Howard-Flanders and Boyce, 1966). The model is illustrated in Fig. 6. The two heteroduplexes are represented on the first line where the position of mis-

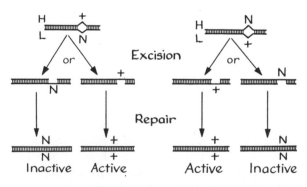

Average activity = 0.5 in each case

FIG. 6. Model for the formation of homoduplexes from heteroduplexes by excision and repair.

match of the bases is indicated by the separation. As with the repair of UV-induced lesions, the first step is imagined to be the excision of one strand or the other in the region of the mismatch. The probability of excising from one strand is considered equal to that for the other. The repair synthesis will then yield one homoduplex or the other with equal probability.

The activity of the heteroduplexes should therefore approach a value that is one-half that exhibited by the wild-type homoduplex as the efficiency of the excision and repair processes increases. This relative activity for the heteroduplex containing the mutant sequence in the template strand should be one-half or less, while for the heteroduplex with a wild-type template strand it could be greater than one-half if transcription of the N gene, and consequent replication of the entire molecule, were to win out against the competing repair mechanism.

On the basis of this model we designed the following experiment. We argued that if we irradiated the cells used in the assay with increasing doses of ultraviolet light, we should be able to create a' sufficient number of lesions in the DNA of the host cell to trap the enzymes involved in repair mechanisms. If the heteroduplex molecules should enter such a cell, the probability of forming the homoduplexes from them by the excision and repair mechanism would be greatly reduced. Consequently, one would predict that the activity of the heteroduplex containing the mutant sequence in the template strand would be greatly reduced compared to that of the heteroduplex with wild-type configuration in that strand.

The predictions from this line of reasoning did in fact obtain when the experiment was performed, as is shown in Fig. 7. In this experiment, the cells were exposed to different doses of ultraviolet light before being infected with helper phage. The survival of competence for the wild-type homoduplex DNA is shown in the upper part of the figure. Competence is a property of the cells that is relatively resistant to ultraviolet light. For example, after 250 sec of irradiation the survival of colony-formation is only 2×10^{-5}, whereas competence is little changed.

Rather than give a survival curve for each DNA, the ratio of the surviving fractions for pairs of DNA's is given in the lower part of the figure. Of primary interest is the survival ratio of the two heteroduplex DNA's, H^+/L^N to H^N/L^+. This ratio does not vary markedly until the dose is increased above 250 sec, at which time there is a dramatic increase such that the ratio reaches a value greater than or equal to 700 to 400 sec. Thus at this dosage there is an essentially qualitative difference between the heteroduplexes—the H^+/L^N is active and the H^N/L^+ is not. We interpret this to mean that the H strand is the template strand for gene N.

That the striking differential effect of ultraviolet irradiation of the cells on genetic activity of the heteroduplexes is specific to the N gene is indicated by a control experiment shown in Fig. 8. The experiment is the same as the previous one, except that the activity of gene R is measured. The helper-phage is an R mutant which can supply the N function if need be. Thus, the only gene that is measured is R, and each of the types of DNA contains the wild-type homoduplex structure for this gene. There is no significant differential effect of irradiation on any of the three DNA's tested, the homoduplex and the two heteroduplexes. Thus the differential effect of ultraviolet irradiation of the cells is not on the whole heteroduplex molecules, but rather only on gene N in those molecules.

Having concluded that the H strand functions as the template during transcription of the N gene, we must now determine whether H is synony-

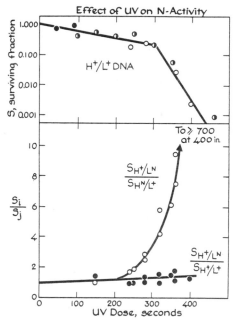

FIG. 7. The effect of ultraviolet irradiation of the recipient cells on N activity. The nonpermissive cells (W3350) were irradiated for the times indicated on the abscissa just before the addition of helper phage (see Table I for the assay of N activity). The source of ultraviolet light was the General Electric germicidal lamp and the intensity at the surface of the culture was 17 ergs·mm^{-2}·sec^{-1}. The upper part of the figure represents the results of three different experiments in which H$^+$/L$^+$ DNA was assayed for N activity. The ordinate represents the surviving fraction of this activity as the recipient cells receive greater doses of ultraviolet light. In the lower part of the figure the ordinate is the ratio of the surviving fraction obtained with one DNA (S_i) to that obtained with another DNA (S_j), both DNA's being assayed with the same irradiated culture.

mous with strand I or II of Fig. 1 before we can identify the orientation of *N*. Two different sets of data indicate that H and strand II (Fig. 1) are the same.

The first set derives from the initial experiments in a series we are doing, which involve the selective degradation of each strand at its 3'- end with the exonuclease I of *E. coli* (Lehman and Nussbaum, 1964). This exonuclease is specific for single strands, which it degrades one residue at a time from the 3'- end. The experiment performed by Doerfler which indicates the 5'- to -3' direction of the H and L strands is summarized in Fig. 9.

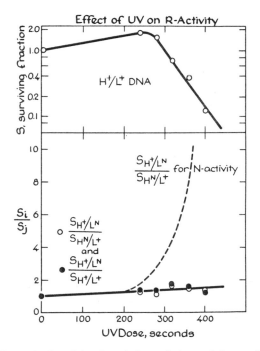

Fig. 8. The effect of ultraviolet irradiation of the recipient cells on *R* activity. The experiment is the same as that indicated in Fig. 7 except that *R* activity was measured.

Each of the isolated strands from wild-type λ DNA was treated with exonuclease I at a concentration, and for a period calculated to remove an average of 2–3% of the residues from the 3'- end of each strand. After stopping the reaction, each treated strand was renatured with its complementary untreated strand by incubation at pH 10.5 and 37°C. The resulting two types of molecules should have opposite ends damaged, as shown by the dotted lines in the figure. However, both molecular types should be active since the assay of fragments indicates that only one cohesive site is necessary for activity. This was found to be the case, whether it was the activity of the *A-B* pair of genes or that of the *R* gene which was being measured. The ratio of the *A-B* to *R* activity was not significantly different for the two molecular types. This is to be expected even if the enzyme degradation did penetrate to these genes, since the untreated complementary strand is wild-type and the function of these genes is not necessary for DNA replication (Brooks, 1965; Joyner *et al.,* 1966).

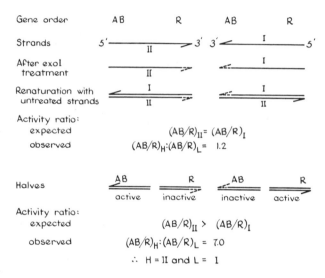

FIG. 9. The 5′-to-3′ direction of the H and L strands. See the text for explanation.

These two types of molecules were next broken near their centers to form the four types of half-molecules shown in Fig. 9. The half-molecules that contain ends in which one strand has been degraded by exonuclease I may be expected to be inactive, since the required cohesive site should have been altered sufficiently to inactivate it (Strack and Kaiser, 1965). Thus the ratio of A-B to R activity in the two populations of half-molecules should exhibit the difference indicated in the figures. The result confirmed the expectation—namely, that the ratio of *A-B* to *R* activity of the half-molecules derived from the form in which the H strand was treated with exonuclease I was sevenfold greater than that in which the L strand was treated. As indicated in the Fig. 9, this result is consistent with H being strand II of Fig. 1.

Very recently, Wu and Kaiser (1967) have obtained more conclusive evidence that H is strand II of Fig. 1. They labeled the duplex λ DNA with ^{32}P phosphate at the 5′-hydroxyl terminus of each strand, using polynucleotide kinase according to the technique described by Weiss and Richardson (1966). By degradation of the isolated strands, they found that the 5′- termini of the H and L strands were G and A, respectively. Since they also found that the ^{32}P was associated with G in the left-half of the duplex, and with A in the right-half, they concluded that the 5′- to -3′ direction of the H strand is from left to right.

It is strand II in Fig. 1 which we therefore identify as the template strand for the transcription of the *N* gene. Accepting the assumption that the mRNA and this template strand are antiparallel, we conclude that the orientation of the *N* gene is from right to left. Thus, in the DNA isolated from λ*dg,* the orientation of the galactose operon and the *N* gene are the same. By the arguments given in the section on gene position, we can also conclude that in the prophage state, *N* has the same orientation as the adjoining galactose operon (Fig. 2).

Conclusion

We may conclude by considering the significance of the preceding information to a determination of the sites in λ DNA at which transcription is initiated.

If the directions of transcription of two genes on the same DNA point away from each other, there must exist at least two points for the initiation of transcription between them. Taken alone, our results do not yet indicate such a case of opposite gene orientation in λ. However, Eisen *et al.* (1966) have recently obtained preliminary evidence for the existence of an operon located to the right of *N* (Fig. 1), which appears to be oriented from left to right. Their genetic map of the relevant region in λ yields the following gene order: -*N-C*$_I$-*X*-*C*$_{II}$-*O*-*P*-, where the *i*λ of Fig. 1 includes *C*$_I$ and *X*, but not *C*$_{II}$ or the mutation sites we have used to identify *N* (i.e., N_7 and N_{53}). Their data suggest the existence of an operon that includes *X, C*$_{II}$ and *O,* but not the other genes. Since some mutations in X inactivate *O* function, while others in this region render the *O* function constitutive (Jacob, personal communication), it would appear that this operon is oriented from left to right.

The orientation of this suspected operon would then be opposite to that for the *N* gene. If this is the case, there must exist two sites for the initiation of transcription in, or very near, the *i*λ region—some 13,000 base pairs from the right end of the DNA. Furthermore, if *N* is part of an operon, its operator locus should lie just to its right, at the left-hand end of the *i*λ-region. One mutant, *t27,* described by Eisen *et al.* (1966) lies within the *i*λ-region, at its extreme left and fails to complement N_7. Perhaps such a mutant defines a region like *X*, but which is operative in the opposite direction. This speculation leads to the view that the *i*λ-region contains two operators at either end, and one or two regulator genes in the center (C_I), whose product(s) control the respective operons.

Opposite orientations for *N* and *O* necessitate the use of different strands as template during transcription. Thus the template strand for *O* should be strand I of Fig. 1. We expect to be able to test this conclusion and to determine if there are changes in the direction of transcription at other positions in λ DNA by finding out if the template strand is different for different regions. We must abandon the heteroduplex technique for all cases in which the function of the relevant gene is not necessary for DNA replication—even one round of replication is too much.

The *O* gene is different from *N* in that mutants of *O* can multiply if the infection is at high multiplicity (Campbell, personal communication), and do yield significant amounts of λ-lysozyme, λ-nuclease, serum-blocking activity, and various parts of the mature phage (observable in the electron-microscope) when induced from the prophage state (Dove, 1966; Protass and Korn, 1966; Eisen *et al.*, 1966). Although the existing data indicate that there is very little if any DNA replication after induction of *O* mutant prophage (Joyner *et al.*, 1966; Eisen *et al.*, 1966), these other characteristics make it dubious that the heteroduplex technique will be applicable to *O*. Indeed, preliminary experiments with *O* heteroduplexes indicate that, with the present assay system, they are as active as the wild-type homoduplex, and that the ratio of activity of the two heteroduplexes does not change significantly after UV irradiation of the recipient cells. We suppose that at the high multiplicities of helper phage used in the assay, at least one round of replication of the heteroduplexes occurs.

A more general method that we are presently adopting consists in (1) the isolation of λ mRNA made *in vivo* which is specific to a given region of the λ DNA, and (2) the determination of which DNA strand will form a hybrid with this mRNA. It is in the isolation of such specific mRNA that we intend to employ the catalogue of λ DNA fragments mentioned at the end of the section on gene position.

Finally, in regard to the general approach given in this article and to these future experiments, it should be emphasized that we have little information on which to base a prediction as to the number of operons in λ DNA, and even less for a prediction of their relative orientations. Thus, at present, no rules exist about the relative orientations of a set of operons, or genes, in viral DNA except that they can be different. Pertinent to this last point is the recent observation by Streisinger *et al.* (1966) that in T4 the N-to-C directions of the structural genes for the lysozyme and head protein are opposite.

REFERENCES

Adler, J., and Kaiser, A. D. (1963). *Virology* **19,** 117.

Adler, J., and Templeton, B. (1963). *J. Mol. Biol.* **7,** 710.

Amati, P., and Meselson, M. (1965). *Genetics* **51,** 369.

Arber, W. (1958). *Arch. Sci. (Geneva)* **11,** 259.

Black, L. W., and Hogness, D. S. (1966). In preparation.

Bode, V. C., and Kaiser, A. D. (1965a). *Virology* **25,** 111.

Bode, V. C., and Kaiser, A. D. (1965b). *J. Mol. Biol.* **14,** 399.

Bremer, H., Konrad, M. N., Gaines, K., and Stent, G. S. (1965). *J. Mol. Biol.* **13,** 540.

Brooks, K. (1965). *Virology* **26,** 489.

Buttin, G. (1963). *J. Mol. Biol.* **7,** 183.

Campbell, A. (1961). *Virology* **14,** 22.

Campbell, A. (1962). *Advan. Genet.* **11,** 101.

Campbell, A. (1964). *Bacteriol. Proc.* p. 140.

Campbell, A. (1965). Personal communication.

del Campillo-Campbell, A., and Campbell, A. (1965). *Biochem. Z.* **342,** 485.

Doerfler, W., and Hogness, D. S. (1965). *Federation Proc.* **24,** 226.

Doerfler, W., and Hogness, D. S. (1966). In preparation.

Dove, W. F. (1966). *J. Mol. Biol.* **19,** 187.

Eisen, H. A., Fuerst, C. R., Siminovitch, L., Thomas, R., Lambert, L., Pereira DaSilva, L., and Jacob, F. (1966). *Virology* **30,** 224.

Hershey, A. D., and Burgi, E. (1965). *Proc. Natl. Acad. Sci. U.S.* **53,** 325.

Hershey, A. D., Burgi, E., and Ingraham, L. (1963). *Proc. Natl. Acad. Sci. U.S.* **49,** 748.

Hogness, D. S., (1966). *J. Gen. Physiol.* **49,** 29.

Hogness, D. S., and Simmons, J. R. (1964). *J. Mol. Biol.* **9,** 411.

Howard-Flanders, P., and Boyce, R. P. (1966). *Radiation Res.* Suppl. 6, p. 156.

Imamoto, F., Ito, J., and Yanofsky, C. (1966). *Cold Spring Harbor Symp. Quant. Biol.* **31** (in press).

Isaacs, L. N., Sly, W. S., and Echols, H. (1965). *J. Mol. Biol.* **13,** 963.

Jacob, F. (1966). Personal communication.

Jacob, F., and Wollman, E. L. (1961). "Sexuality and the Genetics of Bacteria," pp. 295–305. Academic Press, New York.

Joyner, A., Isaacs, L. N., Echols, H., and Sly, S. W. (1966). *J. Mol. Biol.* **19,** 174.

Kaiser, A. D. (1962). *J. Mol. Biol.* **4,** 275.

Kaiser, A. D., and Hogness, D. S. (1960). *J. Mol. Biol.* **2,** 392.

Kaiser, A. D., and Inman, R. B. (1965). *J. Mol. Biol.* **13,** 78.

Lehman, I. R., and Nussbaum, A. L. (1964). *J. Biol. Chem.* **239,** 2628.

Maitra, V., Cohen, S. N., and Hurwitz, J. (1966). *Cold Spring Harbor Symp. Quant. Biol.* **31** (in press).

Protass, J. J., and Korn, D. (1966). *Proc. Natl. Acad. Sci. U.S.* **55,** 1089.

Radding, C. M., and Kaiser, A. D. (1963). *J. Mol. Biol.* **7,** 225.

Rothman, J. L. (1965). *J. Mol. Biol.* **12,** 892.

Salas, M., Smith, M. A., Stanley, W. M., Jr., Wahba, A. J., and Ochoa, S. (1965). *J. Biol. Chem.* **240,** 3988.

Sarabhai, A. S., Stretton, A. O. W., Brenner, S., and Bolle, A. (1964). *Nature* **201,** 13.

Smith, M. G., and Skalka, A. (1966). *J. Gen. Physiol.* **49,** 127.

Strack, H. B., and Kaiser, A. D. (1965). *J. Mol. Biol.* **12,** 36.

Streisinger, G., Okada, Y., Terzaghi, E., Emrich, J., Tsugita, A., and Inonye, M. (1966). *Cold Spring Harbor Symp. Quant. Biol.* **31** (in press).

Stretton, A. O. W., Kaplan, S., and Brenner, S. (1966). *Cold Spring Harbor Symp. Quant. Biol.* **31** (in press).

Studier, F. W. (1965). *J. Mol. Biol.* **11,** 373.

Thach, R. E., Cecere, M. A., Sundararajan, T. A., and Doty, P. (1965). *Proc. Natl. Acad. Sci. U.S.* **54,** 1167.

Weigert, M. G., Gallucci, E., Lanka, E., and Garen, A. (1966). *Cold Spring Harbor Symp. Quant. Biol.* **31** (in press).

Weiss, B., and Richardson, C. C. (1966). *Cold Spring Harbor Symp. Quant. Biol.* **31,** (in press).

Wilson, D. B. (1966). Ph.D. Thesis, Dept. of Biochemistry, Stanford.

Wu, R., and Kaiser, A. D. (1967). *Proc. Natl. Acad. Sci. U.S.* **57,** 170.

Yanofsky, C., Carlton, B. C., Guest, J. R., Helsinki, D. R., and Henning, V. (1964). *Proc. Natl. Acad. Sci. U.S.* **51,** 266.

Young, E. T., and Sinsheimer, R. L. (1964). *J. Mol. Biol.* **10,** 562.

The Synthesis of the λ Chromosome: The Role of the Prophage Termini

William F. Dove

MCARDLE MEMORIAL LABORATORY FOR CANCER RESEARCH
UNIVERSITY OF WISCONSIN, MADISON, WISCONSIN

It is known that the chromosome of the mature lambda phage is a single DNA molecule, a continuous linear duplex of 5×10^4 base-pairs terminated by complementary single-stranded 5′-ended cohesive sites about 10 bases in length (Burgi and Hershey, 1963; Studier, 1965; Hershey *et al.*, 1963; Hershey and Burgi, 1965; Strack and Kaiser, 1965; Wang and Davidson, 1965).

It is known that this phage chromosome can exist as a stable genetic element of its host, *E. coli* K12. This element is called the prophage (see Lwoff, 1953). Physical evidence bearing on the structure of the prophage is limited (Hoffman and Rubenstein, 1966), but genetic evidence of several sorts, one of which will be mentioned here, indicates that the lambda prophage is an intact element inserted into the continuity of the *E. coli* chromosome which is 100 times its length (Campbell, 1963; Franklin *et al.*, 1965; Rothman, 1965; Signer, 1966). The gene order in the prophage is a circular permutation of the order found in the mature phage chromosome. Campbell (1962) has accounted for this structure by proposing that the transition between the free phage chromosome and the prophage takes place via a circular form (Hershey *et al.*, 1963) and reciprocal genetic exchange between a sequence in the center of the mature phage chromosome and a sequence at the prophage attachment site on the host chromosome. The prophage attachment site of lambda lies near the bacterial *Gal* locus; φ80 and all its known hybrids with lambda attach at a site near the bacterial Tryp locus.

In the theory of Campbell, it need not be specified whether the phage sequence and host sequence involved in integration are strictly homologous ("First-order Theory") or quasi-homologous ("Second-order Theories"). In either case, the theory predicts that the prophage is flanked by two sequences which are at least quasi-homologous (see Fig. 2).

Thus the current evidence suggests that both the mature lambda chromosome and the prophage are bounded by terminal redundancies. For the mature chromosome, these termini are single-stranded; for the prophage they are double-stranded. As Thomas (Chapter 2) elaborated the possession of terminal redundancies ranging from 10 to 1000 nucleotides in length is a feature shared by all chromosomes yet studied in the family of double-stranded DNA bacteriophages. This feature permits the facile formation both of circular structures by recombination between the termini of the same molecule, and of end-to-end polymers ("Concatenates") by recombination between the termini of different molecules.

In parallel with the work of Franklin (1966) I have been investigating two roles of the prophage terminal redundancy. First we have studied their role in selection of phage sequences for inclusion in mature virus particles after induction of a prophage. Second, we have investigated the importance of the redundancy in the selective replication of phage sequences after prophage induction.

We shall review Franklin's evidence that both members of the prophage terminal redundancy are necessary for the selection of a homogeneous set of DNA molecules for the mature virus. Second, we shall see that when one prophage terminus is absent, the other directs the selection of sequences after induction. Thus a mechanism exists whereby at least one of the prophage termini can be activated for excision in the absence of homology between the two termini. Third, we shall see that the possession of a terminal redundancy is not essential for the selective replication of phage sequences after induction of a prophage. Phage-specific DNA synthesis occurs, at a somewhat reduced rate, after induction of prophages lacking one of their termini.

To introduce this discussion, let us review briefly what is known about the molecular state of the lambda DNA molecule during its autonomous replication.

Upon injection into the host, the lambda DNA molecule is converted to a form or series of forms which are not infectious in the Kaiser-Hogness (Kaiser, 1962) infectivity system (DNA eclipse; Dove and Weigle, 1965). Late in phage development infectious phage chromosomes are produced (DNA maturation). At least the initial eclipse reaction can be performed in chloramphenicol, implying that it is performed by pre-existing enzymes. DNA maturation is controlled by at least six late-acting phage genes, *A* through *F* (Dove, 1966a; Weigle, 1966).

There is evidence, as yet incomplete, that during DNA eclipse much of the parental phage DNA is in a circular form (Young and Sinsheimer,

1964; Bode and Kaiser, 1965; Weissbach *et al.,* 1966). There is also evidence that during phage replication, DNA is synthesized in a circular form (Young and Sinsheimer, 1964; Lipton and Weissbach, 1966) and/or in a concatenate form (Smith and Skalka, 1966).

The molecular events in DNA maturation are not known. It is probable that at least the production of the cohesive sites and the packaging of the mature chromosomes are involved (see Dove and Weigle, 1965; Weigle, 1966).

We have studied the role of the prophage termini in processes involved in phage chromosome synthesis using a set of deletion mutants derived from the lysogenic strain W1485 ($i^{\phi 80}h_\lambda^+$), described by Franklin *et al.* (1965). We have genetically characterized the phage $i^{\phi 80}h_\lambda^+$ by its ability to complement certain *sus* mutants of lambda. Whereas $\phi 80$ (Matsushiro, 1961; Matsushiro *et al.,* 1962) is $A^-,B^-,C^-,N^+,O^-,P^-,R^+$ in such complementations, the hybrid $i^{\phi 80}h_\lambda^+$ is $A^+,B^+,C^+,N^+,O^-,P^-,R^+$. Since $\phi 80$ possesses cohesive ends homologous with those of lambda (Yamagishi *et al.,* 1965; Burgi, personal communication), we propose the structure shown in Fig. 1, line 3, for the mature chromosome of $i^{\phi 80}h_\lambda^+$. The prophage in the lysogenic strain W1485 ($i^{\phi 80}h_\lambda^+$) has the structure shown in line 4 (see Franklin *et al.,* 1965). This structure is a circular permutation of the presumed structure in line 3.

This prophage structure was established by the existence of a set of deletion mutants which form the material for this discussion. It is a useful fact of natural history that T_1^r mutants at the Tryp locus often are deletion mutants, as if the T_1^s phenotype can be lost only by extensive mutation of

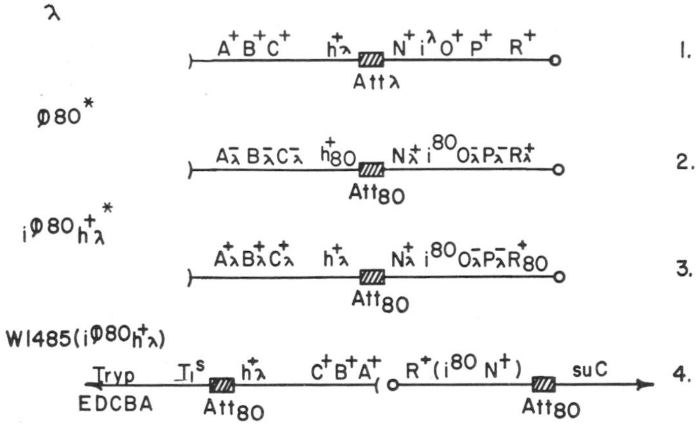

FIG. 1. Genetic composition of λ, $\phi 80$, and $i^{\phi 80}h_\lambda^+$. o,) = complementary cohesive ends; * = presumed gene order; ( ) = not ordered; Att = attachment homology sequence. (Franklin *et al.,* 1965; Signer *et al.,* 1965).

William F. Dove

this gene. Thus many of the T_1^r mutants isolated from W1485 ($i^{\phi80}h_\lambda^+$) are Tryp⁻ deletions (Anderson, 1946; Yanofsky and Lennox, 1959), and many are defective lysogens. Table I shows the genetic characterization of the deletion prophages which will be used for this discussion. The D series was isolated from W1485 ($i^{\phi80}h_\lambda^+$), while the JB5000 series was isolated from W3101 ($i^{\phi80}h_\lambda^+$).

TABLE I

Pᴿᴏᴘʜᴀɢᴇ Dᴇʟᴇᴛɪᴏɴˢ[a]

Strain	Phage yield per cell	Tryp	h_λ^+	J^+	$..F^+..$	A^+	R^+_{80}	$...i^{80}$	N^+_{80}
W1485 ($i^{\phi80}h_\lambda^+$)	10–80	+	+	+	+	+	+	+	+
D1, D3, D11, D12	10^{-2}	−	+	+	+	+	+	+	+
D10	10^{-2}	+	+	+	+	+	+	+	+
D25	$<10^{-7}$	+	−	+	+	+	+	+	+
D24	$<10^{-7}$	−	−	−	+	+	+	+	+
D30	$<10^{-7}$	+	−	−	−	+	+	+	+
D15	$<10^{-7}$	+	−	−	−	−	(+)	(+)	?
D9	$<10^{-7}$	−	−	−	−	−	−	(+)	?
D18	$<10^{-7}$	−	−	−	−	−	−	(+)	?
D19	$<10^{-7}$	−	−	−	−	−	−	−	?
D26	$<10^{-7}$	−	−	−	−	−	−	−	?
W3101 ($i^{\phi80}h_\lambda^+$)	10–80	+	+	+	+	+	+	+	+
JB5062	$<10^{-7}$	+	−	−	+	+	+	+	+

[a] (+) signifies that the presence of the gene cannot be demonstrated by rescue with lambda phages; the R^+ gene is detected by lysozyme synthesis; the $i^{\phi80}$ gene is detected by immunity and by rescue with the homologous phages $i^{\phi80}$ vir h_λ^+.

The selected set shown here demonstrates the basic feature which provides evidence for the linear insertion of the prophage into the bacterial chromosome. Bacterial Tryp and T_1^s characters and prophage characters form a coherent set spanned by the set of deletions. Barring special mechanisms for the origin of these deletions (and they are no more common than T_1^r deletions in a nonlysogenic strain), this feature demonstrates that each prophage marker up to $i^{\phi80}$ is intimately associated with the bacterial chromosome in a linear array. We take this intimate association to be linear insertion of the entire prophage into the bacterial chromosome.

Now let us discuss the role of the prophage termini in the induction, replication, and maturation of the lambda chromosome from the prophage state. First, what is the effect of loss of the h_λ^+ prophage terminus upon the efficiency of formation of a mature phage chromosome after induction? According to the notions we have presented, this formation requires joining between the prophage termini and opening of the *A–R* join followed by packaging of the mature phage chromosome. This event occurs in most cells after induction of W1485 ($i^{\phi 80}h_\lambda^+$), yielding an average burst size of 10 to 80 phages per cell. In the deletion D10, this event occurs in only 4×10^{-3} of the cells after induction, again with an average burst of 10 to 80 phages per cell. These phages, once formed, are capable of normal lytic growth, implying that in deletion D10 no phage genes essential for lytic growth have been deleted. We call D10 a terminal deletion. Within the Campbell theory, D10 has suffered a complete deletion of the h_λ^+ terminal homology segment because no completely normal phage particles are found among the phages yielded by these rare cells. These phages are all incapable of establishing lysogeny (Franklin, 1966).

We conclude from this result that the possession of both prophage termini is essential for the efficient selection of phage sequences for mature phages after prophage induction. We shall see shortly that the inefficient production of phages by D10 is reflected not by inefficient phage DNA replication but by inefficient DNA maturation and packaging. We believe that the step controlled by the termini is the joining between the prophage termini, and that this step is essential for DNA maturation.

Franklin has conducted a study of the distribution of densities of the phages made by D10, which gives further insight into the role of the prophage termini in selection of phage sequences after induction. Whereas the distribution of densities of the phages made by W1485 ($i^{\phi 80}h_\lambda^+$) is narrow and unimodal at $\rho = 1.507$, those rare phages made by D10 possess a broad distribution of densities from $\rho = 1.507$ to $\rho = 1.468$. Both the normal and light density phages breed true with respect to density. Thus the broad distribution of densities of phages made by D10 probably reflects a broad distribution of lengths of DNA selected for packaging in the rare cells which succeed in making phages. This situation will be discussed in more detail elsewhere (Franklin, 1966).

It seems then that the pair of prophage termini are required for efficient and for accurate selection of phage sequences upon induction of the prophage. This is easily understood by the Campbell theory: The selection involves recombination between the prophage termini.

Additional insight into the role of the prophage termini in sequence selection can be gained by asking whether the rare phages made by D10 are formed by recombination events distributed randomly across the prophage. If the sequence selection in normal prophage induction is directed solely by the extensive homology available for recombination between the prophage termini, then one might expect that in the absence of one terminus these recombination events would occur at low probability and would be randomly distributed.

An approximate calculation can be made for this prediction, as diagrammed in Fig. 2.

Let E_L be the length of the region essential for phage growth including h_λ^+ and the A–R join; let E_R be the length of that region between the A–R join and N_{80}^+; and let L be the length of DNA packaged. All lengths are expressed in lambda equivalents. $E_L \geq 0.37$ λ from recombination data (Amati and Meselson, 1965; Jordan and Meselson, 1965); $E_R \geq 0.28$ λ from physical data (Egan and Hogness, 1966). For random excision, with the constraint that the A–R join must be included to allow DNA matura-

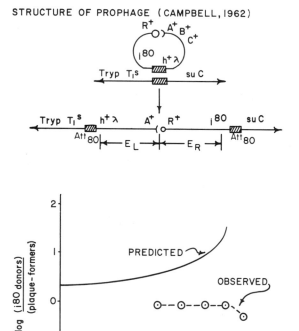

FIG. 2. Random excision of sequences after induction of a terminal deletion. A lysate from D10 was centrifuged to equilibrium in CsCl, $\rho = 1.49$, at 33,000 rpm in an SW39 rotor. Six-drop fractions were collected and titered for $i^{\phi 80}$ donor units and plaque forming units. The log of the ratio of $i^{\phi 80}$ donor units to plaque forming units is plotted as a function of DNA length L for the case of random excision. The observed ratio is plotted in comparison.

tion, we predict that the ratio of probabilities for excision of an i$^{\phi 80}$ donor particle and excision of a plaque - forming particle is

$$\frac{\text{P (i }^{\phi 80}\text{ donor particle)}}{\text{P (plaque former)}} = \frac{L - E_R}{L - E_R - E_L} = \frac{L - 0.28}{L - 0.65}$$

We have plotted $\log \dfrac{\text{(i }^{\phi 80}\text{ donors)}}{\text{(plaque former)}}$ against L in Fig. 2.

The experimental ratio was obtained by analyzing fractions of a density gradient distribution of particles from the D10 burst (a) by adsorbing to C600 (λ), inducing, and plating on a lawn of C600 (λ); and (b) by plating directly on a lawn of C600. Control experiments showed that defective particles of λ*dg* could be quantitatively assayed by procedure (a). The value of L was estimated from the density of the particles by the equation of Weigle (1961).

The observed ratio of $i^{\phi 80}$ donors to plaque formers remains constant as a function of density and actually falls slightly at the light edge of the band, for unknown reasons. Thus the expectation for random excision is not fulfilled, and it seems clear that errors in the values of E_L, E_R, and L cannot be large enough to account for this discrepancy. We conclude therefore that the phages made by D10 are made by selection which ensures that the only particles found contain all phage genes essential for lytic growth.

A trivial explanation for this observation would be that only nondefective phages could be packaged because only in induced bacteria destined to liberate nondefective phages are all essential phage structures synthesized. Although we cannot rigorously rule this out until we can measure the synthesis of all essential phage structures, two arguments make it unlikely. First, in induced D10 cultures all known phage products including lysozyme, tail antigen, and intact phage tails are made with normal efficiency. Second, defective particles (λ*dg*) can appear in a burst from a normal lambda lysogen, presumably because lysogens are multinucleate and carry a normal phage chromosome as well as the defective one after induction.

Aside from the above trivial explanation for this result, one can propose that there exists a system which does not require both termini for the selection of phage sequences. Many elaborate systems can be imagined. One simple scheme requires that one or both of the prophage termini is activated for excision after induction. In the normal prophage this results in recombination between the two termini and, eventually, in the production of phages carrying chromosomes defined in length and sequence. In the terminal delection prophage, this results in recombination between a point fixed in the $i^{\phi 80}$ terminus and a point randomly selected along the chromosome of the lysogen. Were this random point to lie in the $A^+ - {}^+_\lambda$ interval, a defective particle would arise. The absence of such particles is accounted for simply by proposing that there is a minimum length of DNA molecule, near 0.7λ, which can be packaged by lambda.

Thus we propose that upon induction of a lysogen, a specific sequence in the $i^{\phi 80}$ terminus of the prophage is activated for excision. We cannot say whether similar activation of the h^+_λ terminus occurs. The selection of phage sequences is not dictated solely by extensive homology between the prophage termini. This is a logical corollary, for the case of prophage induction, of the proposition of Signer and Beckwith (1966) that a specific "recombinase" directs prophage insertion.

Finally, we want to discuss the role of the prophage termini in allowing the selective synthesis of phage DNA after induction. This is of general interest because of the ubiquitous occurrence of terminally redundant chromosomes and the observation of circular and concatenated replicative forms.

One extreme hypothesis for the role of circular and concatenated forms in viral replication is that only one replication complex ("initiator"; Jacob and Brenner, 1963) is available for such replication, that this complex cannot freely diffuse from template to template, and consequently that manifold replication can be achieved only by the use of circular templates or concatenates. The elements of this hypothesis were first suggested to me by Sinsheimer, and his work on the replication of ϕX-174 (Chapter 11) indicates that it is a cogent hypothesis.

We have tested this extreme hypothesis by asking whether deletion prophages can achieve autonomous manifold replication after induction. The result is that they can, at an overall rate reduced at most by a factor of 4 from the normal rate of autonomous phage DNA replication.

Induced cultures were labeled with tritiated thymidine, either from 55 to 57 min after induction, or continuously from 0 to 70 min in the presence of deoxyadenosine (Boyce and Setlow, 1962). As will be discussed in detail elsewhere (Dove, 1966b), the total synthesis of DNA per induced bacterium did not differ by more than a factor of 2 among the strains being compared by any one procedure. Recovery of material during sonication, purification, or assay was high and nonvarying in such sets. Labeled material was characterized as DNA by alkaline stability, RNase insensitivity, and DNase sensitivity. Such labeled material was assayed for phage-specific DNA by DNA-DNA hybridization in agar matrices (Bolton and McCarthy, 1962).

In Fig. 3 the elution patterns are compared for DNA by W3101 (i$^{\phi 80}$h$^{+}_{\lambda}$), by W3101, and by JB5062; all were labeled from 55 to 57 min after ultraviolet induction. At least 0.69 of the labeled DNA made by the normal lysogen is phage-specific. The background shown by W3101 is 0.05. JB5062 allows a synthesis of at least 0.46 phage-specific DNA at this time.

A measure of the overall rate of phage-specific DNA synthesis is obtained by labeling from 0 to 70 min after induction in the presence of 200 μg/ml deoxyadenosine (Boyce and Setlow, 1962).

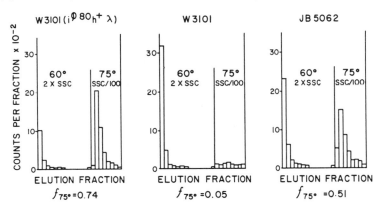

FIG. 3. Hybridization of DNA made from 55 to 57 min by W3101 (i$^{\phi 80}$h$^{+}_{\lambda}$), by W3101, and by JB5062. Cultures were labeled from 55 to 57 min after induction, and DNA isolated, purified, and sonicated. Portions of labeled DNA were hybridized with DNA from purified phage, and fractions eluted at 60°C with 0.30 *M* NaCl, 0.03 *M* sodium citrate ("2 × SSC") and then at 75°C with 0.0015 *M* NaCl, 0.00015 *M* sodium citrate ("SSC/100"). Labeled DNA in these fractions was determined by standard radiochemical methods.

Results of the assay of such labeled material from W1485 (i$^{\phi 80}$h$^{+}_{\lambda}$) and from D10 are shown in Fig. 4. It is calculated that in the normal lysogen at least 36 phage equivalents of phage DNA per induced bacterium are made by 70 min; for D10, at least 15 phage equivalents are made. These estimates are lower limits because the efficiency of hybridization is not 1. It ranges from 0.5 to 0.8.

The most extensive deletion we have found which still permits detectable autonomous phage DNA replication is D15. There is only presumptive genetic evidence (Franklin *et al.,* 1965; Dove, 1966b) that in D15 the *A–R* join has been deleted (Table I). This evidence would dictate that the origin(s) of phage DNA replication lie in the $i^{\phi 80}$-*R*$^{+}$ segment. Furthermore, we have observed in all terminal deletion strains that both lysozyme and phage antigen are made in normal quantities, as shown in

Table II. Since it is known that the synthesis of these late proteins is replication-dependent (Brooks, 1965; Dove, 1966a; Joyner *et al.,* 1966; Echols, Chapter 8), we presume that the genes (*R* and *J,* respectively) for these proteins are replicating after induction of a deletion prophage.

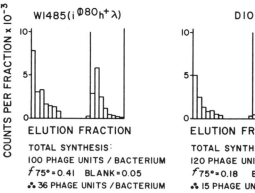

FIG. 4. Hybridization of DNA made from 0 to 70 min by W1485 ($i^{\phi 80}h_\lambda^+$) and by D10. Cultures were labeled from 0 to 70 min after induction, and DNA hybridized and assayed as in the experiments described in Fig. 3.

TABLE IIA

SYNTHESIS OF SERUM-BLOCKING POWER BY INDUCED PROPHAGES[a]

Strain	Maximum synthesis (phage equivalents/ bacterium)	Time of half-maximal synthesis
W1485 ($i^{\phi 80}h_\lambda^+$)	90 ± 20	80 ± 5
D1	70 ± 20	90 ± 5
D10	50 ± 10	90 ± 5
D11	110 ± 60	—
D12	170 ± 90	—
D3	140 ± 50	—

[a] Method described in Dove (1966a).

TABLE IIB

SYNTHESIS OF LYSOZYME[a]

Strain	Maximum level (units/mg protein)	Time, half-maximal synthesis (min)
W1485 ($i^{\phi 80}h_\lambda^+$)	3.8 ± 0.4	69 ± 1
D10	3.5 ± 0.5	82 ± 1
D15	3.8 ± 0.4	83 ± 1

[a] Lysozyme assay: 1 unit = $1.00 \, \Delta \, OD_{540}m\mu/min$. Purified lysozyme (Worthington) = 170 units/mg protein. Method of Sekiguchi and Cohen (1964).

The simplest mode of replication consistent with all the data is that there is an origin of replication at the $i^{\phi 80}$ terminus of the prophage, and that replication proceeds from $i^{\phi 80}$ to h_λ^+. This is our working hypothesis, and it is being directly tested.

In all prophage deletion strains tested we have observed that more nonphage DNA is made than in a normal lysogen. We believe that this reflects replication of bacterial sequences adjoining the prophage under the direction of the phage replication system.

We conclude from these experiments that preferential synthesis of phage DNA can take place after induction of deletion prophages, at a rate reduced from normal by at most a factor of 4. Thus the possession of terminal redundancies is not essential for manifold phage DNA replication in this system. Either, as suggested to me by Stahl, circular structures and/or concatenates can form efficiently by mechanisms not involving the normal termini, or else these structures are not obligate replicative forms after lambda prophage induction.

The fate of the phage DNA synthesized by a deletion lysogenic strain is not normal. Although phage DNA is made at efficiency near 1 in a terminal deletion strain such as D10, infectious DNA is made at efficiency 10^{-3}. Packaging of DNA into particles with densities near that of mature phage cannot be detected at a level of 10^{-2} (See Fig. 5), and mature phages are made at efficiency 10^{-3}, as discussed above. We presume that the replicating molecules cannot be matured, either because

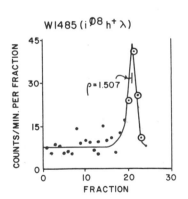 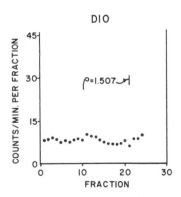

FIG. 5. Packaging of DNA in W1485 ($i^{\phi 80}h_\lambda^+$) and in D10. Induced cultures of W1485 ($i^{\phi 80}h_\lambda^+$) and D10 were labeled with tritiated thymidine at 55 to 57 min after induction, an excess of unlabeled thymidine was added, and incubation was continued until lysis. After incubation of the lysate with 10 μg/ml DNase in 0.1 M MgCl$_2$ at 37°C for 30 min, a portion was centrifuged to equilibrium in CsCl, $\rho = 1.52$, and fractions were collected. Labeled DNA was assayed by standard methods.

excision of DNA near in length to that of the mature phage is essential for DNA maturation and packaging, or because the molecules remain associated with the bacterial chromosome and are not available for maturation for topographical reasons.

In summary, we have reviewed evidence of Franklin that the prophage termini direct the selection of a homogeneous class of molecules for inclusion into phage particles after induction of a lysogen. We presume this selection is by recombination as proposed by Campbell. Second, we have seen that the selection of phage sequences after induction involves a specific system acting at least at the $i^{\phi 80}$ terminus to direct excision even when the h_λ^+ terminus is absent. Thirdly, we have found that prophage structures lacking the h_λ^+ terminus are capable of directing preferential phage DNA synthesis after induction, at a somewhat reduced rate. This result rules out extreme hypotheses of an obligatory role of circular and concatenated forms in lambda DNA replication.

ACKNOWLEDGMENTS

This work was begun in the Department of Biochemistry, Stanford University Medical School, and is being continued at Wisconsin. Throughout this time, the work has been catalyzed by discussion and exchange of information with Dr. Naomi Franklin.

Various parts of this work at Wisconsin were performed with the assistance of Mrs. Mary Walker, Miss Ruthann Zimmer, Mrs. Ermile Hargrove, and Miss Jean Jensen.

At Stanford I was supported by a postdoctoral fellowship of the National Institutes of Health. At Wisconsin, support is provided by the Alexander and Margaret Stewart Trust Fund, and by Program-Project Grant CA-07175 and Training Grant CRTY-5002 of the U.S. Public Health Service.

REFERENCES

Amati, P., and Meselson, M. (1965). *Genetics* **51**, 369.
Anderson, E. H. (1946). *Proc. Natl. Acad. Sci. U.S.* **32**, 120.
Bode, V. C., and Kaiser, A. D. (1965). *J. Mol. Biol.* **14**, 399.
Bolton, E. T., and McCarthy, B. J. (1962). *Proc. Natl. Acad. Sci. U.S.* **48**, 1390.
Boyce, R. P., and Setlow, R. B. (1962). *Biochem. Biophys. Acta* **61**, 618.
Brooks, K. (1965). *Virology* **26**, 489.
Burgi, E., and Hershey, A. D. (1963). *Biophys. J.* **3**, 309.
Campbell, A. (1962). *Advan. Genet.* **11**, 101.
Campbell, A. (1963). *Virology* **20**, 344.
Dove, W. F. (1966a). *J. Mol. Biol.* **19**, 187.
Dove, W. F. (1966b). Manuscript in preparation.
Dove, W. F., and Weigle, J. J. (1965). *J. Mol. Biol.* **12**, 620.
Egan, J. B., and Hogness, D. R. (1966). Cited in Hogness, D. R. (1966). *J. Gen. Physiol.* **49**, 29.
Franklin, N. C. (1966). Manuscript in preparation.

Franklin, N. C., Dove, W. F., and Yanofsky, C. (1965). *Biochem. Biophys. Res. Commun.* **18,** 910.

Hershey, A. D., and Burgi, E. (1965). *Proc. Natl. Acad. Sci. U.S.* **53,** 325.

Hershey, A. D., Burgi, E., and Ingraham, L. (1963). *Proc Natl. Acad. Sci. U.S.* **49,** 748.

Hoffman, D., and Rubenstein, I. (1966). *Abstr. Biophys. Soc.* p. 108.

Jacob, F., and Brenner, S. (1963). *Compt. Rend.* **256,** 298.

Jordan, E., and Meselson, M. (1965). *Genetics* **51,** 77.

Joyner, A., Isaacs, L., Sly, W., and Echols, H. (1966). *J. Mol. Biol.* **19,** 174.

Kaiser, A. D. (1962). *J. Mol. Biol.* **4,** 275.

Lipton, A., and Weissbach, A. (1966). *Biochem. Biophys. Res. Commun.* **23,** 436.

Lwoff, A. (1953). *Bacteriol. Rev.* **17,** 269.

Matsushiro, A. (1961). *Biken's J.* **4,** 133, 137, 139, and 141.

Matsushiro, A., Kida, S., Ito, J., Sato, K., and Imamoto, F. (1962). *Biochem. Biophys. Res. Commun.* **9,** 204.

Rothman, J. L. (1965). *J. Mol. Biol.* **12,** 892.

Sekiguchi, I., and Cohen, S. S. (1964). *J. Mol. Biol.* **8,** 638.

Signer, E. R. (1966). *J. Mol. Biol.* **15,** 243.

Signer, E. R., and Beckwith, J. R. (1966). *J. Mol. Biol.* **22,** 33.

Signer, E. R., Beckwith, J. R., and Brenner, S. (1965). *J. Mol. Biol.* **14,** 153.

Smith, M., and Skalka, A. (1966). *J. Gen. Physiol.* **49,** 127.

Strack, H. B., and Kaiser, A. D. (1965). *J. Mol. Biol.* **12,** 36.

Studier, F. W. (1965). *J. Mol. Biol.* **11,** 373.

Wang, J. C., and Davidson, N. (1965). *J. Mol. Biol.* **12,** 892.

Weigle, J. J. (1961). *J. Mol. Biol.* **3,** 393.

Weigle, J. J. (1966). *Proc. Natl. Acad. Sci. U.S.* **55,** 1462.

Weissbach, A., Lipton, A., and Lisio, A. (1966). *J. Bacteriol.* **91,** 1489.

Yamagishi, H., Nakamura, K., and Ozeki, H. (1965). *Biochem. Biophys. Res. Commun.* **20,** 727.

Yanofsky, C., and Lennox, E. W. (1959). *Virology* **8,** 425.

Young, E. T., II, and Sinsheimer, R. L. (1964). *J. Mol. Biol.* **10,** 562.

The Regulation of Viral Genes and the Uncontrolled Expression of the Galactose Genes During λ Development

H. Echols, B. Butler, A. Joyner, M. Willard, and L. Pilarski

DEPARTMENT OF BIOCHEMISTRY, UNIVERSITY OF WISCONSIN,
MADISON, WISCONSIN

In this chapter, we are concerned with two aspects of viral development in the case of the temperate bacteriophage λ. First, we will consider how the virus controls its own development; then, we will turn to the question of the relationship of viral development to the aberrant expression of certain host genes—the genes of the galactose operon, which bear a rather special relationship to λ.

When wild-type λ infects a sensitive cell, there are two possible outcomes: (1) the lytic response, in which the cell lyses and releases new virus particles, or (2) the lysogenic response, in which the cell survives with the viral genome incorporated into the bacterial genome and replicating with it—the prophage state (see Jacob and Wollman, 1961). The prophage can be induced to enter the vegetative lytic stage by treatment with ultraviolet light, mitomycin C or other agents which interfere with DNA replication. The genome of λ contains genetic functions concerned with both lytic and lysogenic types of response. Viral development during the lytic response requires the expression of the "early" cistrons $N, O, P,$ and Q (Fig. 1) concerned with DNA replication and control of late functions (Joyner *et al.*, 1966; Dove, 1966; Eisen *et al.*, 1966) and the "late cistrons" $A–J$ and R (Fig. 1) concerned with the production of

125

phage structural proteins and other aspects of phage maturation (Mount *et al.*, 1966; Dove, 1966). Results of experiments concerned with control of λ development at the level of DNA transcription into mRNA are presented below.

The lysogenic response requires the repression cistrons c_I and probably c_{II} and c_{III} to achieve the repression of λ functions and the subsequent lysogeny of λ when it infects a nonlysogenic cell (Kaiser, 1957; Bode and Kaiser, 1965). Achievement of the lysogenic response also requires the "lysogeny region" defined by the b_2 deletion (Kellenberger *et al.*, 1961), which includes the region of the λ genome physically necessary as a prerequisite for insertion of the viral genome into the bacterial genome (Campbell, 1965). In the integrated prophage state the only λ phage function which is known to be expressed is the c_I function, which is apparently responsible for maintaining repression of the prophage genome and for immunity of the lysogenic cell to a superinfecting λ phage (see Jacob and Wollman, 1961). We will consider possible relationships between repression and early-late control further on.

Induction of phage λ leads to the production of normal λ particles. In addition, in rare (approximately $\frac{1}{10}^6$) instances, induction can lead to the production of defective λ particles (λ*dg*), which have lost some of the λ genes and acquired the nearby bacterial genes of the galactose (*gal*) operon (see Lederberg, 1960). These defective λ*dg* can thereby transduce cells unable to utilize galactose for growth (*gal*⁻ cells) to galactose utilizing (*gal*⁺) cells. When λ*dg* infects a nonlysogenic bacterial cell, a very rapid synthesis of the galactose enzymes occurs beginning some 15 min after infection (Buttin, 1961). This has been termed "escape" synthesis because enzyme synthesis from the *gal* operon is normally repressed to a low level in the absence of the "inducer" galactose or certain of its analogs, such as D-fucose. As shown by Buttin, infection of lysogenic cells leads to no escape synthesis; however, after infection of lysogenic cells the phage associated *gal* genes can still be turned on by the inducer D-fucose. We will consider here some recent experiments concerned with the phage functions which must be carried out in order for the phage associated *gal* operon to escape the normal cellular control.

Another example of escape synthesis of the *gal* enzymes occurs when λ is induced to develop from the prophage state (Yarmolinsky and Wiesmeyer, 1960; Buttin *et al.*, 1960). Under these conditions, the *gal* operon is substantially derepressed even though λ*dg* particles containing the *gal* operon are found only with the extreme rarity of approximately $\frac{1}{10}^6$. We have recently studied in this case also the phage functions which must be carried out in order for the *gal* enzymes to show escape synthesis.

Regulatory Control of λ Development

THE RELATIONSHIP BETWEEN DNA REPLICATION AND PRODUCTION OF λ PROTEINS

There are two major questions concerning regulatory control of λ development which are now at least partly answered: (1) What λ genes are involved in regulating production of λ proteins—that is, what are the control "rules" in λ development? (2) Are these controls operative at the level of transcription of DNA into mRNA? The most interesting question of all remains at present completely unanswered—*how* do the control mechanisms work in molecular terms?

The regulatory genes for which there is some evidence in λ can be subdivided roughly into three general categories: (1) "developmental" genes regulating production of "late" λ proteins during lytic growth—that is, regulating the expression of the genes *A* through *J,* which appear to be structural genes for proteins involved in the phage particle and its assembly (Mount *et al.,* 1966; Dove, 1966; Weigle, 1966), and gene *R.* which is the structural gene for phage lysozyme (Campbell and del Campillo-Campbell, 1963); (2) "developmental" genes regulating production of some "early" λ proteins during lytic growth—that is, regulating the expression of some of the genes involved in DNA replication and control of late function and thereby providing for a sequential temporal control of the λ genome beginning with perhaps one or two "initiation" functions; (3) repression genes regulating the expression of developmental λ genes during the lysogenic response to infection of nonlysogenic cells, in the prophage state, or in superinfection of lysogenic cells.

The data from a number of laboratories which bear on developmental control of λ are summarized in Table I, which lists the protein and DNA replication phenotypes of λ mutants defective in different λ cistrons. The location of the cistrons are shown in Fig. 1. *N, O, P,* and *Q* mutants show a pleiotropic defect in the late proteins tail antigen (serum-blocking power) and lysozyme (Dove, 1966) and in phage components observable in the electron microscope (Mount *et al.,* 1966). The *T*11 mutation also leads to a pleiotropic defect in late function (Mount *et al.,* 1966), and is characterized further by overproduction of the λ-specific exonuclease (Radding, 1964b)—presumably an early protein (Protass and Korn, 1966b). Mutants in cistron *N* produce neither exonuclease (Radding, 1964a) nor late proteins; this suggests that *N* may be the structural gene for nuclease or alternatively that *N* mutants are blocked in the expression of the (separate) nuclease structural gene, either because *N* has a regulatory (initiation) role in turning on other early functions or because N mutants have a polarity type defect which affects the nuclease gene. The latter two (pleiotropic defect) possibilities have received additional support recently from the finding that another presumptive early λ protein—the β-protein—is not produced after induction of

TABLE I

DNA AND PROTEIN PHENOTYPE OF λ MUTANTS AFTER INDUCTION OF DEFECTIVE LYSOGENS[a]

Prophage	λ DNA Synthesis	λ-Nuclease	Tail antigen	Lysozyme
λ^+	+	+	+	+
R	+	+	+	−
N	−	−	−	−
O	−	+	−	−
P	−	+	−	−
Q	+	+	−	−
T11	−	Hyper	N.A.	−

[a] λ DNA synthesis has been measured by hybridization techniques (Joyner *et al.*, 1966) and by thymidine incorporation (Eisen *et al.*, 1966). λ-Nuclease data is from Radding (1964a,b) and Protass and Korn (1966a). Tail antigen has been measured by Dove (1966) and lysozyme by Dove (1966) and Protass and Korn (1966a). The designation N.A. means that the data were not available.

N mutants (Radding and Shreffler, 1966). An initiation role for N has been proposed recently by Protass and Korn (1966a) and Thomas (1966).

The N, O, P, and $T11$ mutants are defective in DNA replication (Joyner *et al.*, 1966). This shows that one control feature of λ development is that DNA replication is required in order for normal expression of late phage functions to be observed. A similar result has been found

```
                                    c_III      c_I  T11  c_II
 ─────────────────────────────────────────────────────────────────────
 A B C D E F G H M (K L) I J  \____/      N              O   P   Q   R
                               b_2
```

FIG. 1. Approximate location of the λ genetic map of the cistrons A–R described by Campbell (1961) which are required for the production of λ phage in lytic growth; the cistrons c_I, c_{II}, and c_{III} required for repression and lysogeny in the lysogenic response (Kaiser, 1957); the b_2 region (Kellenberger *et al.*, 1961) required for prophage establishment; and the site of the $T11$ mutation (Eisen *et al.*, 1966), which may define a regulatory cistron. Only the order of the markers is shown and their approximate distribution along the vegetative λ map. The vegetative map is linear, running from A to R; the prophage map is also linear, running from c_{III} to J. For recent data on map distances and gene order, see Amati and Meselson (1965) and Franklin *et al.* (1965).

in the case of phage T4 (Epstein *et al.,* 1963). The *Q* cistron mutant appears to show normal DNA replication after induction, but is defective in production of late phage proteins, so that DNA replication is necessary but not sufficient for normal expression of late function. In the case of T4 infection, the DNA negative phage overproduce early enzymes because of a failure of a normal shut-off of enzyme synthesis (Wiberg *et al.,* 1962; Levinthal, this volume). In the case of λ, at least the major overproduction of the early enzyme λ-nuclease after *T*11 induction occurs after λ-nuclease activity ceases to increase in control induction of a $λ^+$ strain (Radding, 1964b); however, there does not appear to be a general correlation between failure of DNA replication and the "hypernuclease" phenotype, because DNA defective *O* and *P* cistron mutants have been found by Radding to be normal in nuclease production.

To summarize the developmental "rules": (1) *N, O, P,* and *Q* functions all appear to be required for normal synthesis of late phage proteins; (2) *N, O,* and *P* are also required for autonomous λ DNA replication to occur; (3) *N* mutants produce no known λ proteins except for the constitutive c_I repressor, so that *N* function may *possibly* be required for expression of other early functions; (4) There may exist a regulatory function (defective in *T*11) required to turn-off nuclease and β-protein production and perhaps other early proteins.

In considering the role of repression functions, it is apparent that in the prophage state or in the case of superinfection of an immune lysogenic cell, the c_I repressor would only have to block one or more critical early functions directly in order to block late functions as well (see discussion in Jacob and Wollman, 1961; Isaacs *et al.,* 1965). The extreme example, suggested by Thomas (1966) mainly on the basis of genetic experiments involving infection by heteroimmune phages, would be that the c_I product blocks only *N* function directly. However, in the case of infection of a nonlysogenic cell, c_I repression does not appear to occur immediately after infection (Kaiser, 1957), and it is possible that the role of the c_{II} and c_{III} products in enhancing lysogeny under these conditions is to provide for an inhibition of late function expression. This might occur through a transient early inhibition of DNA replication, as suggested by Smith and Levine (1964) for phage P22, or through a specific repression of late functions.

GENETIC CONTROL OF EARLY AND LATE λ MRNA

We can then ask if the regulatory processes outlined above are manifested at the level of transcription of DNA to mRNA. This can be done by measuring the fraction of pulse-labeled RNA which forms a

DNA-RNA hybrid with denatured λ DNA and thus the level of λ specific mRNA present in the cell at a given time.

Repression of phage functions in the case of a prophage (Attardi *et al.*, 1963; Sly *et al.*, 1965) or superinfecting phage infecting a lysogenic cell (Sly *et al.*, 1965) has been shown to be accompanied by a repression of λ mRNA to a very low level. Further experiments on mRNA repression utilizing heteroimmune phage (Isaacs *et* al., 1965) have suggested that the c_I product is responsible for the repression of all λ mRNA, in the case of λ superinfection of cells lysogenic for λ and that the genetic regions specifying the receptor sites for the c_I product are all within the "immunity region" between N and c_{II} of Fig. 1. These experiments did not distinguish between two general possibilities: (1) a "direct" repression by the c_I product of all λ mRNA; (2) an "indirect" repression in which only some or all of the mRNA for the early genes would be directly repressed, and the late mRNA would be repressed because an early gene product is needed to initiate effective transcription of the late region (Isaacs *et al.*, 1965). The latter model would now appear much the more likely because of the experiments of Thomas (1966) showing that a repressed phage apparently can supply to some extent all functions except N. Further, as discussed below, a product of the early region (Q) appears to be necessary in order for λ mRNA from the late $A–J$ region to reach normal levels and the N product may be needed for normal transcription of the early region.

The wild-type mRNA phenotype during lytic growth has been studied most thoroughly by Skalka (1966a,b), using half molecules and smaller fragments of λ DNA to hybridize with λ mRNA isolated at different times after infection. She finds that λ mRNA at early times is transcribed mainly from the "right-half" of the phage DNA (see Fig. 1) containing the early cistrons N, O, P, and Q and that transcription from the region of DNA containing the $A–J$ functions does not occur efficiently until approximately half-way through the lytic cycle.

We have recently been engaged in a study of early and late λ mRNA production by λ mutants under lytic growth conditions, either infection by Nc_I, Oc_I, Pc_I, or Qc_I mutants or mitomycin induction of N, O, P, Q, or $T11$ mutants. The object of this work has been to try to establish whether the pleiotropic defect of N, O, P, and Q mutants in late λ proteins was accompanied by a quantitative defectiveness in λ mRNA from the late region of the genome at late times and whether the defect of N mutants in the early protein λ-nuclease was part of a general defectiveness in early λ mRNA production (as might be the case if the N product had an initiator role).

To study specifically the mRNA from the early and late regions of the genome we have carried out DNA-RNA hybridization reactions with DNA isolated from wild-type λ and from the λ*dg* *A–J* deletion phage shown in Fig. 2. The λ*dg* *A–J* has deleted all of the *A–J* region and substituted the epimerase (*E*) and part of the transferase (*T*) gene of the galactose operon (Adler and Templeton, 1963). Thus it cannot hybridize with *A–J* mRNA, and shows only a slight additional hybridization with *E. coli* mRNA over that obtained with λ—an increment of approximately 0.05% in the fraction of *E. coli* RNA hybridized, which is insignificant compared to the λ hybrid in every case except *N* infection.

The results of experiments using infection of nonlysogenic cells by Nc_I, Pc_I, Qc_I, and $+c_I$ are shown in Fig. 3. The results with the $+c_I$ phage are in agreement with the earlier results of Skalka in showing ex-

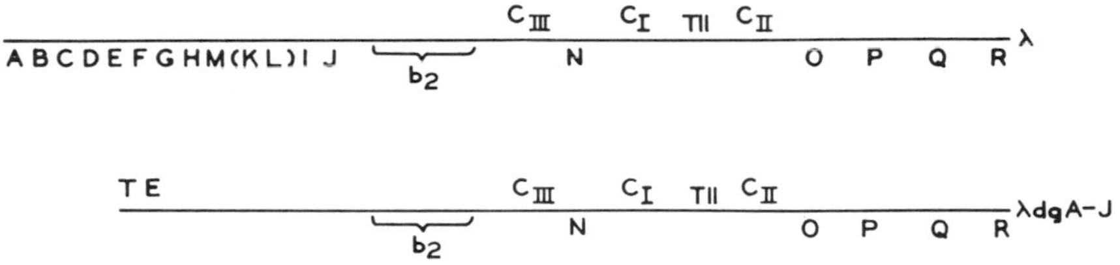

FIG. 2. Comparison of genetic maps of λ and a λ*dg* which has deleted the *A–J* region of λ and substituted the *E* (epimerase) and part of the *T* (transferase) genes of *E. coli*. It is not known whether other *E. coli* genes outside of the *gal* operon are included. The λ*dg* *A–J* genome contains less DNA than λ, as judged by the lower buoyant density of the λ*dg* *A–J* phage in CsCl.

tensive production of *A–J* mRNA at times late in the lytic infection cycle, since the difference between the λ and λ*dg* hybridization represents *A–J* mRNA. Both Pc_I and Qc_I appear to be approximately normal in early mRNA production, but produce much less *A–J* mRNA late in infection than the $+c_I$ control. Nc_I is defective in both early and late mRNA production.

Therefore, the mRNA results show that the pleiotropic defect in synthesis of late proteins by *N, P,* and *Q* mutants is associated with a defectiveness in mRNA production from the region of the DNA that contains most of the genes for these proteins. *P* and *Q* mutants produce approximately 10% the normal level of lysozyme (Dove, 1966; Protass and Korn, 1966a); this is consistent with our finding that some *A–J* mRNA appears to be produced by these mutants. The failure of *N* mutants to produce the early proteins λ-nuclease and β (Radding and Shreffler, 1966) we find to be accompanied by a defectiveness in early mRNA production. Infection data on *O* is still incomplete, but previous mRNA experiments on induced lysogenic cells (Joyner *et al.*, 1966) have shown that *O* is very low in total λ mRNA late in the induction period, which is consistent with the pleiotropic defect in the late proteins.

The simplest interpretation at the present time of our mRNA results and the protein phenotypes of *N, O, P,* and *Q* mutants would appear to be the following: the *N* product is required in order for transcription of the other early genes (*O P Q*) to proceed normally; the *Q* product is required in order for transcription of the late genes (*A–J* and *R*) to proceed normally; and the *O* and *P* genes are required for autonomous λ DNA replication, which in turn is required in order for a normal number of copies of late mRNA and therefore a normal level of late protein to be produced. The requirement for DNA replication may be simply a gene dosage effect; alternatively the DNA may have to be replicated before the *Q* product can act effectively. In the case of *N*, the possibility

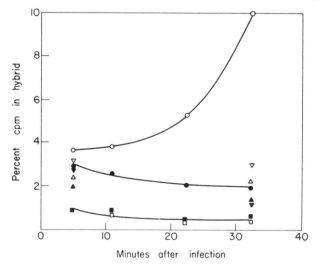

FIG. 3. λ mRNA levels after infection of nonlysogenic strain W3350 *su*⁻ by various *amber* type mutants of λ. Results of infection by the Campbell mutants *sus* N_{53}, *sus* P_{80} and *sus* Q_{21}, are shown here, together with results of infection by a + c_I strain; each of the Campbell mutants also contained a c_I mutation, in order to minimize complications involving the lysogenic response. RNA was pulse-labeled for 2 min with uridine-^{3}H, the incorporation terminated by chilling at the times indicated on the figure, and the RNA was extracted and hybridized with denatured λ DNA (see Sly *et al.*, 1965). Each point on the figure represents the fraction of the ^{3}H-RNA forming an RNase resistant hybrid with denatured λ DNA (open symbols) or λ*dg A–J* DNA (Closed symbols); this percent cpm in hybrid is taken as a measure of the amount of λ mRNA present in the cell, since the fraction of the ^{3}H-RNA which hybridizes with *E. coli* is very nearly the same in all cases. ○, ● represents λ and λ*dg A–J* hybrid, respectively with RNA extracted after + c_I infection; □, ■, $N_{53}c_I$ infection; △, ▲, $Q_{21}c_I$ infection; ▽, ▼, $P_{80}c_I$ infection.

that the protein and mRNA phenotypes result from a polarity defect must still be considered.

One further regulatory function which may affect λ mRNA production is that defective in the $T11$ mutant. The possibility that the $T11$ mutant may overproduce early mRNA at late times and therefore be "hyper" in λ-nuclease (and β-protein) production is suggested by the fact that $T11$ produces much more λ mRNA at times late in the induction period than the other DNA negative (N, O, P) mutants (Joyner *et al.*, 1966) and that this mRNA comes almost entirely from the early region (Skalka *et al.*, 1966). $T11$ may then be defective in a function normally involved in turning off or turning down early mRNA production. Some evidence for such a turn-off mechanism has been provided by the mRNA competition experiments of Skalka (1966a), which show that there appears to be a class of mRNA from the "right-half" of the λ genome which is made early but not late.

Escape Synthesis by the Galactose Enzymes After λ*dg* Infection

When λ*dg* carrying a wild-type *gal* operon infects a nonlysogenic cell, there is a rapid rise in synthesis of the *gal* enzymes, beginning some 15 min after infection (see Buttin, 1963). This phenomenon is shown in Fig. 4 for the epimerase enzyme of the *gal* operon—escape synthesis

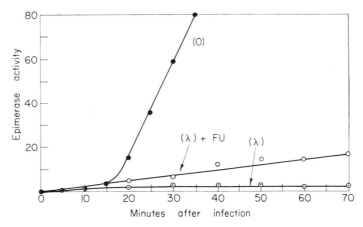

FIG. 4. Epimerase activity as a function of time after infection of epimerase-negative nonlysogenic (○) or lysogenic (λ) cells with and without the inducer D-fucose (FU) by λ*dg* wild-type for DNA replication and Q function. Epimerase activity was measured on lysozyme lysates (Jordan and Yarmolinsky, 1963) and is expressed as △OD/min/0.1 ml of lysate in the standard spectrophotometric assay. ●, infection of nonlysogenic cells; ⊙ infection of lysogenic cells; ○ infection of lysogenic cells in the presence of 10^{-3} M D-fucose.

of epimerase in a nonlysogenic cell and no escape synthesis in a lysogenic cell. We can then ask what λ functions must be carried out in order for the escape synthesis to occur. The location of the *gal* operon in the part of the phage genome normally containing the late functions and the delayed synthesis suggested that the phage-associated *gal* genes might be behaving like late phage genes. If this were the case, escape synthesis by the *gal* genes should require the *Q* function needed for normal mRNA production and protein synthesis from the late region. Alternatively, phage replication might be the critical factor.

To test these possibilities, we have studied epimerase synthesis after infection of nonlysogenic epimerase-negative cells by λ*dg* phage containing *N, P,* and *Q* mutations. The results (Fig. 5) clearly show that DNA replication but not *Q* function is required in order for escape synthesis of epimerase to occur. Thus the phage need only provide for replication of the *gal* operon. Previous experiments concerned with the necessity for replication for escape synthesis have been hampered by the lack of direct measurements of DNA replication by λ*dg* and mutant λ phages which have recently been provided by hybridization analysis (Joyner *et al.*, 1966). There are two obvious ways in which the replication of *gal* operons could lead to escape: (1) the multiplication of operator sites as the phage

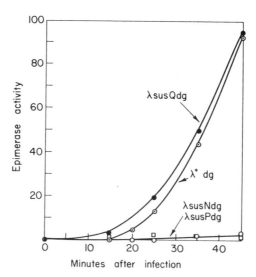

Fig. 5. Epimerase activity as a function of time after infection of nonlysogenic cells by a λ*dg* wild-type for DNA replication and *Q* function (λ⁺*dg*) and λ*dg* mutants defective in the *N* or *P* functions required for DNA replication or the *Q* function required for normal late-function expression. Epimerase was measured as for Fig. 4. ⊙, λ⁺*dg* infection; ●, λ*Q*₂₁*dg* infection; ○, λ*N*₅₃*dg* infection; □, λ*P*₈₀*dg* infection.

replicates may titrate the repressor supply; (2) newly synthesized DNA may be initially insensitive to repression. We hope that further experiments in progress will distinguish between these possibilities.

Escape Synthesis by the Galactose Enzymes After λ Induction

Escape synthesis by *gal* enzymes also occurs when λ is induced from the prophage state, even though the production of λ*dg* phage is a rare event (see Yarmolinsky, 1963; Buttin, 1963). We were interested in determining what λ functions are required in order for this aberrant increased synthesis of the *gal* enzymes to occur. We have used heat induction of *N, O, P, Q,* and *R* mutants containing a temperature sensitive mutation in the c_I *gene* (c_{I857} of Sussman and Jacob, 1962) as a convenient mode of λ induction and measured epimerase activity 50 min after induction to test for escape synthesis of the *gal* enzymes. The results of these experiments (Table II) show that the *N, O,* and *P* functions required for DNA replication must act in order for an increased synthesis of epimerase to be observed. A requirement for λ DNA replication for *gal* enzyme escape synthesis has been inferred by Yarmolinsky (1963) from FUDR inhibition.

Unfortunately the requirement for λ DNA synthesis does not help a great deal in trying to arrive at a molecular model for escape synthesis

TABLE II

EPIMERASE ACTIVITY AFTER PROPHAGE INDUCTION[a]

Prophage	Epimerase activity
N_7	0.6
O_{29}	0.6
P_3	0.7
Q_{21}	3.2
R_{54}	2.8
Nonlysogenic	0.5
Uninduced	0.6

[a] *E. coli* K-12 strain W3101 nonlysogenic or carrying prophages defective in the *N, O, P, Q,* or *R* cistrons was grown at 30°C to 4×10^8 cells/ml, transferred to 42°C for 10 min and then diluted twofold into 37°C medium. Each prophage carries a temperature sensitive mutation in the c_I gene (c_{I857}) so that thermal induction occurs at 42°C. After 50 min at 37°C, the cells were harvested by centrifugation, sonic extracts were prepared, and uridine diphosphogalactose-4-epimerase (epimerase) assayed as described previously (Echols *et al.*, 1963). Units of epimerase are μmoles product/hour/mg protein. The "uninduced" entry refers to prophage containing cells grown at 30°C and not thermally induced. Since *R* mutants are defective only in lysozyme production, we have used the *R* lysogen as the control "normal" example of λ induced epimerase synthesis for comparison to the nonlysing *N, O, P, Q* lysogens.

by the host *gal* operon, since a number of possibilities are left open. The necessity for proximity of the prophage site to the *gal* genes (Buttin, 1963) suggests as one possible model that DNA repair following excision of the prophage may be involved. We have been attracted to DNA repair models because the degradation and resynthesis aspects of DNA repair (see Howard-Flanders and Boyce, 1966) provide two simple mechanisms for aberrant transcription and therefore for escape synthesis: If degradation of one DNA strand and resynthesis of that strand occurs through regions of the genome occupied by the *gal* genes, then the newly resynthesized DNA (assumed to be the transcribed strand) might be insensitive to the normal cellular repression mechanism; alternatively, the degradation of one strand (assumed to be the one not normally transcribed) might lead to uncontrolled transcription of the other DNA strand because of its single-stranded character. A complication in considering repair models is the inability of *O* and *P* to initiate derepression, even though both Lieb and Sly (personal communication) have shown that *O* and *P* are apparently excised following thermal induction as judged by frequent loss of the prophage. The repair capacity of induced lysogens of *O* and *P*, however, may be different from that of induced lysogens of λ^+, where the replicating phage is perhaps using an enzyme needed for repair, such as a polymerase, or making an enzyme which interferes with repair (a "killing" enzyme).

Models of the "multiple copy" variety are also plausible, although perhaps less attractive from a molecular point of view. Escape synthesis could result from titration of the repressor supply as a result of production of many copies of the *gal* operon; the many copies might come from abortive replication of the *E. coli* chromosome at the point where λ is excised or from frequent formation of λ*dg*-like DNA molecules which can replicate but not be packaged into mature phage. Certainly the latter mechanism and possibly the former as well would lead to a requirement for λDNA replication in order for escape of the *gal* operon to occur. Clearly much more information is needed before we can say with any certainty what the mechanism is for the escape synthesis by the host *gal* operon associated with λ induction.

ACKNOWLEDGMENTS

The experiments on regulatory control of λ mRNA reported here comprise part of a collaborative study of the problem being carried out with Ann Skalka. The experiments discussed in this paper have benefited greatly from her ideas and concurrent experiments. We thank William Dove for the gift of many of the strains used in this work and for much helpful discussion, and R. L. Baldwin for suggestions concerning the λ*dg* infection experiments. We also thank Charles Radding,

Allan Campbell, and Julius Adler for strains, and Judy Wagnild for assistance in some of the experiments. This research was supported by U.S. Public Health Service grant GM 08407, U.S. Public Health Service training grant 2 T1 GM 236 BCH, and an institutional grant from the American Cancer Society.

REFERENCES

Adler, J., and Templeton, B. (1963). *J. Mol. Biol.* **7,** 710.

Amati, P., and Meselson, M. (1965). *Genetics* **51,** 369.

Attardi, G., Naono, S., Rouviere, J., Jacob, F., and Gros, F. (1963). *Cold Spring Harb. Symp. Quant. Biol.* **28,** 363.

Bode, V. C., and Kaiser, A. D. (1965). *Virology* **25,** 111.

Buttin, G. (1961). *Cold Spring Harbor Symp. Quant. Biol.* **26,** 213.

Buttin, G. (1963). *J. Mol. Biol.* **7,** 610.

Buttin, G., Jacob, F., and Monod, J. (1960). *Compt. Rend.* **250,** 2471.

Campbell, A. (1961). *Virology* **14,** 22.

Campbell, A. (1965). *Virology* **27,** 340.

Campbell, A., and del Campillo-Campbell, A. (1963). *J. Bacteriol.* **85,** 1202.

Dove, W. F. (1966). *J. Mol. Biol.* **19,** 187.

Echols, H., Reznichek, J., and Adhya, S. (1963). *Proc. Natl. Acad. Sci. U.S.* **50,** 286.

Eisen, H., Fuerst, C. R., Siminovitch, L., Thomas, R., Lambert, L., Pereira da Silva, L., and Jacob, F. (1966). *Virology* **30,** 224.

Epstein, R. H., Bolle, A., Steinberg, C. M., Kellenberger, E., Boy de la Tour, E., Chevalley, R., Edgar, R. S., Susman, M., Denhardt, G. H., and Lielausis, A. (1963). *Cold Spring Harbor Symp. Quant. Biol.* **28,** 375.

Franklin, N. C., Dove, W. F., and Yanofsky, C. (1965). *Biochem. Biophys. Res. Commun.* **18,** 910.

Howard-Flanders, P., and Boyce, R. P. (1966). *Radiation Res. Suppl.* **6** (in press).

Isaacs, L. N., Echols, H., and Sly, W. S. (1965). *J. Mol. Biol.* **13,** 963.

Jacob, F., and Wollman, E. L. (1961). "Sexuality and the Genetics of Bacteria," p. 285. Academic Press, New York.

Jordan, E., and Yarmolinsky, M. (1963). *J. Gen. Microbiol.* **30,** 333.

Joyner, A., Isaacs, L. N., Echols, H., and Sly, W. S. (1966). *J. Mol. Biol.* **19,** 174.

Kaiser, A. D. (1957). *Virology* **3,** 42.

Kellenberger, G., Zichichi, M. L., and Weigle, J. J. (1961). *J. Mol. Biol.* **3,** 399.

Lederberg, E. M. (1960). *Symp. Soc. Gen. Microbiol.* **10,** 115.

Lieb, M., and Sly, W. S. (1966). Personal communication.

Mount, D. W. A., Harris, A., Fuerst, C. R., and Siminovitch, L. (1966). In press.

Protass, J. J., and Korn, D. (1966a). *Proc. Natl. Acad. Sci. U.S.* **55,** 1089.

Protass, J. J., and Korn, D. (1966b). In press.

Radding, C. M. (1964a). *Biochem. Biophys. Res. Commun.* **15,** 8.

Radding, C. M. (1964b). *Proc. Natl. Acad. Sci. U.S.* **52,** 965.

Radding, C. M., and Shreffler, D. (1966). *J. Mol. Biol.* **18,** 251.

Shalka, A. (1966a). *Proc. Natl. Acad. Sci. U.S.* **55,** 1190.

Skalka, A. (1966b). *Cold Spring Harbor Symp. Quant. Biol.* (in press).

Skalka, A., Echols, H., and Butler, B. (1966). Unpublished data.

Sly, W. S., Echols, H., and Adler, J. (1965). *Proc. Natl. Acad. Sci. U.S.* **53,** 378.

Smith, H. O., and Levine, M. (1964). *Proc. Natl. Acad. Sci. U.S.* **52,** 356.

Sussman, R., and Jacob, F. (1962). *Compt. Rend.* **254,** 1517.

Thomas, R. (1966). *J. Mol. Biol.* **22,** 79.

Weigle, J. J. (1966). *Proc. Natl. Acad. Sci. U.S.* **55,** 1462.

Wiberg, J., Dirksen, M. L., Epstein, R. H., Luria, S. E., and Buchanan, J. M. (1962). *Proc. Natl. Acad. Sci. U.S.* **48,** 293.

Yarmolinsky, M. (1963). *In* "Viruses, Nucleic Acids, and Cancer," p. 151. Williams & Wilkins, Baltimore, Maryland.

Yarmolinsky, M., and Wiesmeyer, H. (1960). *Proc. Natl. Acad. Sci. U.S.* **46,** 1626.

Regulation of the Development of the Temperate Phage Lambda

Melvin H. Green

DEPARTMENT OF BIOLOGY, UNIVERSITY OF CALIFORNIA, SAN DIEGO,
LA JOLLA, CALIFORNIA

Upon infection of *E. coli* by the temperate phage, lambda (λ), the host cells may respond in two extremely different ways. A fraction of the population is killed by the virus, which uses the cell for its own replication and development prior to lysis. Most of the remaining fraction survives the infection and becomes lysogenic for λ. Progeny of the lysogenic cells, *E. coli*(λ), grow normally, but are now totally immune to infection by λ. Each genome of the lysogenic cell carries a λ genome (prophage) at a characteristic site.

Two major control mechanisms operate on the prophage genome. The replication of the prophage DNA is under the control of the host replication cycle; it occurs at a time characteristic of its location on the host genome (Nagata, 1963). Certain defective phages, though incapable of replicating their own DNA after infecting *E. coli*, are, nevertheless, replicated while in the prophage state (Jacob *et al.*, 1957). Furthermore, when *E. coli*(λ) is infected by λ, the superinfecting phage DNA enters the immune cell, but does not replicate (Wolf and Meselson, 1963).

A second control is exerted at the level of transcription of the λ genome into RNA. Unlike the replication control, the latter process is governed by the λ genome. The c_I region of λ governs both the immunity specificity of a lysogenic cell and the ability of λ to lysogenize (Kaiser and Jacob, 1957). As first shown by Attardi *et al.* (1963), the c_I product (repressor) is responsible for the repression of the synthesis of λ messenger RNA (mRNA)* in a lysogen. This repression can be overcome

---

* Abbreviations: mRNA, messenger RNA; CAP, chloramphenicol; UV, ultraviolet; MC, mitomycin C; TL, threonine and leucine; (tl), (U32), and (U37) are the lysogenic derivatives of *E. coli* W3110 containing the heat-inducible $\lambda c_{I,tl}$, etc.

by treatments which cause prophage development, such as ultraviolet light (Attardi *et al.,* 1963), mitomycin C (Sly *et al.,* 1965) and heat (Green, 1966).

It would thus appear that the inactivation of the λ repressor releases the prophage from both control mechanisms—that of DNA replication and transcription into mRNA. In this report, experiments will be described which indicate that certain conditions can derepress λ mRNA synthesis without causing subsequent λ maturation. The bearing of these results on our understanding of the mechanisms regulating the development of λ will be discussed.

Inactivation of the Lambda Repressor in the Absence of Induction*

Our studies on the development of bacteriophage lambda began as a result of the following observation by Lieb (1964). *E. coli*$(\lambda c_{I.tl})$ could be heat induced only if protein synthesis occurred at the elevated temperature (41–45°C). If a culture of this lysogen was heated in the presence of CAP, then diluted or washed free of CAP and transferred to 37°C, the cells grew and did not produce phage. Since the c_I mutation was thought to cause the λ repressor to be thermolabile (Jacob and Monod, 1961), one might have expected heat inactivation of the repressor to cause a large fraction of the cells to become induced regardless of the medium in which the cells were heated.

Lieb's results permitted the following interpretations to be considered:

1. The repressor is not directly inactivated by heat. Instead, some protein is synthesized at the elevated temperature that inactivates it, thus causing induction.

2. The repressor is heat-sensitive. However, due to one of several possibilities, it becomes active very rapidly upon lowering the temperature and removing the CAP; thereby induction is prevented. For example, at 45°C in the presence of CAP, a precursor to the repressor might accumulate; this permits the rapid synthesis of repressor at 37°C in the absence of CAP. Or the repressor might be reversibly thermolabile, that is it might become reactivated quickly after the temperature is lowered.

3. The repressor is heat-sensitive, but its inactivation is insufficient cause for induction. Also required is the synthesis of one or more proteins at the elevated temperature.

The correct explanation could be found by employing a direct assay

---

* Terminology: The term "induction" is herein operationally defined as the killing of a lysogenic cell by conditions that would not kill a nonlysogenic cell of the same strain. These conditions must be capable of initiating prophage development and maturation.

for the λ repressor, namely one which detected its ability to inhibit the synthesis of λ mRNA (Jacob and Monod, 1961). As previously demonstrated (Attardi *et al.*, 1963), the DNA-RNA hybridization technique of Hall and Spiegelman (1961) was ideal for this purpose. Cultures were pulse-labeled for 1 min with uracil-^{3}H under a variety of conditions; the RNA was isolated and tested for its ability to hybridize with λ and *E. coli* DNA. The nitrocellulose filtration assay of Nygaard and Hall (1963), as modified by this author (Green, 1963), was used to detect the fraction of labeled *E. coli* and λ mRNA in each preparation. The results are summarized in Table I (Green, 1966).

Samples 1 and 2 are controls showing the repression of λ mRNA synthesis in lysogenic cultures. At 37°C *E. coli*($\lambda c_{I,t1}$) synthesized a very low level of RNA that is homologous to λ DNA (No. 1). *E. coli*(λ), incubated at 45°C for 25 min in the presence of CAP, also remained repressed for λ mRNA synthesis (No. 2). The *E. coli*/λ ratios for the hybridized RNA were 45 and 57, respectively.

The absolute amount of radioactive material in mRNA can vary from one RNA preparation to the next for several reasons. For example, variations in the length of the pulse and the presence or absence of CAP can alter the distribution of label in mRNA, rRNA, and tRNA. Changes in the pool size of RNA precursors affect the specific activity of the pulse-labeled RNA. Assuming that these factors similarly influence the incorporation of uracil-^{3}H into *E. coli* and λ mRNA, the ratio of *E. coli* RNA to λ RNA, obtained by hybridization of a given amount of ^{3}H-RNA with fixed concentrations of *E. coli* and λ DNA, is a more accurate estimate of the extent of λ derepression than is the absolute percentage of input ^{3}H-RNA hybridized.

The other six RNA preparations were obtained from cultures pulse-labeled at 45°C at the indicated times. Each shows a considerably lower *E. coli*/λ hybrid ratio than the controls, indicating derepression of the λ genome. Samples 3 and 4 were isolated from cultures in which no induction (cell death) occurred, even though sample 4 had been infected with λc_I immediately after raising the temperature to 45°C. Both RNA preparations were shown to contain λ mRNA when tested with increasing λ DNA concentrations as described previously (Green, 1963). Although the *E. coli*/λ ratio with RNA from the uninfected culture (sample 3) of *E. coli*($\lambda c_{I,t1}$) was relatively high (8.3) compared to other cases where derepression occurred, it may be concluded that the prophage repressor has been at least partially inactivated at 45°C in the presence of CAP. Additional evidence to support this claim will be presented subsequently. Sample 5 shows that $\lambda c_{I,t1}$ can be transcribed into RNA in a sensitive (nonlysogenic) culture of *E. coli* treated with CAP five min prior to in-

TABLE I

DEREPRESSION OF PROPHAGE $\lambda c_{I,t1}$ AT 45°C[a]

Number	Source of ^{3}H-RNA	Time of 1 min pulse	CPM hybrid with:		% RNA background	$E.\ coli/\lambda$[b]
			λDNA (115 µg/ml)	$E.\ coli$ DNA (440 µg/ml)		
1	$E.\ coli(\lambda c_{I,t1})$	1' (37°)	69 (0.21%)	3122 (9.5%)	1.07	45
2	$E.\ coli(\lambda)$ + CAP	25'	45 (0.13%)	2548 (7.6%)	0.24	57
3	$E.\ coli(\lambda c_{I,t1})$ + CAP	25'	200 (0.57%)	1663 (4.8%)	0.49	8.3
4	$\lambda c_{I,n} + E.\ coli(\lambda c_{I,t1})$ + CAP	25'	521 (1.4%)	2120 (5.5%)	0.52	4.1
5	$\lambda c_{I,t1} + E.\ coli$ + CAP	25'	870 (1.6%)	4027 (7.4%)	0.74	4.6
6	$E.\ coli(\lambda c_{I,t1})$	2'	986 (2.2%)	2372 (5.4%)	0.52	2.4
7	$E.\ coli(\lambda c_{I,t1})$	10'	739 (1.7%)	2374 (5.5%)	0.52	3.2
8	$E.\ coli(\lambda c_{I,t1})$	25'	1497 (5.3%)	2240 (7.9%)	0.81	1.5

[a] Except for sample 1, which was labeled at 37°C, all cultures were pulse-labeled with uracil-^{3}H (0.5 µCi/0.2 µg/ml) at the designated time after transfer to 45°C. The cells were grown in K medium at 37°C to 3–4 × 10^8/ml. CAP (100 µg/ml) was added 5 min prior to transferring the cells to 45°C (Nos. 2–5). Cells were infected at a multiplicity of 5 immediately after raising the temperature to 45°C (Nos. 4 and 5). DNA-RNA hybrids were formed during a 6 hour incubation at 59°C. The data are average values from two identical reactions. The percentage figures in brackets represent the percentage of RNA hybridized based on the acid-precipitable ^{3}H in the reaction mixture.

[b] Ratio of the amounts of $E.\ coli$ RNA and λRNA in a given sample which hybridize with fixed concentrations of $E.\ coli$ and λDNA.

fection. Thus, the results of samples 3, 4, and 5 also indicate that at least part of the λ genome can be transcribed by the *E. coli* RNA polymerase.

Samples 6, 7, and 8 were obtained at various times from heat-induced cultures of *E. coli*(λc$_{I,t1}$) and show low *E. coli*/λ hybrid ratios characteristic of the derepressed prophage. The relatively high rate of λ mRNA synthesis, at 25 to 26 min after induction, is in agreement with the results of Sly *et al.* (1965); he found an increase in the rate of λ mRNA synthesis during the second half of the latent period in the absence of CAP. Of particular importance is the result of sample 6, which shows that the prophage is fully derepressed within 3 min after raising the temperature to 45°C. Kinetic experiments on the rate of induction indicated that under these growth conditions, no induction occurred until after 4 min. It may thus be concluded that derepression of λ mRNA synthesis can occur in the absence of any induction, even if the lysogens are super-infected by λc$_I$ prior to inactivating the repressor. Protein synthesis is not required for the inactivation of the λc$_{I,t1}$ repressor at 45°C.

We next tested the possibility that only a very small fraction of the λ genome (e.g., the c$_I$ region) becomes derepressed under conditions in which there is no induction. This was shown not to be the case by competitive hybridization experiments (Green, 1966) in which RNA from derepressed *E. coli*(λc$_{I,t1}$) cultures (in CAP at 45°C) was shown to compete with a large fraction (~50%) of labeled RNA isolated from a culture undergoing lytic infection by λc$_I$. As a control, RNA isolated from lysogenic cultures in which the λ genome was repressed [*E. coli*(λ) at 45°C; *E. coli*(λc$_{I,t1}$) at 37°C] did not prevent lytic λ mRNA from hybridizing with λ DNA.

Finally, it was of interest to determine whether the absence of induction was due to the rapid synthesis or reactivation of the λ repressor after lowering the temperature to 37°C. *E. coli*(λc$_{I,t1}$) was infected with λc$_I$, aerated 20 min at 45°C in the presence of CAP in order to derepress λ mRNA synthesis, and then return to 37°C. One aliquot of the culture was washed free of CAP; the other was not. Aliquots of each culture were pulse-labeled with uracil-^{3}H at various times. After isolation of the RNA, hybridization experiments were performed with *E. coli* and λ DNA (Green, 1966).

As may be seen from the data presented in Table II, λ mRNA synthesis remained fully derepressed 30 min after transfer to 37°C (Nos. 1–3). After 120 min at 37°C, the cell number of the culture without CAP (No. 6) had increased fourfold, yet λ mRNA synthesis was still not fully repressed. The *E. coli*/λ ratio of 9.5 is significantly lower than that of 45 found for the repressed prophage λ genome (see Table I). However, the presence of CAP at 37°C completely prevented the repression of λ mRNA from becoming reestablished, as seen by comparing the ratios in samples 4–6.

These results permit us to conclude that the $\lambda c_{I,t1}$ repressor was irreversibly inactivated at 45°C in the presence of CAP. The prevention of the synthesis of the λ repressor by CAP lends support to the belief that this repressor is, at least in part, a protein (Thomas and Lambert, 1962; Jacob *et al.*, 1962). After removal of the CAP, repression of λ mRNA was restored relatively slowly in cells growing at 37°C. This result is in

TABLE II

SYNTHESIS OF PROPHAGE λ REPRESSOR[a]

Time of 1 min pulse	Hybrid cpm with:		% RNA back-ground	E. coli/λ
	λ DNA	E. coli DNA		
	At 147 μg/ml	At 424 μg/ml		
(1) 20′, 45° + CAP	475 (1.9%)	1850 (7.5%)	0.88	3.9
(2) 50′, 37° + CAP	894 (3.3%)	2058 (7.5%)	0.43	2.3
(3) 50′, 37° − CAP	738 (2.7%)	1829 (6.6%)	0.54	2.5
	At 130 μg/ml	At 240 μg/ml		
(4) 140′, 45° + CAP	439 (1.7%)	918 (3.6%)	0.39	2.1
(5) 140′, 37° + CAP	375 (1.4%)	966 (3.7%)	0.42	2.6
(6) 140′, 37° − CAP	200 (0.70%)	1893 (6.6%)	0.59	9.5

[a] *E. coli* ($\lambda c_{I,t1}$) in K medium at 37°C were infected with $\lambda c_{I,71}$ at a multiplicity of 5. After 15 min for adsorption, CAP (100 μg/ml) was added, and 5 min later the cells were transferred to 45°C ($t = 0$). After 20 min aeration, the cells were centrifuged, washed, and resuspended in K medium at 37°C. At the indicated times (neglecting the time required for removal of CAP), aliquots were pulse-labeled for 1 min with uracil-^{3}H (0.5 μCi/0.2 μg/ml). DNA-RNA hybrids were formed during an 8 hour incubation at 59°C. The percentage figures in brackets represent the percentage of RNA hybridized based on the acid-precipitable ^{3}H in the reaction mixture.

accord with kinetic studies on the synthesis of the repressor of the *lac* operon (Pardee *et al.*, 1959, Sadler and Novick, 1965). The latter authors suggested that this was due to the repressor being composed of subunits whose rate of aggregation is extremely concentration dependent. This subject will be discussed further in the section on complementation between λc_I mutants.

The cells which grew at 37°C after derepression of λ mRNA synthesis at 45°C in the presence of CAP were shown to be heat-inducible, indicating that the $\lambda c_{I,t1}$ prophage is not lost by this treatment. Although the cells treated in this manner retained their immunity to λc_I, they were not "refractory" since at least 98% of the population could be killed by

$\lambda b5c_I$, a phage of immunity type b5 (Kellenberger *et al.,* 1961). It may be concluded that derepression of λ mRNA synthesis can occur without the loss of the cell's control over prophage replication.

The Repressor-Inducing Protein Hypothesis

Of prime importance to our understanding of the mechanism of induction is the resolution of the problem of whether a particular protein or a certain threshold of many λ proteins causes the induction of a lysogen. The results just described indicate that a point mutation causing the λ repressor to become thermolabile also provoked a requirement for protein synthesis in order to obtain heat induction. Could it be that the repressor itself is the essential "inducing protein"? Since the repressor activity of the c_I product is thermolabile, the need for protein synthesis at 45°C may be due to the necessity for attaining a sufficient level of inducing protein to function before becoming inactivated by heat.

This line of reasoning received strong support when Lieb announced the isolation of a second class of heat-inducible λc_I mutants (Lieb, 1966). Unlike the previous type, which Lieb called class B, the class A* mutants require no protein synthesis or cell metabolism in order for heat induction to occur. Six mutants of each class were characterized and found to map in separate halves of the c_I region. Typical properties of the two classes of λc_I mutants are illustrated in Fig. 1.

In general, heat induction of the class B lysogens is highly dependent on the growth medium, whereas the rate of induction of the class A lysogens is insensitive to the medium. Whereas CAP completely inhibited induction of the class B lysogen, *E. coli*(tl), it had no significant effect on the rate of killing of the class A lysogen, *E. coli*(U32). It is also evident that 0.01 M NaN_3 had no effect on the rate of induction of *E. coli*(U32) at 45°C. This agent was found to reduce the growth rate by tenfold at 37°C. In contrast, *E. coli*(tl) grown in minimal M9 medium were induced at a much lower rate than cells grown in K medium (M9 supplemented with 1.5% casamino acids). The generation time for *E. coli*(tl) in K medium at 37°C was approximately 0.7 times that in M9 medium.

The rate of induction of class A lysogens was also unaltered at 45°C in M9 or in 0.01 M Tris, pH 7.3, 0.005 M $MgSO_4$ (Green, unpublished data). That induction had indeed occurred was seen by lowering the temperature and restoring the cells to complete growth medium. After

---

* The class B mutants were originally called class I (Green, 1966) before the "AB" notation was adopted.

about 90 min, the cultures lysed and phage was released with burst sizes approximating those obtained by UV or mitomycin C induction. Under similar conditions, class B lysogens resume growth and do not produce increased amounts of phage.

Since c_I mutations can affect the activity of both the repressor and the inducing protein, we postulated that the active λ repressor is an inactive form of the inducing protein (Green, 1966). In the case of the class A lysogens, the repressor is converted by heat to the inducing protein, which

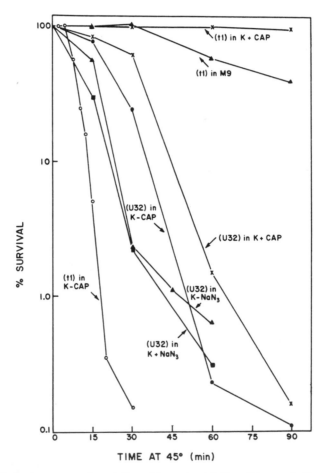

FIG. 1. Induction properties of *E. coli*($\lambda c_{I,t1}$) and *E. coli*($\lambda c_{I,U32}$). Log phase cultures growing in K or M9 medium at 37°C for *E. coli*(tl) and at 32°C for *E. coli*(U32), were diluted into 45°C medium of the original type; but this medium contained the noted additions ($t = 0$). Cell viability was determined by colony forming ability. Plates were incubated overnight at 30°C. When used, CAP was at 100 μg/ml and sodium azide (NaN$_3$) was 0.01 M.

triggers λ maturation and cell death. The repressor of the class B lysogens is inactivated by heat, but is apparently not converted to active inducing protein. At elevated temperatures (e.g., 45°C), the c_I region directs the synthesis of inducing protein rather than repressor, thus causing heat induction. As an alternative explanation, Lieb postulated that the λ repressor interacts with an inducing protein that is coded by some other λ gene in the lysogenic cell (Lieb, 1966). As yet, these alternatives have not been resolved.

Activation of the Class B Inducing Protein at 45°C

The inability of class B lysogens to be heat-induced in the absence of protein synthesis can be explained in one of the following ways: (1) The repressor is heat-inactivated; but, it is not converted to, or does not release, the active inducing protein; or (2) the inducing protein is formed, but it is thermolabile or present in insufficient quantity to function.

In an effort to settle this question, the effect of other common inducing treatments was studied (Green, unpublished data). Mitomycin C (MC) was shown to cause *E. coli*($c_{I,tl}$) to induce rapidly at 45°C + CAP. The following evidence suggests that MC or a cell product formed as a result of MC causes induction by inactivating the λ repressor.

As in the case of UV induction (Lieb, 1964), different c_I temperature mutants exhibited a wide range of sensitivity to MC induction when present as prophage in lysogenic derivatives of the same bacterial strain. The dose response of several lysogens of *E. coli* strain W3110 is shown in Table III; the kinetics of induction of these lysogens is illustrated by Fig. 2. From these data we can arrange the mutants in order of increasing sensitivity to MC: $c_I^+ < c_{I,U32} < c_{I,tl} < c_{I,U37}$. Thus, alterations affect the ability of MC to promote induction.

TABLE III

SENSITIVITY TO MITOMYCIN C[a]

Lysogen	Mitomycin C concentration (μg/ml)				
	0.3	1.0	3.0	10.0	30.0
W3110 (λ)	—	122	46	0.26	0.0015
W3110 (λ_{U32})	130	9.0	0.57	0.0071	—
W3110 (λ_{tl})	170	49	12	4.1	—
W3110 (λ_{U37})	40	0.45	0.11	0.0024	—

[a] The percent survival was determined after 60 min aeration at 32°C in K medium containing the designated concentration of mitomycin C. Viability was determined by the colony count method.

Thymine starvation was found to be another means of inducing class B lysogens at 45°C in the absence of protein synthesis. The thymine requiring strain, CR34, was made lysogenic for $\lambda c_{I,t1}$. This strain also requires threonine and leucine (TL) for growth. As shown in Fig. 3, thymine starvation at 45°C in the presence of CAP or absence of TL provoked cell death after a lag of about 30 min (Nos. 2 and 4). Control cultures containing thymidine with CAP or without TL (Nos. 1 and 3) remained viable at 45°C. When protein synthesis was permitted at 45°C, the absence of thymidine did not cause a lag prior to cell death (Nos. 5 and 6).

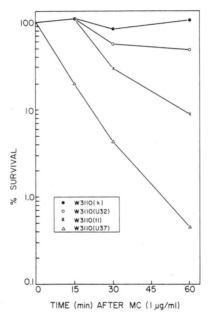

FIG. 2. Sensitivity of various lysogens of *E. coli* W3110 to mitomycin C (MC). Each culture was treated with 1.0 μg/ml MC (Kyowa Hakko Kogyo Co., Ltd., Tokyo, Japan) in K medium with aeration at 32°C. Plates were incubated overnight at 32°C.

In order to show that under these conditions, thymineless death was due to induction of prophage development, the following experiment was done. After thymine starvation for 60 min at 45°C + CAP, the cells were diluted out of CAP and aerated at 37°C in the presence of thymidine. Within an hour the cells lysed and released 27 λ phage per induced cell.

It would thus appear that thymine starvation, as well as MC, can cause the conversion of the class Bλ repressor to the inducing protein at 45°C in the absence of protein synthesis. As in the case of the class A lysogens, induction then proceeds at maximal rate (see Figs. 1 and 3). Once it is formed, the class B inducing protein is therefore stable and active at 45°C. This conclusion is based on the assumptions that MC and thymine starvation cause induction by inactivating the λ repressor and that the activity of the inducing protein is required for λ maturation regardless of the means of induction.

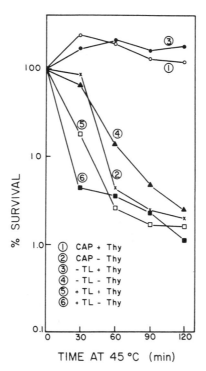

FIG. 3. Induction by thymine deprivation of *E. coli* CR34($\lambda c_{I,t1}$) at 45°C in the absence of protein synthesis. Cells were grown at 37°C in minimal M9 medium supplemented with TL (20 μg/ml each), vitamin B_1 (5 μg/ml), and thymine (10 μg/ml) (M9 + TL thy). The cells were washed on a Millipore filter to remove TL and thymine; then, they were suspended in M9 with or without TL and thy at 37°C, as indicated. CAP (100 μg/ml) was added to samples 1 and 2 five min prior to transfer to 45°C. Samples 3–6 were incubated 15 min at 37°C (± thy, ± TL, as indicated) to exhaust the TL or thymine pools before transferring to 45°C. Plates were incubated overnight at 37°C.

The Effect of Derepression and Induction on the Structure of Intracellular λ DNA

As discussed earlier, the inducing protein is present in an inactive form in the lysogenic cell. Its activity appears after that of the repressor is lost. The only assay for the inducing protein is an indirect one, namely the loss of cell viability. Since class B lysogens could become derepressed for λ mRNA synthesis while retaining their immunity to superinfection (Table I), it seemed that the inducing protein might be involved in some process of λ maturation other than the detachment of the prophage.

Bode and Kaiser (1965) reported that λ DNA assumed a more compact configuration after being injected into an immune lysogen. At neutral pH, the injected DNA sedimented at 1.9 times the rate of uninjected λ DNA. It was suggested that this new form might be that of a twisted circle. The following experiment was performed in order to determine whether the inducing protein alters the structure of the injected λ DNA.

A culture of *E. coli*($\lambda c_{I,r32}$), a class A lysogen, was infected with ^{3}H-$\lambda c_{I,tl}$; *E. coli*($\lambda c_{I,tl}$), a class B lysogen, was infected with ^{32}P-$\lambda c_{I,tl}$, each at a multiplicity of 1. Each culture was grown at low temperature for 30 min to permit the injected λ DNA to assume the compact configuration. Part of each culture was then pooled and harvested (sample B). CAP (100 μg/ml) was added to another portion of each culture and they were aerated for 60 min at 45°C to cause derepression of λ mRNA synthesis. One hundred percent of the *E. coli*($\lambda c_{I,tl}$) cells remained viable whereas 95% of the *E. coli*($\lambda c_{I,r32}$) cells were induced. These two cultures were then pooled and harvested (sample C). Part of the ^{32}P-$\lambda c_{I,tl}$-infected *E. coli*($\lambda c_{I,tl}$) culture that was not treated with CAP was harvested without pooling. These cells were mixed with ^{3}H-$\lambda c_{I,tl}$ just prior to extracting DNA (sample A). DNA was purified according to the procedure of Dove and Weigle (1965), with care taken to prevent sharing. Samples A–C were then subjected to sucrose gradient centrifugation. The sedimentation profiles are shown in Fig. 4, A–C.

The ^{3}H profile (component III) in Fig. 4A represents DNA extracted from λ phage in the presence of a cell extract. It serves as a control to show that little, if any, DNA degradation or aggregation occurred. The ^{32}P profile represents λ DNA that had been injected into immune bacteria. Two new peaks are present, one sedimenting 1.9 times more rapidly than the λ ^{3}H-DNA (component II) and the other ~2.5 times as fast (component I). Component I probably resulted from some type of intracellular association of λ DNA. Component II has the sedimentation rate of the twisted circular form of λ DNA described by Bode and Kaiser (1965).

Samples B and C show nearly identical sedimentation profiles (Fig. 4, B and C). In each case, approximately 60% of the total ^{3}H appeared in the peak sedimenting twice as fast as component III and no ^{3}H peak appeared in the region of component III. Assuming that

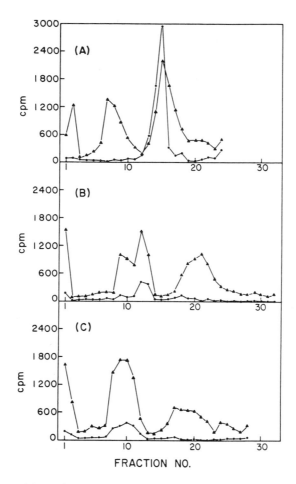

FIG. 4. Effect of induction on the structure of intracellular λ DNA. (A) ³H-DNA from ³H-λc$_{I,t1}$ phage (— ● —); λc$_{I,t1}$ ³²P-DNA 30 min after injection into *E. coli*(tl) at 37°C (— ▲ —), (B) λc$_{I,t1}$ ³H-DNA (— ● —) in *E. coli*(U32) for 30 min at 32°C; λc$_{I,t1}$ ³²P-DNA (— ▲ —) in *E. coli*(tl) for 30 min at 37°C. (C) λc$_{I,t1}$ ³H-DNA (— ● —) in *E. coli*(U32) for 30 min at 32°C followed by 60 min at 45°C + CAP (100 μg/ml); λc$_{I,t1}$ ³²P-DNA (— ▲ —) in *E. coli*(tl) for 30 min at 37°C followed by 60 min at 45°C + CAP (100 μg/ml).

In all cases, the multiplicity of infection was 1. The DNA samples were layered onto 10–30% sucrose gradients and spun for 8 hours at 24,000 rpm at 15°C in the Spinco SW25.1 rotor. Fractions (~1ml each) were collected from the bottom of the centrifuge tubes, TCA-precipitated onto membrane filters, and then counted in a Nuclear-Chicago Liquid Scintillation Spectrometer. The direction of sedimentation is right to left.

the compact DNA (component II) is a twisted circular form, only one break in either of the DNA strands would result in a twofold decrease in sedimentation coefficient. Therefore, neither derepression nor induction by heat in the presence of CAP causes the opening of this circular DNA. No apparent change in the structure of intracellular, parental λ DNA was caused by these conditions.

We are still without a biochemical assay for the inducing protein. The possibility that this protein may be required for λ DNA replication or maturation is currently being tested. It is of interest to note that the c_I region falls within a cluster of cistrons that control the replication of λ DNA (Brooks, 1965).

Complementation between λ c_I Mutants

In view of the postulated bifunctional role of the c_I region, namely, to specify the activity of the λ repressor and the inducing protein, it seemed reasonable to predict that the c_I region might be divided into two cistrons. The clustering of the two classes of temperature mutants into separate halves of the c_I region (Lieb, 1966) lent support to this hypothesis. Further support is herein provided by complementation experiments.

In order to test for complementation between the two classes of λc_I mutants, the following procedure was employed. Cells lysogenic for one class of temperature mutants were infected at low temperature (32° or 37°C) by a mutant of the other class. At this temperature, both the prophage and the superinfecting phage might be expected to synthesize their respective repressors. After a short time (15–45 min), the temperature was raised to 45°C, thereby inactivating the heat-sensitive repressors. However, if heat-stable λ repressors were formed as a result of complementation, several new properties should be exhibited by the superinfected cells: (1) a decrease in the rate of induction (cell death) at 45°C; (2) a delay in the time of lysis at 45°C, since λ-specific endolysin synthesis would be inhibited; (3) a decrease in the rate of λ mRNA synthesis.

The inhibition of the rate of induction of two class A lysogens is shown in Figs. 5 and 6. Superinfection of *E. coli*(U32) by λ, $\lambda c_{I,tl}$, or $\lambda c_{I,71}$ resulted in nearly the same inhibition of induction over a 90 min period at 45°C. It should be noted that $\lambda c_{I,71}$ is a typical "clear" mutant in that it cannot lysogenize *E. coli* at 37°C (Kaiser, 1957). The mutation maps in the B region of c_I; it is further from the A region than is $c_{I,tl}$ (Lieb, 1966). Wild-type λ would be expected to inhibit induction since its repressor is stable at 45°C (Green, 1966).

With *E. coli*(U37), visability assays were performed over a 3 hour period (Fig. 6). It may be seen that the rate of death is greatly inhibited by λ, $\lambda c_{I,t1}$, and $\lambda c_{I,71}$, but only slightly by $\lambda c_{I,r32}$ and $\lambda c_{I,U37}$. Infection by λ even permitted the lysogens to divide at 45°C. In several experiments of this type, colonies were picked from the 1, 2, and 3 hour survivors of the 45°C heat treatment, grown at 32°C, and tested for heat inducibility.

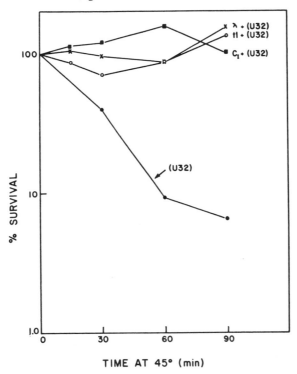

FIG. 5. Inhibition of induction of *E. coli*($\lambda c_{I,U32}$) by λ, $\lambda c_{I,t1}$, and $\lambda c_{I,71}$. Log phase cells (2–4 × 10⁸/ml) were centrifuged and concentrated to ~1 × 10⁹/ml in K medium at 32°C. λ (— × —), $\lambda c_{I,t1}$ (— ○ —), and $\lambda c_{I,71}$ (— ■ —) were added at a multiplicity of 20. After 15 min of phage adsorption, the cells were diluted 100-fold into prewarmed (45°C) K medium (zero time). Samples were removed periodically for viability assays. Plates were incubated overnight at 30°C.

In all cases (~50), those that were lysogenic were still heat-inducible. It would thus appear that the resistance to heat induction is a temporary physiological phenomenon; it is not due to the integration of the λc_I^+ region, either by recombination or lysogenization.

A more direct assay, for the appearance of heat-resistant λ repressor resulting from complementation, was afforded by examining the kinetics of cell lysis at 45°C following superinfection. This assay is a measure of the activity of a λ-specific enzyme, endolysin (Campbell, 1961). Superinfection of *E. coli*(U32) by λc$_{I,tl}$ delayed the time of cell lysis while completely inhibiting induction for at least 90 min at 45°C (Fig. 7).

The most direct assay for complementation between λc$_I$ mutants entailed measuring the rate of λ mRNA synthesis at elevated temperatures. This rate would be expected to decrease if heat-resistant repressors resulted from complementation. To test this prediction, *E. coli*(U32) was infected with λc$_{I,tl}$ at 32°C and aliquots were pulse-labeled with uracil-³H at 45 min and 90 min after transfer to 45°C (see Fig.7). RNA isolated from these cells as well as from an uninfected, heat-induced control, pulse-labeled at 45 min after transfer to 45°C, was tested for hybridization with λ and *E. coli* DNA (see Table IV).

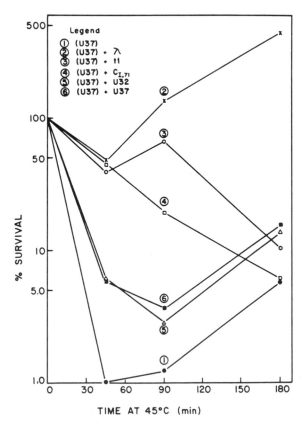

FIG. 6. Inhibition of induction of *E. coli*(λc$_{I,U37}$). Cells were infected as described in Fig. 5 with λ (— × —), λc$_{I,tl}$ (— ○ —), λc$_{I,7l}$ (— □ —), λc$_{I,U32}$ (— △ —), and λc$_{I,U37}$ (— ■ —). After dilution into 45° K medium (t = 0), cell viability was assayed periodically. Plates were incubated overnight at 30°C.

TABLE IV

COMPLEMENTATION BETWEEN λc_I MUTANTS[a]

Source of ³H-RNA	Time of pulse at 45°C (min)	RNA hybridized (cpm) with:		% RNA background	$\lambda/E.\ coli$[b]
		λ DNA (295 μg/ml)	E. coli DNA (400 μg/ml)		
1. E. coli ($\lambda c_{I,U32}$)	45–46	8992 (22.3%)	3224 (8.0%)	0.3	2.79
2. E. coli(U32) + $\lambda c_{I,t1}$	45–46	2044 (4.9%)	7022 (16.8%)	0.5	0.29
3. E. coli(U32) + $\lambda c_{I,t1}$	90–91	4351 (10.9%)	6941 (17.3%)	0.3	0.63

[a] E. coli ($\lambda c_{I,U32}$) were treated as described in Fig. 7. DNA-RNA hybrids were formed during a 4.5 hour incubation at 61°C. The data are average values of two identical reactions. The percentage figures in brackets represent the percentage of RNA hybridized based on the acid-precipitable ³H in the reaction mixture.

[b] Ratio of the amounts of λRNA and E. coli RNA in a given sample which hybridize with fixed concentrations of E. coli and λDNA.

It is apparent that the superinfecting phage, tl, caused a significant reduction in the rate of λ mRNA synthesis. After 45 min at 45°C, there was nearly a tenfold decrease in the $\lambda/E.$ *coli* mRNA ratio. As discussed previously, this ratio is the best index for comparing the relative rates of synthesis of λ and *E. coli* mRNA at different times or under different environmental influences. At 90 min, the $\lambda/E.$ *coli* mRNA ratio increased somewhat, but was still four times less than in the uninfected control. Once again we have an example of λ mRNA synthesis in the absence of cell death. It should be noted that the decreased $\lambda/E.$ *coli* ratio was caused mostly by a decrease in the rate of λ mRNA synthesis. The ability of *E. coli* mRNA to hybridize serves as an internal

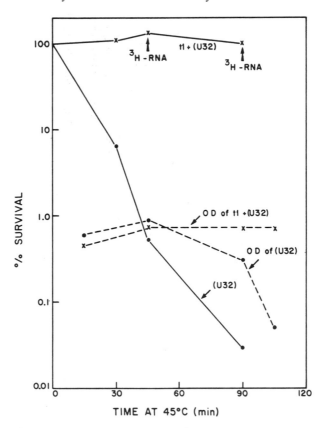

FIG. 7. Inhibition of induction and cell lysis of *E. coli*($\lambda c_{I,U32}$) by $\lambda c_{I,II}$. Conditions were as described in Fig. 5, except that the infected cells were aerated at 32°C for 45 min before diluting into 45°C K medium (zero time). Portions of the infected and uninfected cultures were pulse-labeled at 45°C with uracil-^{3}H (0.5 μCi/ml) at $t = 45$–46 min and 90–91 min. The optical density (OD) of the cultures was measured at 660 mμ.

control which demonstrates that the RNA was not destroyed during isolation.

It is thus apparent that complementation between certain λc_I mutants can give rise to a λ repressor with altered thermal properties. The repressor must be an aggregate, therefore, of one or more products coded by the c_I region. Many more mutants must be examined before we can definitely conclude that the c_I region is comprised of more than one cistron. This finding would enable us to decide whether the λ repressor-inducing protein was an aggregate of the same or different polypeptides.

The fact that the λ repressor is an aggregate helps to account for our observation that it is synthesized slowly with respect to cell growth (see p. 144). Since derepressed class B lysogens may nevertheless retain immunity and not be induced, it is possible that some disaggregated, or otherwise altered form of the λ repressor can interfere with viral development in a way not involving the inhibition of λ mRNA synthesis.

This speculation derives support from the observation of Thomas and Bertani (1964). When *E. coli*(λ) was multiply infected with λ and λ_{VIR}, only infective λ_{VIR} was produced. These authors proposed that the prophage λ repressor can in some way block phage maturation in a cell that contains all the mRNA and proteins needed for the production of λ phage. For example, the repressor might directly block replication of λ DNA or the conversion of noninfective λ DNA to the infective form. Experiments designed to test this hypothesis are in progress.

Summary

The regulation of λ development, as affected by the product(s) of the c_I region, has been shown to be more complex than was first suspected. The c_I region directs the synthesis of a repressor that inhibits the formation of λ mRNA. This activity serves to prevent prophage development and to maintain the immunity of a lysogenic cell. With certain λc_I mutants, the prophage repressor can be inactivated by heat without the concomitant loss of the cell's immunity to superinfection. In these cases, the prophage does not mature; the cell retains its viability and remains lysogenic. We have postulated that in order for prophage development to occur, the repressor must not only be inactivated, but it must also be converted to a substance (the inducing protein) whose activity is required for λ maturation. Complementation studies have indicated that the λ repressor is an aggregate of polypeptides coded by the c_I region. The activity of the c_I product, whether repressor or inducing protein, is thought to depend upon its state of aggregation or conformation.

ACKNOWLEDGMENTS

The author is indebted to Helgi Tarikas and Susan Vaill for their skillful technical assistance, and to the U.S. Public Health Service for research support (Grant No. AI-06013-03).

REFERENCES

Attardi, G., Naono, S., Rouvière, J., Jacob, F., and Gros, F. (1963). *Cold Spring Harbor Symp. Quant. Biol.* **28,** 363.
Bode, V. C., and Kaiser, A. D. (1965). *J. Mol. Biol.* **14,** 399.
Brooks, K. (1965). *Virology* **26,** 489.
Campbell, A. (1961). *Virology* **14,** 22.
Dove, W. F., and Weigle, J. J. (1965). *J. Mol. Biol.* **12,** 620.
Green, M. H. (1963). *Proc. Natl. Acad. Sci. U.S.* **50,** 1177.
Green, M. H. (1966). *J. Mol. Biol.* **16,** 134.
Hall, B. D., and Spiegelman, S. (1961). *Proc. Natl. Acad. Sci. U.S.* **47,** 137.
Jacob, F., and Monod, J. (1961). *J. Mol. Biol.* **3,** 318.
Jacob, F., Fuerst, C. R., and Wollman, E. L. (1957). *Ann. Inst. Pasteur* **93,** 724.
Jacob, F., Sussman, R., and Monod, J. (1962). *Compt. Rend.* **254,** 4214.
Kaiser, A. D. (1957). *Virology* **3,** 42.
Kaiser, A. D., and Jacob, F. (1957). *Virology* **4,** 509.
Kellenberger, G., Zichichi, M. L., and Weigle, J. J. (1961). *Proc. Natl. Acad. Sci. U.S.* **47,** 869.
Lieb, M. (1964). *Science* **145,** 175.
Lieb, M. (1966). *J. Mol. Biol.* **16,** 149.
Nagata, T. (1963). *Proc. Natl. Acad. Sci. U.S.* **49,** 551.
Nygaard, A. P., and Hall, B. D. (1963). *Biochem Biophys. Res. Commun.* **12,** 98.
Pardee, A. B., Jacob, F., and Monod, J. (1959). *J. Mol. Biol.* **1,** 165.
Sadler, J. R., and Novick, A., (1965). *J. Mol. Biol.* **12.** 305.
Sly, W. S., Echols, H., and Adler, J. (1965). *Proc. Natl. Acad. Sci. U.S.* **53,** 378.
Thomas, R., and Bertani, L. E. (1964). *Virology* **24,** 241.
Thomas, R., and Lambert, L. (1962). *J. Mol. Biol.* **5,** 373.
Wolf, B., and Meselson, M. (1963). *J. Mol. Biol.* **7,** 636.

Selectivity of *in Vitro* RNA Synthesis on Lambda DNA Templates[*]

Stanley N. Cohen,† Umadas Maitra, and Jerard Hurwitz

DEPARTMENT OF DEVELOPMENTAL BIOLOGY AND CANCER,
ALBERT EINSTEIN COLLEGE OF MEDICINE,
BRONX, NEW YORK

It has been well established that the expression of genetic information in viruses, as well as in more complex organisms, is subject to temporal control; not all genes are expressed simultaneously (Jacob *et al.*, 1957; Luria, 1962; Epstein *et al.*, 1963). Early and late functioning cistrons have been distinguished in λ, T-even, and other bacteriophages; in certain instances, genetic maps have revealed a physical clustering of genes whose expression is temporally related (Jacob *et al.*, 1957; Campbell, 1961; Epstein *et al.*, 1963). Although the biological mechanism responsible for the sequential expression of genetic information is not known, there is evidence that such regulation may occur at the level of gene transcription (Kano-Sueoka and Spiegelman, 1962; Protass and Korn, 1966; Skalka, 1966).

Experiments by Bremer *et al.* (1965) and by Maitra and Hurwitz (1965) have indicated that transcription of DNA *in vitro* may also be a specifically regulated process. Initiation of RNA chains on a variety of DNA templates occurs predominantly with purine nucleoside triphosphates, which indicates that pyrimidines constitute the principal DNA sites at which such chains are started. Recent kinetic (Berg *et al.*, 1965) and electron microscopy (Crawford *et al.*, 1965) studies of the bind-

[*] This work was supported by grants from the National Institutes of Health, the National Science Foundation and the Public Health Research Council of the City of New York.

† Stanley N. Cohen is the recipient of a Postdoctoral Fellowship grant from the American Cancer Society.

ing of RNA polymerase to DNA templates have shown that only a portion of the template is occupied by enzyme molecules at a given time. However, it has not been clear whether such binding of enzyme to DNA and the *in vitro* RNA synthesis which, under appropriate conditions is its consequence, takes place randomly along the length of the template—or whether it selectively occurs at certain DNA sites.

At least four principal possibilities exist regarding the sites of the DNA template which may be selected by the RNA polymerase for the formation of RNA chains *in vitro* (Fig. 1). Such chains may begin and end randomly along the length of the template, as in A; RNA synthesis may begin at a single site and proceed along the template with newly formed chains terminating at various specified or random sites, as in B; one or more specified segments of the template may be repeatedly copied, as in C; or there may be a preferential, but not exclusive, copying of certain portions of the template while other segments are copied to a lesser degree, as in D.

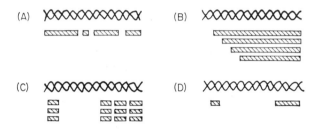

FIG. 1. Possible schemes for transcription of DNA template *in vitro*.

DNA from the bacteriophage λ is an especially useful tool for examining possible specificity of RNA synthesis *in vitro*. Different segments of λDNA are widely divergent in their nucleotide base composition (Hershey and Burgi, 1963; Hershey, 1964); this provides a basis for the physical separation of segments of sheared λDNA by column chromatography (Hogness and Simmons, 1964) or density gradient centrifugation (Hershey *et al.*, 1965; Nandi *et al.*, 1965). Thus transcription from different portions of the DNA template can be studied *in vitro*.

In the present experiments, which have been reported previously in brief (Cohen, 1966). $\lambda c_I b^+$ bacteriophage and 3S λDNA were prepared by modifications of the procedures of Burgi (1963) and of Kaiser and Hogness (1960); the DNA was ascertained to be free of RNase and detectable single strand breaks. The DNA was then sheared into half-length molecules which were separated preparatively, using a modification of the gradient centrifugation procedure developed by Davidson and his associates (Davidson *et al.,* 1965; Nandi *et al.,* 1965). This procedure utilizes the preferential binding of mercuric ions to the AT-rich (right) half of λDNA to increase the buoyant density in Cs_2SO_4 gradients and permit its separation from the GC-rich (left) half.

RNA polymerase products were prepared using whole λDNA and each of its isolated halves as templates; these products were characterized with regard to nearest neighbor frequency, chain length, base composition, and ability to hybridize with various segments of λDNA. Other experiments were performed to determine the effects of altering the DNA template by sonication, denaturation, or covering up the free cohesive ends of linear λDNA on the ability of this DNA to prime and direct RNA synthesis *in vitro.* Evidence has accumulated (Strack and Kaiser, 1965; Kaiser and Inman, 1965) to indicate that the complementary ends of λDNA are important in determining its biological infectivity. Circular λ molecules, aggregates, sheared and cohered "outside-in" molecules and other forms of λDNA were prepared by procedures described elsewhere (Cohen *et al.,* 1967). These altered DNA forms were employed as templates for the synthesis of RNA products.

The RNA polymerase used in these experiments was purified from *E. coli* B by a modification of the procedure of Furth *et al.* (1962). Enzyme preparations were ascertained to be free from detectable DNase and RNase.

Separation and Identification of λDNA Halves

The dripping pattern of a typical Cs_2SO_4 gradient density centrifugation of the Hg^{++} complex of sheared λDNA is shown in Fig. 2. Two major peaks are evident: The small shoulder noted on the first major peak was of the same buoyant density as linearly intact λDNA when rebanded in a CsCl gradient; it may represent either incompletely sheared material or a small amount of reconstituted "outside-in" whole λDNA formed by complementary sequences on the ends of the halves during centrifugation.

The pooled fractions comprising the major peaks shown in Fig. 3 were dialyzed against EDTA to remove the Hg^{++} and were then rebanded in the analytical ultracentrifuge in CsCl gradients using a reference standard consisting of *B. subtilis* phage 2C DNA. A densitometer tracing of such a centrifugation showing separated halves of λDNA is seen in Fig. 3. The observed buoyant densities of the two halves represent $A + T/G + C$ ratios of 55% and 45% (Schildkraut *et al.,* 1962).

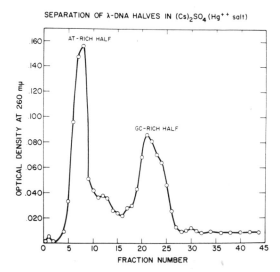

SEPARATION OF λ-DNA HALVES IN (Cs)₂SO₄ (Hg⁺⁺ salt)

FIG. 2. Dripping pattern of Cs_2SO_4 gradient centrifugation of Hg^{++} complex of sheared λDNA. Shearing of whole λDNA into half-length molecules was accomplished with a Fisher steady speed stirrer, fitted with a Vertis stainless steel macro shaft and blades, and was monitored by gradient centrifugation in CsCl (Hershey *et al.,* 1965). One-hundred and seventy mμmoles of sheared λDNA dissolved in 4 ml of 0.1 *M* Na_2SO_4-0.005 *M* sodium-borate buffer, pH 9.0, were mixed with 0.90 ml of 10^{-4} *M* $HgCl_2$ at 0°C, and Cs_2SO_4 was added to the solution to a density of $\rho = 1.508$ at 25°C. The mixture was centrifuged at 4°C in a No.40 fixed angle rotor at 38,000 rpm for 46 hours in a Spinco L2 preparative ultracentrifuge. A hole was pierced in the bottom of the tube, fractions (4 drops each) were collected in the cold, and the adsorbancy at 260 mμ was determined using a Zeiss spectrophotometer.

RNA Synthesis Primed by Various Forms of λDNA

Certain parameters of RNA synthesis primed by whole λDNA and each of its halves is shown in Table I. Although the whole λDNA template used in these experiments was free from single-strand breaks, it should be pointed out that both of its isolated halves and single strand nicks as a result of the shearing process. The S values shown for RNA were determined at 30 min of incubation, i.e., during the period of rapid RNA chain growth. The values shown for RNA synthesis and total initiation are yields, rather than rates; they have been determined at a time (70 min) when initiation is complete and total ribonucleotide incorporation has reached a virtual plateau. It can be seen from this data that marked heterogeneity of chain length is evident in the growing 30-min RNA products synthesized with each of these DNA templates. The average size of completely formed RNA chains (calculated by dividing total ribonucleotide incorporation by γ-^{32}P-ATP and γ-^{32}P-GTP incorporation at termini) synthesized from λ halves was smaller than that found when whole λDNA was used as template. Incorporation of ribonucleotides into RNA products was lowest in reactions primed by the GC-rich half.

The effects of certain structural alterations in the λDNA template on RNA initiation and synthesis are shown in Table II. No substantial differences were observed among native linear λDNA, λ circles, aggregates, sheared λDNA, or sheared and cohered "outside-in" molecules. In contrast, denaturation of the λDNA template resulted in a sharp rise in total initiation, a decrease in synthesis, and a consequent reduction of average RNA chain size, as has been observed previously with other

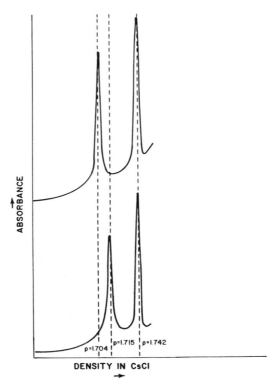

FIG. 3. Densitometer tracing of analytical ultracentrifugation analysis of the λ fractions. The pooled fractions comprising the major peaks shown in Fig. 1 were dialyzed against four 1 liter changes of 0.1 M EDTA, pH 8, followed by four 1 liter changes of 0.15 M NaCl-0.15 M sodium citrate. Aliquots of approximately 1.5 μg of each half were separately combined with 2 μg of *B. subtilis* phage 2 C DNA and were centrifuged in solutions consisting of 0.02 M Tris buffer, pH 8.5, and CsCl (to $\rho = 1.710$) at 44,770 rpm at 25°C for 20 hours. The assumed buoyant density of the 2C DNA reference standard was $\rho = 1.742$, based on a ρ of 1.710 for *E. coli* DNA.

TABLE I

PARAMETERS OF RNA INITIATION AND SYNTHESIS WITH LINEAR WHOLE λDNA AND ITS SEPARATED HALVES[a]

DNA template	Sedimentation range of 30-min RNA product	RNA synthesis (70 min)	Initiation at 70 min			Average RNA chain length at 70 min
			$\gamma\text{-P}^{32}\text{-ATP}$	$\gamma\text{-P}^{32}\text{-GTP}$	Total	
		mμmoles		mμmoles		
Native linear λ	12–>30 S	1.21	0.08	0.21	0.29	4250
AT-rich half	14–20 S	1.17	0.15	0.35	0.50	2350
GC-rich half	12–18 S	0.78	0.08	0.50	0.58	1350

[a] RNA synthesis was carried out in reaction mixtures (0.5 ml) containing 25 μmoles of Tris buffer, pH 8.0, 80 mμmoles each of ATP CTP, and UTP, 40 mμmoles of α-^{32}P-labeled GTP (3.3 × 10^4 cpm/mμmole), 4 μmoles of 2-mercaptoethanol, 0.5 μmole of MnCl$_2$, 2.5 μmoles of MgCl$_2$, 5 units of RNA polymerase, and 4 mμmoles of DNA as indicated. Reaction mixtures were incubated for 30 or 70 min, as indicated.

Sedimentation of 30-min RNA products was carried out for 16 hours at 25,000 rpm at 28°C in 15–30% sucrose gradient solutions containing 0.05 M Tris buffer, pH 7.5, 0.1 M NaCl and 0.2% sodium dodecyl sulfate. Thirty 1 ml fractions were collected through a hole pierced in the bottom of the tube.

Acid-insoluble RNA products were collected on Millipore filters, dried, and counted in a windowless gas flow counter. Initiation experiments were carried out as previously described (Maitra and Hurwitz, 1965), using γ-^{32}P-labeled ATP and γ-^{32}P-GTP (specific activities = 6.2 × 10^9 cpm/μmole and 8 × 10^9 cpm/μmole, respectively) in different reaction tubes. Incubations were at 37°C for 70 min.

TABLE II

PARAMETERS OF RNA SYNTHESIS ON NATIVE AND ALTERED λDNA TEMPLATES[a]

DNA template	RNA synthesis at 70 min	Initiation at 70 min		Average RNA chain length at 70 min
		γ-^{32}P-ATP	γ-^{32}P-GTP	initiation	
	mμmoles		μμmoles		
Native linear λ	1.21	0.08	0.21	0.29	4250
λ circles	0.99	0.07	0.15	0.23	4150
λ aggregates	1.30	0.09	0.21	0.30	5340
Sheared λ	1.12	0.12	0.20	0.32	3500
Sheared-cohered λ	1.05	0.09	0.16	0.25	4100
Sonicated λ	0.70	0.16	0.16	0.32	2200
Denatured λ	0.55	0.27	1.45	1.72	320

[a] The conditions for RNA synthesis and initiation were as described in Table I. The preparation of the various structural forms of λDNA used in these experiments has been described in detail elsewhere. (Cohen *et al.*, 1967).

DNA primers (Maitra and Hurwitz, 1965; Bremer *et al.*, 1966). Sonication of the template also resulted in decreased synthesis; but, no significant increase in initiation sites resulted from this procedure. Average RNA chain length was reduced as a result of the decreased synthetic rate observed with sonicated DNA.

Nearest Neighbor Analysis of RNA Products

Nearest neighbor frequency analysis of the RNA products made from whole λDNA and each of its halves was carried out following alkaline hydrolysis; the ^{32}P-labeled nucleoside monophosphate compounds were separated by high voltage electrophoresis using a pyridinium acetate buffer system. Between six and twelve individual analyses were performed for each of the four labeled α-^{32}P-labeled nucleoside triphosphates incorporated; individual analyses were found to be in close agreement.

The cumulative results of nearest neighbor analyses of 70-min RNA products synthesized using whole λDNA and each of its halves as templates are shown in Table III. These results can be summarized as follows: (1) The products of the two halves can be readily distinguished by differences in the frequencies of individual mononucleotide bases and certain pivotal dinucleotide pairs (e.g., ApA, UpU, GpC, etc.)*. A + U/G + C ratios of 1.252 and 0.777 in the RNA made from isolated AT-rich and GC-rich halves reflect the divergent deoxyribonucleoside monophosphate composition observed in the DNA templates (55% and 45% A + T, respectively). (2) The frequencies of individual bases and pivotal dinucleotide pairs in the RNA formed with whole λDNA as template resemble those formed in the RNA made from the AT-rich half, and differ widely from the frequencies observed in the GC-rich RNA product. (3) The A + U/G + C ratio determined for the whole λ product suggests preferential copying of AT-rich regions of the template; random copying would have produced a ratio which is the average of those observed with the halves.

It appeared from these experiments that transcription of λDNA *in vitro* did not occur randomly along the length of the genome. The inference that certain segments of λDNA are preferentially copied by the RNA polymerase, when the whole λDNA molecule is used as template, was substantiated by the results of DNA-RNA hybrid experiments performed

---

* It should be noted that, in all of these RNA products, the frequencies of complementary mononucleotides (i.e., A and U; G and C) and those of complementary antipolon dinucleotide pairs (e.g., ApA and UpU; GpU and ApC) are not equal. This finding suggests that both strands of the λDNA template may not be copied equally *in vitro* at the same sites.

TABLE III

NEAREST NEIGHBOR ANALYSIS OF RNA PRODUCTS SYNTHESIZED WITH WHOLE λDNA AND EACH λ HALF[a]

Dinucleotide pair	AT-rich half		Whole λDNA		GC-rich half	
ApA UpU	0.0813	0.0910	0.0759	0.0865	0.0489	0.0574
CpA UpG	0.0683	0.0833	0.0720	0.0791	0.0692	0.0774
GpA UpC	0.0604	0.0619	0.0598	0.0578	0.0582	0.0626
CpU ApG	0.0598	0.0530	0.0533	0.0551	0.0556	0.0504
GpU ApC	0.0573	0.0630	0.0582	0.0514	0.0620	0.0535
GpG CpC	0.0489	0.0420	0.0525	0.0503	0.0751	0.0681
UpA	0.0551		0.0512		0.0306	
ApU	0.0826		0.0765		0.0544	
CpG	0.0443		0.0530		0.0828	
GpC	0.0630		0.0693		0.0913	
A	0.265		0.259		0.207	
U	0.291		0.272		0.230	
G	0.230		0.240		0.287	
C	0.214		0.229		0.276	
Ratio A + U/G + C	1.252		1.134		0.777	

[a] The reaction mixtures were as described in Table I. In each case a single α-^{32}P-labeled nucleoside triphosphate (specific activity 4.5 to 8 $\times$ 10⁴ cpm/mμmole) and 10 units of RNA polymerase were added. Following incubation for 70 min at 37°C 1.5 mg of carrier bovine serum albumin was added; the RNA was precipitated with cold 5% trichloracetic acid containing 0.01 M sodium pyrophosphate and was washed twice with cold 2% TCA. The precipitates were dissolved in 0.5 ml of 0.3 N KOH and incubated overnight in tightly sealed tubes at 37°C. A mixture of unlabeled 2′(3′)-ribonucleotide markers was added, the digests were neutralized with Dowex 50-H⁺, and the pH was adjusted to between 6 and 8 with NaOH. The Dowex resin was removed by filtration through glass wool, and the digests were concentrated to dryness in a rotary evaporator at room temperature. They were then dissolved in 0.15 ml of water spotted on Whatman No. 3 paper and subjected to electrophoresis in pyridine : acetic acid : H.O buffer (1 : 10 : 89) pH 3.5 at 6000 volts for 1.5 hours. Electropherograms were cut into 1.5-cm wide strips which were placed in vials and counted in a Packard Tri-Carb liquid scintillation spectrometer. The frequency of dinucleotide pairs was calculated using a Control Data Computer programmed for solution of the simultaneous equations described by Josse et al. (1961).

on nitrocellulose membranes, using the method of Gillespie and Spiegelman (1965).

Hybridization of λDNA with RNA Polymerase Reaction Products

In these experiments, which have been described in greater detail elsewhere (Cohen et al., 1966), alkaline denatured tritiated λDNA was fixed to nitrocellulose membranes; RNA, which was isolated and purified following synthesis in the RNA polymerase reaction, was annealed with this DNA under appropriate conditions. Experiments were performed to determine the specificity of hybridization under various conditions and to ascertain that size differences and divergent base compositions of the DNA templates and the RNA products did not artifactually affect the results. Maximum efficiency of hybridization under the conditions chosen was between 40 and 50%. DNA amounts in excess of saturating levels were used in each instance.

Table IV summarizes the results of certain of these experiments. The RNA product prepared using the AT-rich half of λDNA as template hybridized equally well with whole λDNA and with its own template

TABLE IV

HYBRIDIZATION OF WHOLE λDNA AND EACH OF ITS HALVES WITH RNA PRODUCTS SYNTHESIZED *in Vitro*[a]

RNA synthesized from	Hybridization	
	DNA used	Input RNA hybridization (%)
AT-rich λ half	Whole λ	39–43
	AT λ	38–45
	GC λ	3–6
GC-rich λ half	Whole λ	37–44
	AT λ	3–5
	GC λ	38–43
Whole λDNA	Whole λ	38–44
	AT λ	37–43
	GC λ	10–15

[a] RNA was synthesized in the standard RNA polymerase assay as described in Table I except that Mg^{++} was used in place of Mn^{++} for most experiments, although with either metal ion similar hybridization patterns were obtained. All four ribonucleoside triphosphates employed were labeled with ^{32}P in the α-phosphate group (specific radioactivity of each was 2.5×10^8 cpm/μmole). Following an appropriate period of synthesis DNase was added to a final concentration of 10 μg/ml and the mixture was incubated for an additional 15 min at 37°C. The mixture was heated at 100°C for 3 min and diluted with an equal volume of 2 × SSC containing 0.5% sodium dodecyl sulfate (SDS). The solution containing the RNA product was equilibrated with an equal volume of phenol saturated with 2 × SSC and the aqueous layer was dialyzed for 2 hours each against three successive changes of 2 × SSC containing SDS. Hybridization of this ^{32}P-labeled RNA product with H^3-λDNA or with each of H^3-λDNA halves fixed on nitrocellulose membrane filters was performed at 68°C for 16 hours in 0.5 ml of 2 × SSC, as described by Gillespie and Spiegelman (1965). DNA levels which were in excess of the amount required to saturate the amount of RNA present were used. The values expressed are the range obtained in five different experiments. Input RNA 25 μμmoles.

and showed 4% cross-hybridization with the GC-rich λDNA half. The RNA synthesized using the GC-rich half of λDNA mirrored this hybridization pattern. When RNA made from whole λDNA was hybridized with each of the halves, a nonrandom pattern of homology was observed. This product hybridized almost as well with whole λDNA as with the AT-rich half, but only 10–15% of the input RNA counts annealed with the GC-rich half of the molecule. The preferential homology of the whole λRNA product with the AT-rich DNA half is in keeping with the predominant copying of AT-rich sequences of λDNA demonstrated by nearest neighbor analyses and indicates that *in vitro* transcription of λRNA is selective rather than random.

Effect of Duration of *in Vitro* RNA Synthesis on Selectivity of Transcription

It was of interest to determine whether the threefold preference of the RNA polymerase for synthesizing RNA from the AT-rich half of linear DNA was manifest throughout the duration of·*in vitro* RNA synthesis. In order to determine this, RNA products were isolated and purified following incubation of template with enzyme for various times. Table V shows the effect of the duration of enzymatic RNA synthesis on the annealing characteristics of the RNA product made using whole λDNA as template. It can be seen from this table that efficiency of hybridization was reduced in the product formed early during *in vitro* RNA synthesis, but the proportion of input RNA showing homology with

TABLE V

HYBRIDIZATION OF RNA PRODUCTS SYNTHESIZED AT VARIOUS INCUBATION TIMES[a]

RNA synthesized from whole λDNA		% of input RNA hybridized with:		
Duration of synthesis (min)	mμmoles synthesized	Whole λDNA	AT-rich half	GC-rich half
5	0.20	18.4	16.8	7.1
10	0.38	25.6	27.2	8.3
20	0.64	37.5	37.4	11.8
45	1.10	40.2	38.1	12.4
100	1.46	40.8	42.7	14.9

[a] The conditions of the experiment are as described in Table IV except that incubation was carried out for the time intervals indicated. RNA products were prepared for hybridization experiments as described in Table IV.

each of the λDNA halves remained the same throughout the duration of synthesis. These data indicate that when purified *E. coli* RNA polymerase and linear λDNA are used *in vitro,* there is no temporal shift to increased copying of the GC-rich half, as has been reported by Skalka (1966) to occur *in vivo.*

Effects of Structural Alterations in DNA on Selectivity of Copying

The effects of certain alterations in the DNA template on hybridization of products formed in the RNA polymerase reaction are shown in Table VI. It is evident from these data that preferential transcription, by the

TABLE VI

EFFECTS OF ALTERATIONS OF THE λDNA TEMPLATE ON SELECTIVITY OF COPYING[a]

λDNA template	Average size of 50-min RNA product	% of input (50 min) RNA hybridized with:		
		Whole λDNA	AT-rich half	GC-rich half
Native linear Circles Aggregates Sheared Sheared-cohered	3000–3300	38–44	38–42	10–15
Sonicated	1600	18.7	15.9	8.9
Denatured	240	18.2	18.9	20.0

[a] The conditions of RNA initiation and synthesis are as described in Table I. Isolation of RNA products and hybridization were carried out as described in Table IV. The preparation of various forms of λDNA has been described elsewhere (Cohen *et al.*, 1966).

RNA polymerase of sequences on the AT-rich λ half, is not affected by covering up the free complementary ends present on native linear λDNA by the formation *in vitro* of circles, aggregates, or "outside-in" sheared and cohered molecules. Furthermore, shearing or sonicating the template did not abolish selectivity of transcription. The RNA synthesized on sonicated λDNA showed a decreased efficiency of hybridization, an effect which may be related to the observed decrease in the size of the RNA product made on this template (Table II), but preferential copying of the AT-rich half was still evident. Only denaturation of the λDNA template resulted in elimination of selective copying.

Summary and Conclusions

The studies we have just described indicate that transcription of whole λDNA by the *E. coli* RNA polymerase does not occur randomly along the length of the genome, but rather that the enzyme is directed by the template to preferentially copy certain regions. Sites present on the AT-rich half of the linear λDNA molecule are preferentially transcribed through the duration of *in vitro* RNA synthesis. This contrasts with the temporal change in transcription pattern which occurs *in vivo* at early and late times following λ infection of its host (Skalka, 1966).

The complementary free cohesive ends of linear λDNA do not appear to influence either the priming ability of the template or the selectivity of *in vitro* transcription. Formation *in vitro* of λ circles, aggregates, etc., resulted in no detectable alterations in these parameters.

Denaturation of λDNA resulted in elimination of selective copying by the enzyme, an effect which may be related to the uncovering of new nonspecific sites at which the enzyme binds. The latter is also suggested by the marked increase in RNA chain initiation which results from disruption of the helical secondary structure of the template. In contrast, linearly intact λDNA is not essential for the preferential copying of the AT-rich half *in vitro*; presumably, the template sequences determining selectivity of transcription are small enough to avoid disruption by this procedure, and the creation of new DNA termini by sonication does not affect this mechanism.

REFERENCES

Berg, P., Kornberg, R. D., Francher, H., and Dieckmann, M. (1965). *Biochem. Biophys. Res. Commun.* **18**, 932.

Bremer, H., Konrad, M. W., Gaines, K., and Stent, G. S. (1965). *J. Mol. Biol.* **13**, 540.

Bremer, H., Konrad, M., and Bruner, R. (1966). *J. Mol. Biol.* **16**, 104.

Burgi, E. (1963). *Proc. Natl. Acad. Sci. U.S.* **49**, 151.

Campbell, A. (1961). *Virology* **14**, 22.

Cohen, S. N. (1966). *Federation Proc.* **25**, 651.

Cohen, S. N., Maitra, U., and Hurwitz, J. (1967). *J. Mol. Biol.* (in press).

Crawford, L. V., Crawford, E. M., Richardson, J. P., and Slayter, H. S. (1965). *J. Mol. Biol.* **14**, 593.

Davidson, N., Wedholm, J., Nandi, U. S., Jansen, R., Olevera, B. M., and Wang, J. C. (1965). *Proc. Natl. Acad. Sci. U.S.* **53**, 111.

Epstein, R. H., Bolle, A., Steinberg, C. M., Kellenberger, E., Boy de la Tour, E., Chevalley, R., Edgar, R. S., Susman, M., Denhart, G. H., and Lielausis, A. (1963). *Cold Spring Harbor Symp. Quant. Biol.* **28**, 373.

Furth, J. J., Hurwitz, J., and Anders, M. (1962). *J. Biol. Chem.* **237**, 2611.

Gillespie, D., and Spiegelman, S. (1965). *J. Mol. Biol.* **12**, 289.

Hershey, A. D. (1964). *Carnegie Inst. Wash., Yearbook* **63**, 580.
Hershey, A. D., and Burgi, E. (1963). *Carnegie Inst. Wash., Yearbook* **62**, 482.
Hershey, A. D., Burgi, E., and Davern, C. I. (1965). *Biochem. Biophys. Res. Commun.* **18**, 675.
Hogness, D. S., and Simmons, J. R. (1964). *J. Mol. Biol.* **9**, 411.
Jacob, F., Fuerst, C., and Wollman, E. (1957). *Ann. Inst. Pasteur* **93**, 724.
Josse, J., Kaiser, A. D., and Kornberg, A. (1961). *J. Biol. Chem.* **236**, 864.
Kaiser, A. D., and Hogness, D. S. (1960). *J. Mol. Biol.* **2**, 392.
Kaiser, A. D., and Inman, R. B. (1965). *J. Mol. Biol.* **13**, 78.
Kano-Sueoka, T., and Spiegelman, S. (1962). *Proc. Natl. Acad. Sci. U.S.* **48**, 1942.
Luria, S. (1962). *Science* **136**, 685.
Maitra, U., and Hurwitz, J. (1965). *Proc. Natl. Acad. Sci. U.S.* **54**, 815.
Nandi, U. S., Wang, J. C., and Davidson, N. (1965). *Biochemistry* **4**, 1687.
Protass, J., and Korn, D. (1966). *Proc. Natl. Acad. Sci. U.S.* **55**, 832.
Schildkraut, C. L., Marmur, J., and Doty, P. (1962). *J. Mol. Biol.* **4**, 430.
Skalka, A. (1966). *Proc. Natl. Acad. Sci. U.S.* **55**, 1190.
Strack, H. B., and Kaiser, A. D. (1965). *J. Mol. Biol.* **12**, 36–49.

PART III

SINGLE-STRANDED DNA BACTERIOPHAGES

Bacteriophage φX174: Viral Functions[*]

Robert L. Sinsheimer, Clyde A. Hutchison, III, and Björn H. Lindqvist

DIVISION OF BIOLOGY, CALIFORNIA INSTITUTE OF TECHNOLOGY, PASADENA, CALIFORNIA

Introduction

For some years now, we have been interested in the manner of replication of the small bacterial virus, φX174, which contains as its genetic component a single molecule of single-stranded DNA of molecular weight about 1.7×10^6 (i.e., about 5500 nucleotides) (Sinsheimer, 1959). Since it contains such a limited genetic complement, it seemed evident that this virus could introduce into the host cell only a limited set of functions—which could, potentially, be enumerated and analyzed. It seemed likely, therefore, that the reproduction of this virus must rely to a larger degree upon preexistent processes of the host cell, and that the virus-induced and the normal host functions must be complementary and be intermeshed to a greater degree than is the case with some of the larger, more autonomous viruses.

Successive developments in our and other laboratories have shown that the single-stranded DNA of this virus is circular (Fiers and Sinsheimer, 1962) and that, once inside the cell, it is quickly converted to a double-stranded DNA ring (the replicative form, RF) (Sinsheimer et al., 1962; Kleinschmidt et al., 1963; Chandler et al., 1964), which then replicates by a semiconservative mechanism (Denhardt and Sinsheimer, 1965a). At a later stage of the infectious cycle, single-stranded DNA molecules for the progeny virus particles are formed—very likely by a conservative process—using the double-stranded RF as a template (Denhardt and Sinsheimer, 1965a). It has also been shown, with bacterial

[*] This research was supported in part by U.S. Public Health Service Grants RG-6965 and GM-13554.

spheroplasts, that the free DNA, either viral or RF, is capable of initiating an infection (Guthrie and Sinsheimer, 1963).

More recently we have been interested in an analysis of the interactions between the virus and the host cell—in the interplay of viral and host functions—as a means of learning more about the mechanism of viral replication and of control mechanisms in the bacterial cell.

In this work, in order to resolve the contributions of the virus and of the host cell, we have made extensive use of mutants—of both the virus and of the host—and have made use of varied, often atypical, conditions of infection. As a consequence of these studies, we have reached certain tentative conclusions regarding the mechanism of replication of phage ϕX174, many of which are summarized in Fig. 1.

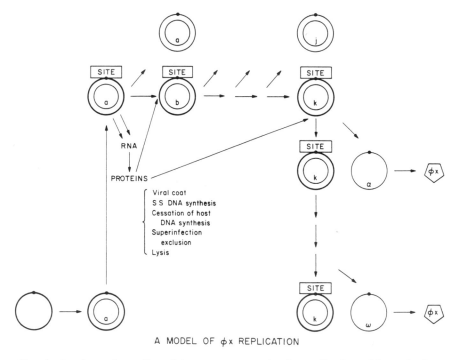

A MODEL OF ϕx REPLICATION

Fig. 1. A schematic outline of important stages in the replication of bacteriophage ϕX174.

The Essential Bacterial Site

The conversion of the parental, single-stranded DNA to the replicative form can be accomplished—apparently at a number of intracellular sites—in the absence of protein synthesis. This step, therefore, does not seem to require the participation of any virus-induced function, but appears to be performed by preexistent host enzymes. (Parenthetically, one might wonder why a cell would have an enzyme capable of catalyzing such a process—that is, capable of converting DNA from a single-stranded to a double-stranded form. We suspect that the enzyme(s) involved is a component of the normal DNA repair system of the cell and that φX174 is a virus which has specialized to take advantage of this repair mechanism.)

Once formed, however, it appears that the RF cannot function—that it cannot initiate further steps in the infection—until it becomes associated with a particular kind of site within the bacterial cell. Moreover, it appears that the number of these sites is limited; and, under certain circumstances there may be just one site. The RF must make this association before it can be transcribed into mRNA, be replicated, or contribute to the formation of progeny DNA. (On the other hand, the RF can apparently participate in recombination even though it has not become associated with such a site.)

At this time, the evidence for this important concept of an essential bacterial site, with which the viral genome must associate and with which or through which it functions, is essentially indirect but persuasive from our viewpoint. We do not know the nature or precise function of this site.

We were first led to think along these lines by the result of an experiment (performed by Dr. David Denhardt) which involved the use of φX particles heavily labeled with ^{32}P (each particle contained several ^{32}P atoms in its DNA) (Denhardt and Sinsheimer, 1965b). When a ^{32}P atom disintegrates inside a φX particle (the half-life of ^{32}P is only 14 days), the DNA chain is broken and the particle is inactivated.

We asked the following question. Suppose that nonradioactive cells in a nonradioactive medium were infected with heavily labeled phage at a low multiplicity of infection. Under these conditions, the only ^{32}P atoms in each infected cell should be those contained in a single parental virus DNA strand. If these cells were then frozen—after various periods of incubation at 37°C—and held to permit the decay of the ^{32}P atoms therein, what would be the effect of this decay on the subsequent success of the infection (as measured by the ability of the cells to form plaques)? It should be pointed out that, in this study, we used starved cells so that the infectious cycle could be synchronized by the addition of nutrient medium.

The results were surprising. It was found that when cells were frozen as long as 9 min after infection—at which time each infected cell contains four or five ^{32}P-free daughter RF molecules—the ability of the infected cells to produce progeny virus particles was still completely inactivated by the decay of the ^{32}P atoms present in the parental virus DNA strands. This can only mean that the parental strand (or rather the RF produced therefrom) has a special function, which cannot be replaced by the daughter RF molecules; this is true despite the fact that the daughter RF molecules, if extracted from the infected cells, are capable of infecting spheroplasts.

This experiment indicated that, under the conditions employed (specifically, using starved cells), the original parental RF plays a special role. Furthermore, the fact that not all RF molecules within a cell are equivalent led us to ask whether more than one kind of ϕX can replicate at the same time in a starved cell. To answer this question, we carried out experiments (which we have termed limited participation experiments) in which several distinguishable ϕX mutants were employed.

Limited Participation Experiments

The mutants used in this study can be distinguished from one another on the basis of their ability (or inability) to grow in various strains of *E. coli,* or of their ability to replicate at various temperatures, or by the morphology of the plaques they produce. Following the ^{32}P suicide experiment, we were led to ask how many of such recognizably different ϕX mutants can replicate in one cell at one time (Yarus and Sinsheimer, 1967). Cells were infected with 3 to 4 particles of each of four distinguishable mutants. The progeny virus particles from single cell bursts were then examined to see how many kinds were represented in each burst. The results are shown in Table I. If normally grown, log phase cells are so infected, all four mutants are frequently seen among the progeny of one cell. If, however, starved cells are infected, the most frequent observation is that only one type is produced in any one cell. Occasionally two, but never more than two mutants are found among the progeny of single, multiply infected, starved cells. We should add that there is no selection of a particular genotype among the four genotypes used. Independent experiments indicate that all of the input single strands are converted to RF, but that in starved cells, there is evidently a limited participation by these RF molecules in the formation of progeny single strands.

Our interpretation of these results is as follows. Having become an RF, the circular DNA molecule must become associated with a particular intracellular site in order to participate further in the infectious process. If it does not become associated with such a site, it can neither replicate nor give rise to progeny DNA molecules. Our data indicate that the number of such sites is limited in log phase cells, and that in starved cells, it is usually one. In the ^{32}P-survival experiment, this one site was occupied by the parental RF, because, until it became associated with the site, there were no progeny RF molecules. In the limited participation experiments it was a matter of chance which of the RF molecules first became associated with the site.

It also appears that an RF which is not associated with such a site is excluded from function—that is, it is not transcribed into mRNA, and thus is not used to specify protein. That this is so was shown by the

TABLE I

LIMITED PARTICIPATION EXPERIMENT[a]

Number of genotypes/burst	Number of single cell bursts			
	Starved cells		Normal cells	
	Observed	Expected (Poisson)	Observed	Expected (Poisson)
1	37	0	15	7
2	14	3	23	21
3	0	20	14	21
4	0	28	3	6

[a] *E. coli* cells log-phase or starved, were infected with a mixture of four distinguishable φX genotypes. The effective multiplicity of infection with each genotype was calculated from the difference of the input and the nonadsorbed phage titers of each type. The number of genotypes present in the progeny of single cell bursts (51 starved cell bursts, 55 log-phase cell bursts) was measured and compared with the distribution to be expected assuming the adsorbed phage were distributed among the infected cells according to a Poisson distribution.

results of experiments in which we undertook to determine whether complementation of function between mutants defective in different cistrons could be observed under conditions of site limitation. If there is only one site, and if only the RF associated with that site can be transcribed, then clearly there can be no complementation.

The results of such an experiment, in which starved cells were multiply infected with particles of two mutants known to contain mutations in different cistrons, are shown in Table II. The mutants used are of the type which will grow in cells at 30°C, but are defective, in different functions, at 40°C. At 30°C, only 11 out of 28 infected starved cells produced both kinds of phage. This is about the proportion (40%) of the starved cells believed to contain two RF-attachment sites. At 40°C *only* those cells which produced both kinds of progeny produced *any* phage. That is, no cell at 40°C produced only one kind of phage. This result indicates that complementation is possible only if one RF of each type becomes associated with a site. Without complementation, no mutant phage can be produced at 40°C. If there is only one RF-attachment site in the cell, there can be no complementation; and at the restrictive high temperature, no progeny virus of either type will be formed.

TABLE II

COMPLEMENTATION IN STARVED CELLS[a]

Cells	Temperature (°C)	$ts\gamma$h + ts9			Frequency of mixed bursts
		Number of cells yielding			
		γh + 9	γh only	9 only	
Starved	30	11	11	6	0.39
Starved	40	12	0	0	1.00

[a] Starved *E. coli* cells were infected with a mixture of the complementary ϕX temperature-sensitive mutants $ts\gamma$h and ts9. (m (γh) = 3.4; m (9) = 4.2). The infected cells were incubated at either 30° or 40°C and the genotypes of the progeny of single-cell bursts at both temperatures were analyzed.

ϕX Mutants

ϕX mutants are of various kinds. There are host range mutants and there are conditional lethal mutants of two kinds: temperature-sensitive mutants, and suppressible mutants. Temperature-sensitive mutants can replicate almost normally at 30°C but, unlike the wild-type, cannot reproduce at 40°C. Suppressible mutants cannot grow in the normal ϕX-sensitive host cell; but they can grow in a related host known to carry a suppressor gene. This gene, which was introduced into a ϕX-sensitive host from the K12 strain CR34, is believed to be a standard amber suppressor of type SU-I; this type of suppressor, by means of an altered transfer RNA molecule, is able to translate a triplet codon (which is nonsense in the normal host) as a serine codon (Capecchi and Gussin, 1965).

Several suppressor-carrying bacterial strains have been isolated by Dr. Alan Garen (Garen *et al.,* 1965). These insert different amino acids in lieu of the nonsense codon. We have tested three of these (as spheroplasts since they are not available in ϕX-sensitive hosts) with our viral mutants; and we have found that most of our mutants are suppressed by any of the three suppressors which insert serine, glutamine, and tyrosine, respectively.

Different viral mutations may be expected to affect different portions of the genome—different cistrons—and thus affect different functions. If a cell is infected with two mutant particles which are defective in different functions, it might be expected that these could complement each other; each mutant would provide the function needed by the other to produce a successful infection. This does happen, and this test divides the various viral mutants into complementation groups. The mutants within one group cannot complement each other—presumably they are all defective in the same function—but they do complement mutants in other complementation groups.

Such complementation is not perfect. The yields of virus produced as a result of complementation are in the range of 10–50% of those observed with wild-type virus. The mutants used in the studies described herein were selected so that the yields of virus in the absence of complementation are, in the lethal condition, less than 2% of that produced by the wild-type virus. Thus, the demonstration of complementation is quite unambiguous. The progeny of the complementation are, of course, with the infrequent exception of recombinants, the two parental genotypes. With one exception, the complementation is symmetric; that is, both genotypes are present in essentially equal proportions among the progeny.

By virtue of these complementation tests, and on the basis of a comparison of reversion frequencies, we believe that all the mutants we have studied are simple, one-locus mutants. A few which appear to be defective in more than one function have been shown, in all probability, to be polar mutations.

DNA Synthesis in φX-Infected Cells

Having become associated with a critical site, the RF is transcribed to form messenger RNA molecules, which are, in turn, translated into proteins. We know several of the functions of these proteins, some of which are indicated in Fig. 1. One of them is essential for the replication of the RF. This point will be considered in what follows. At least two are involved in the formation of the viral coat, and two appear to be involved in lysis of the cell. The formation of progeny single-stranded DNA is dependent, in a complex way, upon the presence of virus-induced protein. A viral-induced protein brings about a cessation of host DNA synthesis at about 10–15 min post-infection, while another protein acts to block superinfection by secondary φX particles. Several of these functions have been assigned to cistrons and have been located on a genetic map.

That a phage function is involved in the formation of infective RF replicas was first demonstrated with mutants of phage S13 by Dr. Ethel Tessman (Tessman, 1966). Similar ϕX mutants have now been characterized. When cells are infected with such mutants, the DNA of the infecting virus is converted to the RF; but, no RF replication ensues.

To demonstrate this phenomenon we have used a new technique, which enables us to study specific viral DNA synthesis in ϕX-infected cells. Previously, our studies had been hampered by the fact that host DNA synthesis continues during the initial stages of ϕX infection and quantitatively overshadows the synthesis of viral DNA. To overcome this difficulty, it is necessary to markedly reduce the synthesis of host DNA, and this can be done by pretreating the cells with mitomycin C. As is well known, this agent reacts with cellular DNA and blocks DNA synthesis. Ordinarily, however, such treated cells will not support the replication of phage ϕX. We have found that it is necessary to use an HCR$^-$ strain (Boyce and Howard-Flanders, 1964) as the host cells in studies of this nature. For reasons which we do not fully understand,

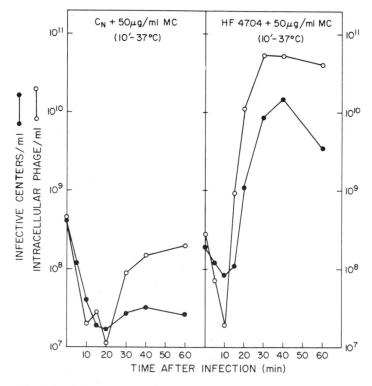

FIG. 2. Growth of ϕX in mitomycin (MC)-treated *E. coli* cells. Left—in an HCR$^+$ strain (C_N); right—in an HCR$^-$ strain (HF4704).

mitomycin-treated HCR⁻ cells will support normal φX development (Fig. 2), even though their DNA synthesis has been reduced to a few percent of normal (Fig. 3) (Lindqvist and Sinsheimer, 1966).

Under these conditions, it is quite possible to follow the incorporation of tritiated thymidine into viral DNA. By velocity centrifugation of the DNA extracted from infected cells through a preformed cesium chloride gradient, one can separate the residual incorporation into host DNA from that into φX single-stranded DNA and into the φX RF. In fact, the two forms of RF—the twisted form with two closed DNA strands and the nicked form with at least one strand open—can be resolved (Fig. 4).

In Fig. 5, a comparison is made of the uptake of tritiated thymidine by mitomycin-treated *E. coli* during the first 12 min of infection with two different φX mutants: φX*am*33, in which RF replication is blocked;

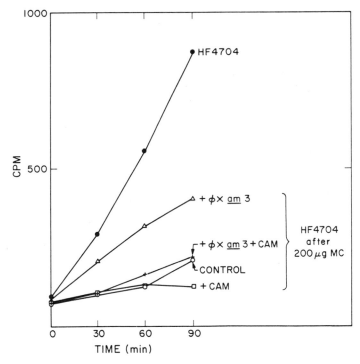

FIG. 3. Incorporation of tritiated thymidine into *E. coli* cells (HCR⁻ strain HF4704). (●) untreated, uninfected, log-phase cells; (○) mitomycin-treated, uninfected cells; (□) mitomycin-treated, uninfected cells plus chloramphenicol (CAM); (△) mitomycin-treated, φX*am*3-infected cells; (+) mitomycin-treated, φX*am*3-infected cells plus chloramphenicol (CAM).

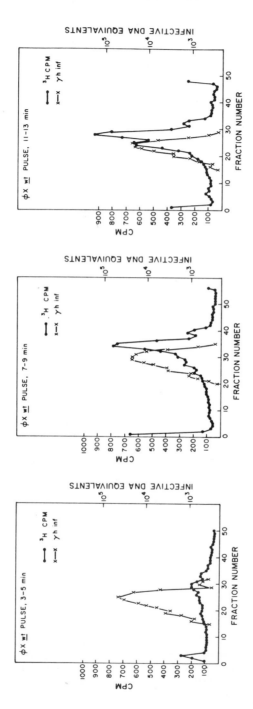

Fɪɢ. 4. Velocity sedimentation (through a preformed CsCl gradient) of DNA extracted at various times from ϕX-infected *E. coli* cells, pretreated with mitomycin C. Left—DNA extracted at 5 min; thymidine-³H supplied during the interval 3–5 min post-infection. Center—DNA extracted at 9 min; thymidine-³H supplied during the interval 7–9 min post-infection. Right—DNA extracted at 13 min; thymidine-³H supplied during the interval 11–13 min post-infection. DNA, isolated from the ϕX mutant, λh, was added to serve as a marker for the position of single-stranded DNA.

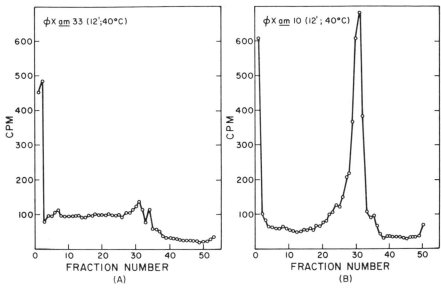

Fɪɢ. 5. Sedimentation analysis (through a preformed CsCl gradient) of DNA extracted after 12 min from øX-infected *E. coli* cells, which had been pretreated with mitomycin C. Thymidine-³H was supplied during the 0–12 min interval. (A)— infection with øXam33. (B)—infection with øXam10.

and *φ*Xam10, in which RF replication proceeds normally, but in which the synthesis of single-stranded DNA is blocked. Incorporation of tritiated thymidine in the case of infection with *am*10 corresponds to synthesis of about 20 RF DNA's per cell, while in the case of infection with *am*33, it is only enough to account for the synthesis of strands complementary to those of the input viral DNA.

Since RF replication had been shown to proceed normally in the presence of 30 μg/ml of chloramphenicol (Sinsheimer *et al.*, 1962), it was surprising to find that a virus-induced function was necessary for the replication of RF. Subsequent studies have shown that RF replication is blocked in an amino acid auxotroph in the absence of the essential amino acid. Hence it is clear that a virus-induced protein (apparently only one), the synthesis of which is relatively resistant to chloramphenicol, is a necessary prerequisite for the replication of RF.

The one RF formed per infecting viral genome in the case of mutants such as *am*33 is sufficient to provide complementation for mutants in several other cistrons, but the *am*33 RF is not rescued in such complementation. Its synthesis cannot be induced by complementation, and it is in such cases that the distribution of genotypes in the progeny of the complementation is asymmetric (Table III). No progeny of mutants in this complementation group are observed in such complementation situations.

These studies are reminiscent of the work of Lark (1966) in which the synthesis of a chloramphenicol-resistant protein—the replicator—was shown to be essential to the initiation of another round of cellular DNA replication. The nature of such chloramphenicol-resistant proteins is, at present, quite obscure.

TABLE III

COMPLEMENTATION IN LOG-PHASE CELLS; PHAGE YIELD AT 40°C RELATIVE TO *wt*

Mutants	Distribution of Progeny		
$ts\gamma$h + ts9	$\dfrac{\gamma\text{h} + \gamma\text{h}}{0.002}$	$\dfrac{9 + 9}{0.006}$	$\dfrac{\gamma\text{h} + 9}{0.094 \ (\gamma\text{h}/9 = 1.1)}$
ts4 + am33	$\dfrac{4 + 4}{0.02}$	$\dfrac{33 + 33}{0.03}$	$\dfrac{4 + 33}{0.64 \ (4/33 = 20)}$

[a] Log-phase *E. coli* cells were infected with two pairs of complementary mutants: $ts\gamma$h plus ts9 or ts4 plus am33. The phage yields obtained under the restrictive conditions were measured and are expressed relative to the yield of ϕX*wt* under the same conditions. The distribution among the progeny of the two imput genotypes was also measured and is expressed as the ratio. As controls, the yields of phage of each mutant type alone under the restrictive conditions are presented.

ϕX Functions and Complementation Groups

The viral functions that have been assigned to definite complementation groups are summarized in Table IV. The defective function in the first complementation group of mutants is clearly concerned with cell lysis. Under restrictive conditions, mature, infective phage particles are made inside the infected cell, but the cell does not lyse. The cells do not divide, but they become elongated, assume a snakelike morphology (Hutchison and Sinsheimer, 1966), and continue to produce phage for 90 min or more. In concentrated cultures, several hundred to 1000 particles may be produced per cell; in dilute cultures, several thousand particles may be produced per cell.

These mutants are useful in that they make it possible to obtain considerably higher yields of phage. They are also useful in that they permit the analysis of various processes over a much longer period of time than is the case with mutants not carrying this defect. With these mutants, complications due to different fractions of the cell population lysing at

TABLE IV

φX Mutants

CG[a]	Function	Mutants	Particle properties		Infective properties				
			Heat stability	Electrical mobility	SBP	SS	RF	Mature phage	Lysis
I	Lysis	am 3, 20	w	—	w	w	w	w	0
II	Coat, lysis	ts 4, 28	w	—	w	w	w	0	<<w
III	Coat	tsγ, 41, 79; am9, 23	w, R, S	≥w	0–w (particle)	0	w	0	w
IV	Viral Protein	ts 6, 9; am 14, 16	w, R, S	~w	w (low s)	0	w	0	w
V	?	am 10, 42	w	—	w	0	w	0	w
VI[b]	RF Replication	am 8, 30, 33	w	—	w	0	<<w	0	late

[a] Symbols: w = wild-type; R = more resistant than *wt*; S = more sensitive than *wt*; SBP = serum blocking power; SS = single-stranded DNA; RF = replicative form DNA; CG = complementation group.
[b] Complementation asymmetric

different times are avoided. Thus, in studying the physiology of several of the other ϕX mutant types, it has been useful to prepare double mutants carrying both this lysis defect and a second mutation; this allows us to study the biochemical effect of the second mutation over a much longer period of time.

It is of interest that net RF synthesis in cells infected with a lysis-defective mutant virtually ceases at about 10–15 min after infection; this is at essentially the same time as in infection with wild-type phage. Only the synthesis of single-stranded DNA is prolonged.

With such mutants, it has been possible to demonstrate that at about 10–15 min after infection, several critical changes take place in the infected cell. Net RF replication ceases, and the synthesis of single-stranded DNA begins. The synthesis of host DNA also ceases at this time, which means that the synthesis of all double-stranded DNA is shut off. Even with mutants which are defective with respect to the initiation of synthesis of single-stranded DNA, RF replication and host DNA synthesis stop (Fig. 6); and, the cells become completely resistant to superinfection by external ϕX.

Superinfection exclusion is most conveniently demonstrated with lysis-defective mutants of complementation Groups I or II. If cells infected with such mutants are superinfected (after 10 min of infection) with wild-type phage, the lytic capacity of the wild-type phage cannot be expressed; and no wild-type phage can be found among the intracellular progeny. Studies with labeled, superinfecting phage indicate that while these phage adsorb and go into eclipse, their DNA is never converted to RF. This exclusion (immunity to superinfection) develops during the first ten min of the infection with the mutant phage.

Mutants in the second complementation group are also defective with respect to lysis, although perhaps not as completely as are those in Group I. Moreover, unlike mutants of the first group which produce mature progeny particles inside the infected cell, the mutants of this group produce defective particles. These particles contain a normal, infective single-stranded DNA, but they cannot adsorb to host cells. They appear to be unstable, and they have altered sedimentation properties (Fig. 7). Instead of the usual 114 S virus particles, particles which contain infective DNA and which have sedimentation velocities of 107 S, 71 S, and 21 S are found.

Further study is required in order to understand the nature of this complex defect. The failure of these mutants to induce cell lysis plus the formation of defective (probably unstable) particles seem to indicate that there may be a protein of the viral coat which also plays an essential role in cell lysis. The possibility that this group might consist of double mutants, or of polar mutants, would seem to be excluded by their ability to complement with the mutants of Group I.

The defect of the mutants in complementation Group III affects a major phage coat protein which reacts with phage antisera. Unlike the members of the other complementation groups, the suppressible mutants and one temperature mutant in this group fail to make serum-blocking power. Phage particles of some members of this group have altered thermal stability; and, several of the mutants with a defect in this cistron have altered electrophoretic mobility.

It is of particular interest (and rather perplexing as well) that cells

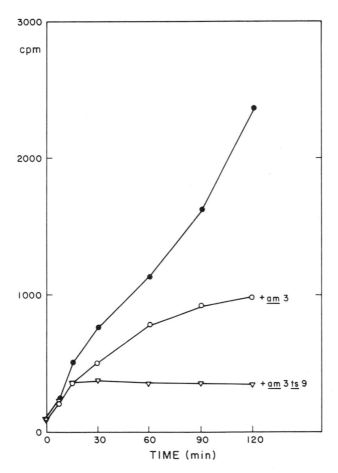

FIG. 6. Incorporation of thymidine-³H into *E. coli* cells. (●) uninfected; (○) infected with φX*am*3, a lysis-defective mutant; (△) infected with φX*am*3*ts*9, a double mutant, both lysis-defective and unable to synthesize progeny single strands.

infected with mutants of this class neither make single-stranded DNA nor continue to accumulate RF. Figure 8 shows that upon infection of cells with a double mutant, $am3ts\gamma$, containing both a Group I mutation to prevent lysis and a Group III mutation to produce defective coat protein, no single-stranded ϕX DNA is made. The RF is made in normal amounts during the first 15 min of the infection, and a small amount of a DNA with a low sedimentation coefficient, of unknown nature, appears. The synthesis of the RF is turned off at the usual time; but, apparently in the absence of a functional coat protein, the syn-

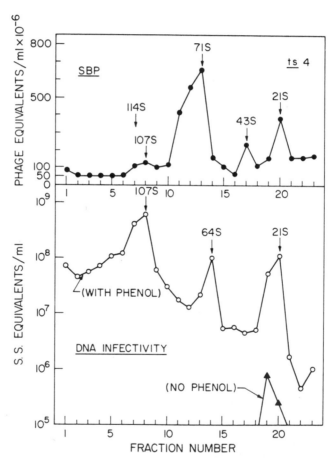

Fig. 7. Distribution, in a sucrose gradient, of serum-blocking power (upper) and infectivity to spheroplasts (lower) of an extract of *E. coli* cells infected with a ϕX mutant (*ts*4) of complementation Group II. In the case of infectivity to spheroplasts, both the infectivity of the extract and the DNA isolated from it with phenol were measured. (Data obtained by Dr. Stanley Krane.)

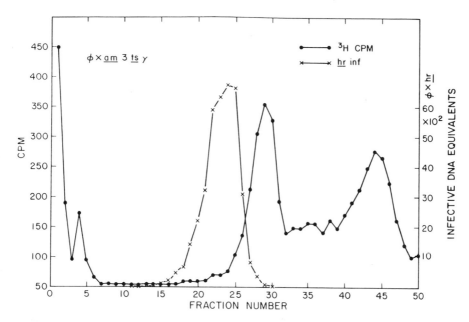

FIG. 8. Sedimentation analysis (through a preformed CₛCl gradient) of DNA extracted, after 90 min, from *E. coli* cells pretreated with mitomycin C and infected with the double mutant øX*am* 3*t*γ. Thymidine-³H was supplied during the 0-90 min interval. DNA isolated from the øX mutant, *hr*, was added to the preparation to serve as a marker for the position of single stranded DNA.

thesis of single-stranded DNA is blocked. At present, the explanation of this result is unclear. It is conceivable that the coat protein plays a role in releasing the single-stranded DNA from its template and that this release is blocked in the presence of a defective coat protein.

Mutants in Group IV lyse the cells normally; they make serum-blocking power; they make and replicate RF; but they do not make single-stranded DNA. The virus particles of some mutants in this group have an altered temperature sensitivity. We interpret this data to mean that this cistron also specifies a viral coat protein, which is, of course, necessary for proper viral assembly. However, we do not know why (as is the case with the mutants of Group III as well) the formation of single-stranded DNA is blocked in the face of a defect in the synthesis of a coat protein.

This analysis has implicated cistrons II, III, and IV in the formation of coat proteins. Apparently the øX particle must be more complex than the simple polyhedral array of identical protein subunits envisioned in some theories concerning the assembly of viral coats.

Concerning complementation Group V, we only know that lysis is normal, that the mutant makes serum-blocking power, that it makes and replicates RF, but that it does not make single-stranded DNA. Perhaps (but by the process of elimination only) the cistron containing the defect is concerned with the synthesis of a DNA polymerase.

The mutants in Group VI, as we mentioned earlier, are those in which the function necessary for the replication of the RF is defective.

A preliminary genetic analysis, based primarily on three factor crosses, suggests that the map order of these cistrons is II, III, I, IV, VI.

At this time, we have no reason to believe that the six complementation groups that we have identified represent the complete set. However, the DNA content of ϕX suggests that ten or twelve cistrons would be an upper limit. It thus appears likely that we have located a majority of the viral functions.

Much remains to be done to fill out the set, to characterize these functions in detail, and, in particular, to understand those interactions of the virus-induced processes and the host structures and processes that lead to the replication of ϕX174.

REFERENCES

Boyce, R. P., and Howard-Flanders, P. (1964). *Z. Verebungslehre* **95**, 345.
Capecchi, M. R., and Gussin, G. N. (1965). *Science* **149**, 417.
Chandler, B., Hayashi, M., Hayashi, M. N., and Spiegelman, S. (1964). *Science* **143**, 47.
Denhardt, D. T., and Sinsheimer, R. L. (1965a). *J. Mol. Biol.* **12**, 647.
Denhardt, D. T., and Sinsheimer, R. L. (1965b). *J. Mol. Biol.* **12**, 663.
Fiers, W., and Sinsheimer, R. L. (1962). *J. Mol. Biol.* **5**, 424.
Garen, A., Garen, S., and Wilhelm, R. C. (1965). *J. Mol. Biol.* **14**, 167.
Guthrie, G. D., and Sinsheimer, R. L. (1963). *Biochim. Biophys. Acta* **72**, 290.
Hutchison, C. A., and Sinsheimer, R. L. (1966). *J. Mol. Biol.* **18**, 429.
Kleinschmidt, A. K., Burton, A., and Sinsheimer, R. L. (1963). *Science* **142**, 961.
Lark, G. (1966). *Bacteriol. Rev.* **30**, 3.
Lindqvist, B., and Sinsheimer, R. L. (1966). *Federation Proc.* **25**, 651.
Sinsheimer, R. L. (1959). *J. Mol. Biol.* **1**, 43.
Sinsheimer, R. L., Starman, B., Nagler, C., and Guthrie, S. (1962). *J. Mol. Biol.* **4**, 142.
Tessman, E. S. (1966). *J. Mol. Biol.* **17**, 218.
Yarus, M. J., and Sinsheimer, R. L. (1967). *J. Virology* **1** (in press).

Gene Function in Phage S13 *

Ethel S. Tessman

DEPARTMENT OF BIOLOGICAL SCIENCES, PURDUE UNIVERSITY,
WEST LAFAYETTE, INDIANA

Introduction

Phage S13 is a small single-stranded DNA phage closely related to ϕX174. Because of its small size, phage S13 appears to be particularly suitable for the determination of the function and the manner of regulation of each phage cistron. Using the conditional-lethal method of analysis for genetic mapping and complementation testing developed by Campbell (1961), six complementation groups have been discovered so far.

The functions of the various groups as known at present are: Group I, formation of a protein coat subunit which is included in the coat region determining host range and adsorption; Group II, unknown; Group IIIa, unknown; Group IIIb, formation of a second protein coat subunit; Group IV, formation of late replicative form DNA; Group V, lysis (E. S. Tessman, 1965, 1966; I. Tessman and Tessman, 1966). [Mutants affecting lysis in ϕX174 have been reported by Hutchison and Sinsheimer (1966).] All the *su* and *t* mutants used were produced by hydroxylamine† mutagenesis, except for mutants of the gene controlling lysis, for which only nitrous acid-induced mutants were used. All the lysis mutants were isolated by I. Tessman. Figure 1 shows the various cistrons and their linkage relationships obtained from two-factor crosses. Groups II and V map to the left of Group I, although their relative order has not yet been established. A few other genes may yet remain to be discovered. As yet, there is no evidence for a circular map despite the known circularity of both the single-stranded and double-stranded DNAs (Fiers and Sinsheimer, 1962; Kleinschmidt *et al.,* 1963; Chandler *et al.,* 1964).

* This work was supported by grant AI-03903 from the National Institutes of Health.

† Abbreviations used: HA, hydroxylamine; RF, replicative form DNA or double-stranded DNA; SS, single-stranded DNA; moi, multiplicity of infection; UV, ultraviolet light; CM, chloramphenicol; MAK, methylated-albumin-kieselguhr.

This chapter is concerned mainly with the Group IV function, which is formation of infectious double-stranded DNA.

Group IV mutants are distinguished by the following characteristics: They fail to lyse the restrictive host, *E. coli* C122, and they yield no infectious particles upon lysozyme lysis. In this respect, they are quite different from Group V mutants, which also fail to lyse but yield a larger than normal burst of infectious particles after lysozyme lysis. *Su* mutants of Groups I, II, IIIa and IIIb all lyse the restrictive host at the normal time though they yield no infectious particles. (Cells infected with Group IV *su* mutants may eventually lyse after about 2 hours at 37°C due to the growth of wild-type revertants.)

For *t* mutants, lysis is not a clear-cut group characteristic. Some *t* mutants of Groups I and IIIb are very much delayed in lysis while others lyse promptly. The two known *t* mutants of Group IIIa are delayed in lysis. Two *t* mutants of gene II were studied and found to lyse at the normal time.

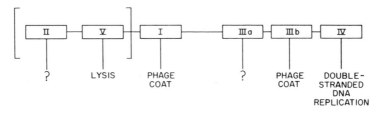

FIG. 1. Sequence of genes on the genetic map of phage S13, and the gene functions. The relative positions of genes V and II are not known. Recombination frequencies establishing position of genes I, IIIa, IIIb, and IV are given in E. S. Tessman (1965). Recombination frequencies establishing positions of genes II and V are unpublished.

Group IV mutants are also blocked in exclusion. Thirty minutes after infection by the Group IV mutant, either an su^+ phage or an *su* mutant of another group can be added and produce a burst of normal size. It has been found that mutants of Groups I, II, IIIa, IIIb, and V exclude superinfecting wild-type phage (E. S. Tessman, unpublished).

Another Group IV property is the poor rescue phenomenon. When the restrictive host is mixedly infected with wild-type and a Group IV mutant, only a small fraction of the particles in the burst have the genotype of the Group IV parent, even though the inputs were equal. Thus in a total average burst of 300 particles, 5 to 10 particles on the average per cell will be Group IV. This poor rescue of the Group IV genome is in contrast to what is observed for mutants of the other genes, which all show complete rescue. Poor rescue of the Group IV genome is also observed in the bursts obtained from complementation tests (asymmetric complementation) although the total burst sizes in these tests are large. Therefore, Group IV mutants complement normally though their genomes are poorly rescued.

Mutants showing some of these properties have been found for other phages. Epstein *et al.* (1963) have found T4 *am* mutants that are blocked in both DNA synthesis and lysis. Poor rescue and asymmetric complementation have been found for the RNA phages MS2 and f2 (Pfeifer *et al.,* 1964, Valentine *et al.,* 1964).

Before presenting the results which show that the function of gene IV is concerned with synthesis of infectious double-stranded DNA, the method of measuring DNA synthesis will be described.

Determination of RF and SS by HA Treatment

In 1962, Sinsheimer *et al.* showed, by means of density gradient centrifugation, that an infectious DNA component having the density and other properties of double-stranded DNA was found in extracts of ϕX-174 infected cells. This component was termed the replicative form or RF. It was also found to be about 10 times more UV resistant than the single-stranded DNA (SS). This component was also purified by Hayashi *et al.* (1963) using a MAK column (Mandell and Hershey, 1960). A DNA component of similar density and UV resistance was found by Howard *et al.* (1966) in extracts of S13-infected cells using CsCl gradients and MAK columns to separate RF from SS. Howard *et al.* developed a method of distinguishing infectious RF from infectious SS. This method consists simply of hydroxylamine treatment and does not involve physical separation of the RF and SS. The RF was found to be very insensitive to HA treatment. After 4 hours in 0.2 M HA, pH 6.0, the infectivity of the RF is still 100%, while in 10 min the infectivity of the SS is reduced to e^{-1}. Figure 2 shows an HA-inactivation curve for DNA in a crude extract of S13-infected cells chilled just before the onset of lysis. The first part of the curve has the slope for SS, while the second part of the curve has the slope for RF. The reason there seems to be relatively so little RF in this preparation is that the RF is at least 20 times less infective than the SS. This conclusion is based on the finding that NaOH treatment increases the RF infectivity by a factor of 20, but does not change SS infectivity. Pouwels and Jansz (1964) first showed for ϕX that RF was 30 to 100 times less infective than SS by this method. Figure 3 shows HA-inactivation curves for DNA from cells infected under conditions giving no SS synthesis. These curves again show that RF survival is still 100% at 4 hours. Curves A and C of Fig. 3 also show that RF formed in 30 μg/ml CM has the same resistance to HA as RF made without CM.

It was found further that the HA-resistant DNA in an extract of S13-infected cells has approximately the same UV-sensitivity that was found by Sinsheimer *et al.* (1962) for ϕX RF.

With the HA method, RF is determined by inactivating the extract for a time such that the SS will make no contribution to the plaque count. The amount of SS in a preparation is determined by subtracting the amount of RF from the total amount of infectious DNA.

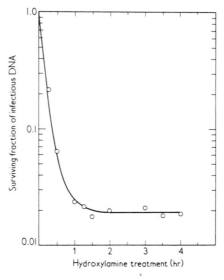

FIG. 2. Hydroxylamine inactivation of the infectivity of unpurified DNA from a culture of *E. coli* C infected with wild-type S13 and chilled at 12.5 min following infection. The extraction and inactivation procedures are described by E. S. Tessman (1966).

Kinetics of Infectious DNA Synthesis by Wild-Type Phage

Infectious RF is already present at the earliest time measured (2 min after infection). The amount of infectious RF at 2 min is proportional to the moi. In broth cultures at 37°C, at an moi of 5.0 or less, the RF increases linearly with time until 12 or 12.5 min; these are the latest times that can be measured before the onset of lysis. (At a high moi, about 15.0, the curve levels off by 10 min.) Figure 4 shows the kinetics of RF and SS formation in bacteria infected with wild-type phage at an moi of 5.0. The data in Fig. 4 are expressed as the ratio of infectious DNA at time *t* min to infectious DNA at 2 min.

While RF synthesis begins very early, SS synthesis does not begin until between 8 and 9 min; then, it increases rapidly.

Since in cells infected with wild-type phage, measurement of DNA synthesis is stopped by the onset of lysis at about 12.5 min, a lysis mutant (Group V) provides a means of measuring kinetics of RF and

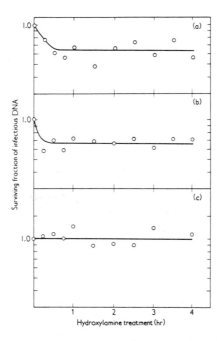

Fig. 3. Hydroxylamine inactivation curves of the infectivity of DNA from *E. coli* C infected under conditions where SS production is blocked. (a) The DNA was obtained from *E. coli* C infected with wild-type S13 and grown in the presence of 30 µg/ml chloramphenicol for 20 min. (b) The DNA was obtained from *E. coli* C infected with *su*129 (Group II) and grown in the absence of chloramphenicol for 12.5 min. (c) The DNA was obtained from *E. coli* C infected with *su*129 and grown in the presence of 30 µg/ml chloramphenicol for 20 min.

SS formation at later times. At 12 min, the lysis mutant studied (*su*N15) had formed the same amount of SS and RF as wild-type. Figure 5 shows that infectious RF synthesis shuts off after 12 min. The data indicate first a shut-off of RF synthesis and then a resumption of synthesis after 35 min to give a value at 65 min which is twice the 12-min value. There is no increase in RF between 65 and 75 min.

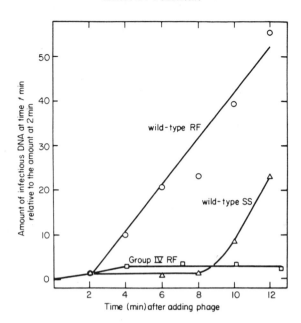

FIG. 4. Kinetics of synthesis of infectious double-stranded DNA (RF) and single-stranded DNA (SS) by wild-type S13 and of RF by a Group IV mutant, *su*100. The amount of RF and of SS at 2 min is set equal to 1.0. The procedure is described by E. S. Tessman (1966). All points for wild-type RF and SS in this figure are from the same experiment. Multiplicity of infection was 5.0, on the basis of the titer obtained by direct plating.

SS synthesis by the lysis mutant is also shown in Fig. 5. A leveling-off of SS synthesis is observed after 45 min.

Synthesis of Infectious RF by *su* and *t* Mutants of Each Group

Synthesis of infectious RF in the restrictive host was measured by comparing the amount of HA-resistant DNA made at 12.5 min after infection with that made at 2 min. Despite the variability between duplicate determinations, it is clear from Table I that mutants of Groups I, II, IIIa, IIIb, and V all yield high values similar to those found with wild-type phage; mutants of Group IV all show very low values. All the mutants shown in Table I are separable by recombination. The amounts of RF made at 2 min are the same for both wild-type and mutants at a given moi. When the kinetics of RF synthesis by a Group IV mutant was determined, it was found that there was no increase in infectious RF after about 2.5 min (Fig. 4). In the permissive host Y,

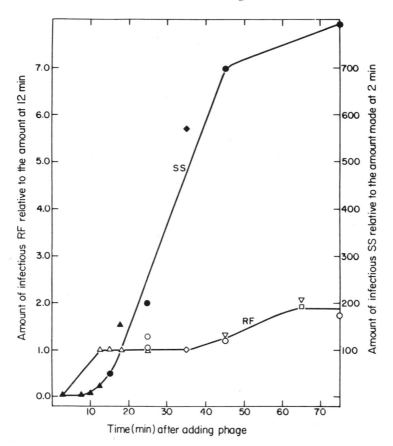

Fig. 5. Kinetics of synthesis of infectious RF and SS by a lysis mutant, *su*N15. Procedures are as described by E. S. Tessman (1966) except that for assay of RF formed at 20 or more min after infection, a 4000 times dilution (rather than the usual 400 times dilution) of the cell extract is required before combining with protoplasts in order to avoid inhibition of the protoplast assay. The inhibition may be caused either by bacterial DNA or single-stranded DNA. Because of the large amounts of SS formed at late times, relatively long periods of HA inactivation (3.5 to 4.0 hours) were used.

however, the amount of RF synthesis at 12 min was the same as that found for wild-type S13.

The *t* mutants of Groups I, II, IIIa, and IIIb tested at high temperatures also showed RF ratios similar to those of wild-type phage. With regard to infectious SS formation, in the interval between 2 and 12.5 min, wild-type shows a 39 to 66 times increase in SS. The *su* and *t* mutants of Group IIIb show a very small increase in SS; at most the increase is a doubling. Mutants of all other groups (except for Group V

TABLE I

Formation of Infectious Replicative Form DNA by *su* and *t* Mutants Grown under Restrictive Conditions[a]

Ratio of hydroxylamine-resistant DNA at 12.5 min after infection to hydroxylamine-resistant DNA at 2 min after infection

Group I		Group II		Group IIIa		Group IIIb		Group IV		Group V	
su39	27	su18	25	su10	19	su44	34	su16	1.2	suN15	55
su39	60	su86	38	su10	42	su63	27	su16	2.0		
su65	9.8	su86	49	su41	27	su66	32	su16	3.1		
su65	25	su129	21	su43	26	su66	66	su34	1.2		
su68	26					su77	35	su45	1.2		
t173[b]	50	t76	70	t163	49	t330	46	su72	2.2		
		wild-type	20					su100	1.0		
		S13	23					su100	1.7		
		(at 37°)	28					su100	1.8		
			56					wild-type	76		
								S13			
								(at 41°)			

[a] Infection, making of cell extracts, HA-treatment and assay of infectious DNA were carried out as described in E. S. Tessman (1966). For all the phages used the moi was 5.0; this calculation was based upon the titer obtained by direct plating.

[b] For *t* mutants at 41°C, 10 min : 2 min ratio is shown.

mutants) show no increase in SS. Possibly infectious SS is made by mutants of some of the other groups, but it may be degraded if not immediately incorporated into phage coats.

Effect of Chloramphenicol on RF Formation

Chloramphenicol at 30 μg/ml reduces protein synthesis in *E. coli* to only a few percent of normal (Kurland and Maaløe, 1962). In ϕX-infected cells (Sinsheimer *et al.*, 1962) and in S13-infected cells (E. S. Tessman, 1966), 30 μg/ml CM block SS synthesis and lysis; this indicates that these functions require formation of phage-specified proteins. In their original work on RF formation in ϕX-infected cells, Sinsheimer *et al.* (1962) found that, at 30 μg/ml CM, RF was produced in large quantity after an initial delay. Since RF synthesis proceeds under conditions in which almost all protein synthesis is inhibited, Denhardt and Sinsheimer (1965) postulated that RF synthesis is under the control of a host enzyme rather than a phage-induced enzyme. However, the present results show that a higher dose of CM will inhibit RF synthesis and that therefore RF synthesis is under phage control.

For cells infected with wild-type S13, it is indeed found that relatively large amounts of RF can be formed in the presence of 30 μg/ml CM. Figure 6 shows the amount of RF formed relative to the normal 12-min value when 30 μg/ml CM was added at 3 min before infection. Although the amount found at 12 min in the presence of CM was only half the

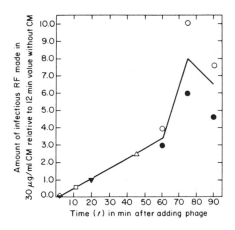

FIG. 6. Synthesis of infectious RF by wild-type S13 in 30 μg/ml CM. The CM was added at 3 min before infection. Points designated by symbols of the same shape are from the same experiment. All values were measured relative to the amount of RF formed by wild-type phage at 12 min without CM.

normal 12-min value, by 75 min there was a very large increase in RF. In one experiment the 75-min value was 10 times the normal 12-min value, and 6 times the normal in another experiment. Both experiments showed an increased amount of synthesis between 60 and 75 min and then a drop by 90 min. (In one experiment where S13-infected cells and ϕX-infected cells were compared, the values obtained after 20 min in CM were found to be the same for both.)

It can be observed directly by the use of a lysis mutant, suN15, that the addition of CM at 30 μg/ml causes a greatly increased production of infectious RF. Figure 7, curve C, shows that the addition of 30 μg/ml CM at minus 3 min results in 7.6 times more RF at 75 min than would be formed without CM.

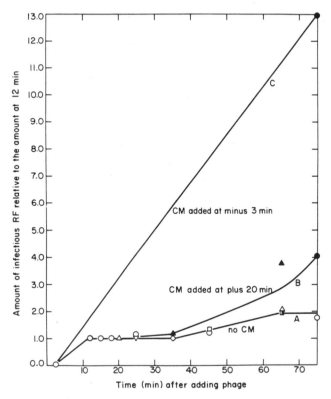

FIG. 7. Formation of infectious RF by a lysis mutant, suN15, in the absence and presence of 30 μg/ml CM. Curve A, no CM added, Curve B, CM added at 20 min after infection, Curve C, CM added at 3 min before infection. Symbols of the same shape are from the same experiment. All black symbols are from cultures with CM; all white symbols are from cultures without CM. For cultures grown without CM, CM was added after chilling to standardize the procedure.

Addition of CM to infected cultures at a late time was also studied. To this end, CM (30 μg/ml) was added to an *su*N15-infected culture at 20 min after addition of phage, a time when RF synthesis has ceased. The results are shown in Fig. 7, curve B. It is seen that the amount of RF at 75 min is double the normal 75-min value, which in itself is twice as great as the initial shut-off value. Thus adding CM at a late time appears to interfere to a limited extent with the shut-off mechanism.

When CM is added at 20 min, no increase in SS DNA is observed between 20 and 75 min after infection.

Although infectious RF synthesis proceeds well in the presence of 30 μg/ml CM, quite different results are observed when the concentration of CM is 100 μg/ml. At 2 min (100 μg/ml CM added at −3 min), the amount of RF is about 50% of a normal 2-min culture, while at 12 min it is only 3% of a normal 12-min culture (Fig. 8). Thus 100 μg/ml CM produces mild inhibition of synthesis of early RF made at

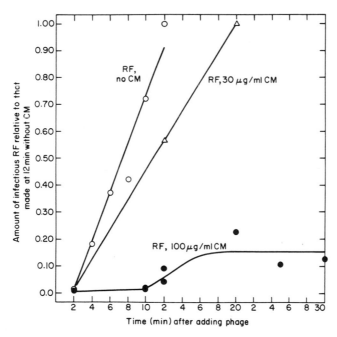

FIG. 8. Formation of infectious RF by *E. coli* C infected with wild-type S13 in 100 μg/ml CM and in 30 μg/ml CM. CM was added at 3 min before infection. The RF ratios are all given relative to the amount of RF which is made in the absence of CM at 12 min. At 2 min the amount of RF made in 100 μg/ml CM is 50% of a control culture lacking CM and in 30 μg/ml CM is 100% of the control.

2 min, but a severe inhibition of synthesis of later RF. At about 20 min there is a partial recovery of synthesis, which may be due to CM breakdown by the cells (Wissemann *et al.* 1954).

The differential effects of CM on RF synthesis are consistent with the finding of phage mutants which initiate the formation of early RF but are unable to promote the synthesis of late RF. Both types of data support the view that early RF is formed by a host-specified enzyme but that late RF is formed by a phage-specified enzyme.

Once DNA synthesis has been initiated, CM at doses greater than 100 μg/ml does not prevent DNA synthesis in T4 (300 μg/ml; Fermi and Stent, 1962) or *E. coli* (200 μg/ml; Friesen and Maaløe, 1965). This indicates that once the enzymes are formed, high doses of CM are not inhibitory.

In an experiment with wild-type phage, 150 μg/ml CM was added at 5.5 min after infection, a time at which some late RF has already been formed. The results are shown in Fig. 9. These results indicate that enzyme made before the addition of CM can continue to act; i.e., that continued enzyme synthesis is not necessary for continued RF formation.

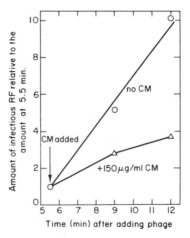

Fig. 9. Effect on the synthesis of infectious RF of adding 150 μg/ml CM at 5.5 min after adding phage. Wild-type S13 was the infecting phage.

Poor Rescue of Group IV RF Synthesis

Mixed infection of the restrictive host with Group IV mutants and wild-type gives little rescue of mature phage particles. In order to see at what stage this failure of rescue occurs, synthesis of infectious RF by these mixedly infected cells was studied. It was found that the presence of wild-type in the cell had little effect on the amount of infectious Group IV-mutant RF synthesized. Table II gives ratios of HA-resistant RF at 12.5 min to that at 2 min for cells infected with the mutant alone and for mixedly infected cells. There is at most a 2 times increase in Group IV RF synthesis over the amount formed in cells infected with the mutant alone. This table also shows, as a control, that RF synthesis by *su* mutants of other groups is not depressed in mixed infection together with wild-type.

In one mixed infection experiment, 30 μg/ml CM was added at 3 min before infection and the culture was allowed to grow for 75 min so as to obtain a much larger amount of wild-type RF than is normally formed at 12 min. This experiment also gave only a twofold increase in mutant RF over the amount formed in unmixed infection. This poor rescue

TABLE II

FORMATION OF INFECTIOUS REPLICATIVE FORM DNA AFTER MIXED INFECTION OF THE RESTRICTIVE HOST WITH WILD-TYPE PHAGE AND AN *su* MUTANT[a]

Phages used	Group of *su* mutant	Ratio of hydroxylamine-resistant DNA at 12.5 min to that at 2 min	
		Wild-type	Mutant
Wild type +*su*16	IV	20	2.5
Wild type +*su*16	—	23	1.6
Wild type +*su*16	—	18	1.7
*su*16 alone	—	—	2.0
Wild type +*su*100	IV	15	2.7
Wild type +*su*100	—	15	2.4
Wild type +*su*100	—	13	1.7
Wild type +*su*100	—	21	3.9
Wild type +*su*100	—	35	1.7
*su*100 alone	—	—	1.8
Wild type +*su*100 (in CM)[b]	—	384	4.6
*su*100 (in CM)[b]	—	—	2.7
Wild type +*su*65	I	19	21
Wild type +*su*86	II	15	16
Wild type +*su*10	IIIa	16	18
Wild type +*su*74	IIIb	14	15

[a] Cells were mixedly infected and extracts prepared and HA-treated as described in E. S. Tessman (1966). In the experiment with CM, 30 μg/ml were added at 3 min before infection.

[b] Seventy-five min: 2 min ratio in CM experiment.

suggests that most of the gene IV product is not freely diffusible in the cell. Possibly the product is synthesized spatially very close to an RF molecule and then is bound to the RF molecule that specified it. The results observed here are at least superficially analogous to the finding by Thomas and Bertani (1964) of lack of rescue of immunity-sensitive phages by immunity-insensitive phages.

The rescue experiments (without CM) provide evidence that the amount of SS formed is proportional to the amount of infectious RF present. The mixedly infected cells contain a large amount of wild-type RF and a small amount of mutant RF. They yield a large number of wild-type SS molecules and mature particles, but a very much smaller number of mutant SS and mature particles.

The complementation results with Group IV mutants show that they complement normally with mutants of all the other groups. Therefore the small amount of infectious RF synthesized by a few Group IV mutant particles per cell can direct the synthesis of as much of a gene product as is needed by a mutant of any other group to give a full'burst.

Discussion

CONCLUSIONS ABOUT GROUP IV FUNCTION

The function of the Group IV gene product is concerned in some way with the formation of infectious progeny RF. Possible functions include a DNA polymerase function, conversion of noninfectious late RF to infectious late RF, or structurally changing early RF so that it becomes capable of replication. The gene product has two exceptional characteristics: It can be formed in 30 μg/ml CM, a concentration at which most other proteins cannot be formed; and, most of the gene product appears not to be freely diffusible, as implied by the poor rescue phenomenon. Some other instances of CM-resistant polypeptides have been discussed by Lark and Lark (1964).

THE SWITCHING-ON OF LATE FUNCTIONS OF GENOMES MUTANT IN GROUP IV

The normal burst sizes obtained in complementation tests between Group IV mutants and mutants of all other groups show that the Group IV mutants can perform all phage functions other than late RF synthesis. However, in unmixed infection, Group IV mutants appear to be blocked in all phage functions. They do not cause exclusion, nor do they lyse the cell. Moreover, they do not form any mature particles,

SS, or cause the formation of serum blocking power. When a nongroup IV mutant is added, some event causes a switching-on of late function in genomes mutant in gene IV. It is possible that onset of late function requires prior translation of gene IV, and gene IV mutants that are nonsense mutants would be expected to be blocked in translation (Sarabhai *et al.*, 1964). The absence of late function might also be due to a block in transcription. In a comparable instance, Attardi *et al.* (1963) have found that O^0 mutants of the *E. coli* lactose operon (which are blocked in synthesis of β-galactosidase, galactoside trans-acetylase and permease) are blocked in production of "lactose" mRNA.

Protass and Korn (1966) have also shown that mutants of λ cistron N, blocked in DNA synthesis and in late functions, are blocked in transcription. Both in these cases and in the presumptive case of Group IV mutants, the block in transcription may not be a primary block but may be a feed-back effect of a block in translation. Experiments to determine whether Group IV mutants form mRNA are in progress.

If the block in onset of late function is due to a block in translation, then mutants of gene IV which are translated would show late functions. It is expected that a missense mutation would be translated. Thus there may exist some *su* and *t* mutants which though blocked in late RF synthesis (because of an inactive gene IV enzyme) yet form some mature particles and lyse. These would appear leaky.

Whether the block is in transcription or translation, it is not understood how the presence of a second phage in the cell that is nonmutant in gene IV could switch on late function of the first phage that is mutant in gene IV. The switching-on of late functions in mixed infection might be due in some way to synthesis of late RF by the nongroup IV phage or to formation of a small amount of late Group IV RF due to the slight degree of rescue.

THE SHUT-OFF OF RF SYNTHESIS

Any explanation for the shut-off of RF synthesis must take into account the finding that RF synthesis continues if a high dose of CM (150 μg/ml) is added after late RF has begun to form (Fig. 9). Since this dose blocks late RF synthesis when added before infection (Fig. 8), pre-sumably this dose prevents the formation of any further RF enzyme when added at 5.5 min. Thus the continued RF synthesis observed after addition of 150 μg/ml CM at 5.5 min indicates that enzyme produced prior to the addition of CM is available for continued synthesis. There-fore, it seems likely that the shut-off of RF synthesis observed after 12 min is not due simply to a cessation of enzyme production.

There appears to be an inverse correlation between RF synthesis and SS synthesis. Onset of SS synthesis is followed by the shut-off of RF synthesis. Around 45 min there is a leveling-off of SS synthesis and a small resumption of RF synthesis. In the presence of CM, there is no SS synthesis, and RF synthesis does not shut off.

A model for shut-off of RF synthesis can be proposed assuming a causal relationship between SS synthesis and RF shut-off. The shut-off may be due to a competition between the RF enzyme and the SS enzyme for the RF molecules. If such a competition occurs, it is clear that the SS enzyme must have the greater affinity for the RF molecule since SS synthesis continues while RF synthesis ceases.

The observed correlations do not provide proof of this model. Absence of infectious SS is not proof that SS has not been formed, because formation of SS as observed in the present assay appears to require concomitant synthesis of other proteins; for example, SS synthesis is observed to stop immediately when CM is added at a later time. Therefore, to confirm the model of shut-off by enzyme competition, it will be necessary to show directly that in the absence of shut-off SS enzyme is not formed.

REFERENCES

Attardi, G., Naono, S., Rouvière, J., Jacob, F., and Gros, F. (1963). *Cold Spring Harbor Symp. Quant. Biol.* **28,** 375.

Campbell, A. (1961). *Virology* **14,** 22.

Chandler, B., Hayashi, M., Hayashi, M. N., and Spiegelman, S. (1964). *Science* **143,** 47.

Denhardt, D. T., and Sinsheimer, R. L. (1965). *J. Mol. Biol.* **12,** 647.

Epstein, R. H., Bolle, A., Steinberg, C. M., Kellenberger, E., Boy de la Tour, E., Chevalley, R., Edgar, R. S., Susman, M., Denhardt, G. H., and Leilausis, A. (1963). *Cold Spring Harbor Symp. Quant. Biol.* **28,** 375.

Fermi, G., and Stent, G. S., (1962). *J. Mol. Biol.* **4,** 179.

Fiers, W., and Sinsheimer, R. L. (1962). *J. Mol. Biol.* **5,** 424.

Friesen, J. D., and Maaløe, O. (1965). *Biochim. Biophys. Acta* **95,** 436.

Hayashi, M., Hayashi, M. N., and Spiegelman, S. (1963). *Science* **140,** 1313.

Howard, B., Tessman, I., and Howard, R. (1967). Manuscript in preparation.

Hutchison, C., and Sinsheimer, R. L. (1966). *J. Mol. Biol.* **18,** 429.

Kleinschmidt, A. K., Burton, A., and Sinsheimer, R. L. (1963). *Science* **142,** 961.

Kurland, C. G., and Maaløe, O. (1962). *J. Mol. Biol.* **4,** 193.

Lark, C., and Lark, K. G. (1964). *J. Mol. Biol.* **10,** 120.

Mandell, J. D., and Hershey, A. D. (1960). *Anal. Biochem.* **1,** 66.

Pfeifer, D., Davis, J. E., and Sinsheimer, R. L. (1964). *J. Mol. Biol.* **10,** 412.

Pouwels, P. H., and Jansz, H. S. (1964). *Biochim. Biophys. Acta* **91,** 177.

Protass, J. J., and Korn, D. (1966). *Proc. Natl. Acad. Sci. U.S.* **55,** 1089.

Sarabhai, A. S., Stretton, A. O. W., Brenner, S., and Bolle, A (1964). *Nature* **201,** 13.

Sinsheimer, R. L., Starman, B., Nagler, C., and Guthrie, S. (1962). *J. Mol. Biol.* **4,** 142.

Tessman, E. S. (1965). *Virology* **25,** 203.

Tessman, E. S. (1966). *J. Mol. Biol.* **17,** 218.

Tessman, E. S., and Shleser, R. (1963). *Virology* **19,** 239.

Tessman, I., and Tessman, E. S. (1966). *Proc. Natl. Acad. Sci. U.S.* **55,** 1459.

Thomas, R., and Bertani, L. E. (1964). *Virology,* **24,** 241.

Valentine, R. C., Engelhardt, D. L., and Zinder, N. D. (1964). *Virology* **23,** 159.

Wissemann, C. L., Smadel, J. E., Hahn, F. E., and Hopps, H. E. (1954). *J. Bacteriol.* **61,** 662.

Heterozygotes of Phage f1

June Rothman Scott and Norton D. Zinder

THE ROCKEFELLER UNIVERSITY,
NEW YORK, NEW YORK

Phage f1, which adsorbs only to donor strains of *E. coli* K12 (Loeb, 1960), is a DNA-containing rod-shaped phage (Zinder *et al.*, 1963) very similar to phage fd (Marvin and Hoffman-Berling, 1963) and to phage M13 (Hofschneider, 1963; Salivar *et al.*, 1964). All three contain single-stranded DNA, and they emerge from the host cell without lysis. We have isolated conditional lethal mutants of phage f1: both temperature-sensitive mutants, which grow at 34°C but not at 43°C, and amber mutants, which grow on a permissive *E. coli* K12 HfrC strain containing the *Su-I* gene but not on the nonpermissive *Su*⁻ strain from which it was derived. All of these mutants, which fall into 6 cistrons on the basis of complementation tests, show a reversion index lower than 10^{-4}. This report describes an unusual type of particle appearing in lysates made by mixed infection with pairs of f1 mutants from different cistrons. From the properties of these particles, it is concluded that they are genetic and physical diploids.

Mutants in one of the 6 cistrons give rise to turbid plaques and such mutants were used in most experiments because of the ease of identifying them visually. In an attempt to observe genetic recombination, the permissive strain was mixedly infected with a turbid amber mutant and a clear amber mutant. To identify wild-type recombinants, the yield from this cross was plated on the nonpermissive strain. It was found that about 2% of the phage produced (as assayed on the permissive strain) formed plaques on the nonpermissive strain. However, the plaques on the nonpermissive strain were much more turbid than wild-type f1 plaques; therefore, they could not have been produced by simple wild-type recombinants. In fact, when these plaques were picked and replated, this time on the permissive strain, they were found to contain, in addition to clear plaque formers, phage that make turbid plaques and do not grow on the *Su*⁻ host. We call these particles that result from mixed infection and form

211

turbid plaques on the nonpermissive host "heterozygotes" because they segregate to yield phage that have the characteristics of both parental types (Notani and Zinder, 1964). Since heterozygotes are formed in mixed infection by all pairs of *am* or *ts* mutants that are in different cistrons, the turbid plaques they produce on the nonpermissive host are probably the result of a process of continuous complementation.

Controls to Rule Out Artifacts

The possibility that heterozygotes were artifacts caused by aggregation of these long sinuous rodlike phages were ruled out in two ways: First, in a reconstruction experiment, they did not appear in mixtures of the two mutants at high concentration. Second, attempts to disaggregate them were unsuccessful. They did not disappear from lysates containing heterozygotes kept at low particle concentration (10^5 per ml) and retitrated at weekly intervals.

Extent of Diploidy: UV

Crosses of double mutants by single mutants also give rise to about 2% heterozygotes. In these cases, the diploid region includes all three markers. Therefore, it seemed likely that in heterozygotes of this small phage, the diploid region is extensive, probably even including the entire genome. The relative size of the genomes of the heterozygous and parental phage should be measurable by their relative sensitivity to ultraviolet inactivation. A single "hit" anywhere on a heterozygous genome should destroy its ability to grow on the nonpermissive host (although it might remain active on the permissive strain). Therefore, if heterozygotes are complete diploids, they should be twice as UV sensitive as haploid phage.

To test this hypothesis, a lysate containing heterozygotes was diluted into saline and subjected to ultraviolet irradiation. Samples were removed at various times and assayed on both the permissive and nonpermissive strains. As can be seen in Fig. 1, the UV target size of a heterozygote is twice that of a parental phage. This supports the hypothesis that the genome of a heterozygote is twice as large as the genome of the parental phage.

Sedimentation

An f1 diploid may be either an end-to-end dimer or a side-to-side dimer. If it is arranged side-to-side, the heterozygote should be heavier and possibly denser than the parental particle, but both should have the same length. If the arrangement is end-to-end, the heterozygote should be heavier, of the same density, and twice as long as the parental particle. A clue to the size and shape of the diploids is provided by their sedimentation behavior. Most of a lysate produced by mixed infection consists of parental type phage with an S of 40, as determined in the analytical ultracentrifuge. But, in sucrose gradients, a shoulder at a higher S value, containing the particles that form plaques on the nonpermissive strain (heterozygotes), can be detected. On the other hand, centrifugation of such a lysate in CsCl reveals that the heterozygotes have the same density as do the parental phage (Notani and Zinder, 1964). The small increase in S (about 5%) of the heterozygotes, as compared to the parental particles, is expected for a rigid rod with twice the length. This difference, and their equivalence in density, support the model of diploid formation by end-to-end attachment of haploid phage.

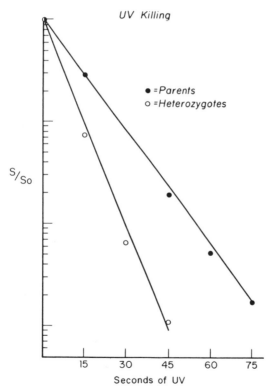

FIG. 1. Inactivation by ultraviolet light. A lysate containing 10^{12}/ml of two amber mutants and 10^{10}/ml heterozygotes was diluted 100-fold into saline-calcium solution; 5 ml was irradiated at 27 cm from a GE germicidal UV lamp. Samples were removed at the indicated times and assayed for parental phage on the permissive host (closed circles) and for heterozygotes on the nonpermissive host (open circles).

Electron Microscopy

Since homozygous diploids are not genetically detectable, it was possible that either all f1 phage are diploid but only a small fraction is heterozygous, or that only a small fraction of f1 particles in a lysate is diploid and the detectable heterozygotes are a random sample of these. The sedimentation data eliminate the former possibility since only a small fraction of the particles in the lysate sediments with the heterozygotes. Hence, it appears that the genetic heterozygotes are a random sample of the physical diploids in a lysate. Therefore, when 2% heterozygotes are present in a lysate, there should be at least another 1% homozygous diploids of each parental type giving a total of 4% diploid particles. Electron micrographs, taken by Dr. Walther Stoeckenius, support the end-to-end model for diploids, since they show that particles with twice the normal length make up about 8% of the total population. Phage particles of three and four times the normal length, as well as diploids, have also been seen by Dr. Lucien Caro in a purified f1 preparation.

Antiserum and Protease Inactivation

Further information on the nature of the diploid particles is provided by treatment with reagents that react with the surface protein components. Neutralization of phage by antiserum usually proceeds by attachment of antibody to the site at which phage attach to bacteria, thereby inhibiting adsorption. Proteolytic enzymes, on the other hand, might react with any of the phage surface components.

When anti-f1 serum is used to inactivate phage, the heterozygotes are found to be twice as sensitive as are parental particles (Fig. 2). Therefore, the diploid particles have either twice the number of serum-sensitive sites or a single site twice the size.

Salivar *et al.* (1964) reported that the similar phage M13 was inactivated by treatment with the proteolytic enzyme subtilisin. An f1 lysate containing diploid particles was treated with subtilisin, papain, ficin, and pronase in turn. For all of these proteases (except pronase, which was without effect) the kinetics of inactivation of both parental and heterozygous particles was the same (Fig. 3).

Phage f1 inactivation by protease is, therefore, different from inactivation by antiserum. The heterozygotes have twice the target size of the parents for serum inactivation; but, for protease inactivation both heterozygotes and parents have the same target size. Although these two facts do not support any unique hypothesis for the structure of the diploid particles, they are both consistent with the suggestion that heterozygotes are end-to-end aggregates of parental phage. If diploids are really twice as long as parental phage, in agreement with the sedimentation and density data presented above; then, since both kinds of particles are equally sensitive to proteolytic enzymes, there must be specific protease sensitive sites on the phage coat. Because of the shape of this phage, it is reasonable that these sites are at the end or ends of the rod. On the other hand, antiserum appears to neutralize f1 by attachment anywhere along the length of the particle in contrast to what is found for other phages. So, although the serum and protease inactivation results reversed our a priori predictions as to which agent would attack the length and which the ends of the phage,

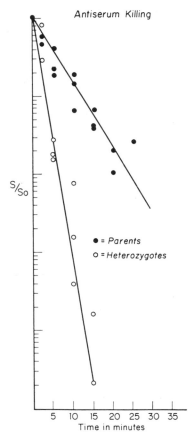

FIG. 2. Inactivation by antiserum. In penassay broth, 10^7 parental phage/ml and 2×10^5 heterozygotes/ml were incubated in an ice bath. Anti-f1 serum was added to this mixture at a final dilution of 10^{-5}; samples were assayed for parental phage on the permissive host (closed circles) and for heterozygotes on the nonpermissive host (open circles) at the indicated times.

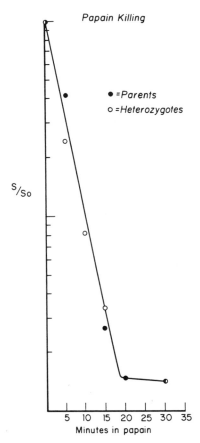

FIG. 3. Inactivation by protease. One-tenth milliliter of a lysate containing 10^{12} parental phage/ml and 2×10^{10} heterozygotes/ml was added to 1 ml of 0.2 *M* pyridine acetate buffer at 37°C. Papain was added to give a final concentration of 0.24 mg/ml. Samples were removed at the indicated times and assayed for parental phage on the permissive host (closed circles) and for heterozygotes on the non-permissive host (open circles).

they are entirely consistent with the hypothesis that the diploids are end-to-end aggregates of parental phage.

Continuity of DNA

Since the f1 heterozygotes consist of two parental particles attached end-to-end, it is important to find out whether the DNA is continuous throughout the double length or whether two completely separate particles are attached externally to form a heterozygote. A preliminary effort to determine this was made by extracting the DNA from a lysate containing heterozygotes and then using this DNA to infect spheroplasts made both from permissive and nonpermissive strains. It was thought that the demonstration of the infectivity of the DNA for the nonpermissive strain would indicate that heterozygotes contain a single dimeric DNA molecule possessing the genes of both parents. However, the spheroplast infectivity system proved so inefficient for f1 DNA that vast excesses of DNA per competent spheroplast had to be employed. Reconstruction experiments demonstrated that multiple infections occurred with the same frequency as infection of the nonpermissive host in the experimental system. The rate of such multiple infections was about 2% of the infectivity for the permissive host; that is, about equal to the frequency of heterozygotes in a phage lysate. Therefore, transformation cannot be employed to detect heterozygous DNA unless some way is first developed to significantly enrich the concentration of diploid particles in a lysate.

Search for DNA Dimers

The two types of experiments performed to detect DNA molecules with twice the normal length produced negative results. First, when DNA from a purified phage preparation containing heterozygotes was run through agar columns, there was never any indication of a separation into two peaks containing molecules of different sizes. Instead, all of the DNA came off in the same fraction. Second, in an experiment performed in collaboration with Mr. Hugh Robertson, when DNA from phage labeled with tritiated thymidine was sedimented in a sucrose gradient, all of the label was recovered in a single peak, which also contained all of the infectivity. This indicates that no labeled or infective component having a higher sedimentation rate was present. This tends to indicate that there are, indeed, no DNA dimers in heterozygous phage particles. Instead, a diploid particle probably consists of two normal phage DNA molecules wrapped up in the same protein coat.

Segregation of Heterozygotes

The idea that heterozygotes contain two DNA molecules is further supported by an experiment which shows that they segregate very early after infection to give both parental types. If the heterozygous structure persisted after infection, as would be expected if the diploid contained a dimeric DNA molecule, then the yield from a cell infected by a diploid should contain more heterozygotes than the yield from a cell mixedly infected with both parental phage. On the contrary, however, cells that have been infected with a single diploid particle never have more than 2% heterozygotes in their yield. Therefore, it seems likely that segregation occurs early after infection with a heterozygote as would be expected if the particle contained two molecules of DNA.

Summary

In summary, mixed infection, with any pair of mutants of phage f1 containing their mutations in different cistrons, gives rise to heterozygotes which grow on the nonpermissive strain by virtue of their ability to complement each other. They appear to be particles made up of two complete f1 phages linked end-to-end; they are produced in the infected cell and not by subsequent aggregation in the lysate. These diploids segregate early after infecting a cell to yield both parental types; this indicates that they do not contain a permanently dimeric DNA molecule. Apparently, heterozygous particles contain two molecules of DNA that have been wrapped up by the cell into the same protein coat.

REFERENCES

Hofschneider, P. H. (1963). *Z. Naturforsch.* **18b,** 203.
Loeb, T. (1960). *Science* **131,** 932.
Marvin, D. A., and Hoffmann-Berling, H. (1963). *Z. Naturforsch.* **18b,** 844.
Notani, G. W., and Zinder, N. D. (1964). *Bacteriol. Proc.* p. 140.
Salivar, W. O., Tzagoloff, H., and Pratt, D. (1964). *Virology* **24,** 359.
Zinder, N. D., Valentine. R. C., Roger. M., and Stoeckenius, W. (1963). *Virology* **20,** 638.

Conditional Lethal Mutants of Coliphage M13[*]

David Pratt, Helen Tzagoloff, William S. Erdahl, and Timothy J. Henry

DEPARTMENT OF BACTERIOLOGY, UNIVERSITY OF WISCONSIN,
MADISON, WISCONSIN

Conditional lethal mutants have been used widely in studying the numbers of cistrons and their functions in the large bacteriophages λ and T4 (Campbell, 1961; Epstein *et al.*, 1963). Similar work has recently been extended to several small viruses containing nucleic acid molecules sufficiently limited in size to make the identification of all of the viral cistrons and their functions feasible. One such virus is the male-specific coliphage M13. M13 is one of several phages, including f1 and fd, which contain single-stranded circular DNA of molecular weight 1.5 to 2×10^6 in filamentous particles 8500 Å long by 60 Å in diameter (Marvin and Hoffmann-Berling, 1963; Salivar *et al.*, 1964; Marvin and Schaller, 1966). The infective cycle of these viruses is distinguished by the fact that progeny phage are released through the wall of the infected cell without lysis and also by the fact that the cell continues to grow and to produce viable infected daughter cells (Hoffmann-Berling *et al.*, 1963; Hofschneider and Preuss, 1963; Salivar *et al.*, 1964).

This chapter will summarize work in press, and in progress, on phage M13 conditional lethal mutants and their use in attempts to identify the viral functions carried out in an infected cell. The full reports on these experiments will be given in other publications.

Genetic Complementation Tests with M13 Temperature Sensitive (*ts*) and Amber (*am*) Mutants

ASSIGNMENT OF MUTANTS TO CISTRONS

Sixty-three *ts* and 33 *am* mutants were isolated from stocks of wild-type phage treated with various mutagens (Pratt *et al.*, 1966). The *ts* mutants

* This work was supported by NIH Grants AI05627 and 5T1GM686.

multiply at 30° to 33 °C but not at 42 °C. The *am* mutants multiply in the Su-1 + permissive host strain K37 (Zinder and Cooper, 1964; Garen *et al.,* 1965) but not in the isogenic Su- host K38.

Genetic complementation tests were carried out by means of spot tests on plates, measurement of phage production in liquid culture, and a third procedure, better suited to M13, which combines features of both types of tests (see legend to Fig. 1; Pratt *et al.,* 1966). On the basis of the complementation tests, the mutants were assigned to 6 cistrons, as shown in Table I. The distribution of mutants in the cistrons is highly nonrandom, with 54 mutants, mostly *ts,* in cistron 2 and only a single, *am,* mutant in cistron 6. With the exception of cistron 6, all cistrons are represented by at least one *am* and one *ts* mutant.

TABLE I

ASSIGNMENT OF M13 CONDITIONAL LETHAL MUTANTS TO CISTRONS

Cistron	Number of mutants		
	ts	*am*	Total
1	7	11	18
2	45	9	54
3	8	5	13
4	2	3	5
5	1	4	5
6	0	1	1
	63	33	96

Rothman (personal communication) has also identified 6 cistrons in phage f1, closely related to M13. Complementation tests between representative f1, and M13 *am* mutants show that the 6 f1 cistrons correspond to the 6 M13 cistrons (unpublished experiments from this laboratory). Tessman (1965, 1966) has found 5 cistrons for phage S13, which has single-stranded DNA of approximately the same molecular weight as M13 DNA.

Isolation, classification, and also genetic mapping of a large number of additional mutants will be required before it can be determined whether phage M13 has cistrons other than the 6 presently identified. In the absence of such information, the total number of cistrons can be estimated by assuming that all of the approximately 6000 nucleotides of the M13 DNA molecule code for amino acids, that there are 3 nucleotides per amino acid, and that each cistron codes for a polypeptide chain of 200 to 300 amino acids. Based on these assumptions, there should be 6 to 10 cistrons in the phage, and therefore, no more than a few cistrons still remain unidentified.

Weak Intercistronic Complementations

Representative ("tester") *ts* and *am* mutants chosen from each of the 6 cistrons were tested against each other for genetic complementation in all pairwise combinations, as described in the legend to Fig. 1. The results are shown in Fig. 1, where the complementation for each pair of mutants is classified as strong, weak, or negative. All of the intracistronic pairs gave negative results, while most of the intercistronic pairs gave strong complementation. However, a few of the mutant pairs classified as intercistronic gave only weak complementation. This will be discussed further, since it raises doubts as to the assignment of mutants to cistrons.

Two of the weak complementations involve the *ts* tester mutant from

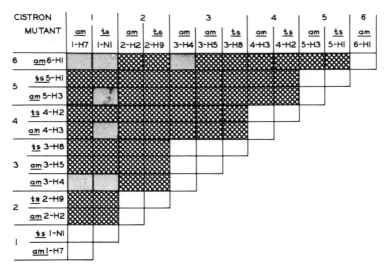

Fig. 1. Results of complementation tests between M13 tester mutants. A culture of the nonpermissive host K38 growing in broth was infected with each of a pair of mutants. Under the conditions used, about 50% of the cells became mixedly infected. The infected culture was diluted before the end of the latent period and plated for plaques on the nonpermissive host. The results were expressed as the ratio of plaques formed to total cells plated. Complementation was classified as strong when the ratio was greater than 0.15, weak when the ratio was between 0.01 and 0.15, and negative when the ratio was less than 0.01. The test and results have been described in detail by Pratt *et al.* (1966). Cross hatched rectangles = strong complementation; stippled rectangles = weak complementation; open rectangles = negative results.

cistron 1 with the cistron 4 and 5 *am* testers. In both cases, the complementation results were on the borderline between weak and strong. On the other hand, the *am* tester mutant from cistron 1 complemented fully with the mutants from cistrons 4 and 5; so, there would seem to be little doubt that cistrons 1, 4, and 5 are distinct.

The remaining weak complementations all involve *am* mutants in cistrons 1, 3, and 6. Not only the tests shown in Fig. 1, but many other intercistronic tests involving mutant pairs from these 3 cistrons were carried out in an attempt to understand the meaning of the weak complementations. All *am* $\times$ *am* mutant pairs except those involving the cistron 3 mutant *am* 3-H5 gave very weak complementation. With *am* 3-H5, all intercistronic complementations were strong. Some *am* $\times$ *ts* pairs from these cistrons gave strong complementation, others very weak. The *ts* $\times$ *ts* pairs gave strong complementation.

Three types of evidence indicate that, despite the weak complementations found with some of the *am* mutants, cistrons 1, 3, and 6 actually are 3 distinct cistrons: (1) Cistron 3 *am* mutant 3-H5 and all *ts* mutants in cistron 3 complement fully with mutants in cistrons 1 and 6; therefore, 3 must be distinct from 1 and 6. (2) Mutations in cistron 6 but not in cistron 1 produce serum-blocking power during infection under nonpermissive conditions (see Table II); therefore, 1 and 6 must be distinct from each other. (3) *Am* type mutants are not known to give intracistronic complementation (Edgar *et al.,* 1964; Garen and Siddiqi, 1962); consequently, any complementations between *am* mutants in cistrons 1, 3, and 6 should be intercistronic.

The genetics literature includes at least two different explanations for weak intercistronic complementations. One explanation, derived from a hypothesis of Lewis (1951), proposes that the weak complementations occur when (1) complementation requires interaction of the products coded by the unmutated copies of the 2 cistrons involved and (2) these products diffuse only slowly from their sites of synthesis in the cell. With the mutations in the *trans* position, the products would be made at different locations in the cell and would not be able to interact fully and therefore would not complement fully. Were this explanation true, it can be seen that the level of complementation in a test should depend on the cistrons involved but not on the particular mutants in those cistrons. The observed results with the M13 mutants in cistrons 1, 3, and 6 are of the opposite kind, with the level of complementation varying from strong to very weak depending on the particular mutant used to represent a cistron. Therefore, it seems unlikely that this explanation would apply to the M13 results. This general type of explanation, involving interaction of products with very limited diffusion, can be modified in various ways to fit the results found with M13. However, the modifications involve additional and unsubstantiated assumptions and seem unlikely.

The second general type of explanation for weak intercistronic complementation finds its basis in work on "polar" mutations, as in the histidine operon in *Salmonella typhimurium* (Ames and Hartman, 1963). A polar mutation prevents the synthesis of the product of the cistron in which it occurs and in addition reduces the levels of products of all cistrons between that in which the mutation occurred and the "distal" end of the cistron cluster making up the operon. Such a polar mutation would not be expected to complement fully with a mutation in one of the distal cistrons of the operon, since in such a complementation test one copy of the distal cistron would itself be mutated and the other copy would be reduced in effectiveness by the polar mutation.

All of the complementation results involving M13 cistrons 1, 3, and 6 can be explained if the 3 cistrons belong to an operon and are susceptible to both polar and nonpolar mutations. The relative positions of the 3 cistrons in the hypothetical operon and the classification of the tester *ts* and *am* mutants as polar or nonpolar are shown in Fig. 2. The figure is

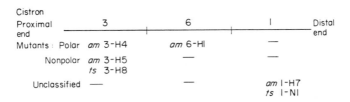

FIG. 2. Operon model for the complementation results for cistrons 1, 3, and 6. The cistrons are ordered in a hypothetical operon. Using the rule that a polar mutant will complement only weakly with any mutant in a cistron distal to it in the operon, the tester mutants are classified as to polarity on the basis of complementation results.

based solely on complementation results; the rule that a polar mutant fails to give strong complementation with mutations in any cistron distal to it in the operon was used. This model fits not only the results for the tester mutants, but also the complementation results for the many other mutant pairs tested for cistrons 1, 3, and 6. However, there is as yet no supporting evidence for the model, such as genetic mapping of the M13 mutants or direct measurement of the effect of a mutation in one cistron on synthesis of the product of another cistron. Despite the lack of supporting evidence, this model appears to provide a more likely explanation of the weak complementation than the model based on limited diffusion of the cistron products.

Tessman (1965) explained complementation results for one mutant of phage S13 on the basis of polarity. It would appear that the weak complementations she found between mutants in S13 groups IIIa and IIIb might also involve polar mutations.

SYMMETRY OF COMPLEMENTATION

Phage yields in liquid culture complementation tests were measured to determine whether the 2 mutants of complementing pairs appeared in equal frequency among the progeny phage. The tests were carried out with mutants from each of the 6 cistrons and showed in all cases that the yields of the 2 phages were approximately equal, that is, symmetrical (Pratt *et al.*, 1966). Asymmetrical complementation has been found by Tessman (1965, 1966) for mutants in one cistron of phage S13, and by Pfeifer *et al.* (1964) and Valentine *et al.* (1964) for some mutants of the RNA phages MS2 and f2. The meaning of asymmetric complementation in these phages is not clear, nor is the meaning of its absence among the M13 mutants so far isolated.

Temperature Shift Experiments

The time at which a cistron functions during a phage infection cycle can be determined, to a first approximation, by temperature shift experiments carried out with *ts* mutants (Edgar and Epstein, 1965). Our interpretation of such experiments will be based on three assumptions: (1) The function specified by a mutated cistron is blocked at the nonpermissive temperature, 42°C, but not at the permissive temperature, 30°C. (2) If a cell infected with a *ts* mutant is incubated first at 42°C, then at 30°C, the infection at 42°C will proceed only to the point at which the function of the mutated cistron is required. On subsequent incubation at 30°C, the infection will continue from that point, with phage production beginning quickly if the block was at a late stage in infection, but only with a longer lag if the block was early in infection. (3) If a cell infected with a *ts* mutant is incubated at 30°C until phage production is underway, and then it is shifted to 42°C, phage production will stop immediately if the function of the mutated cistron is needed continuously for phage production. But, it will proceed at 42°C if the function of the cistron is no longer needed.

Two types of temperature shift experiments were carried out with M13 *ts* mutants, the "shift down" from 42° to 30°C and the "shift up" from 30° to 42°C. The procedures are described in the legends to Figs. 3 and 4.

Results of the shift down experiments for *ts* mutants from cistrons 1 and 2 are shown in Fig. 3A and B. The mutants in cistrons 3, 4, and 5 behaved similarly to the mutant in cistron 1. All gave the results expected if the block at 42°C came late in the latent period: The longer the incubation at 42°C, the shorter the subsequent lag at 30°C before phage production began. In contrast, the cistron 2 mutant gave the results expected if the block came in the early minutes of infection: Phage production at 30°C began only after a considerable lag, even with extended incubation at 42°C. On the basis of these results, cistron 2 is classified as an "early" cistron and cistrons 1, 3, 4, and 5 as "late" cistrons. Cistron 6 could not be classified, since no *ts* mutant in this cistron was isolated.

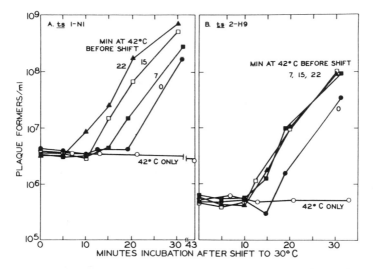

Fig. 3. Shift down experiments. Phage were allowed to attach at 0°C to host cells in 0.08 *M* saline. The infected cultures were diluted into Hershey broth at 42°C to allow phage penetration and any subsequent steps which could be carried out at that temperature. After 0, 7, 15, and 22 min, samples of the 42°C culture were diluted into broth at 30°C for further incubation. All cultures were assayed periodically for total plaque formers. The detailed experimental procedure and control experiments on M13 multiplication at 30°C and 42°C have been described by Pratt *et al.* (1966). (A) Cistron 1 mutant *ts* 1-N1. (B) Cistron 2 mutant *ts* 2-H9.

The results of the shift up experiments for the mutants for cistrons 1 and 2 are shown in Fig. 4A and B. Again, the mutants from cistrons 3, 4, and 5 gave results similar to those for the cistron 1 mutant. For these mutants, phage release stopped immediately after the shift to 42°C. Since M13 phage release occurs simultaneously with phage production (Hofschneider, personal communication), this result means that the functions of cistrons 1, 3, 4, and 5 are needed continuously for phage production. For the cistron 2 mutant, phage production continued at a decreasing rate for about 20 min after the shift up; and then it stopped. This result is neither that expected for a cistron whose function is needed continuously, nor that expected if the function is no longer needed at all. Such a result would be expected for a mutant where the temperature-sensitive product functioned to build up a pool of some temperature-stable second product, such as viral DNA. After the shift to 42°C, phage production could continue until the supply of the second product was exhausted.

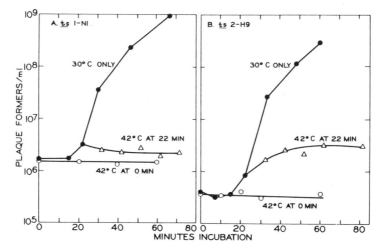

Fɪɢ. 4. Shift-up experiments. Infection was initiated as described in the legend to Fig. 3, except that the post-attachment dilution was into Hershey broth at 30°C. After 0 and 22 min, samples from the 30°C culture were diluted into broth at 42°C for further incubation. Each culture was assayed periodically for total plaque formers. (A) Ciston 1 mutant *ts* 1-N1. (B) Cistron 2 mutant *ts* 2-H9.

Functions Carried Out by *am* Mutants Infecting the Nonpermissive Host

Pʀᴏᴅᴜᴄᴛɪᴏɴ ᴏꜰ Iɴꜰᴇᴄᴛɪᴠᴇ DNA

Cultures of the nonpermissive host strain K38 infected either with an *am* mutant or with the M13 wild-type phage were examined for production of infective DNA. Infected cells were harvested and artificially lysed after 5 and 60 min incubation, as described in the legend to Fig. 5. The lysates were deproteinized with phenol at 70°C and assayed for infectivity on spheroplasts of the permissive host strain K37. The results, expressed as DNA plaque forming units per infected cell, are shown in Fig. 5. The wild-type phage and the *am* tester mutants for cistrons 1, 3, 4, 5, and 6 showed from five- to 200-fold increases in infectivity between 5 and 60 min. These increases presumably represent newly synthesized DNA, although experiments have not yet been carried out to prove this point. There was no increase in infectivity with the cistron 2 *am* mutant shown nor with another cistron 2 *am* mutant.

Samples of the 60-min infected cell extracts were centrifuged in CsCl density gradients (described in the legend to Fig. 6) to determine whether the infectivity banded at the density of single-stranded M13 DNA or at the

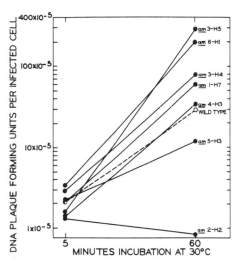

FIG. 5. Infective DNA from nonpermissive host cells infected with M13 *am* Mutants. K38 bacteria were infected in Hershey broth at 30°C. After 5 min incubation, the culture was chilled and washed twice to remove unadsorbed phage. One portion was then concentrated for lysing in 0.033 *M* Tris, pH 8.1. The remainder of the culture was incubated for an additional 55 min, then chilled, washed when necessary, and concentrated for lysing. Lysis was carried out with lysozyme and EDTA; a procedure similar to that described by Denhardt and Sinsheimer (1965) was used. The lysate was deproteinized with phenol at 70°C as described by Denhardt and Sinsheimer (1965). The resulting cell extracts were assayed for infective DNA on spheroplasts of the permissive host K37; the preparation used was a modification of the procedure of Guthrie and Sinsheimer (1963).

lower density which was expected if the infectivity was present in the form of double-stranded DNA molecules. Two DNA's were added to each gradient as density markers: infective single-stranded DNA extracted from particles of a *turbid* (*tu*) plaque type mutant of M13, and T4 phage DNA. The results of a typical gradient, for *am* mutant 3-H4, are shown in Fig. 6. The infectivity from the cell extract banded at a density approximately 0.01 gm/cm³ heavier than the infectivity from the added *tu* phage

DNA. The same result was observed for the infectivity from the 60-min extract for each of the *am* mutant phages which showed an increase in infectivity in the nonpermissive host; it was also found for the wild-type phage. This paradoxical result, that the infectivity from the extracts bands at a density heavier than that of the added single-stranded marker DNA, has not yet been resolved. The marker DNA from the *tu* mutant phage has been shown by analytical centrifugation to band at the same density

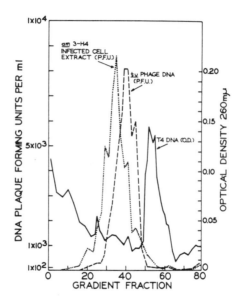

FIG. 6. CsCl density gradient analysis of infective DNA produced in the nonpermissive host by *am* mutant 3-H4. CsCl was dissolved in 0.025 *M* Tris, pH 8.1. A portion of the infected cell extract (see legend to Fig. 5), T4 phage DNA, and infective DNA extracted from phage particles of an M13 *turbid* plaque type mutant were added to the solution. The final density was brought to approximately 1.71. A sample of 2.2 ml was centrifuged at 33,000 rpm for 45 hours in a Spinco SW39 rotor. Two drop fractions were collected into 1 ml lots of 0.025 *M* Tris, pH 8.1. The fractions were used for optical density readings and assays for infective DNA on spheroplasts of the permissive host K37.

in CsCl as DNA extracted from wild-type phage particles; therefore, the *tu* DNA should be a valid marker for the density of single-stranded DNA. It is possible that the infectivity from the cell extracts represents denatured double-stranded circular DNA, as described by Burton and Sinsheimer (1965) and Sinsheimer *et al.* (1965) for ϕX174 phage. DNA denaturation might have occurred during the deproteinization at 70°C. This will be tested by using milder conditions of deproteinization. Alternatively, the infectivity from the extracts may reside in a single-stranded DNA molecule complexed with some other material.

To provide information on the nature of the low level of infectivity present at that early time, samples of the 5-min infected cell extracts for the wild-type phage and one of the *am* mutants were also centrifuged in CsCl. The infectivity was found to band at the density of the single-stranded *tu* mutant marker DNA or slightly heavier. It is not known to what extent the infectivity represents DNA still inside the original phage coat at 5 min and to what extent it represents DNA already engaged in replication.

No peak of infectivity banding at densities lower than that of single-stranded DNA was found in any of the 5 or 60 min gradients. This might be due either to the very low infectivity of double-stranded M13 DNA (Hofschneider, personal communication) or to denaturation of double-stranded material during the hot phenol extraction.

The results obtained thus far show that infective DNA increases between 5 and 60 min independently of the functions of cistrons 1, 3, 4, 5, and 6; but it cannot increase if the function of the "early" cistron, 2, is blocked. This raises the possibility that cistron 2 is involved in some step in DNA synthesis. The role of each cistron in DNA replication will be further investigated by infecting nonpermissive host cells with phages labeled in their DNA and following the fate of the label.

Synthesis of Serum-Blocking Power

Nonpermissive host cells infected with *am* mutants or M13 wild-type phage were examined for the production of serum-blocking power (SBP); that is, antigen capable of blocking the phage-neutralizing ability of anti-M13 serum. Cultures incubated for various lengths of time after infection were harvested, lysed, and assayed for SBP as described in the legend to Fig. 7. The kinetics of SBP production for the *am* tester mutants are shown in Fig. 7. The cistron 3 and 6 mutants formed SBP in the nonpermissive host, while the cistron 1, 2, 4, and 5 mutants did not. SBP production began between 10 and 15 min, which is the time at which phage production and release begin in a normal infection. In experiments carried out with the wild-type phage, SBP production began at a similar time. The level of SBP per cell was severalfold higher for the wild-type phage than for the cistron 3 and 6 mutants; it reached a constant value of approximately 8 phage equivalents per infected cell by 15 min.

The results of these tests show that SBP production can be blocked by mutations in any one of 4 different cistrons. This suggests that the

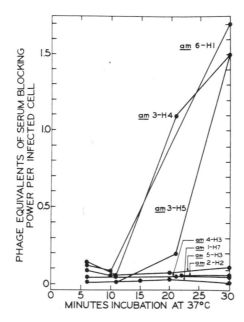

FIG. 7. Production of serum blocking power (SBP) by M13 *am* mutants infecting nonpermissive host cells. Phage were added to K38 cells in Hershey broth at 37°C. After 5 min for adsorption, the culture was diluted in broth for further incubation. Infected cells were collected at the indicated times, chilled, and washed 4 times in phosphate buffer, pH 7, to remove unadsorbed phage. The cells were concentrated in buffer and lysed as described in the legend to Fig. 5. The lysates were serially diluted and SBP determinations were made essentially as described by De Mars (1955). Incubation was carried out for 16 hours at 48°C; then, tester phage, M13 wild-type, were added to 2×10^5/ml. The tester incubation was ended after 2 hours by dilution into cold buffer. SBP values were calculated from the percent of surviving tester phage and reference to a standard curve.

antigen involved is either a complex product with components coded by several cistrons or that its production is indirectly controlled by the products of several cistrons. In either case, the SBP results cannot be used to identify any one cistron as responsible for the production of *the* phage coat protein. The results do show that *am* mutants blocked in the cistron 3 and 6 functions are capable of producing the blocking antigen; this point will be referred to again in considering the function carried out by cistron 3.

KILLING OF HOST CELLS

Cells infected with the wild-type M13 phage survive the infection, as evidenced by their ability to produce colonies on agar plates. The growth rate of the infected cells is slower than that of uninfected cells. When M13 *am* mutants are used to infect the nonpermissive host, the results are very different. Infection with mutants from the early cistron, 2, has absolutely no effect on cell viability or growth rate. In fact, the presence of the cistron 2 mutant in the nonpermissive cell can only be detected by superinfecting with another phage capable of "rescuing" it. In contrast, infection with *am* mutants in cistrons 1, 3, 4, 5, or 6 kills the nonpermissive host cells, stopping their growth after less than one cell doubling. These results are illustrated in Fig. 8; it shows the growth of cultures of the nonpermissive host K38, uninfected and infected with the wild-type phage, a cistron 1 mutant, and a cistron 2 mutant.

Similar results are observed with *ts* mutants infecting cells at the nonpermissive temperature, 42°C. The cistron 2 *ts* mutants have no effect on cell growth and survival, while the cistron 1, 3, 4, and 5 *ts* mutants stop the cell growth. Double mutant *ts* phages, mutated in cistrons 1 + 2, 1 + 3, and 2 + 3 were also tested for their effects on host cells at 42°C. Those containing the cistron 2 mutation had no effect on the

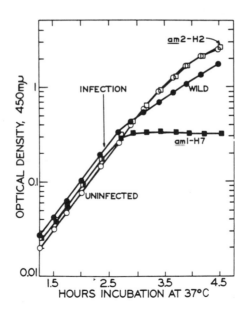

FIG. 8. Growth of infected and uninfected cells of the nonpermissive host K38. Four cultures of K38 bacteria were grown in Hershey broth at 37°C. At the time indicated on the graph, 3 of the cultures were infected with wild-type M13 phage, cistron 1 *am* mutant 1-H7, and cistron 2 *am* mutant 2-H-2. Growth of the cultures, before and after infection, was followed by measuring the increase in optical density at 450 mμ wavelength.

host cells, while the $1 + 3$ double mutant killed the cells. Thus, a double mutant composed of 2 cell-killing mutations, kills the host cells; however, the double mutants containing a nonkilling, cistron 2, mutation do not kill.

In sum, the cell growth and killing experiments show that if all phage cistrons function (wild-type), the infected cells survive but grow at a slower rate than uninfected cells. If cistron 2 functions but any other cistron does not, the cell growth is stopped—that is, the cell is killed. If the function of cistron 2 is blocked, with or without blocks in other cistrons, the infected cell grows as if not infected at all. Therefore, the function of cistron 2 is the primary factor in determining the effect on the host cell growth. The nature of this function and how it affects cell growth are not yet known.

Properties of Mutant Phage Particles Grown Under Permissive Conditions

Conditional lethal mutants grown under permissive conditions would, in general, be expected to produce a protein differing in one amino acid from the corresponding wild-type protein (Edgar *et al.,* 1964; Stretton and Brenner, 1965). If the affected protein was a structural component of the phage particle, properties of the particle could be altered as a result of the amino acid change. Tessman (1965) found this to be true with mutants of phage S13; thus, she was able to identify the cistron controlling the phage coat protein.

M13 mutants from the 6 cistrons were examined for adsorption to the host cell and inactivation *in vitro* by the agents heat, trypsin, nagarse, and anti-M13 serum. None of the mutants tested from cistrons 1, 2, 4, 5, or 6 differed from the wild-type phage in respect to these characteristics. Most of the mutants from cistron 3, on the other hand, did differ from wild-type in one or more ways. All but one of the cistron 3 *ts* and *am* mutants proved to be more labile than the wild-type phage to heat or the proteolytic enzymes, or both. All of the 8 *ts* mutants in cistron 3 had the unexpected property of adsorbing to host cells at a normal rate at 30°C, but only very slowly or not at all at 42°C, where M13 wild-type phage adsorb well. On the other hand, none of the cistron 3 *am* or *ts* mutants was inactivated at an abnormal rate by anti-M13 serum.

These properties of the cistron 3 mutants show that they affect a phage coat protein involved in adsorption to the host cell. There is no direct evidence yet that cistron 3 itself codes for the amino acid sequence of this coat protein rather than controlling its structure through some other means such as an assembly enzyme. However, the mutants in this cistron show as many different degrees of sensitivity to heat and proteolytic enzymes as there are mutants. This seems most readily explained if cistron 3 does code for the protein and each mutant has a protein with a different amino acid sequence.

The likelihood that cistron 3 codes for a phage coat protein must be reconciled with the result mentioned earlier, that cistron 3 *am* mutants infecting the nonpermissive host produce serum-blocking power, which presumably represents either coat protein subunits or assembled phage-like particles. The apparent paradox can be explained if phage M13 has at least 2 coat proteins, with the one which is involved in adsorption not being involved in the serum-blocking antigen. This idea is supported by the finding that at least one of the cistron 3 *am* mutants, 3-H5, not only produces SBP in the nonpermissive host, but also produces a phage-like product which is released into the medium. This product can be purified by the usual differential centrifugation procedure used for M13 phage, contains infective DNA, and has serum-blocking power. The product is abnormal in that it cannot adsorb to the host cell. It seems probable that the major coat protein of the phage is not the same as the adsorption protein controlled by cistron 3. This question is being investigated by "fingerprinting" tryptic digests of phage coat proteins from various mutants.

Genetic Recombination Experiments

Cells infected under permissive conditions with 2 different M13 *am* or *ts* mutants, with revertant levels of less than 10^{-5}, give rise to progeny phage containing up to 1% wild-type. Double mutant phage particles can be found at similar frequencies; thus, the frequency of wild-type phage appears to be a measure of genetic recombination. Several inter- and intracistronic paris of mutants have been crossed in an attempt to construct a genetic map. However, the crosses have so far given inconsistent results and do not allow the construction of a map, either linear or circular.

Summary

Ninety-six amber and temperature sensitive mutants of phage M13 have been assigned to 6 cistrons on the basis of genetic complementation tests. The tests involving mutants in cistrons 1, 3, and 6 gave results of a type expected if the 3 cistrons belonged to one operon.

Temperature shift experiments, carried out for the 5 cistrons for which *ts* mutants were isolated, show that cistron 2 starts to function during the

TABLE II

RESULTS OF INFECTING NONPERMISSIVE HOST CELLS WITH M13 *am* MUTANTS[a]

Result	Wild-type phage	*am* Mutant in cistron					
		1	2	3	4	5	6
Production of infective DNA	+	+	−	+	+	+	+
Production of serum blocking power	+	−	−	+	−	−	+
Survival of infected host cells	+ (reduced growth rate)	−	+ (growth rate unchanged)	−	−	−	−

[a] Experiments were carried out in broth at either 30° or 37°C.

early minutes of infection, while cistrons 1, 3, 4, and 5 start late in the latent period. The functions of both early and late cistrons are needed to maintain phage production once it is underway.

Cultures of a nonpermissive bacterial host strain were infected with either *am* mutants or the wild-type phage and examined for (1) production of infective phage DNA, (2) production of serum-blocking power (SBP), and (3) the effect of the phage on the host cell. The results are summarized in Table II. The wild-type phage produced both infective DNA and SBP. It reduced the growth rate of the host cells but did not kill them. The mutant in cistron 2, the early cistron, was unique in that it produced neither infective DNA nor SBP in the nonpermissive host; in addition, it had no effect on the host cell growth and division. The mutants in each of the other 5 cistrons produced infective DNA; they also killed the nonpermissive host cells, stopping the cell growth after less than one doubling. Mutants in only 2 of these 5 cistrons produced SBP. These results do not permit the assignment of definite functions to any of the 6 cistrons, although they do suggest that cistron 2 is involved at some step in DNA synthesis. They suggest, further, that the serum-blocking antigen is either a complex product coded by several cistrons or a product whose synthesis is controlled in some indirect way by several cistrons. The results show that infection with M13 kills the host cell if cistron 2 is functional; but, the function of any one of the other 5 cistrons is blocked.

Stocks of *ts* and *am* mutants from each of the cistrons were tested for their rate of adsorption to host cells and their rate of inactivation by various agents expected to affect the phage protein coat. The results showed that cistron 3 controls the structure of a coat protein involved in adsorption. It appears that this is not the only coat protein of the virus.

REFERENCES

Ames, B. N., and Hartman, P. E. (1963). *Cold Spring Harbor Symp. Quant. Biol.* **28,** 349.

Burton, A., and Sinsheimer, R. L. (1965). *J. Mol. Biol.* **14,** 327.

Campbell, A. (1961). *Virology* **14,** 22.

De Mars, R. I. (1955). *Virology* **1,** 83.

Denhardt, D. T., and Sinsheimer, R. L. (1965). *J. Mol. Biol.* **12,** 647.

Edgar, R. S., and Epstein, R. H. (1965). *Proc. 11th Intern. Congr. Genet., The Hague, 1963* Vol. 2, p. 1. Pergamon Press, Oxford.

Edgar, R. S., Denhardt, G. H., and Epstein, R. H. (1964). *Genetics* **49,** 635.

Epstein, R. H., Bolle, A., Steinberg, C. M., Kellenberger, E., Boy De La Tour, E., Chevalley, R., Edgar, R. S., Susman, M., Denhardt, G. H., and Lielausis, A. (1963). *Cold Spring Harbor Symp. Quant. Biol.* **28,** 375.

Garen, A., and Siddiqi, O. (1962). *Proc. Natl. Acad. Sci. U.S.* **48,** 1121.

Garen, A., Garen, S., and Wilhelm, R. C. (1965). *J. Mol. Biol.* **14,** 167.

Guthrie, G. D., and Sinsheimer, R. L. (1963). *Biochim. Biophys. Acta* **72,** 290.

Hoffmann-Berling, H., Durwald, H., and Beulke, I. (1963). *Z. Naturforsch.* **18b,** 893.

Hofschneider, P. H. (1966). Personal communication.

Hofschneider, P. H., and Preuss, A. (1963). *J. Mol. Biol.* **7,** 450.

Lewis, E. B. (1951). *Cold Spring Harbor Symp. Quant. Biol.* **16,** 159.

Marvin, D. A., and Hoffmann-Berling, H. (1963). *Z. Naturforsch.* **18b,** 884.

Marvin, D. A., and Schaller, H. (1966). *J. Mol. Biol.* **15,** 1.

Pfeifer, D., Davis, J. E., and Sinsheimer, R. L. (1964). *J. Mol. Biol.* **10,** 412.

Pratt, D., Tzagoloff, H., and Erdahl, W. S. (1966). *Virology* **30,** 397.

Rothman, J. (1966). Personal communication.

Salivar, W. O., Tzagoloff, H., and Pratt, D. (1964). *Virology* **24,** 359.

Sinsheimer, R. L., Lawrence, M., and Nagler, C. (1965). *J. Mol. Biol.* **14,** 348.

Stretton, A. O. W., and Brenner, S. (1965). *J. Mol. Biol.* **12,** 456.

Tessman, E. S. (1965). *Virology* **25,** 303.

Tessman, E. S. (1966). *J. Mol. Biol.* **17,** 218.

Valentine, R. C., Engelhardt, D. L., and Zinder, N. D. (1964). *Virology* **23,** 159.

Zinder, N. D., and Cooper, S. (1964). *Virology* **23,** 152.

Discussion—Part III

Dr. M. L. Fenwick: (Question directed to Dr. E. S. Tessman.) In view of the observed concentration-dependence of the inhibition of RF multiplication by chloramphenicol, is it possible that the original conversion of the parental DNA to RF is also mediated by a relatively chloramphenicol-resistant virus-directed protein?

Answer: The following evidence makes phage control of early RF formation appear unlikely: (1) No phage mutants have been found for which early RF fails to be formed; (2) early RF is formed in the presence of 150 μg/ml chloramphenicol.

Dr. D. Pratt: (Comment following paper by J. Rothman Scott.) William Salivar at Wisconsin has purified "heterozygotes" of phage M13 which appear to be the same as those of phage f1. He discovered that the usual yield of about 1% heterozygotes, relative to total phage, could be increased to at least 10% in infections involving amber (*am*) mutants from M13 cistron 6. When such phage stocks are chromatographed on columns of Biorex 70, heterozygotes reach a maximum frequency of 50%, with the remaining 50% being "diploid" phages which are genetically homozygous. The DNA of the diploid phages, homozygotes and heterozygotes, is made up of 2 separate normal-sized DNA molecules. This was determined both by direct analysis, from sedimentation of DNA extracted from a stock with 40% diploid phages and by indirect analysis, from the shape of UV survival curves for purified diploid phage stocks. Preliminary electron micrographs, of the purified diploids, show phage particles of the normal diameter and twice the normal length. In all of these respects, the results for phage M13 are in accord with Dr. Scott's model for the structure for f1 heterozygotes.

Dr. J. R. Scott: (Comment following paper by Dr. D. Pratt.) With reference to the cell killing effect, we use the loss of ability to form a colony as the criterion of cell death; we find that, under several different conditions that inhibit phage replication, cell killing begins immediately after f1 infection. In agreement with the observations of Dr. Pratt, we find that all of our amber mutants, except those with mutations in cistron 2, kill the nonpermissive host.

Furthermore, wild-type f1 kills its host at 42°C, if 100 μg/ml of chloramphenicol is added during infection, or if the infection is performed in minimal-glucose medium. In the latter instance, the addition of casamino acids prevents cell death. Therefore, it appears that f1 kills its host under conditions that retard or prevent phage protein synthesis.

The commitment to death of these infected cells is a gradual process that can be reversed by altering the causal conditions. For example, a decreasing fraction of infected cells grown at 42°C can be rescued by plating them at a lower temperature. Cells do not lose their ability to form colonies immediately upon infection; instead, an increasing number of cells die as the time after infection increases. Furthermore, f1 infected cells retain their capacity to support the replication of T6 bacteriophage for 30 min after they lose their ability to form colonies (an observation made by Mr. Hugh Robertson in our laboratory). It appears that under all these conditions, cell death is caused by the accumulation of a threshold amount of some material made under the direction of the phage genome. Until this "lethal threshold" is reached, the cells can be rescued.

PART IV

RNA BACTERIOPHAGES

Structure and Function of RNA from Small Phages

Paul Kaesberg

LABORATORY OF BIOPHYSICS AND BIOCHEMISTRY DEPARTMENT,
UNIVERSITY OF WISCONSIN, MADISON, WISCONSIN

Ever since their discovery by Loeb and Zinder (1961), RNA phages have been most useful test objects for the study of a variety of biological and biochemical phenomena. For example, such phages exhibit a close physical and chemical relationship between a particular protein and a particular nucleic acid, so that their study can elucidate the nature of protein-protein and protein-nucleic acid interactions. They provide a small number of genes as well as a gene product in a very small package, whose study can contribute to an explanation of genetic phenomena in structural terms. Furthermore, they are prototypes of disease-inducing viruses, and they may have analogous reproductive processes which we can learn to control. Indeed, the latter is a research goal of many of the authors represented in this volume.

Many of these authors have made important contributions in the above areas. We have also made some research attempts with the same goals in mind, and we have concentrated our efforts on studies of virus structure. I would like to discuss a particular aspect of this work—the structural relationships between RNA of small phages and the proteins encoded in the RNA. I will emphasize the RNA primary structure, even through its tertiary structure is also an important feature of RNA's interaction with the viral coat protein and with the viral RNA replicase.

Let me review briefly some of the structural properties of RNA phages. At least two unrelated RNA phages are known: (a) f2 (and its relatives), discovered by Loeb and Zinder (1961); and (b) Qβ, discovered by Watanabe (1964). There are many reasons to believe, and none to dispute that f2, MS2, fr, R17, M12, and β are virtually identical in physical and chemical properties; that is, their only distinguishing features are due to a few differences in their coat protein and RNA sequences. It should be assumed, in the absence of evidence to the contrary, that these viruses have identical particle weights and that the reported differences are due to experimental shortcomings. Their particle weights thus lie somewhere between 3.6×10^6 daltons, as reported for MS2 (Strauss and Sinsheimer, 1963), and 4.3×10^6 daltons, as reported for fr (Marvin and Hoffmann-Berling, 1963). The weights of their RNA's lie between 1.0×10^6 and 1.3×10^6 (Strauss and Sinsheimer, 1963; Marvin and Hoffmann-Berling, 1963); their coat protein subunits lie between 13,700 for f2 (Konigsberg *et al.,* 1966) and 15,200 for fr (Wittmann-Liebold, 1965). If we take intermediate values for the protein subunit and RNA molecular weights, e.g., those quoted by us for R17 [$M_{\text{protein}} = 14,200$ (Enger and Kaesberg, 1965) and $M_{\text{RNA}} = 1.1 \times 10^6$ (Sinha *et al.,* 1965)] and assume 180 protein subunits, we obtain a virus particle weight of 3.7×10^6 daltons which is in good agreement with the lowest values reported. It is possible that such close agreement is fortuitous. The virus may contain low molecular weight constituents which would contribute to its particle weight, and indeed there may exist a second type of coat protein—for example, one concerned with attachment to the host cell.

The work of Spiegelman and associates (Overby *et al.,* 1966) has shown that the particle weight of phage $Q\beta$ is about 10% greater than that of members of the f2 group. Detailed information about the coat protein subunit and the RNA of this phage has not yet been reported.

RNA phages are not known to have naturally occurring top components (i.e., nucleic acid-free coat protein shells) like those which have been so valuable in elucidating the structure of small plant viruses. However, it is possible to make an artificial top component of R17 (Gullekson and Kaesberg, unpublished data) by brief exposure to high pH. Purified R17 in 0.1 M KCl is mixed with a more concentrated salt solution in proportions such that the virus is dissolved in 1.0 M KCl buffered to pH 11.2. After 8 min at 20°C, the reaction is terminated by bringing the pH to neutrality by the addition of HCl. The resulting solution consists of coat protein and partially degraded RNA. These can be separated easily by digesting the RNA with ribonuclease and then filtering it on G-100 Sephadex (0.15 M KCl, 0.05 M potassium citrate, pH 7). The artificial top component (coat protein) emerges with the first column retention volume, while the degraded RNA is retarded.

Artificial top component has an absorption maximum at 273mμ, and an absorbancy index at that wavelength of 1.09 cm^3/mg. Its sedimentation rate is $s_{20,w} = 41.1$ S. These and other physical parameters lead to the conclusion that it has a nearly spherical protein shell, identical in exterior dimensions to the virus, and that this protein shell is actually the protein portion of the virus. Just as with small plant viruses, it is to be inferred from the structure of the artificial top component that the viral RNA occupies, for the most part, a region in the interior of the protein shell. Although, by analogy with the structure of turnip yellow mosaic virus (Kaper, 1964), it is reasonable to believe that some of the RNA may be embedded to an appreciable extent in the coat protein network. In any event, one must conclude that the RNA in the virus must be packed very efficiently into a very small volume.

It may be pertinent to recall that in solutions of high salt concentration the f2-like phage RNA's are quite compact ($R_G = 230$ Å); and, they have extensive base-base interaction (70% of the RNA is in the "helical" configuration, though not as much as when they are in the virus (80% "helical" configuration) (Mitra *et al.,* 1963). The remarkable finding by Haruna and Spiegelman (1965) that some of these RNA's can be recognized specifically by their own replicases suggests that more work on their structures in solution could be most valuable.

Recently we have been concerned with the primary structures of phage RNA's—that is, with their base sequences. With more than 3000 bases, it would seem probable that attempts to obtain complete sequences with presently known procedures are impractical if not imprudent. What lesser goal, then, can be set that still has some of the utility of a complete sequence determination? It is well known that pancreatic ribonuclease cleaves to produce oligonucleotides terminating in pyrimidines and taka-diastase ribonuclease cleaves to produce oligonucleotides terminating in G. Tener and his associates (Tomlinson and Tener, 1963) and Rushizky *et al.* (1965) showed that it should be possible to separate all small, nonisomeric oligonucleotides from each other. We have analyzed such ribonuclease digests of R17 and M12 RNA's, and we are encouraged by the possibilities that such an approach offers.

So far we have concentrated on digestion with pancreatic ribonuclease; we have succeeded in separating almost all of the products and in determining the base sequences of a number of them. We refer to the collection of digestion products as a *catalog,* and in the case of digestion with pancreatic ribonuclease, we refer to the complete set of pyrimidine-terminated pieces as the pyrimidine catalog. Such a catalog for M12 and R17 RNA is shown in Table I. It is taken from the Ph.D. thesis of J. P. Thirion. The italicized entries indicate those whose sequences we have determined (or whose determination is trivial).

Figure 1 shows a chromatograph of a pancreatic ribonuclease digest of ³²P-labeled M12 RNA, which was carried out according to the method described by Tomlinson and Tener (1963). Such a column (DEAE-cellulose at pH 7.5) fractionates on the basis of oligonucleotide chain length. The numbers 1 through 12 indicate the chain lengths of the fractions invariably found with this RNA. Peak *a* occurs when the digestion conditions are a bit too mild to achieve complete digestion to pyrimidine-terminated pieces. We believe that peak *a* is a unique oligonucleotide, approximately 20 bases long, which contains both U and C. Peak *b* is a heterogeneous collection of incompletely digested fragments and peak *c* represents the remaining material purged from the column by 0.3 *N* NaOH. Table II lists the radioactivities corresponding to the various fractions shown in Fig. 1.

The fractions from such DEAE collulose columns are usually rechromatographed at low pH (e.g., pH 2.8 or 3.4) to achieve separation according to base composition. It is easy to make an accurate comparison

TABLE I

Pyrimidine Catalogs of M12 and R17 RNA

No.	Bases	*M12*	*R17*	No.	Base	*M12*	*R17*	No.	Base	*M12*	*R17*
1.	C	*430*	*441*	5.	A_4C	*2*	*2*	7.	$(A_5G)C$	*1*	1
	U	*418*	*453*		$(A_3G)C$	8	5		$(A_4G_2)C$	*1*	1
					A_4U	*1*	*2*		$(A_5G)U$	*1*	*1*
2.	AC	*92*	*87*		$(A_2G_2)C$	10	12		$(A_3G_3)C$	*1*	1
	GC	*114*	*112*		$(A_3G)U$	4	5		$(A_3G_3)C +$	4	3
	AU	*98*	*94*		$(A_3G)C$	4	4		$(A_4G_2)U$		
	GU	*97*	*105*		$(A_2G_2)U$	15	14		$(A_2G_4)C$	1	1
3.	A_2C	*26*	*27*		G_4C	*1*	*1*		$(A_3G_3)U$	2	2
	AGC	19	*53*		$(AG_3)U$	7	6		$(AG_5)C$	1	1
	GAC	32			G_4U	*1*	*1*		$(A_2G_4)U$	*1*	1
	A_2U	17	*20*	6.	$(A_3G_2)C$	4	4		$(AG_5)U$	1	1
	GGC	*38*	*38*		$(A_4G)U$	0	1	8.	$(A_5G_2)C$	*1*	0
	AGU	15	*46*		$(A_2G_3)C$	7	7		$(A_4G_3)C$	2–3	*1*
	GAU	37			$(A_3G_2)U$	5	3		$(A_5G_2)U$	1	*1*
	G_2U	*31*	*32*		$(AG_4)U$	0	1		$(A_3G_4)C$	1	*1*
					$(A_2G_3)U$	3	3				
4.	A_3C	*11*	*10*		$(AG_4)U$	2	2	9.	$(A_4G_4)C$	*1*	1
	$(A_2G)C$	17	16						$(A_4G_4)U$	*1*	*1*
	A_3U	6	*7*						$(A_3G_5)U$	1–2	1–2
	GGAC	*3*	*14*								
	(GA)GC	11						10.	$(A_4G_5)U$	*1*	0
	AAGU	*4*									
	AGAU	6	21					11.	$(A_6G_4)C$	1	1
	GAAU	8									
	G_3C	6	*7*					12.	$(A_6G_5)C$	*1*	1
	GGAU	6	*16*								
	(GA)GU	9									
	G_3U	*7*	*7*								

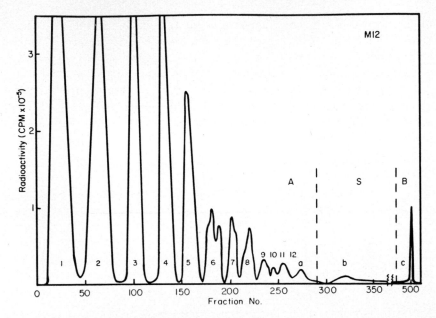

FIG. 1. Chromatography of a pancreatic ribonuclease digest of P³²-labeled M12 RNA on a 1.2 × 105 cm DEAE-cellulose column. A constant gradient, 0.1 to 0.6 M sodium acetate (800 ml of each solution) along with 7 M urea and 0.01 M Tris-HCl, pH 7.5, was used. This was followed by a similar gradient 0.6 to 1.0 M in sodium acetate (250 ml of each solution) and then a wash with 0.3 N NaOH.

TABLE II

COLUMN CHROMATOGRAPHIC ANALYSIS OF M12 RNA

Peak No.	³²P counts (% of total)	Relative No. of bases	Relative No. of pieces
1.	27.56	921.0	921.0
2.	22.47	751.0	375.5
3.	18.15	606.6	202.2
4.	11.34	379.0	94.7
5.	7.90	264.0	52.8
6.	3.81	127.3	21.2
7.	2.60	86.9	12.4
8.	1.51	50.5	6.3
9.	0.75	25.1	2.8
10.	0.38	12.7	1.3
11. and 12.	0.84	28.1	2.4
a	0.71	23.7	—
b	1.51	50.4	—
c	0.47	15.7	—
	100.00	3342.0 bases	

of two similar RNA's by analyzing a mixture in which the viruses are labeled differently, or in which only one of the viral RNA's is labeled with a radioisotope (Figs. 2 and 3). Obviously, then the viruses will have been treated identically with respect to purification, digestion, and analysis. Some inaccuracy is introduced by irreversible adsorption, particularly of larger oligonucleotides, to the columns at low pH (Fig. 4). We had anticipated some preferential adsorption of G-rich pieces, but we have found that oligonucleotides of the same size behave in a similar manner

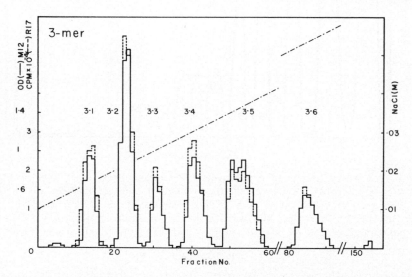

Fig. 2. Rechromatography of the trinucleotide fraction of a pancreatic ribonuclease digest of R17 and M12 RNA's on a 1.2 × 21 cm DEAE-cellulose column. A constant gradient, 0.01 M to 0.10 M NaCl (500 ml of each solution) along with 7 M urea and 0.1 M formic acid, pH 3.3, was used. Essentially no additional radioactivity was released from the column when it was washed with 0.3 N NaOH.

regardless of their composition. When digestion conditions are sufficiently harsh to ensure complete digestion to pyrimidine-terminated oligonucleotides, some destruction of the larger oligonucleotides occurs.

What value do these catalogs have? Let me cite several possibilities.

RNA catalogs can serve as a measure of the "relatedness" of two viruses. For example, apart from any other information, it is obvious from their pyrimidine catalogs that R17 and M12 are very similar to each other. We should be able to tell from pyrimidine and G catalogs whether Qβ and R17 are distantly related (even though they are clearly unrelated serologically). As one might expect, the pyrimidine catalogs from bromegrass mosaic virus RNA and R17 RNA provide no evidence that these two viral agents are related.

RNA catalogs can be a prelude to attempts at base sequence analyses. In a few cases it is possible to obtain naturally occurring pieces of viral RNA; various methods are available for producing pieces. Catalogs for the pieces and for the intact RNA could serve in accounting procedures. For example, BMV RNA is a molecule some 3000 bases long. Two portions of it have been isolated—one contains about 900 bases, the

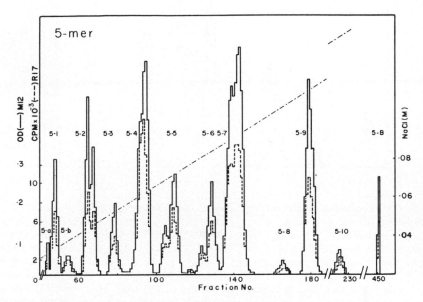

FIG. 3. Rechromatography of the pentanucleotide fraction of a pancreatic ribonuclease digest of R17 and M12 RNA's. Peak 5-B represents material washed from the column with 0.3 N NaOH.

other about 2100 (Bockstahler and Kaesberg, 1965). Table III, taken from the Ph.D. thesis of D. Knoblauch, lists the 7-mer portion of the pyrimidine catalog for BMV RNA. It may be seen from the table, that of the 13 7-mers present in the BMV RNA molecule, eight are located in the region of the large piece, and five are located in the region of the small segment. Some of our cell-free amino acid-incorporation studies suggest that the small piece carries the coat protein gene (Stubbs, 1965). We would suggest that 5 of the 7-mers for intact RNA will be found to code for coat protein sequences, while the remaining 8 will have other coding functions.

The RNA catalogs can be correlated directly with the protein sequences for which they code. Such correlations will become more profitable when coat protein sequences are available and when the other protein products

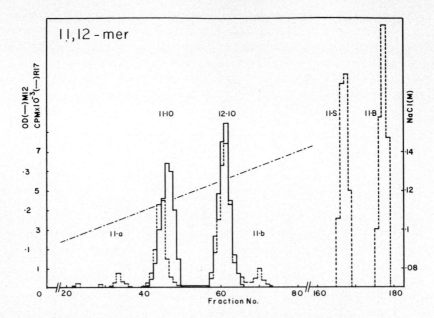

FIG. 4. Rechromatography of the undeca- and dodecanucleotide fractions of a pancreatic ribonuclease digest of R17 and M12 RNA's. Peak 11-S represents radioactivity washed from the column with 1 M NaCl. Peak 11-B represents radioactivity washed from the column with 0.3 N NaOH. The material in these peaks had the same base composition as the average of peaks 11–10 and 12–10. Note that R17 peak 12–10 (the radioactivity peak) superposes M12 peak 12–10 in accord with our finding that their base sequences are identical. R17 peak 11–10 and M12 peak 11–10 do not superpose even though their base compositions are the same. We expect that their base sequences will prove to be different.

TABLE III

7-MER PYRIMIDINE CATALOG OF BMV RNA COMPONENTS

Bases	Intact RNA	Large piece	Small piece
$(A_4G_3)C$	2	1	1
$(A_3G_3)C$	3	2	1
$(A_4G_2)U + (A_2G_4)C$	4	3	1
$(A_3G_3)U$	4	2	2
Total	13	8	5

of infection are better known. So far, the only identified proteins, in addition to coat protein, are the RNA polymerizing enzyme (Haruna and Spiegelman, 1965) and Yamazaki's basic protein (Yamazaki and Kaesberg, 1966). The latter is especially interesting to us because its high lysine content should be reflected in long purine sequences, and thus, in large pyrimidine catalog entries.

At the present time, only fragmentary information regarding the R17 and M12 coat proteins is available. Nevertheless, I will illustrate an obvious type of correlation. In M12 coat protein there exists the sequence, -leu-lys-asp-, in which asp may be either an aspartic acid or an asparagine residue. From a knowledge of the genetic code, only the following pyrimidine catalog entries are possible if asp is asparagine: (1) AAAAAU, (2) AAGAAU, (3) AAAAAC, (4) AAGAAC, (5) AAAAAAU, (6) AAAGAAU, (7) AAAAAAC, (8) AAAGAAC, (9) GAAAAAU, (10) GAAGAAU, (11) GAAAAAC, (12) GAAGAAC.

Of these, Nos. 1, 2, 3, 4, 5, and 7 do not appear in the M12 catalog, and are thus ruled out. There is only one entry whose composition is $(A_5G)U$; its sequence has been determined to be AAAAGAU. Thus Nos. 6 and 9 are eliminated. $(A_5G)C$ appears but once, and it has the sequence AAGAAAC. This eliminates Nos. 8 and 11. $(A_4G_2)C$ appears once and has the sequence AAAGAGC; this eliminates No. 12, $(A_4G_2)U$ appears in the catalog and has not been sequenced. Thus, if the amino acid residue in question is asparagine, the corresponding catalog item *must* have the sequence GAAGAAU. If the residue is aspartic acid, there are 6 unsequenced catalog entries and one known sequence, AAAAGAU, that could correspond.

It seems clear that correlations, such as the one illustrated above, can be extended almost indefinitely once the amino acid sequences of some of the coded proteins are known. Indeed, the easiest way to sequentially order catalog entries (and their larger counterparts produced by limited digestion) may be to locate them by scoring mutants and by examining the corresponding protein sequences.

REFERENCES

Bockstahler, L. E., and Kaesberg, P. (1965). *J. Mol. Biol.* **13,** 127.
Enger, M. D., and Kaesberg, P. (1965). *J. Mol. Biol.* **13,** 260.
Gullekson, G. M., and Kaesberg, P. (1966). Unpublished data.
Haruna, I., and Spiegelman, S. (1965). *Proc. Natl. Acad. Sci. U.S.* **54,** 579.
Kaper, J. M. (1964). *Biochemistry* **3,** 486.
Konigsberg, W., Weber, K., Notani, G., and Zinder, N. (1966). *J. Biol. Chem.* **241,** 2579.
Loeb, T., and Zinder, N. D. (1961). *Proc. Natl. Acad. Sci. U.S.* **47,** 282.
Marvin, D. A., and Hoffmann-Berling, H. (1963). *Z. Naturforsch.* **18b,** 884.
Mitra, S., Enger, M. D., and Kaesberg, P. (1963). *Proc. Natl. Acad. Sci. U.S.* **50,** 68.
Overby, L. R., Barlow, G. H., Doi, R. H., Jacob, M., and Spiegelman, S. (1966). *J. Bacteriol.* **91,** 442.

Rushizky, G. W., Skavenski, I. H., and Sober, H. A. (1965). *J. Biol. Chem.* **240,** 3984.

Sinha, N. K., Fujmura, R. K., and Kaesberg, P. (1965). *J. Mol. Biol.* **11,** 84.

Strauss, J. H., and Sinsheimer, R. L. (1963). *J. Mol. Biol.* **7,** 43.

Stubbs, J. (1965). Ph.D. Thesis, University of Wisconsin.

Tomlinson, R. V., and Tener, G. M. (1963). *Biochemistry* **2,** 697.

Watanabe, I. (1964). *Nippon Rinsho* **22,** 243.

Wittman-Liebold, B. (1965). *Z. Vererbungslehre* **97,** 272.

Yamazaki, H., and Kaesberg, P. (1966). *Proc. Natl. Acad. Sci. U.S.* **56,** 624.

The Interaction of Male-Specific Bacteriophages with F Pili*

Charles C. Brinton, Jr. and Herman Beer†

DEPARTMENT OF MICROBIAL AND MOLECULAR BIOLOGY,
UNIVERSITY OF PITTSBURGH, PITTSBURGH, PENNSYLVANIA

Introduction

The earliest steps in the process of viral infection are the adsorption of the virus particle to a specific cell surface receptor, the release of the viral nucleic acid from its protein coat, and the penetration of the cell by the nucleic acid. The mechanisms by which these necessary steps in viral infection take place have received somewhat less attention than have the mechanisms of intracellular replication and maturation. It is nevertheless important to understand them, since they occur at or near the cell surface, and since, as a consequence, inhibitors of these steps would be less likely to cause cell damage than would inhibitors of the intracellular steps of the infectious process.

In this chapter we will describe some aspects of the adsorption and penetration of the male-specific viruses of gram-negative bacteria. These viruses, first isolated by Loeb (1960), infect only genetic donor cells, although RNA extracted from them will infect spheroplasts of both donor and nondonor cells. The receptors for these viruses are long, thin, bacterial surface appendages, which we have termed F pili. We have proposed that F pili function not only as the adsorption and penetration organelles for both DNA- and RNA-containing, male-specific viruses, but also as conductors of cellular DNA from donor to recipient cells (Brinton *et al.,* 1964; Brinton, 1965, 1966).

The capacity of pili to adsorb male-specific phage was first observed by Crawford and Gesteland (1964) in electron micrographs of mixtures

---

* With electron microscopy by Judith Carnahan.

† Present address: Department of Microbiology, University of Florida School of Medicine, Gainesville, Florida.

of phage and donor cells. Brinton *et al.* (1964), found that only one of the two kinds of pili produced by most donor cells could adsorb RNA phage, and demonstrated their genetic control by genes which are part of the fertility or F factor of gram-negative bacteria.

This virus-host system has several experimental advantages, arising from the fact that the virus-receptor substance is almost entirely external to the cell wall and membrane. The number of receptor sites can, therefore, be visually assayed by electron microscopy, the receptors can be gently isolated and concentrated, and the adsorption process can be studied in the absence of cells. In addition, these receptors can be removed selectively, and virus infection prevented by treatments which do not damage the cells.

Since the properties of pili in general, and F pili in particular, have been the subject of two recent papers (Brinton, 1965, 1966), we will summarize briefly only those aspects which are germane to the problem of male-specific phage adsorption and penetration.

Pili are a widely distributed class of bacterial surface appendages found on nearly all gram-negative bacteria. More than a dozen different kinds have been distinguished on the basis of their diameter, fine structure, serological specificity, virus receptor specificity, and genetic control (Brinton *et al.*, 1954; Brinton, 1959, 1966). The structure of one kind, Type I pili, has been investigated in detail. The Type I pilus is a hollow, rigid, rod, having a diameter of 70 Å, and composed of protein subunits of molecular weight 17,000 assembled in a right-handed helix of pitch distance 24 Å with $3\frac{1}{8}$ subunits per turn of the helix (Brinton, 1965). The axial hole of the Type I pilus is about 20–25 Å in diameter. When Type I piliated bacteria are examined by spreading a bacterial culture with cytochrome C on a water surface, followed by rotary metal shadowing, a macromolecule of about 20–25 Å in diameter can be seen protruding from the end of the Type I pilus. This presumably internal material was first observed by Caro (1966) and has been confirmed for some strains in our own laboratory (Carnahan and Brinton, 1966). The nature of this material is as yet unknown, as is the exact function of Type I pili. We mention this observation because it may be an example of the presence of a macromolecule within a pilus. We shall discuss later the possible presence of DNA within the F pilus.

F pili are rods having a diameter of 85 Å, and an axial hole of about 20–25 Å in diameter (Figs. 1 and 3). They are, in general, longer and more flexible than Type I pili present on the same cell. In electron micrographs of negatively stained preparations, they often exhibit a fine structure similar to that of Type I pili, which suggests that F pili may also be composed of subunits. The sensitivity of F pili to trypsin, and

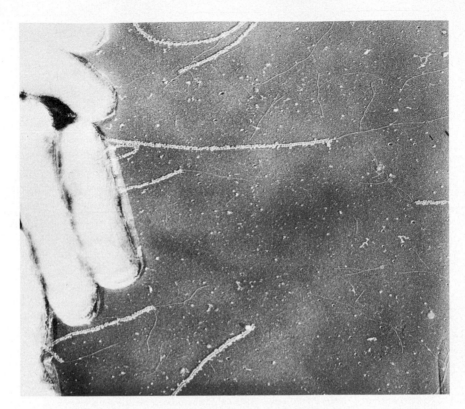

FIG. 1. Metal-shadowed electron micrograph of a mixture of F piliated *E. coli* K12, with spheroidal RNA male phage and rod-shaped DNA male phage added at high multiplicity. The RNA phage adsorbs to the sides of F pili, almost completely covering them. Seven examples of DNA male phage adsorbed to the ends of F pili can be seen. Diameter of F pili = 85 Å.

the UV adsorption spectrum of partially purified F pili preparations indicate that they, like Type I pili, may be composed of protein. Our working hypothesis, the "F pili conduction model," is that F pili are conductors of nucleic acid into or out of the cell across the wall-membrane barrier. Since the demonstration of an F pili requirement for the transfer of cellular nucleic acid renders more plausible the proposal that phage nucleic acid may be carried by the same structure, we will summarize briefly the evidence supporting a role of F pili in the conduction of cellular DNA, as well as of phage RNA and DNA. We should emphasize that the presence of DNA or RNA inside the F pilus, or even directly associated with it, has not yet been observed. However, the indirect evidence and topological arguments currently available provide some convincing evidence in favor of the proposed model.

Evidence and Arguments in Favor of the Hypothesis that Cell-to-Cell Transfer of Chromosomal and Episomal DNA Takes Place via F Pili

1. Both F piliation and fertility are genetically controlled by the F factor. (a) Male strains of bacteria (Hfr, F^+, F') are F piliated and female strains (F^-) are not (Brinton *et al.,* 1964; Crawford and Gesteland, 1964). (b) Genetic elimination of the F factor from F^+ cells by acridine orange treatment eliminates both F piliation and fertility (Brinton *et al.,* 1964). (c) Genetic transfer of the F factor into a wide range of non-F piliated, nonfertile strains confers both fertility and F piliation upon these strains (Brinton *et al.,* 1964). (d) F pili are also present on strains of bacteria transferring both colicinogenic (Brinton, 1966; Caro and Schnös, 1966), and drug resistance factors (Brinton, 1966; Meynell and Datta, 1966). These, like the F factor, are DNA elements, transferable by cell contact.

2. F pili must be phenotypically present for donor cells to be fertile. (a) The phenotypic absence of F pili in Hfr and F^+ strains, produced either by prolonged growth with aeration in complete media, or by aeration in buffer, results in the loss of F piliation and fertility. After inoculation into fresh medium, F piliation and fertility return simultaneously (Brinton, 1965). (b) Mechanical removal of the F pili by an amount of high speed blending which does not affect cell growth or viability results in a loss of fertility, which returns as F pili reappear (Brinton *et al.,* 1964; Brinton, 1965). (c) The fertility of certain F and RTF strains is normally repressed and no F pili are produced. When the repression is relieved such that fertility becomes high F pili are phenotypically expressed (Meynell and Datla, 1966). (d) Both F piliation and fertility of donor cells are repressed and derepressed coordinately by the f_i^+ type of RTF factor (Hirota *et al.,* 1964; Watanabe and Fukasawa, 1962; Watanabe *et al.,* 1962).

3. Light microscopy of mixtures of nonmotile donor and motile recipient cells reveals the presence of pairs of cells connected by a strand of material invisible in the light microscope, the existence of the strand being inferred from the motion of the pairs (Brinton, 1965). Although not in wall-to-wall contact, the pairs remain separated by a distance which never exceeds the length of an F pilus as the motile females move rapidly through the medium. The relative number of these pairs in a mating mixture, and their distance of separation, is consistent with the notion that they are genetic mating pairs connected by F pili.

4. F pili-connected donor-recipient pairs can be seen in electron micrographs of mating mixtures (Brinton, 1965).

5. The presence of an axial hollow region in isolated F pili, the ap-

proximate correspondence of the number of F pili per cell with the number of F factors per cell (when fully expressed), and the ability of donor cells to begin transfer of F factor DNA almost immediately upon contact with the recipient cell, suggests the further possibility that F factor DNA may be present within the axial hollow region of attached F pili. One might therefore suppose that F pili themselves are virus-like particles and that the initiation of cell-to-cell DNA transfer is similar to the initial steps of virus infection.

Evidence in Favor of the Hypothesis That Phage RNA is Transferred via F Pili

1. Male-specific RNA phages infect only cells containing the F factor (Loeb, 1960), which genetically controls F piliation (Brinton, *et al.,* 1964).

2. The inability of RNA phages to infect bacteria lacking the F factor is due to nonattachment of the virus (Crawford and Gesteland, 1964; Dettori *et al.,* 1963; Loeb and Zinder, 1961).

3. The inability of RNA phages to infect bacteria lacking the F factor is not due to their inability to replicate therein, since spheroplasts of non-F piliated female cells have the same susceptibility as do spheroplasts of F piliated male cells to infection by isolated phage RNA (Paranchych, 1963).

4. Male-specific RNA phages can be seen to be adsorbed along the entire length of either free or attached F pili in electron micrographs of F pili mixed with an excess of phage in a medium permitting adsorption (Brinton *et al.,* 1964; Crawford and Gesteland, 1964).

5. The phenotypic presence of visible F pili is necessary for the adsorption of RNA phage to male cells. (a) The capacity of bacterial cells to adsorb RNA phage varies directly with the extent of F piliation during the bacterial growth cycle. (b) Mechanical removal of F pili by blending, centrifugation, and resuspension, under conditions which allow no outgrowth of new F pili, completely removes the ability of F piliated cells to adsorb RNA phage. As F pili reappear, the ability to adsorb RNA phage also reappears (Brinton, 1965). (c) The genetic repression of F piliation effected by the f_i^+ type of RTF factor also represses the adsorption of male-specific phage. Phage-adsorbing ability returns when F pili expression is derepressed (Hirota *et al.,* 1964; Meynell and Datta, 1965; Watanabe and Fukasawa, 1962).

6. The phenotypic presence of visible F pili is also necessary for the infection of male cells. The formation of infected cells does not occur if F pili are absent because of removal by blending (Brinton, 1965),

prolonged aerobic growth, or repression by f_i^+ RTF factors (Hirota *et al.,* 1964; Meynell and Datta, 1965; Watanabe and Fukasawa, 1962).

Evidence in Favor of the Hypothesis That the DNA of Male-Specific Phages is Transferred via F Pili

1. Male-specific DNA phages infect only cells containing the F factor which genetically controls F piliation (Brinton *et al.,* 1964; Hoffmann-Berling *et al.,* 1963; Zinder *et al.,* 1963).
2. The inability of DNA phages to infect bacteria lacking the F factor is due to nonattachment of the virus (Bradley, 1964).
3. Bacteria isolated for their resistance to RNA phages are nearly always resistant to DNA phages and are infertile as well. These cells have lost the F factor and also, therefore, F pili (Bradley, 1964; Raizen, 1966).
4. The number of adsorption sites for the male-specific, DNA phage M13 has been estimated, from the rate and extent of phage adsorption, to be 3 or 4 per cell (Tzagoloff and Pratt, 1964). This number corresponds approximately to the small number of F pili observed per cell (Brinton, 1965), suggesting a single site, or at most a few sites, per F pilus.
5. Male-specific DNA phages can be seen, in electron micrographs, to be adsorbed to the ends of free or attached F pili. This observation, made by Caro and Schnös (1966) has been confirmed in our laboratory (Carnahan and Brinton, 1966) for certain strains (Fig. 1) but not for others (Brinton, 1965).
6. The presence of f_i^+ RTF factors phenotypically represses RNA and DNA male-specific phage sensitivity coordinately (Hirota *et al.,* 1964; Watanabe *et al.,* 1962).

In order to understand the adsorption and penetration mechanisms of male-specific phages, one must first understand the distribution of F pili in a bacterial culture, and the processes of F pili synthesis, assembly, exit and release.

When F pili are removed by mechanical agitation, and the depiliated cells are placed in a complete medium at 37°C, F pili rapidly reappear, reaching a maximum number in about 10 min. Their reappearance is only slightly inhibited by concentrations of chloramphenicol which inhibit new protein synthesis, indicating that F pili, like Type I pili, are produced by the assembly of subunits from a preexisting pool. In all cultures of F piliated bacteria, large numbers of free F pili can be seen. Since the conditions of mechanical agitation commonly used to aerate bacterial cultures are insufficient to detach F pili in large numbers, we must assume that the free F pili are produced mainly by outgrowth and release. The fraction of free F pili observed by electron microscopy, or by assays of the capacity of bacterial cultures to adsorb RNA phage, varies with the strain of bacteria chosen and with the cultural conditions employed. A typical value is about 50%, but the amount of free F pili can be as high as 80–90% of the total F pili present.

The number of attached F pili per cell in fully expressed cultures also varies with the strain of bacteria and with cultural conditions. An average of one is typical, but some strains occasionally have 8 or 10. The fraction of bacteria in fully expressed F piliated cultures which have no visible F pili also varies with the strain and with cultural conditions. Typical values are from 50 to 90%.

The study of the adsorption and penetration of male-specific phages is, therefore, more complicated than that of T phage adsorption because of the peculiar behavior of the receptors for these male phages. For instance, every genetically pure, sensitive culture contains both phenotypically sensitive and phenotypically resistant cells. Resistant cells are continuously and rapidly becoming sensitive due to the appearance of new F pili, and sensitive cells are continuously and rapidly becoming resistant due to the loss of old F pili. A large fraction of the total adsorption sites in the culture are floating free in the medium. In addition, all of these factors vary with the strain of bacteria, cultural conditions, and the presence and activity of F pili repressors.

In order to investigate the interaction of male-specific phages with F pili in greater detail, we have concentrated our present investigation upon two strains of bacteria and one strain of male-specific RNA phage. Using phage whose RNA was radioactively labeled, we have measured phage adsorption to preparations of free F pili, and have examined phage adsorption, and RNA release and penetration in cultures of F piliated bacteria.

Materials and Methods

Bacterial Strains

E. coli K12 Hfr Cavalli W1895 met⁻, having F pili, Type I pili, and flagella. Obtained from L. S. Baron.

E. coli B/r HB11 F lac⁺/lac⁻, having F pili only. Obtained from H. Boyer.

Phage Strains

Male-specific RNA phage R17 isolated by Paranchych and Graham (1962). Obtained from W. Knight.

Male-specific RNA phage M12 isolated by Hofschneider (1963). Obtained from P. H. Hofschneider.

Male-specific DNA phage M13 isolated by Hofschneider (1963). Obtained from P. H. Hofschneider.

Media

Growth and Assay Medium

The "Z" medium of Loeb and Zinder (1961), with 0.002 M CaCl$_2$ added, was used to grow bacterial cultures and to measure adsorption and penetration of phage except where otherwise specified. The medium of Cooper and Zinder (1962) was used for ^{32}P labeling.

Filtration Assay

Adsorption to Bacterial Cultures

The most commonly used method of measuring the adsorption of phage to bacteria is to separate cells from unadsorbed phage by means of centrifugation and to determine the number of plaque forming units (PFU) remaining in the supernatant fluid. For convenience, membrane filters can be used for the same purpose if their pore size is such that phage, but not bacteria, will pass through them. For a filtration assay to be valid, the free phage must not adsorb to the filter material under conditions which permit adsorption of phage to F pili.

We have used a filtration assay developed by Lodish and Zinder (1965). These authors found that the RNA phage f2 would attach to cellulose nitrate filters (Millipore HAWP, 0.45 μ pore size) when suspended in NaCl solutions of 0.10 M or higher concentrations but that they would not attach when suspended in tryptone broth or water. They also found that a fraction of the phage particles present in normal lysates would not attach to the filters when suspended in 0.10 M NaCl, and that these particles had anomalous densities (as determined by cesium chloride density gradient centrifugation). Lodish *et al.* (1965) have also shown that a mutant of phage f2 which does not adsorb to bacteria does not adsorb to the filters in 0.10 M NaCl. It was found that phage populations, homogeneous with respect to density, could be prepared by adsorption of a crude lysate to cellulose nitrate filters in 0.15 M NaCl, followed by the elution of the virus particles therefrom with broth. This method of preparation has the advantages of eliminating nonadsorbing phage particles, and of selecting only those particles which, when suspended in broth, will not attach to the filter, thereby guaranteeing a low background due to the adsorption of free phage to the filter if broth is the assay medium. We have found that Z medium containing 0.002 M Ca^{++} permits the attachment of phage to bacteria, but it also can elute phage from cellulose nitrate filters.

Radioactive R17 phage was prepared by a method similar to that described by Lodish *et al.* (1965). An overnight, aerated culture of strain HB11 F lac$^+$/lac$^-$ in Z medium was diluted 1 : 200 into 100 ml of labeling medium and grown to middle log phase (2×10^8 cells/ml). Ten millicuries of carrier-free P^{32}-phosphoric acid and R17 phage at a PFU/cell multiplicity of 5 were added simultaneously to the culture. Incubation was continued for 2.7 hours, at which time chloroform and lysozyme (100 μg/ml) were added, and the culture was incubated for another 30 min to aid in the release of phage. The crude lysate was clarified by two 40 min centrifugations at 7500 rpm. The clarified, crude lysate, which contained 1.0×10^{12} PFU/ml and 3.9×10^7 cpm/ml, was then diluted with 10 volumes of 0.15 M NaCl and filtered slowly through a 40-mm diameter HA Millipore filter. The filter was eluted for 6 hours with 10 ml of Z medium. The eluate, which was used for adsorption studies, contained 1.0×10^{11} PFU/ml and 2.2×10^6 cpm/ml.

Figure 2 illustrates the results of equilibrium centrifugation (in a cesium chloride gradient) of R17 phage prepared by the method outlined above, and shows the coincidence between PFU and radioactivity. Comparison of these data with those obtained from a comparable centrifugation of a crude lysate showed that the adsorption and elution process—using 10% Z medium-90% saline for adsorption, and 100% Z medium for elution—eliminates nonplaque-forming radioactive material of anomalous density. This is in agreement with the results of Lodish and Zinder (1965) who used saline for adsorption and tryptone broth for elution. Our recovery of phage was considerably lower than that reported by these workers—presumably because of less efficient attachment and/or elution in Z medium, which contains 0.15 M NaCl.

The filtration assay was performed by mixing aliquots of the bacterial culture with phage at zero time in either 1 ml or 5 ml of Z medium and filtering at various times thereafter on 25-mm diameter HA Millipore filters at a rate of about 20 ml/min. The filters were then washed three times with 5 ml volumes of Z medium, dried in an oven, placed in scintillation fluid, and counted in a Packard Automatic Liquid Scintillation Spectrometer. The background radioactivity retained by the filter for phage preparations alone, or for mixtures of phage and non-F piliated bacteria, was of the order of 10^{-4} cpm retained per cpm of phage added. The background was determined in each experiment, and was subtracted from the total counts.

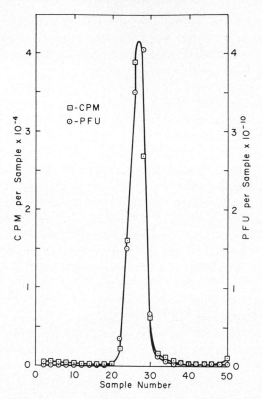

FIG. 2. Cesium chloride density gradient equilibrium centrifugation of R17 RNA phage, purified by adsorption to and elution from cellulose nitrate filters. The ability to form plaques and the RNA radioactivity band together at the buoyant density of the intact phage particles.

ADSORPTION TO BACTERIA-FREE F PILI PREPARATIONS

In some of our earlier experiments (Brinton, 1965), we used the filtration conditions employed by Valentine and Strand (1965) in an attempt to assay the phage adsorbing capacity of preparations of free F pili. The adsorption medium was 0.05 M Tris buffer, pH 7.1, containing 0.0032 M $CaCl_2$. Aliquots of a radioactive phage preparation and an F pili preparation were added to 10 ml of this adsorption medium and the suspension was filtered through Schleicher and Schull B-6 filters. The filters were washed twice with 10 ml volumes of distilled water and then were dried and counted.

Although this form of the filtration assay depends upon RNA phage adsorption to F pili for its response, it is only indirectly an assay for F pili, and is unsatisfactory in several respects. First, the background of radio-activity retained by the filter is very high, and varies with the particular batch of filters being used. Paranchych and Danziger (1966) have shown that Schleicher and Schull B-6 filters have a high affinity for free RNA phage particles in the assay medium of Valentine and Strand, and have thus provided a clear explanation for the high backgrounds characteristic of these assay conditions.

Second, washing the filters with two 10 ml volumes of distilled water, although reducing the very high background level of radioactivity, elutes adsorbed phage from F pili and gives a serious underestimation of the actual number of phage particles adsorbed.

Third, there is no reason to expect that 0.45 μ filters would retain single F pili even when the latter are completely covered with phage, since the diameter of the complex cannot be greater than that of two phages plus one F pilus—which equals about 0.06 μ. We have demonstrated (see later section), that nonaggregated F pili-RNA phage complexes can indeed pass through filters of this pore size. Thus, the assay for adsorption of phage to preparations of free F pili depends upon the presence of clumps of F pili large enough to be retained by the filter. Most F pili preparations contain aggregates, ranging in size from those much larger than bacteria, to single F pili. The filtration assay, therefore, represents an underestimation of the total amount of adsorbed phage, even under ideal conditions of strong phage adsorption to F pili and no phage adsorption to the filter material. One of the dangers of using the filtration assay in the manner outlined by Valentine and Strand (1965) is that treatments of F pili that increase or decrease the amount of radioactivity retained by the filter might be interpreted as affecting the adsorption of phage, whereas they may merely be affecting the aggregation of F pili.

In our more recent measurements of the adsorption of phage to free F pili, we have used the assay conditions described by Lodish and Zinder (with Z medium substituted for tryptone broth), since we have found that these conditions eliminate the first two objections to the assay procedure of Valentine and Strand. The amount of free phage attaching to the filter is low, and washing is carried out with the same medium as that used for adsorption to avoid eluting phage adsorbed to F pili. The third objection to Valentine and Strand's assay is inherent in any filtration or centrifugation assay, but its magnitude can be estimated and a correction made to obtain a more accurate measure of the extent of phage adsorption. An analysis of this and other factors affecting adsorption of phage to preparations of free F pili and to bacterial cultures is made in the next section.

Results

INTERACTION OF PHAGE WITH FREE F PILI

Isolation and Concentration of F Pili

Strain HB11 F lac$^+$/lac$^-$ *E. coli* B/r was selected for the preparation of F pili because it appears, from electron micrographs, to have no flagella, Type I pili, or other appendages to interfere with purification. Thirty liters of bacterial culture were grown at 37°C with aeration in Z medium to a concentration of about 7×10^8 cells/ml, at which time cellular F piliation is maximal in this strain (see next section). The culture was then chilled to stop further growth, and the bacteria were removed by centrifugation in a DeLaval Gyrotester continuous flow centrifuge. Only the clarified supernatant fluid was retained, since 70–90% of the F pili in such a culture are unattached, and blending to remove the attached F pili increases the amount of contaminating "spheroidal" cell wall material (Brinton, 1959). The pH of the clarified supernatant was adjusted to 4.0, and the suspension was allowed to stand in the cold for 24 to 48 hours, by which time a fluffy precipitate formed on the bottom of the vessel. The supernatant liquid was decanted and the precipitate concentrated by low speed centrifugation (7000 rpm for 30 min). The pellets were resuspended in saline, and the pH of the suspension adjusted to 7.0. Material prepared in this way constituted our concentrated F pili preparations. The concentrated preparations were occasionally clarified by low speed centrifugation, but this step usually resulted in a serious loss of F pili due to the removal of large aggregates. A typical, concentrated preparation (Fig. 3) contains many large bundles of F pili rods. F pili aggregates often become extremely large and can be seen easily by phase contrast microscopy. Attempts to disaggregate the F pili clumps without degrading the F pili themselves were unsuccessful. The axial hole characteristic of F pili (Brinton *et al.,* 1964) is particularly prominent in the electron micrograph of Fig. 3.

Kinetics of Phage Adsorption

The time course of adsorption of phage R17 to a preparation of free F pili at three different temperatures is shown in Fig. 4. In each case, 0.05 ml aliquots of F pili and phage suspensions were employed. The fraction of the total radioactivity retained on the filter is plotted as a function of time after the addition of the phage to the F pili. At all temperatures, the amount retained was found to reach a maximum after about 30 min. Three noteworthy features of these curves are: (1) no matter

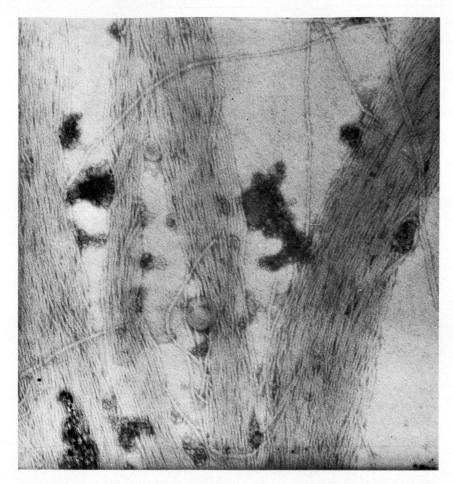

FIG. 3. Electron micrograph of a concentrated free F pili preparation negatively stained with phosphotungstate. The F pili tend to aggregate side-by-side in large bundles. The presence of an axial hollow region is indicated by a central dark line in many of the F pili.

how long the incubation period, only a fraction of the total phage particles is retained by the filter, (2) the maximum fraction retained is characteristic for each temperature, and (3) there is appreciable adsorption of phage at 0°C.

The rate at which adsorption approaches the maximum value is the same for all three temperatures. This is shown in Fig. 5, in which the data from the same experiment are expressed somewhat differently. Here the amount of adsorption at each time interval is plotted as a percent of the amount of adsorption observed after a 30 min incubation period.

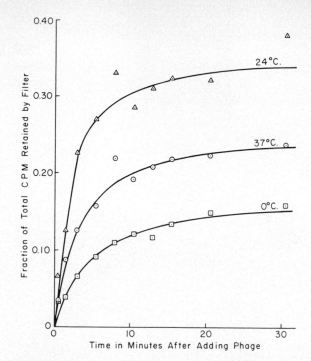

Fig. 4. Kinetics of RNA phage adsorption to a free F pili preparation measured by the filtration assay. The fraction of the total radioactivity retained by the filter is plotted as a function of incubation time for three different temperatures.

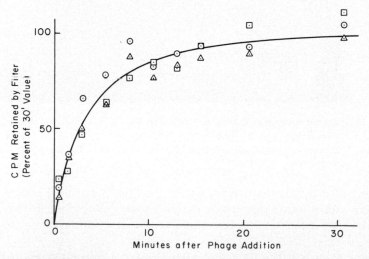

Fig. 5. Kinetics of RNA phage adsorption to a free F pili preparation measured by the filtration assay. The data are the same as in Fig. 4, but radioactivity retained is plotted as percent of the 30 min value for each temperature to show that the maximum value of radioactivity retained is temperature dependent, but the rate at which this value is attained is not.

264

The data fit closely to an exponential function, with 85% and 99.7% of the maximum adsorption being reached at 10 and 30 min, respectively.

Effect of Temperature

The effect of temperature on the fraction of phage adsorbed (retained in the filter) after a 30 min incubation period was determined, using the same preparations of phage and F pili used in the experiment just described. The amount of F pili used (0.10 ml) was twice the amount used in the previous experiment, while the amount of phage used (0.05 ml) was the same. The results are plotted in Fig. 6, along with the 30 min values from the previous experiment. Although there is considerable scatter of the points, both sets may be fitted to empirical curves which differ in position by a constant factor of 0.70, due, undoubtedly to the different concentrations of F pili used in the two experiments.

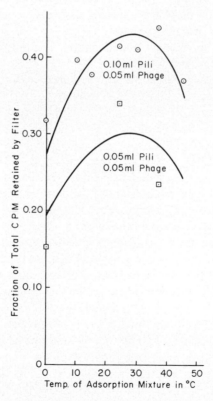

FIG. 6. Effect of temperature on RNA phage adsorption to free F pili measured by the filtration assay. The equilibrium value (30 min of incubation) of radioactivity retained by the filter is plotted as a function of incubation temperature.

The maximum fraction of filter-retained phage shows a broad peak between 20° and 40°C, and the value of 0°C is about 50% of the peak value. Another noteworthy feature of this data is that doubling the amount of F pili in the mixture only increases the amount of adsorbed phage by about 40%.

Effect of Blending of F Pili Suspensions on Binding of Phage

There are several possible explanations for the observation that even after long periods of incubation of a mixture of phage and F pili, not all of the radioactivity contained in the phage is retained by the filter: (a) There is a fraction of phage particles incapable of adsorbing to F pili. (b) The binding forces holding phage to free F pili are weak. The adsorption is, therefore, readily reversible under the assay conditions employed, and an equilibrium is established between free and bound phage. (c) Only very long single F pili or clumps of F pili are retained by the filter. A large fraction of phage particles adsorbed to shorter F pili or to small clumps passes through the filter. (d) All the F pili adsorption sites become saturated with strongly bound phage, and only unadsorbed phage particles pass through the filter.

We shall attempt to demonstrate that a combination of factors (b) and (c) is responsible for the above observation.

An experiment was performed which showed that large amounts of phage adsorbed to shorter F pili pass through 0.45 μ cellulose nitrate filters under our assay conditions. A concentrated F pili preparation was mechanically agitated in a Servall Omni-mixer at a setting of 60 on a 0–100 rheostat for various lengths of time. This intensity of agitation is that used to strip F pili from bacteria without altering the viability or growth rate of the latter (Novotny, 1965). Aliquots of the blended F pili preparations (0.5 ml) were mixed with a constant amount of phage (0.05 ml), incubated for 30 min at 0°C, filtered, washed, and counted. The fraction of counts retained was plotted against blending time, and the results are shown in Fig. 7. After 20 min of blending, all the F pili were small enough to pass through the filter, and no phage particles were retained thereon.

Electron micrographs of the unblended F pili preparation showed many large clumps as well as single F pili. As the amount of blending increased, the clumps decreased in size, and single F pili were broken into shorter lengths. Figure 8 is an electron micrograph of the sample which was blended for 20 min. It shows many short F pili fragments covered with adsorbed RNA phage. These complexes are all capable of passing through the filter, and one must conclude that some of the complexes in the unblended preparation are also capable of so doing. These results provide

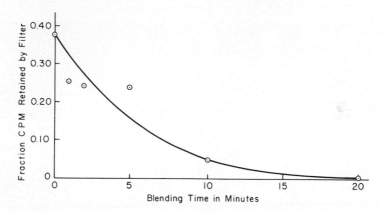

FIG. 7. Effect of the size and state of aggregation of free F pili on the amount of RNA phage radioactivity retained by the filter in the filtration assay. Radioactivity retained by the filter is plotted as a function of time of blending of the free F pili preparation before phage was added. Incubation was for 30 min at 0°C. Blending for 20 min reduces the filtration assay to zero.

one explanation for the observation that there is incomplete retention of radioactivity when ^{32}P-phage-F pili mixtures are filtered.

Equilibrium Studies

An attempt was made to ascertain to what extent each of the four possible factors listed in the previous section were acting to prevent complete filter retention of phage radioactivity. Different amounts of an F pili preparation were mixed with different amounts of a suspension of radioactive phage, the mixtures were incubated for 15 min at 25°C, filtered, washed, and counted. Kinetic studies (see Fig. 5) had shown that, after a 15 min incubation period, 92% as many counts are retained on the filter as are retained after a 30 min incubation period. The results, which are plotted in Fig. 9, may then be considered to approximate closely the equilibrium values.

The lowest curve in Fig. 9 shows that with the smallest amount of phage used (0.01 ml = 304 cpm), the addition of increasing amounts of F pili never allows more than 43% of the total radioactivity to be retained by the filter. Since, in an equilibrium reaction, or in the case of very strong binding, the addition of increasing amounts of F pili should eventually adsorb all the phage, we are left with factors (a) and (c) as possible explanations of the unadsorbed radioactivity.

Electron micrographs of phage-F pili mixtures show a number of phage particles attached to single F pili and to aggregates small enough to pass through the filter. The fraction of phages so attached is of the same order

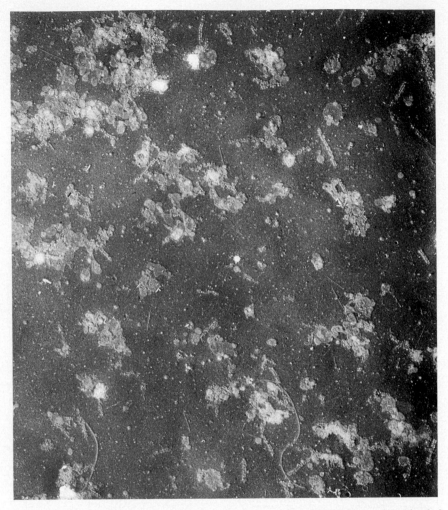

Fig. 8. Metal shadowed electron micrograph of a free F pili preparation blended for 20 min, and mixed with a high multiplicity of RNA phage. Many short pieces of F pili covered with RNA phage can be seen.

of magnitude as the fraction of radioactivity not trapped by the filter when an excess of F pili is present. It seems clear, therefore, that the fact that the total phage radioactivity is not retained by the filter in the presence of an excess of F pili, can be adequately explained on the basis of possibility c.

If it is assumed that the state of aggregation of an F pili preparation is not affected by the presence of phage, the fraction of phage retained by the filter in the experiment illustrated in Fig. 9 will always be 43% of the amount of phage adsorbed to F pili. Thus, if the maximum fraction of phage retained by the filter in the presence of an excess of F pili is known, or can be estimated, this fraction is equivalent to 100% phage adsorption and, therefore, all other values of the retained fraction at all other phage and F pili concentrations can be corrected to give the adsorbed fraction by dividing by this value (0.43 in the case of the data shown in Fig. 9).

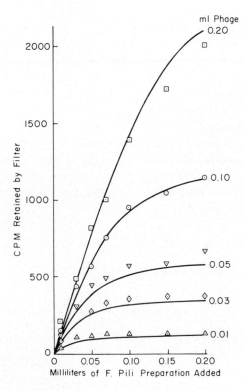

FIG. 9. Effect of F pili concentration and RNA phage concentration on phage adsorption to free F pili. Incubation was for 15 min at 25°C. Curves are theoretical, and are computed from equation (7).

By applying this correction to the data shown in Fig. 9, we were able to obtain reasonably good agreement between simple equilibrium theory and the entire set of data. The familiar equilibrium equation was used to describe the concentration ratios of the three components in a reversible phage adsorption reaction. It may be written

$$\phi + P \rightleftarrows \phi P \tag{1}$$

where $\phi = $ phage particles, $P = $ pili adsorption sites, $\phi P = $ phage — site complexes. The equilibrium equation is

$$\frac{[\phi P]}{[\phi] \, [P]} = K \tag{2}$$

where $K = $ equilibrium constant. Also

$$\frac{[\phi P]}{[\phi]} = K[P] = \frac{\text{Bound phage}}{\text{Free phage}} = \frac{f}{1 - f} \tag{3}$$

where $f = $ fraction of bound phage.

Since neither an accurate count of the number of phage particles nor an accurate estimate of the amount of F pili in our preparations was available, the phage concentration was expressed in counts per minute per milliliter, and the F pili concentration in phage site equivalents is also expressed in counts per minute per milliliter.

Let $V = $ phage/ml in cpm/ml, $S = $ sites/ml in cpm/ml, $V_\phi = $ volume of phage suspension added to reaction mixture, $V_P = $ volume of pili preparation added to reaction mixture, $V_R = $ volume of reaction mixture. Then, total phage $ = VV_\phi$, total pili $ = SV_P$, total phage concentration $ = (VV_\phi/V_R)$, total pili concentration $ = SV_P/V_R$ and, free phage/ml $ = $ total phage/ml $ - $ bound phage/ml or,

$$[\phi] = \frac{VV_\phi}{V_R} - f \frac{VV_\phi}{V_R} = (1 - f) \frac{VV_\phi}{V_R} \tag{4}$$

Also, free sites/ml $ = $ total sites/ml $ - $ bound sites/ml or,

$$[P] = \frac{SV_P}{V_R} - f \frac{VV_\phi}{V_R} \tag{5}$$

(since bound sites are equal to bound phage). Substituting this value of $[P]$ in Eq. 3 one obtains

$$\frac{f}{1 - f} = \frac{K}{V_R} (SV_P - fVV_\phi). \tag{6}$$

Solving this quadratic equation, one obtains

$$f = \frac{\left(\dfrac{V_R}{K} + VV_\phi + SV_P \right) - \left[\left(\dfrac{V_R}{K} + VV_\phi + SV_P \right)^2 - 4VV_\phi SV_P \right]^{1/2}}{2VV_\phi} \tag{7}$$

For the experiment the results of which are plotted in Fig. 9, $V = 30,400$ cpm/ml and $V_R = 5$ ml. The curves drawn in Fig. 9 are computed from Eq. 7 using $S = 44,100$ cpm/ml, $K = 5.55 \times 10^{-3}$ ml/cpm, the fraction of radioactivity retained by the filter being 0.43. The reasonable fit of

Eq. 7 to our data using only three arbitrary constants is evidence in favor of an equilibrium between free phage and bound phage.

An estimate of the equilibrium constant, expressed as milliliters per PFU, can be obtained, since the number of PFU per cpm for the phage preparation used here was 1×10^5 PFU/cpm. Thus:

$$K = \frac{5.5 \times 10^{-3} \text{ ml/cpm}}{1.0 \times 10^5 \text{ PFU/cpm}} = 5.5 \times 10^{-8} \text{ ml/PFU}$$

Since the number of phage particles is usually about five to twenty times the number of PFU's for typical phage preparations, the true value of the constant, K, would be about 4×10^{-9} to 1×10^{-8}.

Equation 7 provides a means of using the filtration assay to measure, in an unambiguous way, the number of phage adsorption sites contained in an F pili preparation. The assay is carried out at several concentrations of phage or pili, and Eq. 7 is fitted to the data using the known value of K for the given assay conditions, and adjusting for the best fit the constants S (sites/ml) and the fraction of pili retained. "Adsorption sites" in this sense means the space on an F pilus occupied by one phage particle, since we have assumed in the definition that one phage blocks one site. Since one phage might block more than one site, the term "phage equivalent site" might be more appropriate.

We consider as unlikely the possibility that the fraction of phage passing through the filter is due to a nonadsorbing fraction of particles. Equilibrium centrifugation of our radioactive phage preparation in a cesium chloride gradient shows a nearly perfect coincidence between PFU's and radioactivity. Therefore, any hypothetical nonadsorbing particles would have the same bouyant density as do the plaque forming particles. Lodish *et al.* (1965) have shown, by means of competition experiments that particles which adsorb to filters also adsorb to bacteria and that particles which do not adsorb to filters do not adsorb to bacteria. By selecting, for use in our filtration assays, only those phage particles which will adsorb and elute from filters, we have effectively eliminated nonadsorbing particles.

Effect of Enzymes and Organic Solvents on Free F Pili

Attempts were made to break up aggregates of F pili by solvent and enzyme treatment. Incubation with Tris buffer-EDTA-lysozyme, which is capable of lysing the host bacteria, was found to have no effect on aggregates or single F pili as judged by electron microscopy. Trypsin, which does not affect Type I pili, was found to digest F pili. During trypsin digestion, an intermediate form, consisting of much shorter and thicker rods, appears. The final product(s) of trypsinization is not yet known. An electron micrograph of a trypsinized F pili preparation, with RNA phage added, is shown in Fig. 10.

F pili are solubilized by organic solvents. The concentration of organic solvent needed depends upon the relative "hydrophobicity" of the solvent. Approximately 50% ethylene glycol, 50% ethanol, water 25% saturated with benzene, or water 25% saturated with chloroform (at room temperature) are sufficient to completely disrupt F pili as judged by electron microscopy. This solubility could mean either that F pili are lipoproteins and therefore related to the cell membrane, or that they are composed of a high proportion of amino acids with hydrocarbon side chains.

Treatment of F pili with chloroform renders them unable to either interact with phage or to be retained by the filter. Filtration assays of a

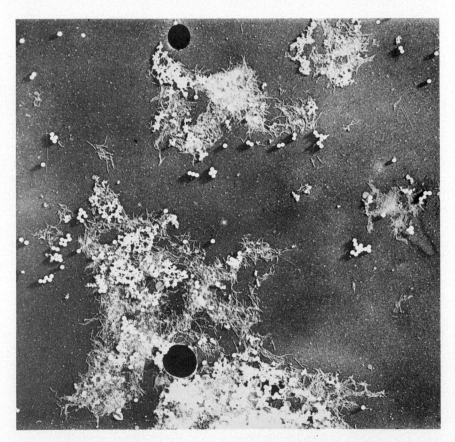

FIG. 10. Effect of trypsin on free F pili. Metal shadowed electron micrograph of a trypsinized preparation of free F pili with RNA phage and polystyrene latex particles added. Shorter fragments of F pili remain which do not adsorb the phage.

chloroformed and unchloroformed preparation are shown in Fig. 11. Since, in electron micrographs of chloroform treated suspensions, one sees no particles large enough to be retained by the filter, it is impossible to state whether or not the fragments produced by disruption of the F pili can interact with the phage.

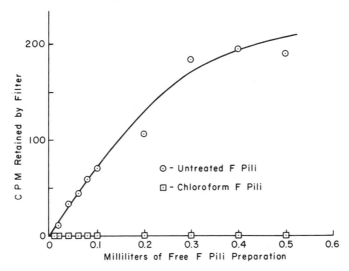

FIG. 11. Effect of chloroform on the retention of phage radioactivity by filters in mixtures of free F pili and phage. The F pili preparation was pretreated with chloroform saturated water at room temperature for 15 minutes. After removal of chloroform, the F pili were mixed with phage, incubated, and filtered.

INTERACTION WITH BACTERIAL CULTURES

The study of the interaction of RNA phage with bacteria involves at least three initial steps: adsorption to the F pilus, release of RNA (resulting in loss of phage viability), and penetration of RNA into the cell. One might also suppose that F pili attached to bacteria behave differently in these processes than do free F pili. In fact, if the F pilus itself is the site of the release and penetration of RNA, the attached F pilus must be different, since adsorption of phage to free F pili does not lead to a loss of plaque forming units.

An important question to ask is whether or not the process of adsorption to F pili attached to bacteria is identical to the process of adsorption to free F pili. This question can be answered for at least one temperature, since adsorption can take place at $0°C$, but penetration cannot. A bacterial culture should adsorb phage at $0°C$ in basically the same way as does a preparation of free concentrated F pili, except that the fraction of phage retained (as measured by the filtration assay) will depend upon what fraction of F pili are attached to the bacteria, and what fraction of free F pili in the culture are in large enough aggregates to be retained by the filter.

Kinetics of Adsorption of Phage R17 to Bacterial Cultures at 0°C

The time course of adsorption of RNA phage to a culture of HB11 F lac$^+$/lac$^-$ grown with aeration in Z medium to a concentration of 7×10^8 cells/ml is plotted in Fig. 12. The phage concentration was

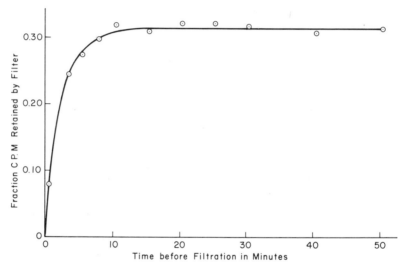

FIG. 12. Kinetics of RNA phage adsorption to a bacterial culture measured by the filtration assay. Cell concentration was 7×10^8 per ml, phage concentration was 7×10^7 PFU per ml, and incubation was at 0°C. Bacteria were *E. coli* B/r HB11 F-lac$^+$/lac$^-$ grown to late log phase in 'Z' medium.

7×10^7 PFU/ml and adsorption was at 0°C. The maximum fraction of phage retained by the filter was 0.31, and the rate at which absorption approached the maximum value was about twice that observed in the case of adsorption to the concentrated preparation of free F pili. Eighty-five percent of the maximum value was reached in 4.5 and 10 min with the bacterial culture and suspension of free F pili, respectively.

Dependence of Adsorption to Bacteria at 0°C on Presence of F Pili

We have shown previously that F pili do not reappear at 0°C after their removal by blending (Brinton *et al.,* 1964), that bacteria depiliated by blending, centrifugation, and resuspension at 0°C do not adsorb phage (Brinton *et al.,* 1964), and that the ability to adsorb phage returns as

F pili regenerate. We have repeated these experiments using the new assay conditions, and have confirmed our previous findings. An aerated culture of Hfr C W1895 in Z medium, containing 7×10^8 cells/ml, was diluted to 5×10^7 cells/ml and chilled to 0°C. Seventy milliliters of the culture were blended for 2 min at 0°C in an Omni-mixer at a rheostat setting of 60. Two aliquots of the blended culture were centrifuged in the cold to sediment the bacteria. One of the pellets was resuspended in cold Z medium to measure the effects of blending on adsorption, and the other was resuspended in the same volume of Z medium at 37°C to measure the effects of F pili regeneration on adsorption.

The results, plotted in Fig. 13, demonstrated that one blending and washing cycle reduced the amount of phage retained by the filter to 0.00053 of that retained by the unblended culture. Regeneration of F pili at 37°C (observable in the electron microscope) allows phage adsorption to return to 0.30 of the control value (unblended cells) after 10 min.

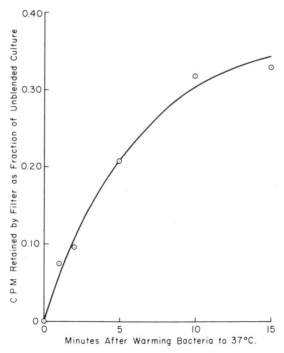

FIG. 13. Effect of F pili removal and reappearance on the adsorption of RNA phage to bacteria as measured by the filtration assay. F pili were removed by one cycle of blending, centrifugation, and resuspension at 0°C. Incubation with phage was for 30 min at 0°C. Regeneration was studied by resuspension at 37°C. At various times regeneration was stopped by chilling to 0°C.

Dependence of Phage Adsorption to Bacteria at 0°C on Bacterial Strain and Cultural Conditions

The amount of F piliation per cell varies widely with the strain of bacteria and the cultural conditions (Brinton *et al.,* 1964). Overnight, aerated cultures of some Hfr and F+ strains become infertile (F− "phenocopies"), and they also lose F pili (Brinton *et al.,* 1964).

We have used the filtration assay to compare the phage-adsorbing ability of HB11 F lac+/lac− and Hfr C at different phases of the culture cycle. Growth of bacteria was in Z medium at 37°C with vigorous aeration. Aliquots of the cultures were chilled to 0°C, and adsorption of

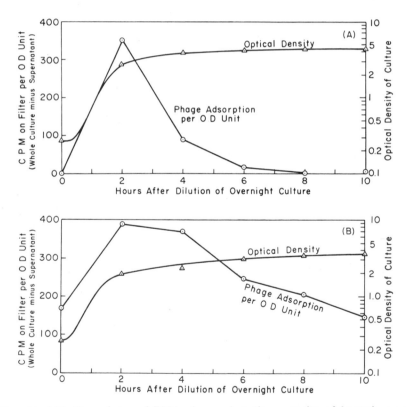

Fig. 14. (A) Dependence of RNA phage adsorption capacity of bacteria on the phase of the culture cycle for *E. coli* K12 Hfr Cavalli. Growth was in vigorously aerated Z medium. Phage adsorption capacity was proportional to F piliation assayed by electron microscopy.

(B) Dependence of RNA phage adsorption capacity of bacteria on the phase of the culture cycle for *E. coli* B/r HB11 F lac+/lac−. Growth was in vigorously aerated Z medium. Phage adsorption capacity was proportional to F piliation assayed by electron microscopy.

phage (30 min incubation period) was measured. Other aliquots were centrifuged and the capacity of the supernatant fluids to adsorb phage was assayed. The value obtained with the supernatant fluid was subtracted from the value obtained with the whole culture to obtain a value more closely representing the adsorption capacity of the bacteria. This value will be in error by the amount of adsorption to F pili clumps that are sedimentable with bacteria. However, the number of these clumps in a culture is small compared to the number in a concentrated preparation of free F pili.

The results of these experiments are plotted in Figs. 14a and 14b. In both strains, phage adsorbing capacity is maximal at the beginning of the stationary phase, when F piliation is maximal. Cells of strain Hfr C subsequently lose all F pili and phage adsorbing capacity in 6 hours, whereas the ability of HB11 F lac$^+$/lac$^-$ cells to adsorb phage only declines to about 50% of the maximum value in 8 hours and remains at this level overnight. These results indicate that in different bacterial strains different systems may exist for controlling the phenotypic expression of F piliation and, therefore, of fertility and phage adsorption. In all adsorption experiments described in other sections of this manuscript, bacterial cultures at the beginning of the stationary phase (about 7×10^8 cells/ml) were used in order to maximize F piliation.

Phage Adsorption Equilibrium Studies at 0°C with Bacterial Cultures

In order to test the validity of the equilibrium Eq. 7 and the filtration assay procedure for bacterial cultures, the amount of ^{32}P-labeled R17 phage retained by the filter was studied as a function of cell concentration at constant phage concentration and also as a function of phage concentration at constant cell concentration. The results are plotted in Figs. 15A and 15B. Adsorption was for 30 min at 0°C, using aliquots of a culture of HB11 F lac$^+$/lac$^-$ grown in aerated Z medium to a concentration of 7×10^8 cells/ml. The reaction volume was 1 ml.

The curve drawn in Fig. 15A is plotted from Eq. 7 using $V_R = 1$ ml, $K = 5 \times 10^{-8}$ ml/PFU, VV_ϕ (total PFU) $= 1.0 \times 10^9$, SV_P (total PFU-equivalent pili sites) $= 38$ sites/cell $\times$ cells/ml, and the fraction of pili retained by the filter $= 0.35$. The number of sites per cell, and the value of the equilibrium constant K, were chosen to fit the data. The value of $K = 5 \times 10^{-8}$ ml/PFU is approximately equal to the value of $K = 5.5 \times 10^{-8}$ ml/PFU, calculated from the equilibrium data obtained from adsorption studies with free F pili. The number of sites per cell is expressed as PFU equivalents. If we assume, that the number of plaque forming units is 5–20% of the number of adsorbing phage particles, (and this is usually the case), the number of adsorption sites per cell would

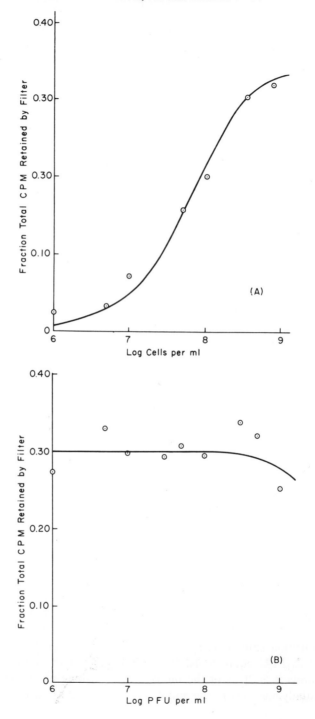

(A)

(B)

be about 200 to 800, a value in reasonable agreement with our electron microscope observations. It must be kept in mind that the number of sites per cell includes both free and attached F pili.

The curve drawn in Fig. 15B is also plotted from Eq. 7 using the same values for all parameters except that for the fraction of F pili retained by the filter, which was 0.30 in this experiment. These data fit well to the theory, although they do not cover the entire range of the dissociation curve.

Thus, knowledge of the parameters of phage adsorption equilibrium permits the extension of the filtration assay to bacterial cultures. The total number of phage adsorption sites is determined in the same way as for a preparation of free F pili if adsorption is carried out at 0°C—at which temperature both phage penetration and gain or loss of F pili by outgrowth are prevented.

The true average multiplicity of infection, which equals the average number of PFU adsorbed per cell, could also be calculated if the fraction of phage adsorbed to attached F pili were known. Phage adsorbed to attached F pili might be estimated by centrifugation of the culture to remove bacteria, and by carrying out a filtration assay on the supernatant. This value would then be subtracted from that obtained from a filtration assay using the whole culture. This is an unsatisfactory method of determining the fraction of free F pili in a culture, since clumps of F pili large enough to be retained by the filter are also removable by centrifugation. Electron microscopy can provide a reasonably reliable estimate of the fraction of attached F pili, and a careful count of well separated Hfr C cells, grown with aeration in Z medium to a concentration of 7×10^8 cells/ml, showed that only 10–20% of the total F pili were attached to cells. Since f in Eq. 7 is the fraction of phage adsorbed to both free and attached F pili, the total amount of phage adsorbed to attached F pili equals $f \times$ the fraction F pili attached to cells $\times$ total PFU

FIG. 15. (A) Equilibrium dissociation curve for adsorption of RNA phage to an F piliated bacterial culture. Fraction of phage radioactivity retained by the filter after 30 min incubation at 0°C is plotted as a function of cell concentration. The curve is theoretical and is calculated from equation (7) using $K = 5 \times 10^{-8}$ ml per PFU, average number of adsorption sites per cell $= 38$, and fraction of radioactivity trapped by the filter $= 0.35$.

(B) A portion of the equilibrium dissociation curve for adsorption of RNA phage to an F piliated bacterial culture. Fraction of phage radioactivity retained by the filter after 30 min incubation at 0°C is plotted as a function of phage concentration. The curve is theoretical and is calculated from equation (7) using the same parameters as in Fig. 15A. Bacterial concentration was 7×10^8 cells/ml.

added. Dividing this value by total cells gives the true average multiplicity of infection. Using 15% as the fraction of F pili attached to cells in HB11 F lac$^+$/lac$^-$ cultures, the true multiplicity of infection at 1×10^8 cells/ml and 1×10^9 PFU/ml $= 0.62 \times 0.15 \times 10^9/10^8 = 0.93$. Thus, because of the equilibrium between free and adsorbed phage, and the relatively large fraction of free phage receptors, the true average multiplicity of infection in the above example is about one when the phage-to-cell ratio is ten.

The Kinetics of Penetration of Phage RNA, and the Effect of Multiplicity of Infection on Penetration

A method of estimating the true multiplicity of infection makes it possible to design experiments to answer the question: "Do all the phage particles adsorbed to a single F pilus, and that are potentially capable of infecting the cell, do so, or, owing to some type of exclusion mechanism, does only one or at most a few phage genomes enter the cell?"

In an attempt to answer this and similar questions, we have measured the penetration of phage RNA into bacteria as a function of time and multiplicity of infection. *E. coli* HB11 F lac$^+$/lac$^-$ was grown in aerated Z medium to a concentration of 7×10^8 cells/ml. Phage was added at zero time at the desired PFU to cell ratio, and the mixture was aerated at 37°C. Samples were removed at various times and diluted rapidly into 50 ml of cold Z medium to stop further penetration. The diluted samples were then blended for 2 min in a Servall Omni-mixer maintained in an ice-water bath to remove F pili and unpenetrated phage. The depiliated bacteria were centrifuged in the cold at 7500 rpm for 30 min, and resuspended in 5 ml of cold Z medium. The suspension was then filtered and the filters were washed three times with cold Z medium and counted.

This experiment is a direct measure of penetration, since chilling to 0°C completely stops further penetration, and the subsequent blending, centrifugation, and resuspension removes unadsorbed phage in addition to adsorbed but unpenetrated phage, and phage adsorbed to free F pili. The fractions of total phage radioactivity that have penetrated after various periods of incubation are plotted in Fig. 16 for five different PFU to cell ratios. The values shown in this figure represent the sum effect of three processes—adsorption, release of viral RNA, and penetration of cells by the RNA.

The first important feature of the curves in Fig. 16 is that RNA penetration is reasonably efficient at low PFU to cell ratios, but it is very inefficient at high ones. The second feature to note is that the overall process appears to take place in two phases: a phase of rapid phage adsorption, release of RNA, and penetration of the viral genome into the cell, followed by a second phase in which there is a slower linear increase in the amount of RNA which has entered the cells.

Since the mixture is incubated at 37°C, the growth of bacteria with a concomitant loss of existing attached F pili and the production of new attached F pili must be expected to occur. Our interpretation of the curves is that the first phase represents adsorption of phage to F pili which are attached to bacteria at time zero, followed by the release and penetration of some of the RNA, and that the second phase represents the adsorption, RNA release, and penetration of phage RNA into new F pili formed by outgrowth during the incubation period. Since the initial adsorption, release, and penetration processes are complete in about 10 min—which is about the time needed for completion of adsorption alone —we conclude that the release and penetration steps occur immediately or within one or two min after adsorption of viable phage to F pili.

For the purpose of determining the effect of multiplicity of infection on the penetration of cells by phage RNA, we have assumed that the penetration that occurs during the rapid initial phase reflects the initial multiplicity conditions. During the second phase the multiplicity will be changing continuously and in a complex manner. The amount of RNA injection which occurs via F pili existing at zero time has been estimated by extrapolating the linear portions of the curves to zero time.

The term multiplicity must be redefined when only RNA penetration is measured, since only a fraction of the RNA molecules may be able to form plaques. This multiplicity is equal to the average number of phages capable of injecting their RNA that are adsorbed to competent F pili, divided by the number of competent F pili. By competent F pili we mean those capable of bringing about the release and penetration of the RNA of adsorbed phage. We do not assume a priori that both competent and noncompetent attached F pili exist. It is possible that all attached F pili are competent. Similarly, all phage particles may be capable of injecting their RNA, although we know that not all phages can form plaques.

An inspection of the adsorption-release-penetration curves at the low PFU/cell ratios shows that, at 30 min, 16% of the total added phage radioactivity has penetrated the cells, and that the fraction is still increasing. Thus, at least 16% of the phage population is capable of injecting its RNA. We have estimated from electron micrographs of individual cells, that, in cultures of HB11 F lac$^+$/lac$^-$, about 10–15% of the F pili are attached, while 85 to 90% are free. From Eq. 7 one can calculate that at the cell and phage concentrations existing at all PFU to cell ratios except 100, almost 100% of the phage particles are adsorbed to either free, or attached F pili. Since, at the lowest multiplicity, 13% of the total

phage radioactivity (or nearly 100% of the radioactivity of phage adsorbed to attached F pili) has penetrated at 10 min, the simplest assumptions consistent with our data is that all phage particles are capable of
RNA-injection, all attached F pili are competent, and every adsorption
at low multiplicity leads to a successful penetration.

These assumptions would permit us to calculate the penetration multiplicity, if we knew the total number of F pili and the total number of
phage particles. Since the phage particles in the particular preparation
used in this experiment were not counted in the electron microscope or
estimated chemically, we can only guess at the particle count from the
past experience of ourselves and others. That is, about 5 to 20% of the
particles are capable of forming plaques. Assuming a value of 10%, the
total number of phage particles is 10 times the number of PFU's. The
total number of individual F pili estimated from electron micrographs
of HB11 F lac$^+$/lac$^-$ cultures is approximately one per cell. Therefore,
the true multiplicity of penetration is equal to approximately ten times
the PFU to cell ratio.

Inspection of the curves in Fig. 16 shows clearly that penetration is a
function of multiplicity. Since adsorption to F pili was the same (almost
100%) for all PFU to cell ratios except 100 (where adsorption was
35% because of the lower cell density), the reduction in the amount of
RNA which penetrates the cells at higher multiplicities must be due to
the fact that the RNA of some of the adsorbed phage is either not
released or does not penetrate. Since the particle multiplicity is known
approximately, we can test the hypothesis that the RNA of one and only
one phage particle per F pilus can penetrate the cell. This statement
can be expressed mathematically. Let the average multiplicity $= m$, and
number of F pili $= N$. Then the number of injected viral genomes will
be equal to the number of F pili that have adsorbed one or more phage.

By the Poisson formula this will be $N(1 - e^{-m})$. Since the total number of phage particles is mN, the fraction, F, of adsorbed phage that
have injected RNA will be

$$F = \frac{(1 - e^{-m})\,N}{mN} = \frac{1 - e^{-m}}{m}\,.$$

The theoretical values are compared with the experimental values for
various multiplicities in Table I. Particle to F pilus multiplicities were
estimated as described above. The experimental values for the fraction
of adsorbed phage that have injected their RNA are estimated by assuming that 15% of the F pili are attached, and that, therefore, 0.15 of the
total radioactivity is adsorbed to attached F pili. The amount of radioactivity which has penetrated the cells via these F pili is determined by
extrapolation as described above and is shown in Fig. 16. The fraction

penetrated is obtained by dividing the extrapolated value by 0.15. The agreement between experiment and theory is reasonable, considering the approximations involved, and we conclude that RNA from only one or a few virus particles adsorbed to the same F pilus succeeds in pene-

TABLE I

PENETRATION OF RNA OF ADSORBED PHAGE AS A FUNCTION OF MULTIPLICITY

PFU/cell ratio	Particle/F pilus ratio $= m^a$	$F = \dfrac{1 - e^{-m}}{m}$	Experimental value of F^b
0.015	0.15	0.93	0.77
0.10	1.0	0.57	0.65
1.0	10	0.10	0.25
10	100	0.010	0.0060
100	350^c	0.0029	0.0013

^{a} Based on the assumption that all phage particles are capable of injecting RNA, that 100% of the phage particles are adsorbed to F pili, and that the fraction of particles that form plaques is 0.10.

^{b} Based on the assumption that 15% of F pili are attached to cells.

^{c} Same assumptions as in footnote ^{a} except that only 35% of the phage particles are adsorbed to F pili because of a lower cell concentration (7×10^7 cells/ml instead of 7×10^8).

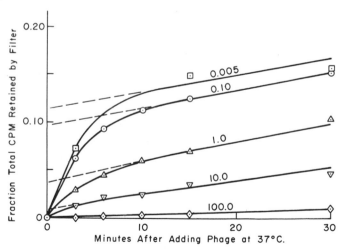

FIG. 16. Kinetics of phage RNA penetration into bacteria as a function of multiplicity of infection. Multiplicity shown is the PFU to cell ratio. Cell concentrations were 7×10^8 except for multiplicity $= 100$ where the cell concentration was 7×10^7. Phage were added at zero time to aerated bacterial cultures of *E. coli* B/r HB11 F lac$^+$/lac$^-$ growing in Z medium. Penetration was stopped by chilling to $0°C$ at various times and free phage, phage adsorbed to free F pili, and phage adsorbed to attached F pili whose RNA had not penetrated were removed by blending, centrifugation, and resuspension at $0°C$.

trating the cell, even though all phages are capable of adsorption and RNA injection, and all attached F pili are competent. If our conclusions are correct, the existence of an exclusion mechanism, which allows only one or a few RNA phages to inject RNA via the same F pilus may be postulated.

Discussion

It is useful at this point to examine critically the possibility that F pili may *not* be the organelles of male-specific phage adsorption, release, and penetration leading to productive infection, but merely structures that for some trivial reason adsorb male-specific phage. The structure responsible for the adsorption of male-specific phage has been shown experimentally to have the following properties:

1. Controlled by genes associated with F factor DNA.
2. Activity repressed by genes repressing F factor functions.
3. Completely removed by blending.
4. Produced by the cell under the same physiological conditions under which F pili are produced.
5. Total amount varies coordinately with the amount of visible F pili.

We believe the existence of a structure other than F pili, but having these properties in common with F pili, to be highly unlikely. Although serious attempts have been made in several laboratories to detect a chemical difference between the cell wall material of donor and nondonor bacteria, no consistent differences, clearly linked to the presence of F factor, have been found. However, since phage resistance or sensitivity can depend upon the presence or absence of a single kind of molecule in the very complex cell wall structure, this negative result is not particularly significant. In any case, critics of the F pili hypothesis are charged with the responsibility of finding another structure having all the properties listed above.

Valentine and Strand (1965) have questioned the hypothesis that F pili have a role in phage infection, their objection being based on their observations that interaction of phage with preparations of free F pili neither causes a loss of plaque forming units nor renders the F pili-phage complex sensitive to RNase. On the other hand, interaction of phage with bacterial cultures renders the phage RNA sensitive to RNase (Zinder, 1963). These observations do not exclude a role of F pili in release and penetration, but they do require the additional assumption that free F pili are somehow different from attached F pili.

The nature of this difference is as yet unknown. We have suggested previously (Brinton, 1965) that attached F pili might contain DNA, which is lost when the F pilus detaches itself by outgrowth or is removed from the cell by blending. An obvious difference between free and attached F pili is that attached F pili are connected to the bacterial cell wall and membrane system. Release of and penetration of cells by phage RNA might require a cooperative reaction between the phage, the F pilus, and the cell wall. If this were true, the F pilus alone would be unable to cause release of phage RNA. The phage particle might arrive at the base of the F pilus by moving along the F pilus rod, and the weak binding of RNA phage to the F pilus could well permit such a process. At whatever site release and penetration occur, all phage RNA becomes accessible to hydrolysis by RNase (Zinder, 1963) or to binding by streptomycin (Brock, 1962; Brock and Wooley, 1963), since both these agents can completely prevent phage infection.

Valentine and Strand (1965) have claimed that divalent cations are required for attachment of f2 phage to free F pili, although Loeb and Zinder (1961) had reported earlier that calcium ions were not required for attachment, but only for a productive infection. Paranchych (1966) has made a careful study of the ionic requirements of the adsorption, release, and penetration steps of R17 infection, and he found that Mg^{++}, Ca^{++}, Sr^{++}, or Ba^{++} are required only for the penetration step of infection. Monovalent ions were shown to suffice for the attachment and eclipse (release) steps of the infectious process. Paranchych therefore questioned the premise that F pili play a role in phage infection on the basis of Valentine and Strand's claim that divalent cations are required for phage attachment to free F pili. In a later study, Paranchych and Danziger (1966) resolved this paradox by demonstrating that monovalent ions would suffice for the attachment of R17 to free F pili.

Raizen (1966) has recently completed a study of the relationships between the attachment sites of donor cells for the RNA and DNA male-specific phages, M12 and M13, and for recipient bacteria. Her studies showed that no rigorous distinction could be made between the three attachment sites by any of the following treatments: blending, heating, change in pH, periodate inactivation, F-phenocopy formation, periodate-soluble conjugation inhibitor treatment, M12 preattachment, M13 preattachment, and female cell preattachment. There were, however, quantitative differences in the reactions of the sites to these treatments. She concluded that all three sites are on the F pilus.

If the F pilus, like the Type I pilus, is composed of identical protein subunits assembled in a helical form with a small number of subunits per turn of the helix, each protein subunit would have at least six different active sites. Naming the faces of the subunit as distal (to the cell), lateral (to the F pilus), and proximal (to the cell), the distal face would contain the site for DNA male phage adsorption, recipient cell adsorption, and polymerization to the proximal face of the subunit next to it. The lateral face would contain the site for adsorption of RNA male-specific phage and for specific aggregation with another F pilus. The proximal face would contain the site for polymerization to the distal face of the subunit next to it. These properties would explain why the RNA male-specific phage can adsorb to the entire length of the F pilus (Brinton *et al.,* 1964), and why usually one, but sometimes two (and more rarely three) DNA male-specific phages can adsorb to only one end of any free F pilus (Caro and Schnöss, 1966). Considering that the same structure is probably involved in the release of viral RNA and DNA, in addition to the conduction of viral RNA and DNA and cellular DNA, the subunit of the F pilus must rank as one of the most talented molecules synthesized by the cell.

Summary

Adsorption to F pili, the organelles of intracellular DNA transfer, is an obligatory step in the infection of male bacteria by both RNA and DNA male-specific phages. Bacterial cultures contain both free and attached F pili, both of which can adsorb phage. Many bacterial cultures have more free than attached F pili, the free F pili arising by outgrowth and release from the cell. F pili are hollow rods 85 Å in diameter, and it is very likely that they are assembled from protein subunits. They are very hydrophobic, and form longitudinal aggregates spontaneously. F pili are degraded by organic solvents and by trypsin. Phenotypically phage-resistant cells having no F pili are found in cultures of male bacteria. The total amount of F pili, the relative proportions of free and attached F pili, and the relative proportions of F piliated and non-F piliated bacteria depend upon the bacterial strains chosen, the cultural conditions employed, and the presence and activity of repressors (Meynell and Datta, 1965; Watanabe *et al.,* 1962).

RNA·male-specific phage particles are weakly and reversibly bound to the sides of free or attached F pili, with an equilibrium constant of about 5×10^{-9} milliliters per particle at $0°C$ in Z medium. The equilibrium constant is only slightly temperature-dependent, being about twice as large at temperatures between $20°$ and $40°C$ as it is at $0°C$. Adsorption does not require divalent cations, but has a nonspecific ionic strength requirement (Paranchych, 1966; Paranchych and Danziger, 1966). It does not require metabolic activity by the cell.

Measurements of adsorption of phage to bacteria may be made by using filters to separate bacteria from unadsorbed phage and from phage adsorbed to free F pili, if a correction is made for the aggregates of free F pili retained by the filter.

The amount of F pili in a bacteria-free preparation can be estimated by measuring the amount of radioactive RNA phage retained after filtration of the phage-F pilus mixture through a filter of pore size 0.45 μ. In using this assay quantitatively, one must take into account the equilibrium between free and bound phage, which depends upon the concentrations of phage and F pili, and the fraction of phage-F pili complexes which are retained by the filter. The results of experiments based on the filtration assay must be interpreted with extreme caution if the state of aggregation of the F pili preparation has not been taken into account.

The RNA of phage particles adsorbed to F pili attached to bacteria can penetrate the cell at 37° but not at 0°C. Penetration requires that the cell be metabolically active and is dependent upon the presence of the divalent cations (Mg^{++}, Ca^{++}, Sr^{++}, or Ba^{++}) (Paranchych, 1966). Only the RNA of the phage penetrates the cell, the protein remaining outside (Edgell and Ginoza, 1965). Only one, or at most a few of several phages adsorbed to the same attached F pilus succeed in injecting RNA into the cell, even though all phages (in the purified preparations used in our experiments) are potentially capable of adsorbing and injecting RNA.

Phages which adsorb but do not inject RNA are easily eluted from attached F pili. Some may be eluted without loss of activity, but many of them lose their ability to form plaques (Paranchych and Graham, 1962). This eclipse of activity does not require the presence of divalent cations, and is therefore distinct from the penetration step of infection (Paranchych, 1966). It presumably represents either the release of RNA from the phage particle, or a rearrangement of the protein coat rendering the phage particle inactive. The RNA of adsorbed phage which has not penetrated the cell becomes sensitive to hydrolysis by RNase (Zinder, 1963). Since RNase can completely prevent phage infection, the release (eclipse) step is indispensable to the infectious process, and must occur on or outside the cell wall.

Since free F pili do not inactivate RNA phage particles adsorbed to them (Valentine and Strand, 1965) and since they do not render phage RNA sensitive to hydrolysis by RNase (Valentine and Strand, 1965), free F pili must be different from attached F pili. Attached F pili may contain DNA, which free F pili have lost, or the release (eclipse) step may require the cooperation of an element of the cell wall.

Although the majority of phage particles in properly purified preparations can adsorb and inject RNA, usually only a fraction of them can form plaques. The inability to form plaques must therefore be due to defects in the phage RNA which affect intracellular functions.

DNA male-specific phages adsorb to the ends (Brinton, 1966; Caro and Schnöss, 1966) of attached F pili, and to one end only of free F pili (Caro and Schnöss, 1966). The rate of adsorption of DNA phage is about one hundredth the rate of adsorption of RNA phage at the same cell concentrations (Tzagoloff and Pratt, 1964). The slower rate can be understood on the basis of a limited number of adsorption sites for the DNA phage, thus lowering the probability of a fruitful collision. DNase has no effect on infection with DNA male-specific phages (Tzagoloff and Pratt, 1964), indicating that with these agents, the viral genome is not exposed during penetration, as is the case with the male-specific RNA phages. Other characteristics of infection with DNA phages are similar to those of infection with RNA phages (Tzagoloff and Pratt, 1964). The protein portion of the phage remains outside the cell, attachment does not require divalent cations but it has a nonspecific ionic strength requirement of about 0.1. Attachment is not temperature sensitive and can occur at 0°C. Penetration is highly temperature sensitive and depends upon host cell metabolism.

ACKNOWLEDGMENTS

The authors thank Dr. William Paranchych for helpful discussions and for making available his unpublished results. The advice and assistance of Dr. William Knight and Dr. Charles Novotny was extremely valuable. This research was supported by grants from the American Heart Association and the National Institute of Allergy and Infectious Disease of the U.S. Public Health Service. Dr. Beer was supported by a Training Grant from the U.S. Public Health Service to the Department of Microbiology, Medical School, University of Pittsburgh.

REFERENCES

Bradley, D. E. (1964). *J. Gen. Microbiol.* **35**, 471–482.
Brinton, C. C., Jr. (1959). *Nature* **183**, 782–786.
Brinton, C. C., Jr. (1965). *Trans. N.Y. Acad. Sci.* [2] **27**, 1003–1053.
Brinton, C. C., Jr. (1966). *In* "The Specificity of Cell Surfaces" (B. D. Davis and L. Warren, eds.), pp. 37–70. Prentice-Hall, Englewood Cliffs, New Jersey.
Brinton, C. C., Jr., Buzzell, A., and Lauffer, M. A. (1954). *Biochim. Biophys. Acta* **15**, 533–542.
Brinton, C. C., Jr., Gemski, P., Jr., and Carnhan, J. (1964). *Proc. Natl. Acad. Sci. U.S.* **52**, 776–783.
Brock, T. D. (1962). *Biochem. Biophys. Res. Commun.* **9**, 184–187.
Brock, T. D., and Wooley, S. O. (1963). *Science* **131**, 1065–1067.
Caro, L. G. (1966). Personal communication.

Caro, L. G., and Schnös, M. (1966). *Proc. Natl. Acad. Sci. U.S.* **56,** 126–132.

Cooper, S., and Zinder, N. D. (1962). *Virology* **20,** 605–612.

Crawford, E. M., and Gesteland, R. F. (1964). *Virology* **22,** 165–167.

Dettori, R., Maccacaro, G. A., and Turri, M. (1963). *Giorn. Microbiol.* **11,** 15–34.

Edgell, M. H., and Ginoza, W. (1965). *Virology* **27,** 23–27.

Hirota, Y., Nishimura, Y., Ørskov, F., and Ørskov, I. (1964). *J. Bacteriol.* **87,** 341–351.

Hofmann-Berling, H., Dürwald, H., and Beulke, I. (1963). *Z. Naturforsch.* **18b,** 893–898.

Hofschneider, P. H. (1963). *Z. Naturforsch.* **18b,** 203.

Lodish, H. F., and Zinder, N. D. (1965). *Biochem. Biophys. Res. Commun.* **19,** 269–278.

Lodish, H. F., Horiuchi, K., and Zinder, N. D. (1965). *Virology* **27,** 139–155.

Loeb, T. (1960). *Science* **131,** 932–933.

Loeb, T., and Zinder, N. D. (1961). *Proc. Natl. Acad. Sci. U.S.* **47,** 282–289.

Meynell, E., and Datta, N. (1965). *Nature* **207,** 884–885.

Meynell, E., and Datta, N. (1966). *Genet. Res.* **7,** 134–140.

Novotny, C. (1965). Unpublished results.

Paranchych, W. (1963). *Biochem. Biophys. Res. Commun.* **11,** 28–33.

Paranchych, W. (1966). *Virology* **28,** 90–99.

Paranchych, W., and Danziger, R. E. (1966). Unpublished manuscript.

Paranchych, W., and Graham, A. F. (1962). *J. Cellular Comp. Physiol.* **60,** 199–208.

Raizen, C. E. (1966). Ph.D. Thesis, University of Wisconsin.

Tzagoloff, H., and Pratt, D. (1964). *Virology* **24,** 372–380.

Valentine, R. C., and Strand, M. (1965). *Science* **148,** 511.

Watanabe, T., and Fukasawa, T. (1962). *J. Bacteriol.* **83,** 727–735.

Watanabe, T., Fukasawa, T., and Takano, T. (1962). *Virology* **17,** 217–219.

Zinder, N. D. (1963). *Perspectives Virol.* **3,** 58–67.

Zinder, N. D., Valentine, R. C., Roger, M., and Stoeckenius, W. (1963). *Virology* **20,** 638.

Viral "Minus" Strands and the Replication of RNA Phages

Charles Weissmann

DEPARTMENT OF BIOCHEMISTRY, NEW YORK UNIVERSITY
SCHOOL OF MEDICINE, NEW YORK, NEW YORK

Introduction

RNA phages are small polyhedral viruses about 200 Å in diameter, having a particle weight of about $3.5–4 \times 10^6$ (Loeb and Zinder, 1961). They consist of about 180 capsomeres of a molecular weight of 17,000 each and a RNA molecule of about 10^6 molecular weight (Loeb and Zinder, 1961; Zinder, 1963; Strauss and Sinsheimer, 1963; Mitra *et al.,* 1963; Gesteland and Boedtker, 1964; Enger and Kaesberg, 1965). A number of phages have been isolated and described in some detail, such as f2 (Loeb and Zinder, 1961), MS2 (Davis *et al.,* 1961), Qβ (Watanabe, 1964), and many others. [For further references, see, Zinder (1963) and Weissmann and Ochoa (1967).]

RNA phages specifically infect male *E. coli* (Loeb and Zinder, 1961) or bacteria containing the *E. coli* F agent (Brinton *et al.,* 1964). Following attachment of the phage to the male-specific F pili, the RNA is introduced into the host, the coat protein remaining outside (Edgell and Ginoza, 1965). An eclipse period of 15 min ensues, following which infectious particles accumulate within the cell. Lysis of infected bacteria occurs between 22 and 60 min, and each cell may yield up to 40,000 virus particles, of which, however, only a fraction (5–50%) is viable, depending on the phage and the growth conditions (Loeb and Zinder, 1961; Paranchych and Graham, 1962; Cooper and Zinder, 1963). The formation of RNA phages is independent of host DNA synthesis (Cooper and Zinder, 1962; Hofschneider, 1963), and no base sequence homology between the phage RNA and the DNA of either infected or noninfected host cells (Doi and Spiegelman, 1962), has been detected. Exposure of infected *E. coli* spheroplasts to actinomycin D at concentrations that completely inhibit host cell RNA synthesis, does not abolish

RNA phage synthesis. These findings speak against the participation of host DNA in the synthesis of RNA phages.

Protein synthesis is required, not only for the formation of the coat protein, but also during the early infective period, for synthesis of viral RNA (Cooper and Zinder, 1963; Paranchych and Ellis, 1964). These findings are in accordance with the subsequent discovery that one (or two) virus-specific RNA synthesizing enzymes are produced in the host within minutes after infection (Weissmann *et al.*, 1963b; August *et al.*, 1963; Haruna *et al.*, 1963). This implies that one of the early functions of the viral RNA is that of a messenger, and, in fact, after penetration into the cell, parental viral RNA is found associated with the polysome fractions (Godson and Sinsheimer, 1966). *In vitro* experiments (Nathans *et al.*, 1962; Nathans, 1965) have shown that viral RNA can indeed function as a messenger in the cell-free system of protein synthesis derived from uninfected *E. coli*. The protein synthesized under the direction of viral messenger RNA consists largely of the viral coat protein. However, other virus-specific proteins appear to be formed as well (Ohtaka and Spiegelman, 1963).

The earliest event that can be related to the synthesis of viral RNA is the conversion of parental RNA (plus) strands to a double-stranded form* by the synthesis of a complementary (minus) strand (Weissmann and Borst, 1963; Weissmann *et al.*, 1964a). Subsequently, in each infected cell, several hundred additional minus strands are formed, most of which can be isolated as double strands (Fig. 1). The double-stranded RNA, after an isolation procedure involving RNase, has been characterized as a double helix consisting of equivalent amounts of plus and minus strands (Weissmann *et al.*, 1964a; Langridge *et al.*, 1964). At least part of it appears to be derived from a more complex structure—the replicative intermediate (Erikson *et al.*, 1964; Fenwick *et al.*, 1964). This intermediate has been shown to turn over during RNA synthesis (Fenwick *et al.*, 1964; Billeter *et al.*, 1966a).

Several enzyme preparations have been purified from phage-infected *E. coli*. Their action *in vitro* gives rise to viral RNA which has been characterized either by annealing techniques (Weissmann *et al.*, 1964b; Weissmann, 1965), or by infectivity assays (Spiegelman *et al.*, 1965). In all cases, minus strands are either present in the enzyme preparation, or are formed during the early stage of synthesis (Weissmann *et al.*, 1964b; Weissmann and Feix, 1966).

---

* It is possible that the plus and minus strands within the replicating complex, as it occurs *in vivo*, may not have the tightly hydrogen bonded, RNase-resistant structure that is observed after isolation by standard methods involving phenol extraction (Borst and Weissmann, 1965). This reservation must be borne in mind whenever mention is made of the involvement of double-stranded RNA in replication.

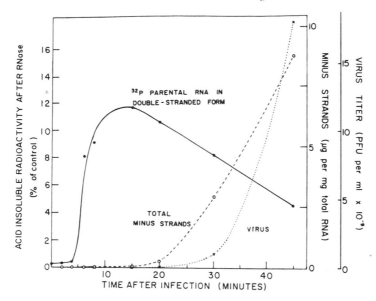

Fig. 1. Formation of double-stranded MS2 RNA and minus strands following infection of *E. coli* with phage MS2. The continuous curve (^{32}P-labeled parental RNA) illustrates the formation and subsequent decrease of RNase-resistant double-stranded RNA containing ^{32}P-labeled parental RNA strands. The dashed curve shows that minus strands (determined by an annealing procedure) increase throughout infection. Most minus strands are recovered in a double-stranded form. Note that different ordinates are used for the two curves. The dotted line shows the time course of intracellular MS2 phage production (Weissmann *et al.*, 1964a).

In vivo, completed RNA strands are incorporated into virus particles with a delay of 10 to 15 min (Cooper and Zinder, 1963; Ellis and Paranchych, 1963). Lodish *et al.* (1965) have demonstrated the existence of a genetically determined maturation step in the late phase of virus synthesis.

Four viral functions have been clearly defined by biochemical and genetic criteria: (1) synthesis of minus strands, (2) synthesis of plus strands, (3) synthesis of coat protein and (4) a maturation step. Functions (1) and (2) are associated with the formation or one or two virus-specific polymerases. An analysis of MS2-infected, actinomycin treated *E. coli* spheroplasts suggests that six different virus-specific proteins are produced in the infected host (Haywood and Sinsheimer, 1965).

In Vivo Studies

FATE OF PARENTAL RNA

Davis and Sinsheimer (1963) demonstrated that less than 3% of the radioactivity of labeled, parental RNA is recovered in the progeny phage.

The parental strand is conserved in the host cell in which it initiated
infection, and it can be recovered intact at the end of the infectious
cycle (Doi and Spiegelman, 1963). Sedimentation analysis of the contents of host cells infected with ^{32}P-labeled phage showed that, within
min after infection, the parental RNA strand is found in association with
the polysome fraction (Godson and Sinsheimer, 1966). Sedimentation
analysis of the deproteinized RNA from cells infected with ^{32}P-labeled
phage revealed a striking change in the physical state of the labeled parental
RNA strand shortly after infection. Part of the labeled RNA (original
$s_{20,\ w} = 27$ to 28 S) sedimented broadly around 16 S (Kelly and Sinsheimer, 1964; Erikson *et al.*, 1964) (Fig. 2). A considerable fraction of
the 16 S RNA was RNase-resistant. The conversion of the parental strand
into what has been called the replicative intermediate (Erikson *et al.*,
1964) can be followed by the appearance of ribonuclease-resistant parental RNA (Weissmann and Borst, 1963; Weissmann *et al.*, 1964a;
Kelly and Sinsheimer, 1964). RNase-resistant parental RNA is first detectable around 5 min after infection (Fig. 1). After 12 min, 12% of the
input phage RNA is resistant to RNase. Since only 20% or less of the
labeled phage was viable, 12% may represent a substantial fraction of
the RNA actually initiating an infection. After 12 min, the RNase-

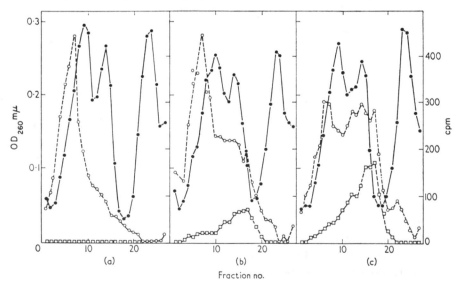

FIG. 2. Conversion of parental ^{32}P-labeled R17 RNA into the replicative intermediate. RNA was isolated from *E. coli* infected with ^{32}P-R17 at (*a*) 0, (*b*) 6, and
(*c*) 12 min after infection, and analyzed by sucrose density gradient centrifugation. —○———○—, total radioactivity; —□———□—, RNase-resistant radioactivity;
—●———●—, absorbance (Erikson *et al.*, 1964).

resistant fraction again decreased. As shown by Geidushek *et al.* (1962), resistance to RNase is a striking property of double-stranded RNA, and, in fact, the isolated RNase-resistant radioactive RNA showed all the properties expected of a double-strand—namely (*a*) sharp thermal transition to a RNase-sensitive state (103°C in 0.15 *M* NaCl, 0.015 *M* sodium citrate) (Weissmann *et al.* 1964a); (*b*) sedimentation with an $s_{20, w}$ of about 13.4 S (Erikson *et al.,* 1965), a value compatible with double-helical RNA of molecular weight of about 2×10^6; and (*c*) buoyant density of about 0.02 gm/cm^3 less than that of single-stranded RNA (Weissmann *et al.,* 1964a). Moreover, the specific dilution assay showed that the radioactive component of the double-strand was indeed a viral plus strand (Weissmann *et al.,* 1964a,b).

Erikson *et al.* (1965) isolated the fraction of parental-labeled RNA sedimenting at 16 *S* and demonstrated that after heat denaturation a considerable amount of the radioactive RNA sediments at 27 S, the characteristic value of single-stranded viral RNA. This showed that parental RNA is conserved intact within the replicative intermediate, and can be released from the duplex by denaturation. The appearance of the radioactive, parental RNA in a double-stranded form is considered to be due to the synthesis of a complementary (or minus) strand with the use of the plus strand as template (Weissmann *et al.,* 1964a). This conversion is prevented by the addition of chloramphenicol to the bacterial culture immediately preceding infection, suggesting that the synthesis of a virus-specific enzyme is required for this step (Erikson *et al.,* 1964; Kelly and Sinsheimer, 1964). The decrease in the amount of labeled, double-stranded RNA during the further course of replication (Fig. 1) can be explained by assuming either (*a*) that the double-stranded RNA is degraded or separated into two single strands, or (*b*) that the parental strand is displaced from the duplex by a newly formed plus strand, as indicated in the scheme of semiconservative asymmetric replication mechanism in Fig. 10.

The kinetics of the process, in particular the slow rate of displacement, could be accounted for if the parental strand, after being displaced from the duplex as a single strand, is again converted to a double-stranded form, and passes through several such cycles until it is finally trapped in a single-stranded state (possibly as the consequence of some discrete damage). The curve in Fig. 1 would then represent the result of nonsynchronous cycling of a large population of parental strands (Ochoa *et al.,* 1964). Several observations can be explained by the latter interpretation. Lodish and Zinder (1966b) infected host cells with ^{32}P-labeled *sus*-3, an amber mutant of f2 that, under nonpermissive conditions, causes the formation of excessive amounts of the enzyme responsible for the synthesis of minus strands. In the nonpermissive host, the amount of parental ^{32}P-RNA converted to a double-stranded form increased throughout the period of observation and reached 50% of the input after 1 hour. Since the double-stranded RNA was defective (6–7 S), these results might be explained by the hypothesis that, in this case, the parental RNA was preferentially trapped in double-strands which could no longer participate in replication because of their defectiveness. It is obvious that extensive intracellular degradation of double-stranded RNA to acid-soluble fragments, as called for by hypothesis (*a*), did not occur. Lodish and Zinder (1966a) isolated a temperature-sensitive mutant of f2, *ts*-6, that neither showed parental strand conversion nor caused viral RNA replication within the host at 43°C, although adsorption of the virus and penetration of the RNA took place normally. After transferring to 34°C, both the conversion to a double-stranded form and, in due course, viral RNA synthesis occurred. This finding suggests that the processes associated with the conversion of the parental strand to 'a duplex are related to early stages of RNA replication. Further experiments showed that, in the case of this mutant, formation of minus strands, but not that of plus strands, was blocked at high temperature (Lodish and Zinder, 1966a). When cells infected with ^{32}P-labeled *ts*-6 were grown at 34°, allowing double-stranded ^{32}P-labeled RNA to form, and were then transferred to 43°C, the label in double-stranded RNA disappeared almost entirely within 3 min (Lodish, 1966). In terms of hypothesis (*b*) this means that under conditions where minus, but not plus strand synthesis was inhibited, parental ^{32}P-labeled strands were displaced from the duplex and were no longer reconverted to double-strands.

VIRUS-SPECIFIC RNA SPECIES FORMED IN BACTERIA INFECTED WITH RNA PHAGES

Minus Strands and Replicative Intermediate

As already mentioned, ^{32}P-labeled, double-stranded RNA is detected some 5 min after infection of *E. coli* with ^{32}P-labeled MS2 phage, showing that minus strands, i.e., strands with a base sequence complementary to that of the parental RNA, are formed. The relative rates of synthesis of viral plus and minus strands are shown in Fig. 3. Synthesis of viral minus strands is detected prior to that of plus strands. Whereas the rate of minus strand synthesis becomes constant about 15 min after infection, the rate of synthesis of plus strands continues to increase up to about 22 min (Billeter *et al.,* 1966a). This suggests the occurrence of separate control mechanisms for the two processes. Fifteen minutes after infection, an MS2-infected cell may contain about 30, and at 45 min after infection, as many as 500 equivalents of the viral minus strands, most of which are double-stranded (Weissmann *et al.,* 1964a). The bulk of the double-stranded RNA sediments at 14–16 S, and, while some of it sediments more rapidly, very little sediments more slowly (Fig. 4). If host-specific RNA synthesis in infected cells is inhibited by UV irradiation or by actinomycin, uracil-^{14}C is seen to be incorporated into RNA, part of which is RNase-resistant, and to sediment in the 14–16 S region

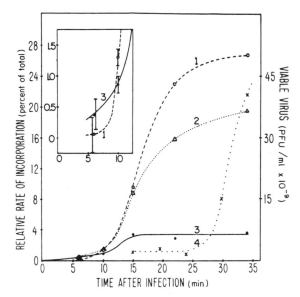

FIG. 3. Rate of labeling of plus, minus, and double MS2 RNA strands as a function of time after infection. Cells were labeled for 30 sec with uracil-^{14}C at the times indicated. Radioactive plus strands (curve 1) and minus strands (curve 3) were determined by the double isotope specific dilution assay (cf. Fig. 11) and radioactive double-stranded RNA (curve 2) by RNase-resistance. In the inset, the values for plus (curve 1) and minus (curve 3) strands, 6 and 10 min after infection, are plotted on an eight-fold enlarged ordinate scale. The standard error, from two independent determinations, is shown. Curve 4, viable virus (Billeter *et al.,* 1966a).

(Fenwick *et al.,* 1964; Kelly *et al.,* 1965). This structure is analogous to the one in which parental RNA is detected after initiating infection— the so-called replicative intermediate (Fig. 2).

Treatment of the radioactive replicative intermediate with RNase at low concentrations causes part of the labeled RNA to become acid-soluble; the remainder sediments in a narrow band at about 12–13 S and is entirely RNase-resistant (Fenwick *et al.,* 1964; Billeter *et al.,* 1966b). The sedimentation coefficient expected of a duplex consisting of two full-length virus strands is 13.4 S (Billeter *et al.,* 1966b). RNase at higher concentrations cleaves the double-stranded RNA to fragments sedimenting rather homogeneously at 8–9 S. It is thought (Fenwick *et al.,* 1964) that the replicative intermediate sedimenting at 14 S and above may consist of double-stranded RNA with one or more growing single plus strands attached (Fig. 5a). Mild RNase treatment prior to sucrose gradient centrifugation would hydrolyze the single strands, thereby making the double-stranded RNA sediment more homogeneously and with a lower sedimentation coefficient (Fig. 5b). More rigorous digestion would lead to cleavage of the double-stranded RNA itself (Fig. 5c). Evidence has been presented, suggesting that the virus-specific 14–16 S RNA may, in fact, consist of two components: (*a*) A double-stranded molecule without single strands attached, sedimenting at about

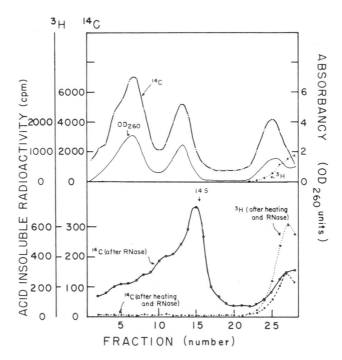

Fig. 4. Sedimentation profile of RNA from MS2-infected *E. coli*. MS2 infected *E. coli* was labeled with thymidine-³H and uridine-¹⁴C. The nucleic acids were treated with DNase and centrifuged through a sucrose gradient. The acid-insoluble ¹⁴C-radioactivity of each fraction was determined directly (———●———), after RNase treatment (—○————○—) and after heat-denaturation and RNase treatment (×——×). Some ³H-labeled DNA fragments (+····+), remaining on the top of the gradient, are partly acid-insoluble and RNase-resistant before and after heating. Apparently some ¹⁴C-labeled precursor has been incorporated into DNA and behaves analogously (Billeter and Weissmann, unpublished experiment).

14 S, and (*b*) a double-stranded molecule with attached single strands sedimenting at 16 S and above. Radioactive component (*a*) would predominate after long periods of labeling and may represent double-stranded material not engaged in replication. Radioactive component (*b*) would be predominant after a short labeling period and may represent the actual replicating intermediate (Lodish and Zinder, 1966b).

The question has been raised (Spiegelman, 1965) as to whether the minus strand in the duplex is continuous and of the same length as the plus. strand, or whether it consists of a series of short segments base-paired with a plus strand, with the implication that such a defective double-

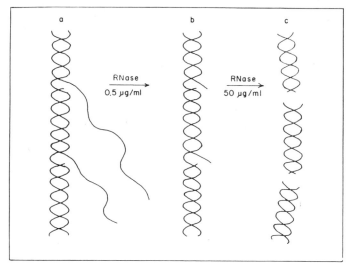

FIG. 5. Schematic representation of the replicative intermediate. (A) As isolated after deproteinization (modified from Fenwick *et al.,* 1964); (*b*) after mild treatment with RNase, the "tails" are cleaved off; (*c*) after intensive RNase digestion, the duplex is fragmented.

strand is but an incidental side-product of replication. In order to decide between these alternatives, labeled 14–16 S RNA (containing replicative intermediate) was isolated and sedimented through a sucrose gradient, either before or after heat denaturation. Labeled minus strands were localized by annealing each fraction after denaturation with an excess of nonlabeled plus strands and determining the acid-insoluble radioactivity. As shown in Fig. 6, minus strands were found in the 14–16 S region in the case of nondenatured samples and in the 27 S region after denaturation. This shows that at least part of the minus strands within the duplex are full-length viral strands (Colthart and Weissmann, unpublished experiments).

Under certain conditions, such as conversion of infected cells to spheroplasts, followed by treatment with actinomycin (Kelly *et al.,* 1965) or after UV irradiation of the host cell (Fenwick *et al.,* 1964), considerable amounts of double-stranded RNA with a sedimentation coefficient of about 6–7 S may be found. In view of the circumstances under which this material is found, its formation may reflect some abnormal process.

Purified Double-Stranded RNA

As already mentioned, part of the so-called replicative intermediate consists of RNase-resistant RNA. Large-scale purification of this fraction is usually carried out after phenol extraction followed by treatment with DNase and pancreatic RNase in solutions of moderate ionic strength. The RNase-resistant RNA is then isolated by exclusion chromatography

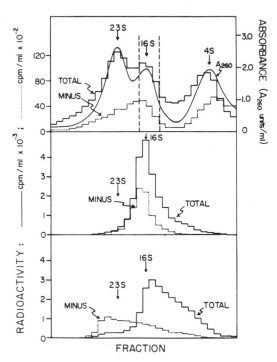

FIG. 6. Demonstration that the replicative intermediate contains a full-length minus strand. Labeled RNA from MS2 infected cells was centrifuged through a sucrose density gradient (top) and the fractions sedimenting around 16 S containing the replicative intermediate, were pooled and recentrifuged directly (middle) or after heat denaturation (bottom). Total = total radioactive nucleic acids; Minus = minus strands, as detected by annealing with excess unlabeled MS2 RNA. Heat denaturation of replicative intermediate ($s_{20,\ w} \sim 16$ S) liberates minus strands sedimenting around 27 S. (Colthart and Weissmann, unpublished experiment).

on Sephadex G-200 (Montagnier and Sanders, 1963; Billeter *et al.*, 1966b; Langridge *et al.*, 1964; Kaerner and Hoffmann-Berling, 1964). An analytically pure preparation of double-stranded MS2 RNA has been described in detail (Billeter *et al.*, 1966b). Table I shows some of the analytical data obtained with this material. The base composition is that expected of an RNA duplex consisting of an MS2 RNA strand and its complement, and it is very different from that of the host cell RNA. The molar absorbance, ϵ (P), is lower than that of single-stranded RNA, and the absolute and relative (thermal) hypochromicity is higher. The T_m of the helix-coil transition is dependent on the ionic strength, and is substantially higher than that of single-stranded RNA under comparable conditions, and higher than that of native DNA with a similar base composition. The X-ray diffraction pattern is well resolved and characteristic of a double-helical structure (Langridge *et al.*, 1964). The molecular weight of double-stranded RNA isolated after a purification procedure involving extensive treatment with RNase is in the order of 4.5×10^5 to 7.5×10^5 as determined by sedimentation behavior and by electron microscopy (Billeter *et al.*, 1966b).

Ammann *et al.* (1964) have developed a purification procedure which does not involve the use of RNase. This procedure yielded a large proportion of filaments with a length of about 1.14 μ, corresponding to a molecular weight of about 2×10^6. The preparation was noninfectious in its native state, but, upon thermal denaturation, infectivity was demonstrable in the spheroplast assay. In conjunction with the experiments described above, one may conclude that double-stranded RNA isolated without RNase treatment contains full-length viral plus and minus strands.

The resistance of purified, double-stranded RNA to pancreatic RNase A and B, as well as to RNase T_1, is very pronounced (although not absolute) in 0.15 M NaCl, but not at lower salt concentrations. After heat denaturation, double-stranded RNA is as susceptible to pancreatic RNase as is single-stranded RNA. Reannealing occurs readily and almost quantitatively at 80°–90°C, at NaCl concentrations above 0.15 M (Billeter *et al.*, 1966b). The reannealing reaction is highly specific (Table II) and is the basis for a quantitative assay of viral plus and minus strands (Weissmann *et al.*, 1966).

SHORT-TERM LABELING EXPERIMENTS

It is postulated that a structure consisting of enzyme, viral minus, and viral plus strands occurs as an intermediate in RNA replication. After isolation involving deproteinization, this structure is recovered as

TABLE I

COMPARISON OF MS2 RNA, DOUBLE-STRANDED MS2 RNA, AND RIBOSOMAL RNA[a]

Parameter	MS2 RNA or R17 RNA	Double-Stranded MS2 RNA	E. coli ribosomal RNA
Phosphorus (%)	—	9.58 (calc. 9.46)	—
Orcinol test (A_{665}-units per μmole of purine)	—	11.8	11.8
Cysteine test	—	Less than 2% DNA	—
$\epsilon\,(P)_{260\,m\mu}$	8600[b]	6670	7450[c]
Base composition (%) A	23.7[b]	24.6 (calc. 24.05)	25.1
U	24.4	23.7 (calc. 24.05)	20.4
G	27.1	25.9 (calc. 25.95)	32.6
C	24.8	25.8 (calc. 25.95)	21.9
Buoyant density in Cs_2SO_4 (gm/cm³)	1.626	1.609	1.646[c]
$s_{20,w}$ (high ionic strength)	26 S (0.2 M NaCl)[d]	8.45 S (SSC)	24 S, 18 S (0.1 M KCl, 0.05 M Tris)[c]
$s_{20,w}$ (low ionic strength)	20.4 S (0.02 M NaCl)[d]	8.1 S (0.01 SSC)	19 S, 15 S, (0.005 M KCl, 0.0025 M Tris)[c]
T_m (high ionic strength)	47° (SSC)	103° (SSC)	54° (0.1 M sodium phosphate buffer)[d]
T_m (low ionic strength)	—	87° (0.1 × SSC)	—
Hypochromicity	18%	26.5%	21.9%
X-ray diffraction pattern	Not oriented	Highly oriented[e]	Not oriented

[a] Billeter et al. (1966b).
[b] Strauss and Sinsheimer (1963).
[c] Stanley (1963).
[d] Gesteland and Boedtker (1964).
[e] Langridge et al. (1964).

TABLE II

SPECIFICITY OF ANNEALING ASSAY[a]

Labeled RNA	Nonlabeled RNA	RNase-resistant radioactivity after annealing[b]	
		cpm over control	% of input
1. ³²P-MS2 RNA (4 µg, 11,630 cpm)	MS2 RNA (400 µg)	11	0.1
	Ribosomal RNA (400 µg)	11	0.1
	Soluble RNA (400 µg)	11	0.1
	TMV RNA (400 µg)	16	0.14
	Normal *E. coli* RNA (400 µg)	0	0
	Infected *E. coli* RNA (400 µg)	1095	9.4
	Partially purified double-stranded MS2 RNA (88 µg)	3006	26
2. ³²P-MS2 RNA (0.24 µg, 9867 cpm)	Purified double-stranded MS2 RNA (8.7 µg)	8209	83.1
	Purified double-stranded Qβ RNA (6.2 µg)	0	0
3. ³²P-Qβ RNA (0.24 µg, 11,375 cpm)	Purified double-stranded Qβ RNA (6.2 µg)	9050	79.5
	Purified double-stranded MS2 RNA (8.7 µg)	10	0.1
4. ³²P-TMV RNA (1.88 µg, 11,600 cpm)	Purified double-stranded MS2 RNA (2.5 µg)	30	0.25
	Normal tobacco leaf RNA (0.9 mg)[c]	54	0.46
	TMV-infected tobacco leaf RNA (1.3 mg)[c]	2243	19.3
5. ³²P-*E. coli* RNA (1.12 µg, 3515 cpm)	TMV-infected tobacco leaf RNA (0.9 mg)[c]	0	0
6. P³²-*E. coli* RNA (1.3 µg, 6600 cpm)	Partially purified double-stranded MS2 RNA (400 µg)	0	0

[a] From Weissmann *et al.* (1964a), Burdon *et al.* (1964), and Colthart and Weissmann (1966).

[b] The mixtures were heated to 120°C for 3 min and annealed at 80°C for 30–60 min. The control is the RNase-resistant radioactivity of the nonheated mixture. Controls: Experiment 1: 36–100 cpm; Experiment 2: 45 cpm; Experiment 3: 110 cpm; Experiment 4: 153 cpm; Experiment 5: 96 cpm; Experiment 6: 300 cpm.

[c] The leaf RNA was first treated with RNase and the RNase was removed by phenol extraction.

the so-called replicative intermediate described above, which may consist of a double-strand of RNA with single strands attached (Fig. 5a). Several lines of evidence suggest that double-stranded RNA, or a structure giving rise to double-stranded RNA upon isolation, is involved in viral RNA replication. (*a*) In MS2 infected bacteria, the formation of viral minus strands precedes that of plus strands (Billeter *et al.,* 1966a). (*b*) As described in a previous section, the radioactive parental RNA is converted to a double-stranded form within a few minutes after entering the host cell. (*c*) A complex, involving minus strands appears to be involved in the *in vitro* synthesis of viral plus strands by RNA synthetase (Weissmann *et al.,* 1964b; Weissmann, 1965). Findings with the *in vitro* system, and the fact that radioactive parental RNA, after conversion to a double-strand is subsequently displaced from the duplex (Weissmann *et al.,* 1964a), led to the suggestion that double-stranded RNA, or a structure giving rise to double-stranded RNA upon isolation, is an intermediate in viral RNA replication (Weissmann *et al.,* 1964b). Short-term labeling experiments lend strong support to this idea. Fenwick *et al.* (1964) labeled (for 10 seconds with radioactive uridine) R17-infected bacteria in which host RNA synthesis had been inhibited by irradiation with ultraviolet light. Upon sucrose gradient centrifugation of the isolated RNA, the label was found almost exclusively in the 14–16 S region, corresponding to the replicative intermediate. If cells thus labeled were subsequently incubated in unlabeled medium (chase), there was a marked decline of RNase-resistant label within the 16 S fraction, which originally was 70% RNase-resistant. Concomitantly, radioactive 27 S RNA, characteristic of viral RNA, appeared. This suggested a precursor-product relationship between the 16 S and the 27 S RNA. These experiments were confirmed and extended by Billeter *et al.* (1966a). Following exposure of MS2-infected *E. coli* to ^{14}C-labeled guanine for 8 seconds, the distribution of radioactivity between the total and the RNase-resistant RNA was determined, both immediately after the pulse, and at various times after termination of the pulse by addition of excess unlabeled precursor. Whereas, immediately after the pulse, about 25% of the total acid-insoluble radioactivity was in RNase-resistant RNA (it should be noted that host RNA synthesis proceeds actively in these cells), this value dropped to about 9.0% during the first 2 min after termination of the pulse and to 6.8% after 7 min (Fig. 7). The loss of radioactivity from pulse-labeled, double-stranded RNA was not due to breakdown of the duplex, since the sedimentation pattern of the RNase-resistant RNA did not change at any time after the pulse, nor was it caused by separation of the two strands of the duplex because, as shown below, the radioactive minus strands were conserved within the double-stranded structure. The loss of radioactivity was due to displacement of labeled plus strands from the duplex.

The proportion of radioactive viral plus and minus strands in double-stranded RNA was determined with the use of the specific dilution assay (Weissmann *et al.,* 1964b) (*a*) immediately after an 8 sec labeling period, and (*b*) at various times after dilution of the label. The results of the dilution assays are shown in Fig. 8. It may be seen that (*a*) after brief exposure to labeled precursors, about 87% of the RNase-resistant ^{14}C-radioactivity was in viral plus strands, and (*b*) upon dilution of the label, this value fell to about 60% and 40% after 2 and 12 min, respectively, indicating displacement of the plus strand from the duplex.

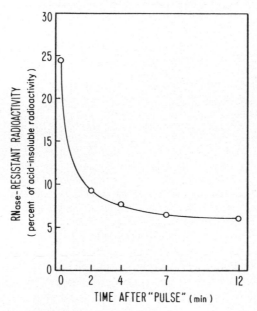

FIG. 7. Decrease of labeled double-stranded RNA formed during a short pulse of guanine-^{14}C. MS2-infected cells were exposed to guanine-^{14}C for 8 sec. RNA was prepared from samples taken immediately after the pulse and at different times after dilution of the label. Double-stranded RNA was determined as RNase-resistant radioactivity (Billeter *et al.,* 1966a).

For further analysis, the proportion of viral plus strands present both in double- and in single-stranded RNA at various times after the pulse of radioactive precursors was determined by means of the double isotope specific dilution assay (Weissmann *et al.,* 1966). At the termination of the pulse, some 60% of the labeled plus strands were in double-stranded RNA. This value dropped to about 8% in the following 4 min, as the

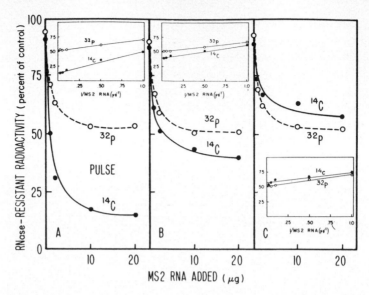

Fig. 8. Distribution of radioactivity between the plus and minus strands of double-stranded RNA immediately after a pulse of guanine-^{14}C, and at various times after dilution with cold guanine. ^{14}C-labeled double-stranded RNA was prepared from MS2-infected *E. coli,* at (*A*) 0, (*B*) 2 min and (*C*) 12 min after labeling for 8 sec with guanine-^{14}C, and was analyzed by the specific dilution test. Samples were heated and renatured in the presence of increasing amounts of unlabeled MS2 RNA. The radioactive RNA remaining RNase-resistant at "infinite" concentrations of MS2 RNA (see extrapolation in inset) is due to labeled minus strands; that converted to a RNase-sensitive form is due to plus strands. Double-stranded RNA, uniformly labeled with ^{32}P, i.e. containing equal amounts of label in plus and minus strands, was added as an internal standard to each sample (Billeter *et al.,* 1966a).

proportion of labeled plus strands in double-stranded RNA decreased. Minus strands were synthesized during the labeling period and appeared largely in double-stranded RNA. However, in contrast to the plus strands, the minus strands showed no significant turnover. The ratio of total plus to total minus strands synthesized was about 10. These findings are illustrated in Fig. 9.

These findings are compatible with the mechanism outlined in Fig. 10, where it is suggested that replication of phage RNA proceeds in 2 steps: (*A*) Synthesis of minus strands using plus strands as templates and (*B*) synthesis of progeny plus strands by an asymmetric, semiconservative process, in which a replicating complex is an intermediate. If a strictly semiconservative mechanism were operative in step (*B*), all of the viral RNA labeled upon exposure of infected cells to radioactive RNA precursors for a time shorter than that required to displace the equivalent of one viral plus strand from the complex, should be recovered in double-stranded RNA. In other words, there should be no free, labeled

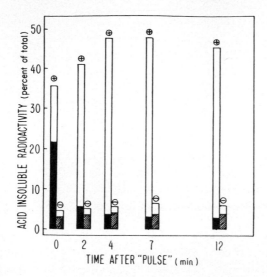

Fɪɢ. 9. Distribution of radioactivity between plus and minus MS2 RNA strands immediately after a pulse of guanine-^{14}C, and at various times after dilution of the label. Solid or shaded areas, radioactivity in double-stranded RNA; nonshaded areas, radioactivity in single-stranded RNA. Immediately after the pulse, 60% of the plus strands are found in double-stranded molecules. Subsequently, plus strands are displaced from the duplex and less than 6% remain double-stranded. The proportion of minus strands in a double-stranded form remains constant (Billeter *et al.*, 1966a).

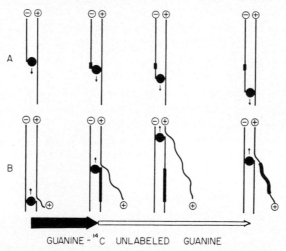

Fɪɢ. 10. Scheme of viral RNA synthesis and the fate of labeled precursors. Two steps in synthesis are postulated: (*A*) Synthesis of a minus strand, leading to the formation of a replicating complex; (*B*) Synthesis of plus strands by the replicating complex, by a mechanism in which newly formed strands displace their counterparts from the complex. A pulse of guanine-^{14}C is incorporated into plus and minus strands in the ratio 10 : 1. After dilution of the radioactive label, labeled plus strands, but not labeled minus strands are displaced from the complex.

plus strands after a sufficiently short pulse. However, reduction of the labeling time to 2 sec gave essentially the same results as did labeling for 8 sec. The proportion of plus strands in double-stranded RNA did not exceed 50–60%. These findings could be explained if some of the newly formed plus strands in step (*B*) displace their counterparts from the duplex while some do not. If the component strands of the duplex dissociate as synthesis of a new plus strand proceeds through base pairing with a minus strand, subsequent reformation of the duplex on completion of the new plus strand might occur randomly with equal chance for displacement of either the old or the new strand. Alternatively, if hydrogen-bonding between the plus and the minus strands of the replicating complex occurs only during the isolation procedure, the randomization may occur at this point.

Another feature deserving comment is that the displacement of the radioactive plus strand from the duplex was found to be incomplete even 12 min after termination of the pulse. It is assumed that this is due to the conversion of some of the released, labeled plus strands to double-stranded material through the continuing synthesis of new minus strands. Experiments by Lodish and Zinder (1966b) support this interpretation. As already mentioned, *E. coli* infected with *ts*-6, a mutant of the RNA phage f2, synthesize double-stranded RNA, viral RNA, and virus at 35°C but not at 43°C. However, once viral replication is initiated, and some double-stranded RNA is formed at 35°C, viral plus strand synthesis but not synthesis of double-stranded RNA occurs also at the higher temperature. This mutant then appears to induce the synthesis of a temperature-sensitive step one enzyme activity, and a stable step two enzyme activity. Short-term labeling experiments carried out at 35°C showed incomplete displacement of radioactivity from double-stranded RNA as in the experiments described above. However, at 43°C under conditions where no further minus strands were synthesized, displacement of the radioactivity from double-stranded RNA was virtually complete.

Enzymatic Studies

Formation of Viral RNA Polymerase(s) after Phage Infection

Following infection with RNA phages, a new RNA synthesizing activity is detectable in the host cell about 6 min after infection, and reaches a maximum at about 30 min (Weissmann *et al.,* 1963b; August *et al.,* 1963). Several amber mutants of RNA phages (*sus*-3, *sus*-11, *Mu*-9) cause the formation of 5–15 times the normal amount of both viral polymerase activity and of double-stranded RNA in certain nonpermissive strains of *E. coli* (Lodish *et al.,* 1964; Lodish and Zinder, 1966b; Feix *et al.,* 1966). In these cases, the rate of enzyme formation is increased to a variable degree and above all, synthesis continues beyond 60 min after infection, whereas it normally ceases after 25–30 min. It is remarkable that the lesion of these mutants is localized in the coat protein cistron (Lodish and Zinder, 1966a; Notani *et al.,* 1965). Since the mutant is considered to contain a single lesion only, this implies that the increased enzymatic activity observed is due to disturbed control, either of the synthesis of the enzyme or of its activity, while the actual enzyme protein is normal. It has been suggested (Lodish *et al.,* 1964), on the basis of experiments in which viral RNA synthesis was inhibited by precursor deprivation, that in the case of wild-type phage, only the parental strand (and perhaps some early progeny strands) are utilized as messengers for the synthesis of the polymerase(s), while in the case of the hyperproducing mutants, late progeny RNA strands also serve this function; this would lead to an excessive production of enzyme(s).

PROPERTIES OF VIRAL POLYMERASE(S)

Three laboratories have reported on the isolation and partial purification of phage-induced RNA polymerases. The properties of these preparations are different, so that each preparation is described separately. In what follows an attempt is made to correlate these findings.

RNA Synthetase

Weissmann *et al.* (1963a, 1964b) isolated an RNA synthesizing enzyme from *E. coli* infected with MS2 phage. After partial purification (about twentyfold) the preparation was free of DNA-dependent RNA nucleotidyl transferase. The preparation contained substantial amounts of endogenous RNA, and, in particular, viral minus strands, most of which were presumed to occur in a double-stranded form. There was no stimulation of enzyme activity by the addition of MS2 RNA or of any other RNA and treatment with RNase and subsequent removal of the nuclease led to irreversible inactivation of the enzyme. It was therefore tentatively concluded that the enzyme contained an endogenous template, presumably the viral minus strand.

Radioactive RNA synthesized with the partially purified enzyme was analyzed by the double isotope specific dilution assay. The product proved to be almost entirely virus-specific, consisting of about 90% plus and about 8% minus strands (Weissmann, 1965) (Fig. 11). Only a small

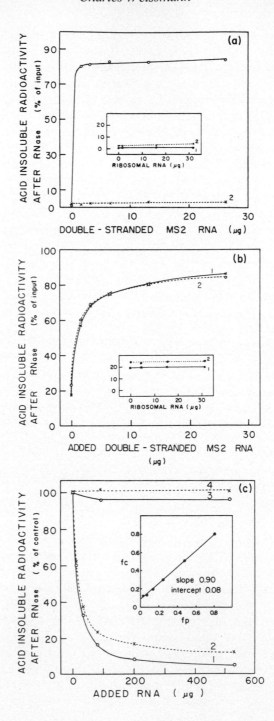

part of the product of RNA synthetase was resistant to ribonuclease when tested prior to deproteinization. However, following extraction with phenol or treatment with sodium dodecylsulfate, over 50% of the product was RNase-resistant (Borst and Weissmann, 1965). The RNase-resistant fraction of the enzymatic product, isolated after phenol extraction, was indistinguishable from the MS2-specific, double-stranded RNA found *in vivo* in regard to its RNase-resistance, T_m, buoyant density and annealing behavior. Analysis by the specific dilution assay showed that the radioactivity in the duplex was almost entirely due to plus strands (Weissmann *et al.*, 1964b). It was therefore concluded that the nonlabeled minus strands found in the duplex were present in the enzyme preparation prior to the incubation with radioactive triphosphates.

The results suggest that RNA synthetase is associated with an endogenous template, namely viral minus strands either in a single- or double-stranded form, and that it catalyzes the synthesis of labeled viral plus strands on incubation with radioactive nucleoside triphosphates. The fact that some of the newly formed, RNase-sensitive plus strands are recovered in a double-stranded state following phenol extraction suggests that the newly formed plus strands and the minus strands are held in proximity, possibly by the enzyme, so that annealing of the complementary strands is greatly facilitated (Borst and Weissmann, 1965).

FIG. 11. Identification of the radioactive product of RNA synthetase as viral RNA of the parental type (double isotope specific dilution assay). (*a*) When a mixture of ^{32}P-MS2 RNA and an excess of unlabeled double-stranded MS2 RNA was heated and reannealed, most of the ^{32}P-RNA was converted to an RNase-resistant form (curve 1). ^{14}C-labeled product of RNA polymerase, primed with *E. coli* DNA, was also present in the mixture and remained RNase-sensitive (curve 2). Ribosomal RNA, in place of double-stranded MS2 RNA, had no effect (inset). (*b*) when the experiment was carried out as above, however, using ^{14}C-labeled product of RNA synthetase, both the ^{32}P (curve 1) and the ^{14}C-labeled RNA (curve 2) were converted to a double-stranded form to an equal extent. (Some radioactive RNA is RNase-resistant without addition of double-stranded RNA, due to unlabeled minus strands derived from the enzyme preparation.) (*c*) A mixture of ^{32}P-MS2 RNA and ^{14}C-labeled product of RNA synthetase was converted into a double-stranded form as described under (*b*) and subjected to the specific dilution assay. It was heated and reannealed in the presence of increasing amounts of MS2 RNA. ^{32}P-labeled (curve 1) and ^{14}C-labeled (curve 2) RNA was displaced from the duplex and converted to an RNase-sensitive form, identifying it as MS2 plus strands. There was no effect of ribosomal RNA (curves 3 and 4). The combination of the steps of Figs. 11b and 11c constitutes the double isotope specific dilution assay. When the fraction of RNase-resistant ^{14}C-radioactivity (f_c) is plotted against the corresponding value for ^{32}P (f_p) a straight line results; the fraction of radioactivity in plus strands is given by the slope, that in minus strands by the intercept (Weissmann, 1965).

Viral RNA Polymerase

August *et al.* (1965) utilized a nonpermissive strain of *E. coli* infected with the amber mutant *sus*-11 as a source of enzyme. This host-phage system does not yield viable virus particles, however it produces 5–10 times the normal amount of viral polymerase activity, and of virus-specific double-stranded RNA (Lodish *et al.*, 1964; Lodish and Zinder, 1966b). The 100-fold purified preparation required RNA as a primer. A variety of RNAs, such as f2, TMV, ribosomal or soluble RNA, was about equally active but homopolynucleotides were not active. The nearest neighbor frequency of the product was different for TMV or f2 primer, suggesting that the added RNA determined the nucleotide sequence of the product. Its base composition, and the finding that after deproteinization more than half of the newly formed RNA was resistant to RNase (Shapiro and August, 1965) led to the suggestion that it was a complementary or minus copy of the template and that the enzyme functions *in vivo* to catalyze the synthesis of a minus strand complementary to the viral plus strand.

Mu-9, an amber mutant of phage MS2, induced, like *sus*-11, the production of large amounts of polymerase and double-stranded RNA, when grown on a nonpermissive host such as *E. coli* Hfr 3000 (Feix *et al.*, 1966). Annealing assays showed that as much as 60% of the product formed *in vitro* by extracts of *E. coli* Hfr 3000 infected with *M*u-9 consisted of viral minus strands, in marked contrast to the results obtained with crude extracts of cells infected with the wild-type phage (MS2), where the corresponding value was about 15%. These findings support the conclusion of Shapiro and August (1965).

RNA Replicase

Haruna *et al.* (1963, 1965a) isolated, from *E. coli* infected with phage MS2, an enzyme that under appropriate ionic conditions specifically required the addition of MS2 RNA for activity. Ribosomal, soluble, and TMV RNA were ineffective, but TYMV RNA caused some stimulation. Significantly, the RNA of phage Qβ, which shows no base sequence homology with MS2 RNA (Weissmann, 1966), did not stimulate the MS2 replicase. Moreover, a replicase isolated from *E. coli* infected with phage Qβ was stimulated by intact Qβ RNA but not by MS2 RNA. In contrast to the two enzyme preparations described in the preceding sections, Qβ replicase was found to catalyze the synthesis of RNA over a period of several hours, producing many times the amount of input RNA. It was also found to be saturated at relatively low ratios of RNA to protein (1 μg/40 μg of protein). After long incubation periods, when the amount of RNA synthesized greatly exceeded that originally added as template, the isolated product was (*a*) as effective a template for Qβ-induced replicase as Qβ RNA itself (Haruna and Spiegelman, 1965b) and (*b*) infective for *E. coli* spheroplasts (Spiegelman *et al.,* 1965). In contrast to the reaction just described, an abnormal synthesis results when Qβ replicase is primed with partially degraded Qβ RNA—the rate of synthesis is lower, its extent is limited and the product is largely double-stranded (Haruna and Spiegelman, 1965c, 1966).

It is clear from the above results, that the replicase system is capable of synthesizing viral RNA of the parental type. The question therefore arises as to whether this replication involves preliminary formation of viral minus strands as suggested by the *in vivo* experiments and by the results with other enzyme preparations mentioned earlier. Annealing assays of the radioactive product isolated after various times of incubation of Qβ replicase with intact Qβ RNA (Weissmann and Feix, 1966) showed that, during the first few minutes, minus strands were predominantly, if not exclusively, synthesized, but that more plus than minus strands were formed later on (Fig. 12). About 20% of the product was RNase-resistant at any time during incubation, but the bulk of the minus strands was RNase-sensitive, and demonstrable only through annealing with Qβ RNA.

Haruna and Spiegelman (1966) have suggested a different interpretation of these results. They postulate that the parental viral RNA has a

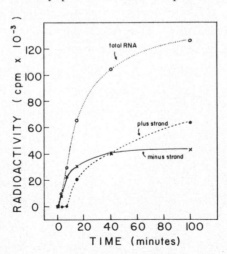

FIG. 12. Analysis of the ¹⁴C-labeled product of Qβ replicase. Qβ replicase was incubated with Qβ RNA and radioactive nucleoside triphosphates. Qβ plus (●— — —●) and minus (×...×) strands were determined by the double isotope specific dilution assay (cf. Fig. 11) (Weissmann and Feix, 1966).

TABLE III

FAILURE OF Qβ PLUS STRANDS TO SELF-ANNEAL[a]

Labeled RNA	Addition	Heated	Total acid-insoluble radioactivity (cpm)	Acid-insoluble radioactivity after RNase (cpm)
1. ³²P-Qβ RNA (heterogeneous)[b] (0.3 µg)	None	No	13,180	111 (0.83%)
	None	Yes	—	117 (0.89%)
	Qβ RNA (5µg)	Yes	—	126 (0.95%)
	Qβ RNA (10µg)	Yes	—	124 (0.94%)
	Qβ RNA (20 µg)	Yes	—	117 (0.89%)
2. ³²P-Qβ RNA (27 S)[c] (0.2 µg)	None	No	5810	8 (0.14%)
	Qβ RNA (5µg)	Yes	—	7 (0.12%)
	Qβ RNA (10 µg)	Yes	—	9 (0.15%)
	Qβ RNA (20 µg)	Yes	—	9 (0.15%)
3. ³²P-Qβ RNA (fragmented, 4–8 S)[d] (0.1 µg)	None	No	2494	3 (0.12%)
	None	Yes	—	3 (0.12%)
	Qβ RNA (3 µg)	Yes	—	2 (0.08%)
	Qβ RNA (10 µg)	Yes	—	2 (0.08%)
	Qβ RNA (16 µg)	Yes	—	1 (0.04%)
	double-stranded Qβ RNA (10 µg)	Yes[e]	—	870 (35%)

[a] Samples were dissolved in 0.05 ml 2.5 × SSC (0.375 M NaCl, 0.0375 M sodium citrate). Heating was at 85°C for 1 hour in a sealed tube. The samples were diluted to 2 ml and a final concentration of 1 × SSC. Acid insoluble radioactivity was determined following incubation with 50 µg/ml of pancreatic RNase and 25 units/ml of RNase T1 for 30 min at 25°C. All determinations were carried out in duplicate.

[b] ³²P-labeled Qβ-RNA was prepared by phenol extraction of ³²P-labeled Qβ phage, purified as described elsewhere (Weissmann and Feix, 1966). About half of the labeled RNA had an $s_{20, w}$ of 27 S, the remainder sedimented more slowly.

[c] The RNA preparation described under (b) was centrifuged through a sucrose density gradient and the fractions corresponding to an $s_{20, w}$ of 27 S were pooled.

[d] The purified 27 S ³²P-Qβ RNA, described under (c), was heated at 100°C for 15 min and centrifuged through a sucrose density gradient. The fractions corresponding to an $s_{20, w}$ of 4 to 8 S were pooled.

[e] The samples were first heated at 120°C for 3 min in order to denature the double-stranded, unlabeled RNA, and then annealed for 1 hour at 85°C.

beginning sequence rich in adenine, and a terminal, complementary sequence rich in uracil. In support of this view, they report that 6% of ^{3}H-labeled Qβ RNA (plus strands) can be converted to a RNase-resistant state by heating with excess unlabeled Qβ plus strands. They propose then, that RNA synthesis by Qβ replicase leads to the formation of plus strands only, and that the large proportion of annealable RNA found is due not to synthesis of minus strands, but to that of A-rich beginning sequences presumed to be present in plus strands. Three objections to this interpretation may be raised. (1) Using ^{32}P-Qβ RNA prepared from carefully purified Qβ virus, we have been unable to demonstrate any self-annealing whatsoever under conditions similar to those reported by Haruna and Spiegelman (1966) (Table III). Analogous negative results have been reported earlier for ^{32}P-labeled MS2 RNA (Weissmann *et al.*, 1964a). The evidence for the postulated self-complementarity cannot be confirmed by us. (2) Even if Qβ could anneal with itself, the proportion of annealable RNA formed by Qβ-replicase at a time when, according to Haruna and Spiegelman (1966), the middle sequences of plus strands are being formed, is far in excess of the 6% which were attributed to self-complementary regions. (3) After a long-term incubation, 30% of the radioactive RNA sedimenting in the 27 S region (corresponding to full-length single-stranded viral RNA) was annealable with plus strands (Fig. 13). If this behavior of the RNA were due to self-complementarity, then viral RNA itself should be self-complementary to the same extent, which it is not. Other arguments of Haruna and Spiegelman in favor of their views

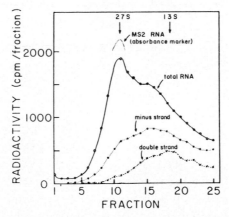

FIG. 13. Sedimentation analysis of the ^{14}C-labeled product of Qβ replicase. The product was centrifuged through a sucrose gradient and the fractions were analyzed for total radioactivity (—○—○—) and RNase-resistant radioactivity (.. × .. × ..). Minus strands (.. ● ... ● ..) were determined by annealing with unlabeled Qβ RNA (Weissmann and Feix, 1966).

are based on the assumption that the *in vitro* reaction proceeds synchronously and that all Qβ RNA added as a template is in fact utilized as such. Since no evidence was offered in support of either of these assumptions, the conclusions based on them cannot be judged at present.

Haruna and Spiegelman (1966) furthermore claim that no double-stranded RNA is formed during RNA synthesis by Qβ-replicase, although from 12 to 20% (Haruna and Spiegelman, 1966; Weissmann and Feix, 1966) of the radioactive RNA formed is RNase-resistant. This conclusion is based on their failure to render the RNase-resistant RNA sensitive to RNase by heating it to 100°C in SSC (0.15 M NaCl, 0.015 M sodium citrate solution). However, since the Tm of double-stranded Qβ RNA is 97.5°C in SSC (see Fig. 14), and since complete denatura-

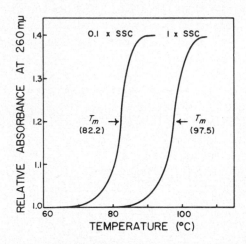

FIG. 14. Thermal transition of Qβ RNA. Qβ RNA was heated in 0.1 × SSC or in 1 × SSC at 2.4°C/min; temperature and absorbance were recorded automatically.

tion requires heating to about 10°C above the T_m, the conditions required for denaturation were not met. By contrast, it has been shown that if the product of Qβ replicase is heated at 100°C in 0.01 × SSC, more than 99% of the product of Qβ replicase becomes RNase-sensitive (Weissmann and Feix, 1966). By these criteria there is little doubt that some double-stranded RNA is formed by Qβ replicase even when intact template is used. Its role in the *in vitro* reaction remains to be elucidated.

Taken as a whole, the results discussed above suggest that Qβ replicase preparations catalyze the formation of 2 products, namely viral plus and viral minus strands, some of which occur in a double-stranded form; it is not known whether one or two separable enzymes are involved.

Conclusion and Summary

The evidence discussed in the preceding sections is compatible with the following sequence of events. Following adsorption of the phage to the male-specific pili of the host, the viral RNA strand penetrates the cell and is used as messenger to direct the synthesis of one or possibly two viral RNA polymerases. Within a few minutes after infection, the parental RNA is found to participate in a replicating complex. Synthesis of both viral minus and plus strands ensues and a large number of additional replicating complexes are formed. Whereas, at first about equal amounts of plus and minus strands are formed, this ratio soon becomes about 10 : 1 in favor of the plus strands.

The structure of the replicating complex is not known. It may consist of one or more viral RNA polymerase molecules as well as of minus and plus strands, and possibly ribosomes attached to the latter. In particular, it is not clear if, or to what extent, minus and plus strands form a tight double-helical structure in the native complex. After deproteinization, the so-called replicative intermediate is isolated, which consists of both double- and single-stranded RNA and contains a full-length minus strand. It may have the structure outlined in Fig. 5. Precursor experiments suggest that synthesis of viral RNA involves a mechanism whereby newly formed plus strands displace preexisting ones from the replicating complex while the minus strands are conserved (semi-conservative asymmetric replication). The newly formed plus strands may (a) undergo maturation and give rise to progeny virus, (b) function as messenger for the formation of virus-specific proteins or (c) participate in a new replicating complex. The results of *in vitro* studies of viral RNA synthesis are compatible with a mechanism involving a minus strand as a template for the synthesis of progeny plus strands, although an alternative interpretation of the relevant data has been attempted by Haruna and Spiegelman (1966). Further work will be necessary to substantiate their suggestion that full-length minus strands are neither formed nor required for viral RNA replication (Haruna and Spiegelman, 1966). A careful analysis of the replicase system will undoubtedly yield a clear answer to this question as well as to the problem as to whether one or two separable enzymes are involved in viral RNA synthesis.

REFERENCES

Ammann, J., Delius, H., and Hofschneider, P. H. (1964). *J. Mol. Biol.* **10,** 557.
August, J. T., Cooper, S., Shapiro, L., and Zinder, N. D. (1963). *Cold Spring Harbor Symp. Quant. Biol.* **28,** 95.

August, J. T., Shapiro, L., and Eoyang, L. (1965). *J. Mol. Biol.* **11,** 257.
Billeter, M. A., and Weissmann, C. (1965). Unpublished experiment.
Billeter, M. A., Libonati, M., Viñuela, E., and Weissmann, C. (1966a). *J. Biol. Chem.* **241,** 4750.
Billeter, M. A., Weissmann, C., and Warner, R. C. (1966b). *J. Mol. Biol.* **17,** 145.
Borst, P., and Weissmann, C. (1965). *Proc. Natl. Acad. Sci. U.S.* **54,** 982.
Brinton, C. C., Jr., Gemski, P., Jr., and Carnahan, J. (1964). *Proc. Natl. Acad. Sci. U.S.* **52,** 776.
Burdon, R. H., Billeter, M. A., Weissmann, C., Warner, R. C., Ochoa, S., and Knight, C. A. (1964). *Proc. Natl. Acad. Sci. U.S.* **52,** 768.
Colthart, L., and Weissmann, C. (1966). Unpublished experiments.
Cooper, S., and Zinder, N. D. (1962). *Virology* **18,** 405.
Cooper, S., and Zinder, N. D. (1963). *Virology* **20,** 605.
Davis, J. E., and Sinsheimer, R. L. (1963). *J. Mol. Biol.* **6,** 203.
Davis, J. E., Strauss, J. H., Jr., and Sinsheimer, R. L. (1961). *Science* **134,** 1427.
Doi, R. H., and Spiegelman, S. (1962). *Science* **138,** 1270.
Doi, R. H., and Spiegelman, S. (1963). *Proc. Natl. Acad. Sci. U.S.* **49,** 353.
Edgell, M. H., and Ginoza, W. (1965). *Virology* **27,** 23.
Ellis, D. B., and Paranchych, W. (1963). *J. Cellular Comp. Physiol.* **62,** 207.
Enger, M. D., and Kaesberg, P. (1965). *J. Mol. Biol.* **13,** 260.
Erikson, R. L., Fenwick, M. L., and Franklin, R. M. (1964). *J. Mol. Biol.* **10,** 519.
Erikson, R. L., Fenwick, M. L., and Franklin, R. M. (1965). *J. Mol. Biol.* **13,** 399.
Feix, G., Davern, C. I., and Weissmann, C. (1966). Unpublished results.
Fenwick, M. L., Erikson, R. L., and Franklin, R. M. (1964). *Science* **146,** 527.
Geidushek, E. P., Moohr, J. W., and Weiss, S. B. (1962). *Proc. Natl. Acad. Sci. U.S.* **48,** 1078.
Gesteland, R., and Boedtker, H. (1964). *J. Mol. Biol.* **8,** 496.
Godson, G. N., and Sinsheimer, R. L. (1966). *J. Mol. Biol.* (in press).
Haruna, I., and Spiegelman, S. (1965a). *Proc. Natl. Acad. Sci. U.S.* **54,** 579.
Haruna, I., and Spiegelman, S. (1965b). *Science* **150,** 884.
Haruna, I., and Spiegelman, S. (1965c). *Proc. Natl. Acad. Sci. U.S.* **54,** 1189.
Haruna, I., and Spiegelman, S. (1966). *Proc. Natl. Acad. Sci. U.S.* **55,** 1256.
Haruna, I., Nozu, K., Ohtaka, Y., and Spiegelman, S. (1963). *Proc. Natl. Acad. Sci. U.S.* **50,** 905.
Haywood, A. M., and Sinsheimer, R. L. (1965). *J. Mol. Biol.* **14,** 305.
Hofschneider, P. H. (1963). *Z. Naturforsch.* **18b,** 203.
Kaerner, H. C., and Hoffmann-Berling, H. (1964). *Z. Naturforsch.* **19b,** 593.
Kelly, R. B., and Sinsheimer, R. L. (1964). *J. Mol Biol.* **8,** 602.
Kelly, R. B., Gould, J. L., and Sinsheimer, R. L. (1965). *J. Mol. Biol.* **11,** 562.
Langridge, R., Billeter, M. A., Borst, P., Burdon, R. H., and Weissmann, C. (1964). *Proc. Natl. Acad. Sci. U.S.* **52,** 114.
Lodish, H. F. (1966). Personal communication.
Lodish, H. F., and Zinder, N. D. (1966a). *Science* **152,** 372.
Lodish, H. F., and Zinder, N. D. (1966b). *J. Mol. Biol.* **19,** 333.
Lodish, H. F., Cooper, S., and Zinder, N. D. (1964). *Virology* **24,** 60.
Lodish, H. F., Horiuchi, K., and Zinder, N. D. (1965). *Virology* **27,** 139.
Loeb, T., and Zinder, N. D. (1961). *Proc. Natl. Acad. Sci. U.S.* **47,** 282.
Mitra, S., Enger, M. D., and Kaesberg, P. (1963). *Proc. Natl. Acad. Sci. U.S.* **58,** 68.

Montagnier, L., and Sanders, F. K. (1963). *Nature* **199**, 664.

Nathans, D. (1965). *J. Mol. Biol.* **13**, 521.

Nathans, D., Notani, G., Schwartz, J. H., and Zinder, N. D. (1962). *Proc. Natl. Acad. Sci. U.S.* **48**, 1424.

Notani, G. W., Engelhardt, D. L., Konigsberg, W., and Zinder, N. D. (1965). *J. Mol. Biol.* **12**, 439.

Ochoa, S., Weissmann, C., Borst, P., Burdon, R. H., and Billeter, M. A. (1964). *Federation Proc.* **23**, 1285.

Ohtaka, Y., and Spiegelman, S. (1963). *Science* **142**, 493.

Paranchych, W., and Ellis, D. B. (1964). *Virology* **14**, 635.

Paranchych, W., and Graham, A. F. (1962). *J. Cellular Comp. Physiol.* **60**, 199.

Shapiro, L., and August, J. T. (1965). *J. Mol. Biol.* **11**, 272.

Spiegelman, S. (1965). "Symposium on Macromolecular Metabolism." N.Y. Heart Assoc., New York.

Spiegelman, S., Haruna, I., Holland, I. B., Beaudreau, G., and Mills, D. (1965). *Proc. Natl. Acad. Sci. U.S.* **54**, 919.

Stanley, W. M. (1963). Thesis, University of Wisconsin.

Strauss, J. H., Jr., and Sinsheimer, R. L. (1963). *J. Mol. Biol.* **7**, 43.

Watanabe, I. (1964). *Nippon Rinsho* **22**, 243.

Weissmann, C. (1965). *Proc. Natl. Acad. Sci. U.S.* **54**, 202.

Weissmann, C. (1966). *Federation Proc.* **25**, 785.

Weissmann, C., and Borst, P. (1963). *Science* **142**, 1188.

Weissmann, C., and Feix, G. (1966). *Proc. Natl. Acad. Sci. U.S.* **55**, 1264.

Weissmann, C., Ochoa, S. (1967). *Progr. Nucleic Acid Res. Mol. Biol.* **6**, 353.

Weissmann, C., Simon, L., and Ochoa, S. (1963a). *Proc. Natl. Acad. Sci. U.S.* **49**, 407.

Weissmann, C., Simon, L., Borst, P., and Ochoa, S. (1963b). *Cold Spring Harbor Symp. Quant. Biol.* **28**, 99.

Weissmann, C., Borst, P., Burdon, R. H., Billeter, M. A., and Ochoa, S. (1964). *Proc. Natl. Acad. Sci. U.S.* **51**, 682.

Weissmann, C., Borst, P., Burdon, R. H., Billeter, M. A., and Ochoa, S. (1964b). *Proc. Natl. Acad. Sci. U.S.* **51**, 890.

Weissmann, C., Colthart, L., and Libonati, M. (1966). In preparation.

Zinder, N. D. (1963). *Perspectives Virol.* **3**, 58.

Infectious M12 Phage Replicative Form RNA: A Tool for Studying Viral RNA Replication

P. H. Hofschneider, J. Ammann, and B. Francke

MAX-PLANCK-INSTITUT FOR BIOCHEMISTRY, MUNICH, GERMANY

Since 1964, when several laboratories reported on phage specific double-stranded RNA in RNA phage-infected cells (references as cited by Ammann *et al.*, 1964b, and Francke and Hofschneider, 1966), different studies were performed on the function of this RNA species in replicating viral RNA.

Today, double-stranded RNA is widely assumed to be an intermediate in the process of replicating single-stranded phage RNA of the f2 phage class. Controversial results exist only for the replication of phage Qβ RNA (Haruna and Spiegelman, 1966; Weissmann and Feix, 1966). According to the most probable model, double-stranded molecules (replicative form, RF) replicate phage type RNA (plus strands) semiconservatively and asymmetrically (Weissmann *et al.*, 1964; Fenwick *et al.*, 1964). The intact complementary minus strand in the RF serves as a template on which plus strands are synthesized. The first plus strand synthesized is thought to be displaced by the synthesis of the following one. According to this hypothesis a heterogeneous population of double-stranded molecules should exist which is composed of two classes: pure double-stranded molecules and double-stranded molecules with nascent single-stranded tails of different length (the latter class termed replicative intermediate by Fenwick *et al.*, 1964).

The model described is based mainly on two lines of arguments. (1) In pulse-labeling experiments one part of the pulse passes at first through an RNase-resistant, i.e., double-stranded RNA, and is later found in single-stranded plus strands (Fenwick *et al.*, 1964; Weissmann, 1965; Kelly *et al.*, 1965; Lodish and Zinder, 1966). Another part of the pulse stays in the RNase resistant complexes; it is incorporated into minus strand material as revealed by hybridization experiments (Billeter *et al.*,

1967). (2) The sedimentation constants of the RNase-resistant complexes vary widely indicating heterogeneity of the material (Fenwick *et al.*, 1964; Billeter *et al.*, 1967; Erikson and Gordon, 1966; Franklin, 1966). The relationship between the molecular weight and the sedimentation constant of DNA was used for the calculation of the molecular weight of double-stranded RNA. The molecular weight of some of the RNase-resistant RNA exceeded twice the molecular weight of single-stranded phage RNA. Thus the complexes could contain complete plus and minus strands.

While these arguments are plausible, they are not necessarily convincing. It is not yet established that the calculation of molecular weights with the formula for DNA is also applicable for double-stranded RNA. One can argue that these molecules actually have smaller molecular weights. They could then be nonfunctional fragments composed of plus strands being complemented with minus counterparts of different length. But even if one assumes the molecular weight calculation to be correct and in good agreement with the postulated size of RF, the continuity of the strands is not established, i.e., single-stranded breaks in the double-stranded molecule could not be excluded.

For a definite proof of the model, it is not sufficient to study the properties of RNase-resistant materials, only. It must be established that these materials contain both the plus and minus strands in a biologically intact form. Intact RF should contain a plus strand which is as infectious as the mature phage RNA. Therefore, the demonstration of its infectivity is one of the possibilities to prove the intactness of, at least, one of the RF strands.

Thus, having found the way to isolate and to assay infectious RF (Ammann *et al.*, 1964b; Delius and Hofschneider, 1967) of RNA phage M12 (Hofschneider, 1963), studies were performed on its formation, size, and strand integrity. The results support the findings mentioned above as well as the corresponding ones in animal virus systems (Hausen, 1965; Katz and Penman, 1966; Darnell *et al.*, 1966; Bishop *et al.*, 1966).

At first, a method for preparative isolation of infectious M12 RF and attempts to resolve this RF into two further fractions will be outlined (results, sections 1 and 2). A technique will be described which allows the following of the time kinetics of RF formation in M12 infected cells. As it was found, infectious M12 RF is synthesized prior to the appearance of progeny single-stranded phage RNA. It is not accumulated late in infection as is RNase resistant material (Weissmann *et al.*, 1964).

After we selected chromatographically for a completely double-stranded material, an infectious RF was obtained sedimenting homogeneously with 15 S. When the purification procedure was modified to avoid removal of partly single-stranded RF, another class of infectious RF was detected which sedimented with approximately 17–27 S. In the infectious 15 S RF, the minus strand has the same sedimentation constant as the infectious plus strand. Consequently, the faster sedimenting infectious RF (17–27 S) is probably identical with the replicative intermediate; and, it possesses single-stranded tails of "plus" type RNA (section 4). These findings appear to exclude the possibility that RF molecules are only artifacts such as RNase resistant complexes of fragmented plus strands and complementary counterparts. On the contrary, the observations support, for 15 S RF, the integrity of the plus as well as of the minus strand. Thus, a mandatory morphological prerequisite of models which involve double-stranded complementary RNA in the replication of single-stranded phage RNA is met.

Preparative Isolation and Properties of M12 RF

In principle, M12 RF was isolated from the nucleic acid fraction of cells infected for 30 min. DNase-digestion was followed by (1) specific denaturation of DNA, (2) by precipitation of ss RNA by addition of 10% NaCl, and (3) by final chromatography on MAK columns (Ammann *et al.*, 1964b). A typical elution pattern is shown in Fig. 1.

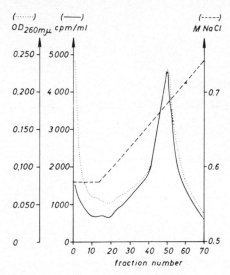

Fig. 1. Final purification of M12 RF on a MAK column. M12 RF is eluted with a linear NaCl gradient from 0.58–0.78 M at 0.69 M. ———— Eluted material measured as ^{32}P cpm/ml; · · · · · by absorption at 260 mμ; – – – – – NaCl gradient (moles/liter). MAK columns were prepared as described by Mandell and Hershey (1960).

The methods yield RF material of high RNase resistance and of a buoyant density of 1.614 g/ml in cesium sulfate which is 0.02 g/ml less than the buoyant density of M12 ss RNA. In the electron microscope, M12 RF looks like double-stranded DNA (Fig. 2). Besides some shorter segments, the molecules are mostly 11,300 to 11,500 Å long. Since this length corresponds to a molecular weight of approximately 2.4×10^6 which is double the molecular weight of ss phage RNA, the major part of the preparation should consist of intact RF molecules.

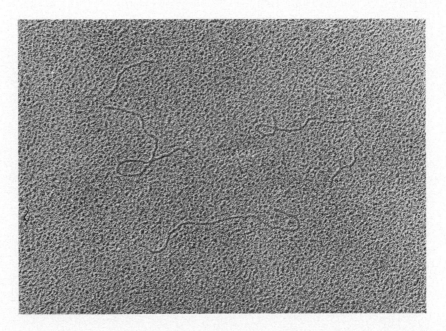

FIG. 2. Electron micrograph of M12 RF (RF I): total magnification 50,000-fold ± three percent. Three molecules reach lengths between 11,300 and 11,500 A. Prepared according to Kleinschmidt and Zahn (1959).

The best evidence for the integrity of at least one strand of the RF molecule comes from the infectivity of the denatured RF for *E. coli* spheroplasts. The ambient temperature melting curve (absorption at 260 mμ) is paralleled by the increase of infectivity obtained when samples taken at ambient temperatures were chilled quickly and tested for infectivity.

Three facts rule out a possible contamination by ss phage RNA or the dissociation of noninfectious RNA aggregates to an infectious form: (a) the sharpness of the transition which takes place over a range of 6 degrees centigrade, (b) the similar T_M values of both biological and absorbance curves (89°C and 91°C, respectively, in 0.001 *M* NaCl, 0.001 *M* Tris, pH 7), and (c) the reproducible increase in infectivity of at least 10,000-fold.

RNase treatment of RF before heat denaturation (1 μg RNase/ml for 5 min at 37°C) inactivates its infectivity at least 1000-fold. However, no more than 15% of the RF is converted to acid-soluble nucleotides by extensive treatment with RNase (100 μg RNase/ml). This observation explains why no infectious RF can be obtained, when RNase-digestion is involved in the purification procedure.

Preparative Segregation of Infectious M12 RF

By repeated chromatography on MAK columns M12 RF can be segregated into at least 2 fractions. Reproducible results, however, can only be obtained when any pretreatment (see the first section) of the crude nucleic acid fraction is avoided.

In Fig. 3, three successive chromatographic runs (1, 2, and 3) are superimposed. After having collected and rechromatographed the fractions 50–145 of run 1, pattern 2 was obtained. RF infectivity was eluted as expected from Fig. 1 at 0.69 *M* in fractions 90–110 (not shown in Fig. 3). Further infectivity was found in fractions 130–160. Rechromatography of the fractions 130–160 of run 2 gave curve 3.

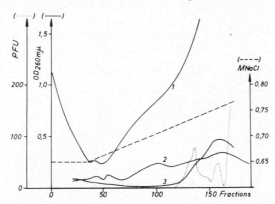

FIG. 3. Segregation of RF II from RF I by MAK column chromatography. (————), Eluted material measured by light absorption at 260 mμ. Crude nucleic acids of *E. coli* infected for 30 min with M12 phages (1); rechromatography of fractions 50–145 of curve 1 (2); rechromatography of fractions 130–160 of curve 2 (3). (– – – – –), NaCl gradient (moles/liter). (· · · · ·), infectivity after denaturation against *E. coli* spheroplasts measured as plaque forming units (PFU). (The infectivity of RF I between fractions 90 and 110 is not shown in this figure). For details on MAK columns and nucleic acid preparation see Hofschneider *et al.* (1967).

A pattern of infectivity after denaturation was found still ranging from fraction 120 to fraction 160 of run 3 with mainly two maxima eluted at NaCl concentrations of about 0.74 *M* and 0.75 *M,* respectively. Thus, it becomes obvious that RF can be chromatographically segregated into two main fractions. Both fractions are infectious only after denaturation. The first one (RF I) is eluted at a low salt concentration (0.69 *M* NaCl) similar to the elution concentration for double-stranded DNA; the second fraction (RF II) is eluted at a much higher salt concentration (0.74–0.75 *M* NaCl) close to the point where infectious single-stranded M12 RNA comes off the column. Further studies on this line are in progress. Taking into account the present results, RF I is assumed to consist of completely double-stranded molecules whereas RF II, according to its chromatographical single-strandlike behavior, could be partly single-stranded. It is probably identical with 22 S RF. Results on this will be presented on p. 332.

Formation of Infectious RF in M12 Infected Cells

To follow the formation of RF in the infected cells during replication of RNA phages, several techniques have been applied. However, it was not determined to what degree the RF, i.e. the RNase resistant material studied, was biologically intact. As reported in the first section M12 RF is not infectious for *E. coli* spheroplasts unless it has been denatured to single strands. Using this type of infectivity as a criterion, it seemed possible to determine the formation of biologically intact RF. Since in infected cells an excess of phage RNA is present which causes infectivity before melting, the problem was to find a reproducible technique for separating RF quantitatively from infectious single strands.

SEPARATION OF RF FROM SS PHAGE BY SELECTIVE ADSORPTION

Benzinger *et al.* (1967) developed a selective adsorption procedure to separate, in a single step, RF of the DNA phage φX174 from single-stranded φX DNA. They mixed the total nucleic acids from the infected cells with methylated albumin coated silicic acid (MAS) in buffer which had a salt concentration at which the single strands adsorb to the MAS and the RF stays in solution. After sedimentation of the MAS, enriched RF is found in the supernatant.

A modification of this method (Francke and Hofschneider, 1966) allows purification of infectious M12 RF from ss phage RNA. Separations of total infectious nucleic acids of M12 infected cells are shown in Fig. 4A–D for 10, 20, 60, and 120 min after infection. At the times mentioned, nucleic acid samples were prepared from an infected 5 liter culture, and each one was split into 12 portions. After precipitation with 3 volumes ethanol, these portions were adsorbed to MAS at 12 different molarities of NaCl, between 0.8 *M* and 1.5 *M;* then, MAS was removed by sedimentation. The adsorption step was repeated twice. The supernatant was assayed for infective RNA before and after denaturation as described by Francke and Hofschneider (1966).

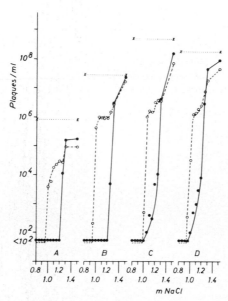

Fig. 4. Separation of total infectious intracellular RNA by selective adsorption to MAS. *E. coli* cells were harvested at 10 min (A), 20 min (B), 60 min (C), and 120 min (D) after infection, and the nucleic-acid fractions were tested for total infectious RNA (x·····x). After adsorption to MAS at different NaCl molarities of sodium chloride-sodium acetate buffer (0.02 *M* sodium acetate, pH 5.0) and after sedimentation of the MAS the supernatant was tested for infectivity (PFU/ml supernatant) before (○———○) and after (○– – – –○) heat-denaturation (1 min, 100°C in 0.003 *M* EDTA). For all technical details, see Francke and Hofschneider (1966).

In all 4 curves of Fig. 4A–D, no infectivity is found in the supernatant up to a molarity of 0.9 *M* NaCl. At 1.0 *M,* the infectivity after denaturation (RF infectivity) greatly increases and reaches a plateau between 1.0 *M* and 1.2 *M.* Above 1.2 *M* increasing amounts of phage RNA are found that show a decrease of infectivity after denaturation, typical for ss phage RNA (ss infectivity).

Two controls have been made to evaluate the question of how much RF is lost in each separation step. Table I shows the biological titer of

the total RNA (30 min after infection) and of the RNA in the supernatant after one and after two separation steps at 1.15 *M*. In the second
step, only single strands (infectivity before melting) are removed;
whereas, the RF titer (infectivity after melting) stays the same. This
shows that in the second and further steps no RF is lost. In the first
step the conditions are different. The large amount of bacterial nucleic
acids adsorbing to the MAS could include RF mechanically. These conditions were reconstructed by mixing the nucleic acids extracted from
4×10^{10} uninfected cells with already purified RF. After one adsorption
step 80% of the biological activity of the input RF was recovered.

TABLE I

Infectivity of a Deproteinized Lysate (30 Min after Infection) in the Supernatant after One and Two Adsorption Steps (at 1.15 M NaCl) to MAS

RNA	PFU/ml	
	Before denaturation	After denaturation
Total infectious RNA	5.3×10^7	2.7×10^7
Infectious RNA after one adsorption step	5.2×10^5	1.4×10^6
Infectious RNA after two adsorption steps	10^2	1.3×10^6

These results show that an infectious RF can be separated with good
recoveries and almost quantitatively from ss phage RNA by the MAS
method between 1.0–1.2 *M* NaCl: At a recovery of 80%, RF infectivity
increases by a factor of up to 10^6 after separation.

The question as to whether there are other forms of RF that appear
in MAS supernatants at molarities higher than 1.2 *M* NaCl and close
to those for elution of ss phage RNA will be answered in section 4.

TIME KINETICS OF RF AND SS PHAGE RNA FORMATION

Having determined the conditions for separating infectious RF, the
MAS technique was applied to study the time kinetics of RF and ss
phage RNA formation in infected cells. The results are shown in Fig. 5.
The main increase of infectious progeny RNA (curve 3) is found to be
10 min before that of intact intracellular phage (curve 2). The majority
of RF is synthesized between 10 and 20 min after infection (curve 4),
which is again 10 min before the appearance of free ss phage RNA.

After 20 min, the level of RF stays practically constant; it starts decreasing not earlier than at 90 min, at which time the cells have started lysing.

The data in Fig. 5 for infectious RNA are phage equivalents of single-stranded RNA. Therefore, the data for the infectivity before denaturation give the actual amount of ss phage RNA in the total lysate (curve 3). To obtain the actual amounts of RF present, a correction must be made for heat inactivation. Assuming that RF loses 40% of its infectivity by heat denaturation as does ss phage RNA and that the plating efficiency of RF is the same as for ss phage RNA (see section 4) then it follows that there is only one infectious RF molecule per cell. In other experi-

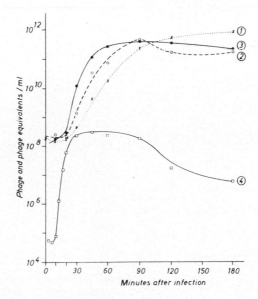

FIG. 5. Time kinetics and M12 phage, ss phage RNA, and RF formation. From an infected 2.5 liter culture 200 ml samples (5×10^{10} cells) were removed at different times and the nucleic acid fractions were prepared. The adsorption to MAS was done at 1.15 M NaCl and repeated once. Extracellular phage (1) was tested from the supernatant after sedimentation of the cells. Intracellular phage (2) was tested after lysis of the sedimented cells with 1% sodium dodecyl sulphate. After phenol extraction and ethanol precipitation a test was made for total infectious RNA (3). Following adsorption of ss phage RNA to MAS at 1.15 M NaCl, RF was assayed before and after denaturation. Only infectivity after denaturation is given (4), since in no case was the infectivity before denaturation more than 1%. The data given for infectious RNA are presented as phage equivalents for single-stranded RNA per ml of original culture. In different experiments one PFU was obtained for each 10^4 to 10^5 phage equivalents of a standard RNA preparation. For more technical details, see legend to Fig. 4.

ments up to ten infectious molecules per cell were found. The total number of RF molecules may be still higher due to other RF forms, which are not infectious or which are lost by the separation technique used.

As to the function of RF it is the main result of these studies that infectious RF accumulates early in infection as it would be expected for a precursor of ss phage RNA. It is synthesized between 10 and 20 min after infection, after which time its amount stays constant until lysis. The maximum amount of infectious RF is reached at the time when the progeny ss phage RNA starts to appear.

This time kinetics for the formation of infectious RF is comparable to the rate of incorporation of ^{32}P-labeled parental phage RNA into the RF fraction as obtained by the MAS separation at 1.0 M NaCl (Fig. 6).

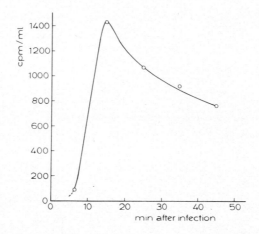

FIG. 6. Incorporation of parental phage RNA in the RF fraction and their subsequent release. *E. coli* cells were infected with ^{32}P-labeled M12 phages and nucleic acids, fractions being prepared at the times indicated. Following adsorption of ss phage RNA to MAS at 1.0 M NaCl (compare Fig. 4), the supernatants were tested for total acid precipitable counts. For technical detail, see Ammann *et al.* (1964a).

However, whereas the labeled parental RNA gradually becomes removed from the RF fraction shortly after the maximum is reached, the infectious RF remains at its maximum for a longer period of time. The results obtained by Kelly *et al.* (1965), who find a periodic incorporation of labeled parental MS2-RNA into RNase-resistant material, suggest that RF, formed at early times after infection, is subjected to a turnover. This could account for the difference described above.

By assaying for the overall synthesis of complementary strands, however, Weissmann *et al.* (1964) find a completely different time kinetics.

Amounts of complementary strand, which can be measured by their reannealing technique, continue increasing at least until 40 min after infection. Therefore, the question arises whether, besides the infectious RF separated by selective adsorption, there are other forms of RF or RF-like fragments accumulating in the cell.

The Two Classes of Infectious RF

As outlined in the introduction two classes of functional intact RF molecules are postulated: completely double-stranded molecules, and molecules with the complete plus strand being partly displaced from the minus strand by a nascent plus strand (replicative intermediate or "tailed" RF). Assuming this postulate mainly based on turnover and sedimentation studies of RNase-resistant complexes to be true, also two classes of infectious RF should be detectable. To separate both "normal" and tailed RF, the selective adsorption to MAS and band-sedimentations were used.

Fractionation of RF by Selective Adsorption to MAS

The ^{32}P-labeled nucleic acids of 6×10^{11} cells were divided into 11 samples and precipitated with three volumes of ethanol. The precipitates were dissolved at 10 different salt concentrations (0.9–2.0 M NaCl) as indicated in Fig. 7. While sample 11 served as a control, samples 7–10 were fractionated by selective adsorption to MAS. As already described on p. 327, infectious ss phage RNA adsorbs to MAS at 1.1 M NaCl and can be removed by sedimentation, whereas, infectious RF stays in solution. At high salt concentration (1.3 M NaCl), ss phage RNA also stays in the supernatant. A partly double-, partly single-stranded molecule, such as tailed RF, is expected to appear in the supernatant at an intermediate salt concentration. If both, a completely double-stranded and a tailed RF should be present in the preparation, the former alone should be obtained at 1.1 M NaCl and a mixture of both at a salt concentration closer to 1.3 M NaCl.

The results are given in Fig. 7. The infectivity curves show that both RF and ss phage RNA are removed below 1.15 M NaCl (MAS fraction 1 and 2); between 1.15–1.25 M NaCl (MAS fraction 3–6) only RF-infectivity is present, while ss infectivity is observed not before 1.35 M NaCl (fraction 8). A comparison of fractions 9 and 10 with the control (fraction 11) shows that MAS does not remove ss infectivity above 1.4 M NaCl. On the other hand more than 50% of the ^{32}P-labeled material is removed; this indicates a purification of infectious RNA from host RNA even at high salt concentrations. A small shoulder in the RF

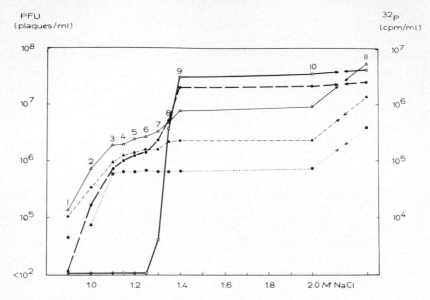

FIG. 7. Fractionation of ^{32}P-labeled nucleic acids of M12 infected cells by selective adsorption of ss RNA to MAS (Francke and Hofschneider, 1966; Benzinger *et al.*, 1967) at different salt concentrations (0.9–2.0 M NaCl, 0.02–sodium acetate, pH 5). Nucleic acids were prepared 30 min after infection (Francke and Hofschneider, 1966) and distributed into 10 MAS tubes and one control tube. After adsorption at the indicated salt concentration, MAS was removed from the samples by sedimentation. This step was repeated twice. The ten final supernatant fractions (1–10) and the control sample (11) were assayed for infectivity (Francke and Hofschneider, 1966) before O————O and after O– – – –O denaturation (PFU/ml supernatant), for total acid-precipitable counts ————, RNase resistant counts – – – – (for RNase digestion, see Ammann *et al.*, 1964b), and alkali resistant counts due to DNA – – – – (alkali hydrolyzation in 0.3 M KOH, 37°C, 16 hours). The curve for infectivity before denaturation represents selectively the presence of infectious ss RNA (ss infectivity). An infectivity increase after denaturation (1 min, 100°C in 0.003 M EDTA), as seen in MAS fractions 2–7 for some orders of magnitude, is due to RF (RF infectivity). In fraction 9 and 10 the sensitivity to heat-treatment of ss RNA infectivity results in a lowered infectivity following denaturation.

infectivity curve (MAS fraction 7), before the main increase of ss infectivity (MAS fraction 8 and 9), suggests an inhomogeneity of RF possibly due to tailed RF. To establish this possibility band sedimentations were performed with MAS-fractions 3 to 10.

SEDIMENTATION CONSTANTS OF INFECTIOUS RF

The sedimentation pattern of MAS-fraction 3, the first one to contain reasonable amounts of RF, is demonstrated in Fig. 8. Equal quantities of the material were sedimented before (Fig. 8, top) and after (Fig. 8, middle) heat-denaturation. Native RF sediments after a broad band of host-DNA in a homogeneous, sharp infectivity-peak with the maximum in fraction 18. The sedimentation constant, calculated from the marker RNA (27 S, Fig. 8, bottom), is 15 S. Some more slowly sedimenting material is found which is also partly RNase resistant but not infectious in any of the experiments shown, probably represents nonfunctional double-stranded RNA or preparation artifacts. By heat-denaturation (Fig. 8, middle) the 15 S RF becomes completely denatured: No material

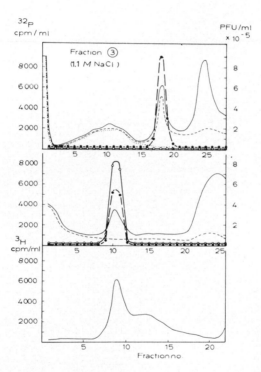

FIG. 8. Band sedimentation of MAS fraction 3. A sucrose gradient 5–20% (weight per volume) was used in 0.1 M Tris, 0.003 M EDTA, pH 7.0. The nucleic acids of MAS fraction 3 were precipitated with 3 volumes ethanol and redissolved in 0.003 M EDTA. The same quantities of native and denatured material were layered on two gradients (Fig. 8, top and middle picture, respectively). Single-stranded phage RNA as 27 S marker was layered on a third gradient (Fig. 8, bottom). It was prepared from purified uridine-3H labeled phages as cited for Fig. 4. Samples were centrifuged for 14 hours at 24,000 rpm in a spinco SW 25 rotor at 5°C. The fractions collected from the gradient were tested for infectivity before O———O and after O––––O denaturation (PFU/ml), for total acid precipitable counts (———), for RNase resistant counts (– – – –), and for alkali-resistant counts due to DNA (····). For better clearness no dots are given for the radioactivity assays.

can be redetected at the 15 S position. A new peak of ss infectivity is found, instead, at the 27 S position, containing 60% of the radioactivity of the 15 S peak. This result shows that the 15 S RF contains at least one intact strand, being infectious and, with respect to sedimentation, identical with mature ss phage RNA.

No change in the sedimentation pattern of RF is found for MAS fractions 4–7 (1.15–1.3 M NaCl). As depicted for MAS fraction 7 in Fig. 9, top, RF still sediments homogeneously. However, as soon as the first traces of ss infectivity can be detected (MAS fraction 8, Fig. 9, middle) a faster sedimenting shoulder of RF infectivity is also found (Fig. 9, middle, fractions 27–31).

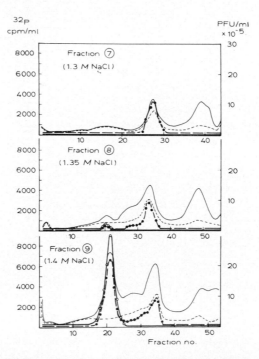

Fig. 9. Band sedimentation of MAS fraction 7 (Fig. 9, top), 8 (Fig. 9, middle) and 9 (Fig. 9, bottom). For technical explanations, see legend of Fig. 8.

In MAS fraction 9 (1.4 M NaCl), the amount of both infectious ss infectivity and infectious "fast" RF material has increased with respect to the 15 S RF peak (Fig. 9, bottom, and 10, top). To rule out the possibility that the observed phenomenon is caused by a nonspecific aggregation of concentrated ss RNA and RF, the gradient fractions containing the "fast" and "slow" RF materials (A and B material in Fig. 10, top) were combined in two pools, concentrated by ethanol precipitation, and rebanded. Fast and slow RF are redetected at their original positions. While A material forms a broad band of RF infectivity ranging from 17 S to 27 S with the maximum at 22 S (Fig. 11, top), B material sediments sharply with 15 S (Fig. 12, top). When A and B materials were denatured before being banded one sharp ss infectivity band is formed in both cases (Fig. 11, middle, and 12, middle). As already proven for denatured 15 S RF, these bands are at the 27 S position. Thus, not only the 15 S but also the fast 22 S RF must contain at least one intact infectious strand sedimenting as does mature ss phage RNA.

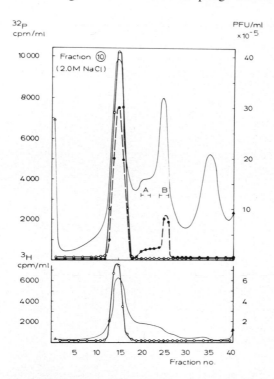

FIG. 10. Band sedimentation of MAS fraction 10 (10, top) and of ss phage RNA as 27 S marker (10, bottom). For technical explanation, see legend of Fig. 8.

In contrary to the infectivity assays, the ^{32}P-assays depict several peaks in the range of 15 to 25 S as seen in Fig. 11, top, Fig. 11, middle, and Fig. 12, middle. As demonstrated in Figs. 11a,b and 12a,b all these materials are completely RNase-sensitive. Most probably they represent ribosomal RNA which had not been removed by the selective adsorption at 2.0 *M* NaCl. They may also include RF degraded by denaturation

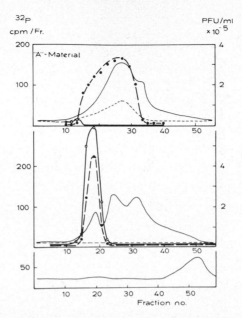

FIG. 11. Band sedimentation of native (Fig. 11, top) and denatured (Fig. 11, middle) A material from Fig. 10, top. The material in Fig. 11, bottom, is the same as in Fig. 11, middle, but digested with RNase before the sedimentation. For technical explanation, see legend of Fig. 8.

and, as far as A material is concerned, incomplete plus strands (see next paragraph).

It follows from these results that two distinct infectious RF classes exist, which sediment with 15 S and with 22 S (17–27 S). Fifteen S RF behaves chromatographically in the MAS procedure in a way similar to double-stranded DNA, whereas, 22 S RF appears in the supernatant practically at the same salt concentration as ss phage RNA. Therefore, it is concluded that infectious 15 S RF (RNase resistant to 85%, compare Table II) is constituted of completely double-stranded molecules having twice the molecular weight of mature phage RNA; in addition, 22 S RF (17–27 S) possesses single-stranded tails as just now demonstrated electronmicroscopically for phage R17 replicative intermediate (Granboulan and Franklin, 1967). This conclusion is further supported by the segregation of infectious RF into two fractions (RF I and II) as described earlier (p. 325).

Another possible explanation, however, would be that both forms contain only one intact strand, i.e., the infectious plus strand, and that they differ in the length of their incomplete minus-counterparts. Should this be true, 15 S RF would contain a minus strand smaller than an

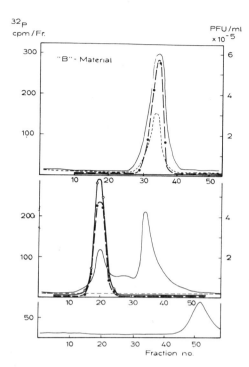

FIG. 12. Band sedimentation of native (Fig. 12, top) and denatured (Fig. 12, middle) B material from Fig. 10, top. The material in Fig. 12, bottom, is the same as in Fig. 12, middle, but digested with RNase before sedimentation. For technical explanation, see legend of Fig. 8.

intact plus strand. To eliminate the latter possibility, experiments were performed to establish the integrity of both strands of the 15 S RF.

STRAND INTEGRITY

Two results, already presented suggest the integrity of both 15 S RF strands. (1) If infectious 15 S RF contains the plus and minus strand in intact form, denatured and rebanded 15 S RF material should sediment quantitatively with 27 S, the sedimentation constant of mature ss phage RNA. At least 60% of 15 S RF could be redetected at the 27 S position (compare Fig. 8, top, and 8, middle). (2) The specific infectivity of 15 S RF (calculated as PFU/^{32}P cpm and corrected for heat inactivation during the melting procedure) is 0.3 to 0.5 that of mature ss phage RNA. Accepting this value and assuming that the minus strand is shorter than the plus strand, one would expect an increase of the specific infectivity by a factor of two for denatured 27 S material since shorter minus strands should not be found at the 27 S position. On the contrary, as calculated from Fig. 8 (top and middle pictures), for instance, the specific infectivity stays constant. This is to be expected only if both strands are intact.

Along these lines, an annealing assay was performed with denatured 15 S RF and with 27 S material, derived from denatured 15 S RF. Both the materials were obtained and prepared for the annealing assay as described in legend to Fig. 7. The yield of 27 S material can probably be increased by using a milder denaturation procedure (Katz and Penman, 1966).

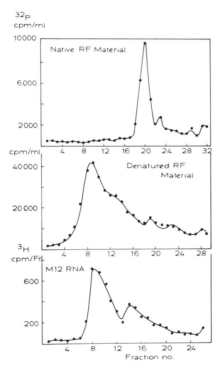

FIG. 13. Preparation of 15 S RF and of 27 S material originating from denatured 15 S RF. In a first step (not shown) infectious 15 S RF was isolated from a gradient as described in Fig. 8, top, by collecting the peak fractions. The peak material was precipitated with 3 volumes ethanol and redissolved in 0.003 M EDTA. For further purification of 15 S RF, one part was rebanded without further treatment (Fig. 13, top). For obtaining purified 27 S material, another part was denatured and then rebanded (Fig. 13, middle). Single stranded M12 phage RNA was layered on a third gradient as 27 S marker. Following the sedimentation (for conditions, see legend of Fig. 8) the 15 S peak material (fractions 19–21) of Fig. 13, top, and the 27 S peak material (fractions 8–10) of Fig. 13, middle, respectively was collected, and dialyzed against 0.015 M NaCl and 0.0015 M sodium-citrate.

TABLE II

RENATURATION OF DENATURED 15 S RF AND OF 27 S MATERIAL AS REVEALED BY INCREASE OF RNase RESISTANCE[a]

	15 S RF (%)			27 S material (%)		SS phage RNA (control) (%)	
	Native	Denatured	Renatured	Denatured	Renatured	"Denatured"	"Renatured"
Acid-precipitable counts							
Before RNase digestion	100	93	98	100	107	100	92
After RNase digestion	85	9	65	11	51	1	3
			100 ————		79 ————		

[a] Samples were diluted to 0.0015 M NaCl, 0.00015 M sodium-citrate and, when desired, denatured by heating to 100°C for 1 min and chilling in an icebath. Then the samples were freeze-dried. Renaturation was performed in a total volume of 0.04 ml by annealing in 0.375 M NaCl, 0.0375 M sodium citrate for 1 hour at 85°C (Weissmann, 1965). RNase resistance was assayed by digestion with 20 µg RNase/ml for 30 min at 37°C in 0.15 M NaCl, 0.015 M sodium citrate. 100% is adequate to 400–4000 cpm or 0.1–1 µg RNA, approximately.

If infectious 15 S RF contains both strands intact, comparable amounts of renaturation should be obtained by reannealing denatured 15 S RF as well as the 27 S material (Fig. 7b). The results are given in Table II. Native 15 S RF is RNase resistant to 85% (measured as acid precipitable counts before and after RNase digestion). After denaturation, RNase resistance drops to 9%. It can be reconstituted again to 65% by renaturation. The RNase resistance of the 27 S material is 11%, and it can be increased to 51% by renaturation.

The control assay with ss phage RNA (10 times as concentrated as RF in the RF assays) does not show any increase of RNase resistance. Thus the results obtained with RF cannot be explained as an artifact caused by intramolecular base-pairing of the plus strand. Setting the amount of renaturation obtained with denatured 15 S RF as 100% under the actual experimental conditions, a renaturation of 79% can be calculated comparably for 27 S material. Thus, 15 S RF contains similar amounts of plus and minus strands both sedimenting like infectious ss phage RNA. At least a part, possibly all, infectious 15 S RF molecules must be composed of an intact plus as well as an intact minus strand. Thus, intact plus and minus strands are very likely the physiological building units of native, infectious 15 S RF of phage M12.

ACKNOWLEDGMENTS

The authors are indebted to Professor A. Butenandt for his continuous support, to Dr. H. Delius for his collaboration in the earlier parts of this work, to Dr. D. S. Ray for stimulating discussions, and to Miss Antje Wirtz for help in translating this manuscript. Thanks also go to Miss A. Preuss, Miss Ch. Seufert, and Miss I. Mayer for their able technical assistance and to the Deutsche Forschungsgemeinschaft and the Studienstiftung des Deutschen Volkes for financial support.

REFERENCES

Ammann, J., Delius, H., and Hofschneider, P. H. (1964a). *Ber. deut. Bunsenges.* **68,** 729.

Ammann, J., Delius, H., and Hofschneider, P. H. (1964b). *J. Mol. Biol.* **10,** 557.

Benzinger, R., Jaenisch, R., and Hofschneider, P. H. (1966). *J. Mol. Biol.* **21,** 493.

Billeter, M., Libonati, M., Vinuela, E., and Weissmann, C. (1967). *J. Biol. Chem.* **241,** 4750.

Bishop, J., Koch, G., and Levintow, L. (1966). *In* "The Molecular Biology of Viruses," pp. 355–373. Symp. Faculty Med., Univ. Alberta, Academic Press, New York.

Darnell, J. E., Girad, M., Baltimore, D., Summers, D. R., and Maizel, J. V. (1966). *In* "The Molecular Biology of Viruses," pp. 375–401. Symp. Faculty, Med., Univ. Alberta, Academic Press, New York.

Delius, H., and Hofschneider, P. H. (1967). *In* "Methods in Enzymology" (S. P. Colowick and N. O. Kaplan, eds.). Academic Press, New York (in press).

Erikson, R. L., and Gordon, J. A. (1966). *Biochem. Biophys. Res. Commun.* **23,** 422.

Fenwick, M. L., Erikson, R. L., and Franklin, R. M. (1964). *Science* **146,** 527.

Francke, B., and Hofschneider, P. H. (1966). *J. Mol. Biol.* **16,** 544.

Franklin, R. M. (1966). *Proc. Natl. Acad. Sci. U.S.* **55,** 1504.

Granboulan, N., and Franklin, R. M. (1966). *J. Mol. Biol.* **22,** 173.

Haruna, I., and Spiegelman, S. (1966). *Proc. Natl. Acad. Sci. U.S.* **55,** 1256.

Hausen, P. (1965). *Virology* **25,** 523.

Hofschneider, P. H. (1963). *Z. Naturforsch.* **18b,** 203.

Hofschneider, P. H., Ammann, J., and Francke, B. (1967). *In* "Methods in Enzymology" (S. P. Colowick and N. O. Kaplan, eds.). Academic Press, New York (in press).

Jaenisch, R., Hofschneider, P. H., and Preuss, A. (1966). *J. Mol. Biol.* **21,** 501.

Katz, L., and Penman, S. (1966). *Biochem. Biophys. Res. Commun.* **23,** 557.

Kelly, R. B., Gould, J. L., and Sinsheimer, R. L. (1965). *J. Mol. Biol.* **11,** 562.

Kleinschmidt, A., and Zahn, R. K. (1959). *Z. Naturforsch.* **14b,** 770.

Lodish, H., and Zinder, N. (1966). *Science* **152,** 372.

Mandell, J. D., and Hershey, A. D. (1960). *Anal Biochem.* **1,** 66.

Weissmann, C. (1965). *Proc. Natl. Acad. Sci. U.S.* **54,** 203.

Weissmann, C., and Feix, G. (1966). *Proc. Natl. Acad. Sci. U.S.* **55,** 1264.

Weissmann, C., Borst, P., Burton, R. H., Billeter, M. A., and Ochoa, S. (1964). *Proc. Natl. Acad. Sci. U.S.* **51,** 682.

RNA Phage-Specific RNA Synthesis in *Escherichia Coli**

Mamoru Watanabe† and *J. Thomas August***

DEPARTMENT OF MOLECULAR BIOLOGY,
ALBERT EINSTEIN COLLEGE OF MEDICINE,
BRONX, NEW YORK

The small size of the known RNA bacteriophages suggests that these agents may have little effect on metabolism of the host bacterium. The RNA phage genome contains approximately 3000 nucleotides, and hence may code for only 3 to 5 proteins. Only two phage-specific proteins have been identified—a phage-specific RNA polymerase, and the viral coat protein. In spite of this limited functional capacity, however, it will be shown that infection by some RNA bacteriophage profoundly inhibits the synthesis of host RNA.

The extent to which phage and host RNA synthesis proceeds concomitantly appears to vary with different RNA bacteriophages. Infection by phage f2 does not significantly affect the synthesis of bacterial RNA, and only 5 to 10% of the RNA synthesized after infection appears in progeny phage particles (Zinder, 1965; Lodish and Zinder, 1966). With R17 infected bacteria, Ellis and Paranchych (1963) observed an 80% inhibition of ribosomal RNA synthesis and estimated that at least 30% of the RNA synthesized after infection was phage RNA. A new phage, R23, recently isolated in our laboratory, completely inhibits RNA synthesis of the host cell.

* This work was supported in part by grants from the National Institutes of Health (GM-11936 and GM-11301), and the National Science Foundation (GB-5082). This is Communication No. 69 from the Joan and Lester Avnet Institute of Molecular Biology.

† Recipient of the Research Fellowship of the American College of Physicians, 1964–1967.

** Recipient of an Investigatorship of the Health Research Council of New York City under contract I-346.

Phage R23 resembles f2 in its physical, chemical, and serological characteristics. Bacteria infected with R23, however, have greater phage-directed RNA synthesis, phage RNA polymerase activity, and phage yield than do other known RNA bacteriophages. The studies reported here are concerned with the extent of phage-directed RNA synthesis in cells infected with R23. The incorporation of uracil-^{14}C into both phage RNA and into phage double-stranded RNA was investigated. Phage RNA was defined as the amount of ^{14}C-labeled RNA recovered in purified phage particles. The loss of virus particles incurred in the purification procedure was estimated by monitoring the recovery of ^{3}H-labeled RNA phage added to the phage lysate prior to purification. The amount of ^{14}C-labeled RNA resistant to ribonuclease was determined following the isolation of RNA by phenol extraction. Kinetic analysis of RNA synthesis was made possible by use of the antibiotic phleomycin,* which has recently been shown to inhibit RNA phage growth and RNA-directed RNA synthesis (Watanabe and August, 1966). Phage maturation is not affected by phleomycin. Phage RNA synthesized prior to the addition of the drug is recovered in mature phage particles.

TABLE I

QUANTITATION OF PHAGE RNA SYNTHESIS[a]

	Incorporation of ^{14}C	
	cpm/ml	%
Total RNA	74,631	100
Phage RNA	41,586	55.7

[a] *E. coli* K38 was grown in LB medium (Loeb and Zinder, 1961) to a density of 2×10^8 cells/ml, and R23 was added at a multiplicity of infection of 35. Uracil-^{14}C was added to a final concentration of 1 μCi/ml (17 $\mu\mu$mole/ml). Two hours after phage infection, total RNA was determined by measuring the amount of uracil-^{14}C which had been incorporated into acid-insoluble material. Phage RNA was determined following purification of the virus (Watanabe and August, 1967).

In R23 infected cells, 56% of the RNA synthesized after infection was recovered in phage particles (Table I). The rate at which phage RNA is synthesized during the infectious cycle was established by the use of phleomycin (Table II). When RNA synthesis was stopped after the first 15 min of infection, 5% of newly synthesized RNA was re-covered in phage particles. During the next 15 min, 53% of the RNA

* Phleomycin was kindly supplied by Dr. K. E. Price of Bristol Laboratories, Syracuse, New York.

synthesized was subsequently recovered in phage particles; from 30 min post-infection on, 60% of the RNA synthesized was phage RNA. In similar experiments, it was found that with either f2- or Qβ-infected cells only 25% of the RNA synthesized after infection was recovered in progeny phage particles.

TABLE II

RATE OF SYNTHESIS OF PHAGE RNA[a]

Sample No.	Time interval (min)	Total RNA (cpm/ml)	Phage RNA (cpm/ml)	Phage RNA (% of total RNA synthesized)
1.	0– 15	5303	284	5.3
2.	15– 30	6674	3555	53.3
3.	30– 45	11,072	6750	60.9
4.	45–120	51,582	30,986	60.1

[a] Bacteria were grown and infected with phage R23 in the presence of uracil-^{14}C as described in Table I. In samples 1, 2 and 3, RNA synthesis was inhibited by the addition of 20 μg/ml phleomycin at 15, 30, and 45 min post-infection, respectively. In all samples, the amount of ^{14}C in the total RNA and in phage particles was determined 120 min after infection. The values in columns 3, 4, and 5 refer to the RNA synthesized during the corresponding time interval.

A second form of phage-specific RNA has been identified on the basis of its resistance to RNase (Weissmann *et al.*, 1964; Kaerner and Hoffmann-Berling, 1965; Kelly and Sinsheimer, 1964; Nonoyama and Ikeda, 1964; Erikson *et al.*, 1965). The amount of RNase-resistant RNA found in *E. coli* K38 cells following R23 infection was greater than that found in bacteria infected with other RNA bacteriophages (Weissmann *et al.*, 1964; Lodish and Zinder, 1966). Ten percent of the RNA synthesized during the first 15 min after R23 infection was found to be resistant to RNase. Thus, early after infection, the ratio of RNase-resistant RNA to phage RNA was 2 : 1 (Table III). About 30% of the RNA synthesized after 15 min was found to be RNase-resistant and the ratio became 2 : 1 in favor of phage RNA.

The sequence of events which occurs after R23 phage infection is summarized in Fig. 1. Shortly after phage infection a viral RNA polymerase activity, which is not seen in uninfected cultures, is observed. The synthesis of RNase-resistant RNA is first detected 6 min after infection. Phage RNA is synthesized in significant quantities from 15 min post-infection. Thereafter, phage RNA accounts for 60–70% and RNase-resistant RNA for 30–40% of the total RNA synthesized in the

TABLE III

Time after infection (min)	% of total RNA synthesized		
	Phage RNA	RNase-resistant RNA	RNase-resistant RNA/phage RNA
5	5.3	10.5	1.98
10	5.3	8.8	1.67
15	5.3	10.4	1.97
20	53.3	28.8	0.54
30	53.3	28.8	0.54
40	60.9	25.3	0.42
50	60.9	24.5	0.40
60	60.1	25.0	0.42
75	60.1	31.2	0.52
90	60.1	26.0	0.43
120	59.6	41.0	0.68

[a] Bacteria were grown and infected in the presence of uracil-^{14}C as described in Table I. At various intervals after phage infection, RNase-resistant RNA was determined as described by Lodish and Zinder (1966), except that pancreatic ribonuclease A was used in a concentration of 1 μg/ml in 0.15 M NaCl, 0.15 M sodium citrate, 0.001 M MgCl$_2$ and 0.01 M Tris, pH 7.4.

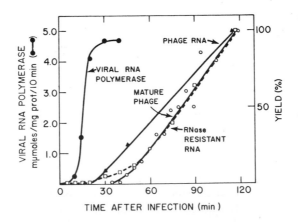

Fig. 1. Intracellular events following RNA phage infection. Viral RNA polymerase activity was determined as described elsewhere (August and Eoyang, 1967). Phage RNA and RNase-resistant RNA were determined as described in Tables I and II. Intracellular phage was assayed as plaque forming units by the agar overlay technique described by Adams (1959).

infected cell. Therefore, in contrast to other RNA bacteriophages, R23 assumes almost complete control of RNA synthesis in the infected cell. Both phage RNA and RNase-resistant RNA are synthesized continuously throughout the course of the infectious cycle. The first intracellular phage can be detected shortly after 30 min post-infection. Ten to 15 min are required between the synthesis of phage RNA and the encapsulation of this RNA into phage particles. The late accumulation of RNase-resistant RNA parallels the appearance of mature phage particles.

An RNA bacteriophage that completely inhibits host RNA synthesis has certain advantages over those that do not in studies of RNA-directed RNA synthesis *in vivo*. This characteristic, in combination with the techniques used in the experiments described in the preceding section and with the technique of pulse-labeling, has made it possible to examine certain mechanisms of phage replication in the R23-*E. coli* K38 infected (Fig. 2). The sequential appearance of uracil-^{14}C in RNase-resistant RNA and in RNA encapsulated in phage particles has been determined. Twenty minutes after infection uracil-^{14}C was added to the bacterial culture. Thirty seconds later, unlabeled uridine

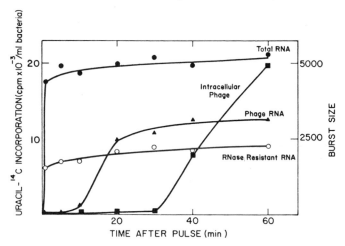

Fig. 2. Analysis of phage-specific RNA following a uracil-[14]C "pulse." *Escherichia coli* K38 was grown in medium A (Davern, 1964) to a density of 2×10^8 cells/ml and infected with R23 at a multiplicity of infection of 50. Uracil-^{14}C was added 20 min after infection to a final concentration of 1 μCi/ml. Thirty seconds later unlabeled uridine was added (200μg/ml) and the bacteria were harvested by centrifugation and resuspended in fresh medium containing 500 μg/ml of unlabeled uridine. At various intervals thereafter, ^{14}C radioactivity present in total RNA, RNase-resistant RNA, and phage RNA was determined. Intracellular phage were determined following lysis of the cells.

was added, and the bacteria were harvested by centrifugation and re-suspended in fresh medium containing unlabeled uridine. At different times thereafter, the localization of ^{14}C in RNase-resistant and phage RNA was determined. During the 60-min period, phage RNA consistently represented 60% of total, newly synthesized RNA. The radio-activity present in RNase-resistant RNA (40% of the total) neither diminished nor entered phage particles, despite the continued production of intracellular phage.

The results of the "pulse-chase" experiment allow a quantitative interpretation of the mechanism of phage RNA replication. It can be concluded that the majority (at least 90%) of the RNase-resistant RNA cannot be involved in a semiconservative mechanism of RNA replication. Models of RNA phage replication must take into account the fact that RNA is not transferred from the bulk of the RNase-resistant RNA to progeny virus particles. If semiconservative replication does play a role, only a small fraction (less than 10%) of the RNase-resistant material could be involved, since greater amounts would have been detected as a chase in the pulse-chase experiment. The functional role, if any, of the remaining RNase-resistant material remains to be determined. These results may also be explained on the basis of other hypothetical models of RNA phage replication. One is that RNase-resistant RNA may not be an intermediate in the synthesis of viral RNA. Another model is that the progeny strand arises from the "replicative intermediate" by a conservative mechanism. In such a case, the replicative form would be an obligatory intermediate. However, the radioactivity present in the double-stranded form would not be expected to enter mature phage particles. A previous observation which provides support for a model of conservative replication is that the RNA of parental phage is not transferred to progeny phage particles (Davis and Sinsheimer, 1963; Doi and Spiegelman, 1963).

Summary

The extent to which phage RNA and host RNA is synthesized after infection varies with different RNA bacteriophages. A strain recently isolated in our laboratory, R23, completely dominates the capacity of the host cell for RNA synthesis. About 60% of the RNA synthesized after R23 infection is recovered in phage particles, while 30 to 40% is present as RNase-resistant material. An analysis of the fate of newly synthesized RNA in a pulse-chase experiment demonstrated that in the case of R23-infected cells the radioactivity present in double-stranded form is not displaced into phage particles, despite the continued production of intracellular phage. It can be concluded that the majority of the RNase-resistant RNA is not involved in a semi-conservative mechanism of RNA phage replication.

REFERENCES

Adams, M. H. (1959). "Bacteriophages." Wiley (Interscience), New York.

August, J. T., and Eoyang, L. (1967). *In* "Methods in Virology" (K. Maramorosch and H. Koprowski, eds.), Vol. II. Academic Press, New York (in press).

Davern, C. I. (1964). *Australian J. Biol. Sci.* **17,** 719.

Davis, J. E., and Sinsheimer, R. L. (1963). *J. Mol. Biol.* **6,** 203.

Doi, R. H., and Spiegelman, S. (1963). *Proc. Natl. Acad. Sci. U.S.* **49,** 353.

Ellis, D. B., and Paranchych, W. (1963). *J. Cellular Comp. Physiol.* **62,** 207.

Erikson, R. L., Fenwick, M. L., and Franklin, R. M. (1965). *J. Mol. Biol.* **13,** 399.

Kaerner, H. C., and Hoffman-Berling, H. (1964). *Nature* **202,** 1012.

Kelly, R. B., and Sinsheimer, R. L. (1964). *J. Mol. Biol.* **8,** 602.

Lodish, H. F., and Zinder, N. D. (1966). *Science* **152,** 372.

Loeb, T., and Zinder, N. D. (1961). *Proc. Natl. Acad. Sci. U.S.* **47,** 282.

Nonoyama, M., and Ikeda, Y. (1964). *J. Mol. Biol.* **9,** 763.

Watanabe, M., and August, J. T. (1966). *Bacteriol. Proc.* p. 115.

Watanabe, M., and August, J. T. (1967). *In* "Methods in Virology (K. Maramorosch and H. Koprowski, eds.), Vol. II. Academic Press, New York (in press).

Weissmann, C., Borst, P., Burdon, R. H., Billeter, M. A., and Ochoa, S. (1964). *Proc. Natl. Acad. Sci. U.S.* **51,** 682.

Zinder, N. D. (1965). *Ann. Rev. Microbiol.* **19,** 455.

Discussion—Part IV

Dr. J. Wodak: (Question directed to Dr. C. Weissmann.) Does fragmentation of the "replicative complex" by ribonuclease lead to the production of double-stranded fragments having any regularity with respect to size? If so, this would imply that the synthesis of new "plus strands" starts at a fixed time after that of the previous strand, if your model is correct.

Answer: Treatment of the replicative intermediate with RNase at 50 μg/ml leads to the production of fragments of double-stranded RNA of rather uniform size with a molecular weight of about 300,000. This would be in agreement with your expectation regarding the spacing of enzymes on the template.

Dr. B. W. Burge: (Question directed to Dr. C. Weissmann.) Does labeled TS-6 parental RNA become ribonuclease resistant when infecting cells at a nonpermissive temperature? Are there any temperature-sensitive mutants, defective in RNA synthetase, which are able to enter the ribonuclease-resistant structure, but which proceed no further at nonpermissive temperatures?

Answer: As shown by H. Lodish and N. Zinder, TS-6 parental RNA does become RNase-resistant as long as the cells are kept at a nonpermissive temperature. The mutant you inquire about has, unfortunately, not yet been found.

Dr. C. Weissmann: (Questions directed to Dr. M. Watanabe.) (1) It has been noted by Billeter *et al.** that turnover of virus specific double-stranded RNA is difficult to detect in the early phase of phage MS2 replication; this is probably because most of the newly formed "plus" strands are recycled into double-stranded RNA. Have you tried to pulse and chase at later times after infection? (2) Under certain conditions, newly labeled double-stranded RNA may be degraded intracellularly to material sedimenting around 6 S, which is no longer capable of turnover. Have you examined the sedimentation profile of the pulse-labeled double-stranded RNA at different times after the pulse?

Answers: (1) We have not examined the fate of pulse-labeled double-stranded RNA at later times after infection. In our experiment the "pulse" was given 20 min after infection with R23. This is at a time when 60% of the RNA being synthesized was phage RNA (Table II); therefore,

---

* M. Billeter, M. Libonati, E. Vinuela and C. Weissmann, *J. Biol. Chem.* (in press).

we do not consider it an early period of replication. Furthermore, the chase period covered a total of 60 min. The bulk of the phage RNA was synthesized between 20 and 60 min (Fig. 1); therefore, even if labeled "plus" strands were recycled, the labeled "plus" strands should have had an adequate opportunity to be displaced by nonlabeled "plus" strands during this period of maximal phage RNA synthesis. (2) We have not yet examined the sedimentation profile of the double-stranded RNA at different times after a pulse. The problem of heterogeneity of double-stranded RNA is being investigated with respect to size, function, and synthesis.

Dr. M. L. Fenwick: (Question directed to Dr. M. Watanabe.) Isn't a possible explanation of the lack of chasing out of the ribonuclease resistant label after a 30 sec pulse (which amount effectively to at least a 1 min pulse, including the time for dilution of precursor pools) simply that it is largely in minus strands? In other words, the relative rates of synthesis of plus strands and of double strands from plus strands are such that 60% of the pulse label would be found in plus strands (the majority of which has already been displaced from the duplex by the end of a 1 min pulse) and 40% of the label is in minus strands being built into new duplexes. The latter would be ribonuclease resistant and would remain so during a semiconservative displacement mechanism of RNA synthesis. If so, one would expect to find a measurable chase (up to 60%) of ribonuclease-resistant labeled plus strands into virus particles following a much shorter pulse, say 2 sec.

Answer: We have not examined pulses shorter than 30 sec. Libonati and his co-workers* have reported that after a 30 sec pulse, about 50% of the radioactive "plus" strands was in double-stranded RNA. Shortening the pulse duration to 8 or 2 sec did not increase this value.

* Libonati, M., Vinuela, E., Billeter, M. A., and Ochoa, S. (1966). *Federation Proc.* **25**, 651.

PART V

MAMMALIAN RNA VIRUSES

Biological and Physicochemical Aspects of Poliovirus-Induced Double-Stranded RNA*

J. Michael Bishop(a), Gebhard Koch(a)† and Leon Levintow(b)

(a) LABORATORY OF BIOLOGY OF VIRUSES, NATIONAL INSTITUTE OF ALLERGY AND INFECTIOUS DISEASES, NATIONAL INSTITUTES OF HEALTH, BETHESDA, MARYLAND, AND (b) DEPARTMENT OF MICROBIOLOGY, UNIVERSITY OF CALIFORNIA SCHOOL OF MEDICINE, SAN FRANCISCO, CALIFORNIA

The small RNA-containing animal viruses, typified by poliovirus, provide a logical bridge at the halfway point in this volume, since they are similar to the RNA phages with respect to both size and architecture and have roughly similar complements of single-stranded RNA. Moreover, virus replication in both cases is typically unaffected by actinomycin, and in both cases infection induces one or more new RNA polymerases. The appearance of virus-specific, RNase-resistant RNA in the infected cell is a characteristic feature common to both groups of viruses. These similarities, among others, strongly argue for the view that replication of both groups proceeds by mechanisms which are similar, at least in essential details. If this is indeed a valid point of view, the conclusions we have drawn from the study of RNase-resistant RNA of poliovirus may also be relevant to the mechanism of replication of the RNA phages.

A single outstanding difference between the animal virus and phage systems must be mentioned; namely, the fact that the RNase-resistant RNA is demonstrably infectious in the former case and not in the latter. Whether or not this situation reflects an important distinction between the two systems is not presently clear, but it seems more reasonable to assume

---

* Abbreviations employed: SSC = standard saline citrate = 0.15 M NaCl, 0.015 M sodium citrate; SDS = sodium dodecyl sulfate; PBS = phosphate-buffered saline; PFU = plaque forming units; TCA = trichloroacetic acid; PSM = physiological saline with 10^{-3} M MgCl$_2$; RSB = reticulocyte standard buffer = 10^{-2} M tris-HCl pH 7.4, 10^{-2} M KCl, 1.5×10^{-3} M MgCl$_2$; MAK = methyl-esterified serum albumin-kieselguhr; RNase = ribonuclease.

† Present address: Heinrich Pette-Institut, Hamburg, Germany.

that the difference will be shown to have a relatively trivial explanation. In any event, we would like to begin by briefly summarizing the conclusions we have drawn from our studies with poliovirus, and then, hopefully, validating these conclusions with appropriate data.

The RNase-resistant RNA extracted with phenol from poliovirus-infected cells is a double-stranded, base-paired structure consisting of a strand of viral RNA and a complementary strand. Although the base-paired duplex molecule probably does not represent the functional replicative form of viral RNA, it is nevertheless biologically active—it is intrinsically infectious. The fact that the material is infectious implies that it embodies the entire information content of the genome in a usable form. The most reasonable explanation for the biological activity is the presence of an intact molecule of viral RNA. If one assumes that this biologically active molecule is in fact a portion of the active replicative structure, the presence of a complete strand of viral RNA imposes certain constraints on permissible models for replication.

Preparation of Double-Stranded RNA

The property of RNase-resistance provides a convenient tool for both the characterization and isolation of double-stranded RNA, but the fact is that resistance to enzymatic attack is far from absolute, even at high concentrations of salt. It therefore seemed advantageous to develop a preparative procedure which did not involve treatment with RNase. We have utilized the familiar techniques of salting out and chromatography on MAK columns, in a procedure similar to that described by Ammann et al. (1964) for phage-specific RNA. The cells were infected in the presence of actinomycin D and radioactive precursors of RNA, and the conditions of chromatography were such that the only labeled species of cellular RNA, that is, the transfer RNA's, were eliminated during loading of the column.

As shown in Fig. 1, application of the chromatographic procedure to unfractionated RNA differentiates two major peaks of labeled viral RNA, one of which is cleanly separated from ribosomal RNA. If the RNA is

Fig. 1. MAK column chromatography, (Mandel and Hershey, 1960; Sueoka and Cheng, 1962) before and after differential NaCl precipitation, of RNA extracted from poliovirus-infected HeLa cells. 1.5×10^9 cells, preincubated for 1 hour with actinomycin at 5 μg/ml, were infected with 100 PFU per cell of Mahoney Strain poliovirus (see Bishop et al., 1955); replication was allowed to proceed for 5 hours in the presence of ^{32}P-orthophosphate (25 μCi/ml). Phenol extraction was performed at 60°C and pH 7.2 in PBS, yielding 34 mg of RNA after one ethanol precipitation and a single wash of the precipitate with a 2 : 1 mixture of ethanol: 0.15 M NaCl at 4°C. The RNA was redissolved in PBS at a concentration of 1

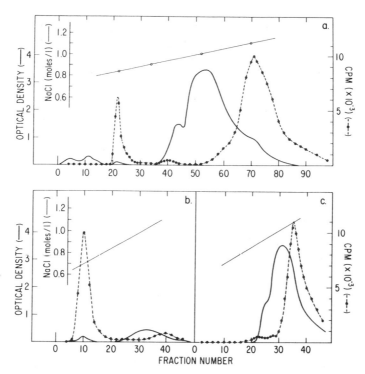

mg/ml and divided into two equal parts. One was chromatographed directly (a); the other was first adjusted to 10% weight volume NaCl, allowed to freeze overnight at −20°C, then thawed, and the resulting precipitate sedimented at 2000 rpm for 20 min. The precipitate was washed once with cold 10% NaCl solution, the wash pooled with the original supernatant, and the precipitate redissolved in PBS. After this procedure, approximately 20% of the original optical density at 260 mμ remained in the supernatant. Supernatant (b) and precipitate (c) were chromatographed separately. The columns consisted of 20 gm celite beds divided into small upper and lower "neutral layers" and a single intervening albumin-containing layer.

The RNA was loaded at concentrations never exceeding 0.5 mg/ml in 0.6 M NaCl, 0.05 M sodium phosphate buffer, pH 6.8. At this NaCl concentration, acid soluble nucleotides, inorganic ^{32}P and over 95% of HeLa cell sRNA pass through the column without adsorbing. Elution of high molecular weight RNA's was accomplished at 35°C (Kubinski *et al.*, 1962) with linear NaCl gradients produced by adding 1.6 M NaCl dropwise to 0.6 M NaCl. Absorbance of column eluates at 260 mμ was recorded continuously, using a Cary spectrophotometer fitted with a flow cell with a 1 mm light path. Fractions of 2 ml were collected and 20 μl samples were taken for plating and counting on a gas-flow low background counter. Generally, radioactivity was monitored continuously by passing the eluate through a coil of narrow gauge polyethylene tubing, over which an end-window Geiger-Mueller tube was mounted. However, the count profile was always verified by plating and counting as described. The NaCl concentration of fractions was determined from the refractive index, which was measured on an Abbe refractometer and converted to NaCl concentration, using an empirically derived standard curve.

fractionated by treatment with 10% NaCl prior to chromatography, the late-eluting species of viral RNA, and the bulk of the ribosomal RNA are precipitated. The other species of virus-specific RNA is soluble in 10% NaCl, and it thus can conveniently be obtained in pure form by salting out followed by chromatography. The larger peak of viral RNA is susceptible to RNase, and represents the single-stranded RNA found in mature virions. The smaller, soluble peak of RNA is resistant to RNase, and, as demonstrated below, fulfills the criteria for a base-paired double-stranded molecule. A third, small peak of labeled RNA appears to be virus-specific and salt-precipitable, but it is not cleanly separated from cellular RNA and we are not ready to offer any good evidence on its constitution.

Figure 2 depicts a representative chromatographic run on a preparaparative scale. In this example 500 μg of double-stranded RNA were isolated from a total of 500 mg of RNA extracted from infected cells.

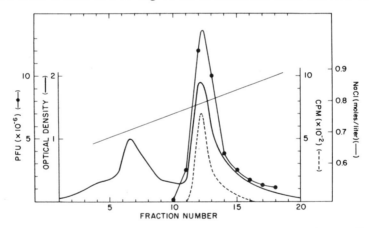

FIG. 2. Preparative chromatography of double-stranded RNA from polio-virus-infected HeLa cells. A total of 500 mg of RNA, extracted as described in Fig. 1 from 1.8×10^{10} cells 5 hours after infection, was precipitated with 1 M NaCl. The supernatant was adjusted to 0.6 M NaCl, and a small quantity of ^{32}P-labeled, purified double-stranded RNA was added to serve as a marker. Chromatography was performed as described in Fig. 1. Samples from each fraction were diluted appropriately for determination of infectivity, using an agar-cell suspension assay (Koch *et al.*, 1966).

Physical and Chemical Properties of Double-Stranded RNA

Figure 3 depicts the densitometer tracings of patterns obtained during band sedimentation of purified double-stranded RNA according to Vinograd *et al.* (1963). The configuration of the bands is indicative of a largely homogeneous population of molecules, with perhaps a small degree of heterogeneity with respect to size. There is a small contaminant, no more than 10% of the total UV-absorbing material, which probably represents some transfer RNA. The progressive broadening of the leading edge of the major band with time is characteristic of concentration-dependent sedimentation, and is reminiscent of the behavior of native double-stranded DNA under similar conditions. The result is therefore in accord with the conclusion that the RNA is in fact a base-paired structure. An uncorrected sedimentation coefficient of 18.4 S has been calculated for the major component from boundary sedimentation experiments (not shown) at pH 6.8 in 0.05 M phosphate buffer and 0.4 M NaCl.

Another characteristic of the double-stranded RNA is its low buoyant density observed by equilibrium centrifugation in Cs_2SO_4, relative to the density of either ribosomal RNA or single-stranded viral RNA (Fig. 4). Table I illustrates the effect of salt concentration on the susceptibility of poliovirus double-stranded RNA to attack by pancreatic RNase, and demonstrates that relative resistance is conferred by either 0.15 M NaCl

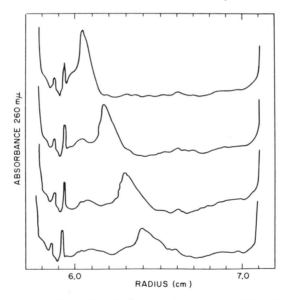

FIG. 3. Band sedimentation of poliovirus double-stranded RNA. Purified double-stranded RNA in 0.05 M sodium phosphate buffer, pH 6.8, was examined in the Model E analytical centrifuge according to the technique of Vinograd *et al.* (1963), using a lamellar solution volume of 20 μl with $A_{260} = 1.8$. The bulk solution consisted of 1.0 M NaCl, 0.2 M sodium phosphate buffer, pH 7.8, in D_2O. Centrifugation was at 44,770 rpm, 7.9°C. Using ultraviolet optics, photographs were taken at 8 min intervals. The tracings displayed in the figure were made with a Beckman Analytrol densitometer; they represent 16 min intervals.

TABLE I

EFFECT OF SALT CONCENTRATION ON SUSCEPTIBILITY TO RIBONUCLEASE[a]

Conditions of treatment	Precipitated cpm	Relative resistance to enzymatic digestion
Untreated control	300	100
$2 \times$ SSC	280	90
SSC	280	90
0.1 SSC	8	3
0.01 SSC	10	3
10^{-2} M MgCl$_2$, 0.01 SSC	280	90
10^{-3} M MgCl$_2$, 0.01 SSC	90	30
10^{-4} M MgCl$_2$, 0.01 SSC	10	3

[a] 8×10^8 cells were infected in the presence of 10 mCi of ^{32}P-orthophosphate, treated with SDS-phenol at 5 hours after infection, and RNase-resistant RNA was isolated. Equal portions of the preparation were diluted so as to obtain the designated salt concentrations in a volume of 0.5 ml and were treated at pH 7.0 with 80 μg of pancreatic RNase per ml for 15 min at 37°C. The reaction mixtures were chilled, 0.5 mg of yeast RNA and TCA to 5% were added in immediate succession and acid-precipitable radioactivity was measured in the low background counter.

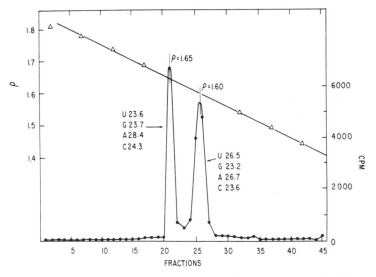

FIG. 4. Equilibrium centrifugation of phenol-SDS-extracted RNA in Cs$_2$SO$_4$. Two $\times$ 10^8 cells were infected in the presence of 3.8 mCi of ^{32}P-orthophosphate, and RNA was extracted at 5 hours with SDS-phenol. The RNA preparation was sedimented through a sucrose gradient (see Fig. 12 for details) and the material between 14 and 20 S was isolated. After dialysis of the pooled fractions for 4 hours against 0.15 M NaCl, 5 $\times$ 10^{-3} M Tris- HCl pH 7.4, one-fifth of this material was mixed with Cs$_2$SO$_4$ solution, centrifuged and analyzed, with the bottom of the gradient to the left. The density (gm/ml, calculated from the refractive index of the Cs$_2$SO$_4$) and base composition of each of the two peaks of labeled RNA are indicated on the diagram. The preparation was not treated with RNase at any time.

or 10^{-2} M MgCl$_2$. The criterion for resistance to RNase in this experiment is precipitability of labeled RNA by TCA, a convenient technique applicable to very small amounts of material.

This criterion can also be utilized to demonstrate the thermal denaturation of double-stranded RNA. A sharp transition to RNase susceptibility is observed at 87°C in 0.01 × SSC (Fig. 5). Under similar conditions, the customary optical procedure yields a T_m of 82°C. The presence of

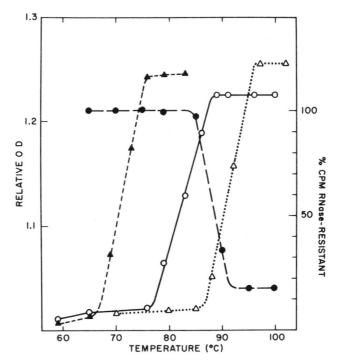

FIG. 5. Thermal Denaturation of Double-Stranded RNA. (*a*) Thermal transition of RNase-resistance. ^{32}P-labeled purified double-stranded RNA was diluted 1000-fold into 0.01 × SSC. Duplicate aliquots of 1 ml were maintained at the indicated temperatures for 15 min, then quenched in a 0°C bath. The salt concentration was adjusted to 2 × SSC and pancreatic RNase (10 µg/ml) added. After incubation at 37°C for 30 min, the samples were precipitated by rapid sequential addition of 1 mg yeast RNA and TCA to a final concentration of 5%. The precipitates were washed with 5% TCA, dissolved in 1% NH$_4$OH, and radioactivity (●) was measured in an end window, low background counter. (*b*) Determination of optical melting curves. Double-stranded RNA, prepared as described in Fig. 2 and diluted to an A$_{260}$ of 0.5, was dialyzed to desired electrolyte conditions. Heating was carried out in microcuvettes in a Gilford spectrophotometer equipped to record the A$_{260}$ and the temperature of the samples. Melting curves in 0.01 × SSC (0), 50% methanol—0.01 × SSC (▲), and 1.5 × 10^{-3} M NaCl (△) are shown. The data are presented without correction for solvent expansion.

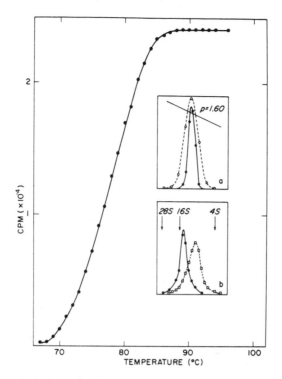

Fig. 6. Thermal elution of poliovirus double-stranded RNA from hydroxyapatite. ^{32}P-labeled, purified double-stranded RNA, in 0.6 M NaCl, 0.05 M potassium phosphate buffer, pH 6.8, was sheared at 21,000 psi with a steel pressure cell. Small portions were examined by zonal centrifugation (5–20% sucrose in 0.15 M NaCl, 5 mM Tris-HCl, pH 7.5; 25,000 rpm, SW 25, 4°C, 12 hours) and equilibrium density gradient centrifugation in Cs_2SO_4 (33,000 rpm, SW 39, 4 days, 4°C). The lower inset portrays (a) the distribution of labeled, sheared RNA (broken line) in the sucrose gradient compared to that of unsheared RNA (solid line) run simultaneously in a separate bucket, both preparations having been centrifuged in the presence of unlabeled, phenol-extracted HeLa cell RNA. The upper inset demonstrates the band widths in Cs_2SO_4 of sheared and unsheared RNA, again centrifuged simultaneously in separate buckets.

The bulk of the sheared material was passed through a jacketed column (1 ml bed volume) of hydroxyapatite in 0.03 M potassium phosphate buffer, pH 6.8; under these conditions, all high molecular weight RNA adsorbs to the hydroxyapatite. The temperature of the column was then raised to 67°C in a linear fashion 0.2°C/min and the column thoroughly washed with 0.08 M phosphate buffer to remove any single-stranded RNA that would be present if any denaturation occured during the shearing procedure. The temperature of the column was then slowly raised in a linear fashion, and a continuous wash with 0.08 M phosphate buffer maintained. Fractions of 18 ml of the eluate were collected and treated with TCA. The data are presented as an integral plot of cumulative count eluation against the temperature, recorded by a thermosensor probe in the buffer immediately above the column surface.

50% methanol lowers the melting temperature to about 72°C; in dilute salt without sodium citrate, the observed transition temperature was raised to 92°C.

Further evidence along the same lines was obtained from a study of the behavior of the RNA on hydroxyapatite columns according to the procedure of Miyazawa and Thomas (1965). These authors showed that fragments of native, double-stranded DNA adsorb to hydroxyapatite, and as the temperature of the column is raised, the DNA is eluted as it is denatured. Figure 6 illustrates an experiment of this sort with double-stranded poliovirus-specific RNA. The RNA was fragmented, albeit not very effectively, in a pressure cell at 21,000 psi. Comparison of zonal and equilibrium centrifugation behavior before and after treatment suggests that the breakage which occurred yielded predominantly half molecules. These fragments adsorbed to the hydroxyapatite column, and were eluted as the temperature was raised.

This procedure yields a value of about 78°C for the T_m. Preliminary results indicate that the first RNA to elute as the temperature is raised is relatively richer in adenine and uracil than is the RNA which elutes last (Martin and Bishop, 1966). This finding may at least point the way toward methods for separating the poliovirus genome into distinct, defined fragments.

Figure 7 illustrates some other aspects of the thermal denaturation of

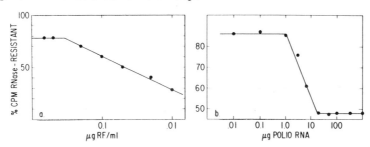

Fig. 7. Annealing studies of double-stranded poliovirus RNA. (*a*) Concentration-dependence of self-annealing of double-stranded RNA. ^{32}P-labeled purified double-stranded RNA (=RF) at varying concentrations in 0.01 × SSC was held at 100°C for 15 minutes. The salt concentration was then adjusted to 2 × SSC and the samples cooled to room temperature over 12 hours. TCA-precipitable counts were determined after digestion with RNase (10 μg/ml) for 30 min. at 37°C. (*b*) Displacement of radioactivity from ^{32}P-labeled, double-stranded RNA by denaturation and annealing in the presence of excess unlabeled single-stranded poliovirus RNA. Varying quantities of single-stranded RNA, isolated from purified poliovirus by phenol extraction, were added to 0.5 ml aliquots of ^{32}P-labeled double-stranded RNA at a concentration of approximately 0.15 μg/ml in 0.01 SSC. Heat denaturation, slow cooling, and determination of counts, acid precipitable after RNase digestion were performed as described above.

the double-stranded RNA. In the experiment shown on the left, samples of labeled RNA were thermally denatured in dilute SSC and slowly cooled in concentrated SSC. As the concentration of RNA was increased, a larger fraction was rendered RNase-resistant by the annealing procedure, up to a plateau of about 80%. Under conditions which favor this phenomenon of renaturation, the addition of increasing amounts of unlabeled RNA from purified, mature poliovirus (Fig. 7, right) progressively reduced the label present in the renatured complex to an end point where half of the label was displaced. A relatively huge excess of single-stranded RNA —on the order of 1000 to 1—is required to reach the end point. The renaturation is manifested not only by restoration of RNase resistance, but also by restoration of the buoyant density of the native material (Fig. 8).

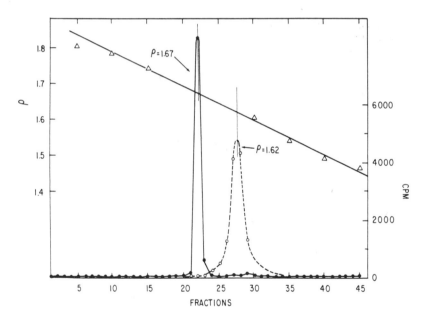

FIG. 8. Equilibrium centrifugation in Cs_2SO_4 of denatured double-stranded RNA and of a similar preparation after heating and slow cooling. A sample of double-stranded RNA was diluted in 0.01 × SSC, held at 100°C for 5 min, and quenched in an ice bath. A second sample was similarly treated at 100°C, and then slowly cooled in 2 × SSC as described in Fig. 7. The two samples were subjected to equilibrium centrifugation in Cs_2SO_4 in separate buckets of the same rotor. The analyses of the two Cs_2SO_4 gradients are superimposed. The peak to the left (unbroken line) represents the heated and quenched RNA, the peak to the right (dashed line), the slowly cooled material. The observed densities for denatured and renatured RNA differ from those given in Fig. 4, doubtless because of imperfect temperature control in the Model L ultracentrifuge.

Table II summarizes the results of a number of determinations of the base composition of RNA from mature virions, purified double-stranded RNA, and renatured double-stranded RNA from which the labeled viral or plus strand has been displaced by unlabeled viral RNA. The base composition was determined from the distribution of ^{32}P in nucleotides produced by alkaline hydrolysis of the RNA, that is to say, the results refer only to labeled species of RNA. The composition of the purified, RNase-resistant RNA is in close agreement with the calculated values for a base-paired, molecule consisting of a molecule of viral ("plus") RNA and a complementary ("minus") strand. Moreover, if the labeled "plus" strand is displaced from the RNase-resistant complex by unlabeled viral RNA, the base composition of the remaining labeled RNA approximates the calculated values for a minus complementary strand. These results provide strong support for the view that the RNase-resistant RNA is indeed a base-paired, double-stranded structure.

TABLE II

CALCULATED AND OBSERVED BASE COMPOSITIONS[a]

Bases	Mature poliovirus observed	Complement (−) calculated	Double (+, −) calculated	Double observed	Complement observed
U	23.8 ± 0.2	29.3	26.5	26.2 ± 0.1	28.0 ± 0.1
G	23.2 ± 0.3	23.8	23.5	23.1 ± 0.0	23.7 ± 0.1
A	29.3 ± 0.1	23.8	26.5	26.7 ± 0.1	23.9 ± 0.2
C	23.8 ± 0.2	23.2	23.5	24.0 ± 0.1	24.4 ± 0.1

[a] Base composition determined by measuring the relative amounts of isotope in the nucleotide products of alkaline hydrolysis. See text and Bishop *et al.* (1965) for details.

Biological Properties of Double-Stranded RNA

As mentioned in the introduction, double-stranded poliovirus-specific RNA is capable of initiating the infective cycle in susceptible cells and giving rise to complete, infectious progeny. It should be emphasized that it is the native, double-stranded molecule which is biologically active; the infectivity of double-stranded phage RNA, as discussed by Hofschneider, in this volume, is apparent only after denaturation. In Table III, the specific infectivity of the double-stranded RNA is compared with that of single-stranded viral RNA. Infectivity has been determined by a new procedure with HeLa cells suspended in agar in the presence of one or more polybasic substances (Koch *et al.*, 1966). The presence of these compounds enhances the specific infectivity of both species of viral RNA but the conditions necessary in each case for maximum titers are

TABLE III

SPECIFIC INFECTIVITY OF SINGLE- AND DOUBLE-STRANDED RNA[a]

	Titer (PFU/μg RNA)	
Polycation added	Single-stranded RNA	Double-stranded RNA
None	1.6×10^{-1}	3.1×10^{1}
Poly-ornithine	6.7×10^{4}	2.3×10^{6}
DEAE-dextran	1.6×10^{2}	1.1×10^{6}
Poly-ornithine and DEAE-dextran	6.7×10^{4}	2.3×10^{6}

[a] See text for details.

somewhat different. The best titers obtainable with double-stranded RNA are more than tenfold higher than those observed with single-stranded RNA, but this difference may simply reflect the relative RNase-resistance of the double-stranded molecule. This property of the double-stranded RNA is illustrated in Table IV. In agreement with other criteria for RNase resistance, the presence of 0.15 M NaCl is required to protect infectivity against attack by pancreatic RNase. Moreover, even at high salt concentrations the protection is not absolute, and the infectivity can be abolished by treatment with higher concentrations of enzyme. In contrast to the at least relative resistance of double-stranded RNA, the in-

TABLE IV

EFFECT OF SALT CONCENTRATION ON INACTIVATION OF INFECTIVITY BY RIBONUCLEASE[a]

Conditions of treatment	RNase conc. μg/ml	Precipitated cpm	Relative resistance	PFU/ml
Untreated control	0	1076	100	3.6×10^{4}
0.01 SSC	0.1	745	69	$<3 \times 10^{2}$
	1.0	232	22	$<3 \times 10^{2}$
	100	80	7	$<3 \times 10^{2}$
2 × SSC	0.1	1028	96	1.2×10^{4}
	1.0	998	93	$<3 \times 10^{2}$
	100	911	85	$<3 \times 10^{2}$

[a] Chromatographically purified, ^{32}P-labeled, double-stranded RNA was adjusted to appropriate salt concentrations, and 1 ml volumes treated with pancreatic RNase at indicated concentrations for 15 min at 37° C. Aliquots were then withdrawn and diluted in PSM for assay of infectivity, and the remainder of the samples precipitated with 5% TCA, using 1 mg of yeast RNA as carrier. The precipitates were washed with 5% TCA, dissolved in 1% NH$_4$OH and plated for counting.

fectivity of single-stranded RNA is completely abolished by 1000-fold less enzyme, either in high or low salt. The effect of enzymes on infectious double-stranded RNA is further demonstrated in Fig. 9, in which time courses of inactivation are depicted. A relatively high concentration of pancreatic RNase in SSC abolishes infectivity with single-hit kinetics, but there is a sharp change of slope after 90% inactivation. The significance

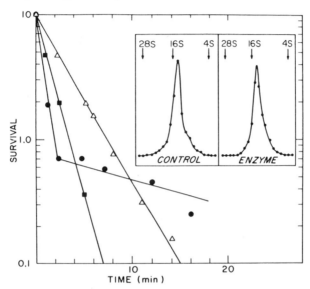

FIG. 9. Enzymatic inactivation of infectious double-stranded RNA. Chromatographically purified RNA with a titer of 2×10^6 PFU/ml was used throughout. Conditions of incubation with enzymes were as follows: (a) pancreatic RNase (●) (0.5 µg/ml) in SSC at 37°C; *E. coli* RNase II (potassium-activated phosphodiesterase), 2 units/ml (△) and 4 units/ml (■) in 0.1 *M* Tris-HCl, pH 7.5, 0.1 *M* KCl, and 10^{-3} *M* MgCl₂ at 22°C. At the indicated time points, samples were withdrawn and diluted into PSM containing 100 µg/ml of DEAE-dextran in the case of pancreatic RNase, and into 0.02 *M* sodium phosphate buffer, pH 7.2, with 10^{-3} *M* EDTA and 100 µg/ml DEAE dextran in the case of RNase II. These procedures sufficiently reduce the activity of the respective RNase to allow reproducible results in determining kinetics of inactivation. The inset depicts the zonal centrifugation in a 5–20% sucrose gradient of a preparation of double-stranded RNA before and after treatment with RNase II.

of the relatively resistant tail of infectivity is unexplained; it is not observed when single-stranded poliovirus RNA is treated with RNase. Also shown on Fig. 9 is the course of inactivation of double-stranded RNA by two concentrations of RNase II, a potassium-activated exonuclease from *E. coli** specific for single-stranded RNA (Singer and Tolbert,

* We are indebted to Dr. Maxine Singer for providing a sample of this enzyme.

1965). The one-hit inactivation of infectivity is somewhat unexpected, considering the specificity of the enzyme and the substantial evidence for the double-stranded nature of at least the bulk of the substrate. The reason for this result is not clear. If one postulates that single-stranded extensions of the double-stranded molecule are required for infectivity as in the case of phage λ DNA (Strack and Kaiser, 1965), it is necessary to make the further assumption that their composition is such that they are not readily attacked by pancreatic RNase. Whatever the site of attack by RNase II, inactivation is effected with the release of no more than 1% acid-soluble material and without demonstrable change in the sedimentation coefficient (Fig. 9, inset). This experiment incidentally shows that the sedimentation coefficient of double-stranded RNA obtained by zonal centrifugation in a sucrose gradient (in 0.15

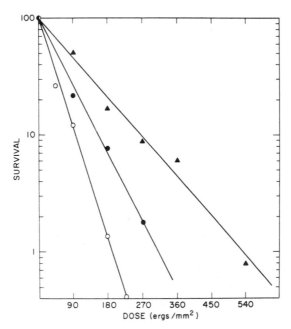

FIG. 10. Inactivation of infectivity by ultraviolet irradiation of whole virus (○), single-stranded RNA (●), and double-stranded RNA (▲). A General Electric G15T8 germicidal lamp was used to irradiate 1 ml aliquots of appropriate virus or RNA dilutions, made in 0.15 M NaCl, 0.02 M sodium phosphate buffer, pH 7.2, and contained in 60 mm plastic petri dishes which were gently rocked during irradiation. The dose administered was determined as described by Latarjet *et al.* (1953). Virus and RNA infectivity titrations were performed with an agar-cell suspension plaque assay (Koch *et al.*, 1966). Infectious, double-stranded RNA was purified as described in the legend for Fig. 2, single-stranded RNA was extracted from purified poliovirus with phenol at 60°C.

M NaCl) is about 14 S, in distinction to the value of 18.4 S calculated from boundary sedimentation (in 0.4 M NaCl) in the analytical ultracentrifuge.

Figure 10 illustrates the inactivation by UV irradiation of the infectivity of intact poliovirus, and single- and double-stranded RNA. One-hit inactivation was observed in each of the three cases, but the rate of the reaction was different in each instance. In agreement with previous observations (Norman, 1960), intact virus was somewhat more sensitive to irradiation than its free, single-stranded genome. The single-stranded RNA was in turn considerably more sensitive than double-stranded RNA, a finding which is in accord with the conclusion that the infectivity of the latter is a distinct, intrinsic property of the molecule. This conclusion is confirmed by the experiment shown in Fig. 11, which demonstrates that,

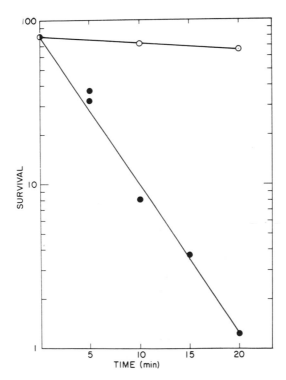

Fig. 11. Effect of formaldehyde on the infectivity of single-stranded (●) and double-stranded (○) poliovirus RNA. Preparations of RNA were treated with 1% formaldehyde (Baker reagent grade) in 0.02 M sodium phosphate, 0.004 M sodium citrate buffer, pH 7.1, containing 0.03 M NaCl (Lozeron and Szybalski, 1966) at 22°C. Aliquots were withdrawn at indicated time points, and diluted no less than 10-fold into PSM for assay of infectivity.

unlike the case of single-stranded RNA, the infectivity of double-stranded RNA is resistant to inactivation by formaldehyde.

Replication of Poliovirus RNA

The fact that poliovirus-specific, double-stranded RNA possesses intrinsic biological activity offers some assurance that it is an authentic product of the process of virus replication, rather than being merely an artifact of isolation or an incidental by-product. It seems unlikely, however, that it represents the active replicative form of viral RNA. Our evidence on this point is as follows. As shown in Table V, isolation of RNase-resistant RNA by treatment of cytoplasmic particles at room temperature with SDS alone yields material with a composition different from that calculated for a simple base-paired duplex molecule.

TABLE V

BASE COMPOSITION OF VIRUS-SPECIFIC RNA's[a]

Bases	Mature poliovirus (+) observed	Double (+, −) calculated	Phenol-SDS extracted observed	Triple (2+, −1) calculated	SDS-extracted observed
U	23.8 ± 0.2	26.5	26.2 ± 0.1	25.6	25.3 ± 0.2
G	23.2 ± 0.3	23.5	23.1 ± 0.0	23.4	23.2 ± 0.2
A	29.3 ± 0.1	26.5	26.7 ± 0.1	27.4	27.4 ± 0.1
C	23.8 ± 0.2	23.5	24.0 ± 0.1	23.6	24.0 ± 0.3

[a] See text and Bishop *et al.* (1965) for details.

The observed values, however, are in close agreement with those calculated for a structure with one minus and two plus strands. Values corresponding to a triple-stranded structure were obtained whether or not the material had been treated with RNase during the course of isolation. It should be pointed out that, although the measured values are clearly inconsistent with a simple duplex molecule, the results are not sufficiently precise to differentiate between a structure with two plus strands and one with, for example, one and a half.

The sedimentation analysis of SDS-extracted, virus-specific RNA is shown in Fig. 12. The preparation had been doubly-labeled with uridine-^{14}C for a long interval and uridine-^{3}H for a short pulse. The radioactivity profiles before and after treatment of RNase are indicated, with the absorbancy profile of cellular RNA as points of reference. The ^{14}C-labeled, RNase-resistant RNA has a sedimentation coefficient of 14–16

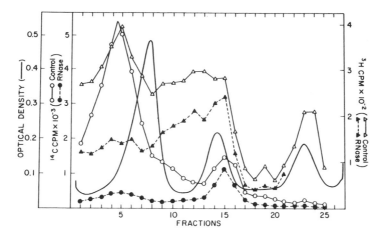

FIG. 12. Sedimentation analysis of doubly labeled RNA extracted from the large particle fraction of poliovirus-infected HeLa cell cytoplasm. Six $\times$ 10^7 cells in concentrated suspension (4×10^6 cells/ml) were incubated with 5 μg/ml of actinomycin for 1 hour, then they were infected with poliovirus at 100 PFU/cell. At the time of infection, uridine-^{14}C (specific activity was 24 mCi/mM; 1×10^{-5} M) was added, and four hours after infection, uridine-^{3}H (23,000 mCi/mM; 2.9×10^{-6} M) was added. After 2.5 min, the cell suspension was poured onto an equal volume of pulverized, frozen medium. The cells were sedimented by centrifugation (1400 rpm, 4 min, International PR 2), washed once with 100 ml of PBS at 4°C, then resuspended in 15 ml of cold RSB. After 5 min at 0°C, the cells were broken with 15 strokes in a Dounce homogenizer. Cell lysis exceeded 90% as checked by phase microscopy, and the extent of nuclear breakage appeared minimal. Cell debris and nuclei were sedimented (1600 rpm for 15 min) and the supernatant was centrifuged at 40,000 for 20 min. The resulting pellet was suspended in 3 ml of .15 M NaCl 5 mμ Tris-HCl, pH 7.5 and treated with 0.5% SDS. The RNA present in such a preparation amounts to approximately 0.5 mg. The entire solution was layered over a 15–30% gradient (used because it more effectively retards contaminating free isotope than does 5–20%) of sucrose in the above buffer without SDS. After centrifugation at 25,000 rpm for 20 hours in a Spinco SW25 rotor at 4°C, the gradient was collected in 1 ml fractions, monitoring A$_{260}$. Duplicate, 0.25 ml samples of each fraction were adjusted to 1.0 ml volume and 0.3 M NaCl, after which one was precipitated with TCA (2.5%) directly, the other being treated first with RNase, 0.2 μg/ml, at 37°C for 15 min. TCA precipitation was carried out without the addition of any carrier material. The precipitated samples were filtered through Gellman GA-6 membrane filters, which were then washed twice with 10 ml volumes of 2.5% TCA. The filters were dried in counting vials at 60°C, after which the acid precipitates were eluted by soaking each filter in 1-ml hyamine hydroxide-methanol. Liquifluor-toluene (15 ml) was added to each vial, and both isotopes were counted simultaneously in a Packard scintillation spectrometer, calibrated so that 0.1% of a suitably quenched ^{3}H standard was counted in the ^{14}C-channel, and 10% of a quenched ^{14}C-standard counted in the ^{3}H channel.

S, while the pulse-labeled RNase-resistant RNA, as first noted by Baltimore and Girard (personal communication; see also Darnell *et al.,* this Symposium) was distributed in a broad zone throughout the lower half of the tube. We conclude that the pulse-labeled material represents the functional replicative RNA, and that it consists of double-stranded RNA with an additional or part of an additional plus strand.

Discussion and Conclusions

Some special mechanism must be invoked to explain the initiation of infection by double-stranded RNA. In the usual infection, the incoming single-stranded RNA probably participates in the organization of a polyribosome and thus directs the immediate synthesis of virus-specific proteins. In the case of double-stranded RNA, three alternative initial events can be envisaged. (1) Direct translation into polypeptide of the incoming RNA. Effort to demonstrate messenger activity *in vitro* for double-stranded polyribonucleotides have, however, been unsuccessful (Nirenberg and Matthaei, 1961; Miura and Muto, 1965). (2) Strand separation. No model of the enzymatic reaction required for this alternative is presently known. (3) Transcription of double-stranded RNA into single-stranded RNA with host-cell RNA polymerase. Recent findings (Shatkin, 1965) indicate, however, that double-stranded reovirus RNA is inactive as a primer for the RNA polymerase of *E. coli.*

Clearly, the question of the initial step in the infectious cycle is unresolved. Study of the replication of reovirus (for example, see Shatkin, this volume) would appear to be a more promising approach to this question, since that experimental system has a higher efficiency of infection and thus permits the synchronous infection of many cells and a detailed analysis of the infectious cycle.

With regard to the mechanism of replication of poliovirus RNA, it now appears that there is a replicative intermediate which is appreciably labeled only by brief pulses of radioactive precursor, and which is characterized by partial RNase resistance, heterogeneous sedimentation behavior, buoyant density intermediate between single- and double-stranded RNA, and solubility in 2 *M* LiCl. We assume that the phenol-extracted, double-stranded poliovirus-specific RNA is a portion of this active complex. Granting this assumption, the fact that the double-stranded RNA is infectious and that it yields intact single strands on solvent denaturation (Katz and Penman, 1966) limits the permissible hypotheses on the nature of the active complex and the detailed mechanism of RNA replication. A widely discussed model involves the sequential synthesis of plus strands against a stable complementary single strand, implying that several partially completed strands are attached to the template. A double-stranded RNA derived from such a replicative complex would presumably have gaps in the plus strand. Granting that infectivity is the property of an intact plus strand, some mechanism for repair of the gaps with covalent links is necessary to account for the observed results.

Our results can be more easily reconciled with an alternative model: a stable template, consisting of intact plus and minus strands, to which a nascent plus strand, or part of one, is attached. The attachment of the plus strand is such that, under certain conditions, it is relatively resistant to attack by RNase.

It should be pointed out, however, that if the double-stranded RNA represents an inactive template rather than a portion of the active replicative complex, our results are not then in conflict with the semiconservative model for RNA replication. Whatever the relationship of the double-stranded RNA to replication may be, the issue is far from settled (see Haruna and Spiegelman, 1966; Weissmann and Feix, 1966), and the fact that the double-stranded RNA is intrinsically infectious lends considerable interest to the study of its properties.

ACKNOWLEDGMENT

We gratefully acknowledge the advice and assistance of C. Hiatt (UV inactivation experiments) and F. DeFilippes and A. Shatkin (optical ultracentrifuge experiments).

Part of this work was supported by Research Grant No. AI-06862 from the U.S. Public Health Service.

REFERENCES

Ammann, J., Delius, H., and Hofschneider, P. H. (1964). *J. Mol. Biol.* **10,** 557.
Baltimore, D., and Girard, M. (1966). Personal communication.
Bishop, J. M., Summers, D. F., and Levintow, L. (1965). *Proc. Natl. Acad. Sci.* **54,** 1237.
Haruna, I., and Spiegelman, S. (1966). *Proc. Natl. Acad. Sci. U.S.* **55,** 1256.
Katz, L., and Penman, S. (1966). *Biochem. Biophys. Res. Commun.* **23,** 557.
Koch, G., Quintrell, N., and Bishop, J. M. (1966). In press.
Kubinski, H., Koch, G., and Drees, O. (1962). *Biochim. Biophys. Acta* **61,** 332.
Latarjet, R., Morenne, P., and Berger, R. (1953). *Ann. Inst. Pasteur* **85,** 174.
Lozeron, H. A., and Szybalski, W. (1966). Personal communication.
Mandel, J. D., and Hershey, A. D. (1960). *Anal. Biochem.* **1,** 66.
Martin, M., and Bishop, J. M. (1966). Unpublished observations.
Miura, K. J., and Muto, A. (1965). *Biochim. Biophys. Acta* **108,** 707.
Miyazawa, Y., and Thomas, C. A. (1965). *J. Mol. Biol.* **11,** 223.
Nirenberg, M. W., and Matthaei, J. H. (1961). *Proc. Natl. Acad. Sci. U.S.* **47,** 1588.

Norman, A. (1960). *Virology* **10**, 384.

Shatkin, A. J. (1965). *Proc. Natl. Acad. Sci. U.S.* **54**, 1721.

Singer, M. F., and Tolbert, G. (1965). *Biochemistry* **4**, 1319.

Strack, H. B., and Kaiser, A. D. (1965). *J. Mol. Biol.* **12**, 36.

Sueoka, N., and Cheng, T. Y. (1962). *J. Mol. Biol.* **4**, 161.

Vinograd, J., Bruner, R., Kent, R., and Weigle, J, (1963). *Pro. Natl. Acad. Sci. U.S.* **49**, 902.

Weissmann, C., and Feix, G. (1966). *Proc. Natl. Acad. Sci. U.S.* **55**, 1264.

The Synthesis and Translation of Poliovirus RNA*

J. E. Darnell, M. Girard,† D. Baltimore,** D. F. Summers,††
and J. V. Maizel****

DEPARTMENTS OF BIOCHEMISTRY AND CELL BIOLOGY, ALBERT EINSTEIN COLLEGE OF
MEDICINE, BRONX, NEW YORK

The small, RNA-containing viruses have been popular subjects in molecular biology because of the potentiality of identifying and studying the function of the limited number of gene products which can be encoded by these viruses. Poliovirus growth in HeLa cells has been explored in our laboratory for some years now with the foregoing hopes in view.

This paper will describe various areas of research in which poliovirus has been a central object of study. Since some of the material has been prepared for publication elsewhere, it will not be fully described here.

The basic techniques employed have been fully described in the past. These include the growth, labeling, and harvesting of infected HeLa cells (Eagle, 1959; Scherrer and Darnell, 1962) and the subsequent fractionation of these cells into cytoplasmic and nuclear fractions (Penman *et al.*, 1963; Penman, 1966). Most of the experiments have utilized zonal centrifugation of either cellular extracts to observe polyribosomes, single ribosomes, and various ribonucleoprotein particles on the pathway to functioning ribosomes, or deproteinized RNA solutions (phenol-SDS or SDS extraction) to observe the sedimentation profile of viral RNA, replicative forms of viral RNA, newly synthesized cytoplasmic messenger RNA and ribosomal RNA, and various types of nuclear RNA.

* Supported by funds from the United States Public Health Service (CA 07592 and CA 07861) and the National Science Foundation (GB 2477).

† Fellow of the Committee of Molecular Biology, General Delegation for Scientific and Technical Research, Paris, France. Present address: Department of Microbe Physiology, Pasteur Institute, Paris, France.

** Present address: The Salk Institute for Biological Studies, San Diego, California.

†† Career Research Scientists of the City of New York.

*** Recipient of Career Development Award, USPHS.

Kinetics of Poliovirus RNA Synthesis

Early studies on the time course of synthesis of poliovirus RNA revealed that the majority (perhaps 85%) of viral RNA is formed at a linear rate between about 3 and 4.5 hours after infection (Darnell *et al.,* 1961). These studies did not provide satisfactory evidence on the time of initiation of viral RNA synthesis. The discovery that actinomycin D has no effect on the multiplication of poliovirus RNA (Reich *et al.,* 1961), while almost completely stopping incorporation of nucleoside precursors into cellular RNA, provided conditions necessary to detect the earliest viral RNA synthesis. The characteristic sedimentation profile of RNA formed in poliovirus-infected, actinomycin-treated cells shows a predominant 35 S peak—the same as that of RNA recovered from purified virus (Darnell, 1962). By observing the type of newly labeled RNA which accumulates in such cells, it is possible to detect 35 S RNA as early as 60 min after infection. This type of RNA accumulates exponentially for 2.5–3 hours, and then the rate of accumulation becomes approximately linear (Baltimore *et al.,* 1966). Since it is known that the infected cell accumulates about $2–3 \times 10^5$ viral RNA molecules by the close of infection (Levintow and Darnell, 1960; Joklik and Darnell, 1961), the total radioactive RNA accumulated can be equated to $2–3 \times 10^5$ viral RNA molecules, and the number of molecules present in an infected cell at any time can be estimated. By back extrapolation of the course of exponential increase of viral RNA, it is found that the infected cell has about 30 new molecules within 30 min after infection. Since the multiplicity of infection in these experiments was about 30 PFU*/cell, it appears that viral RNA replication is definitely underway within a half hour after infection (Baltimore *et al.,* 1966).

In order to investigate the possibility that the exponential rate of increase of viral RNA was due to the participation of progeny molecules, and not due to an asynchronous entry into the replicative process of parental RNA as the only *templates* for virus production, the following experiment was carried out. Actinomycin-treated cells were infected at various virus/cell multiplicities, including some less than one. By following uridine-^{14}C incorporation, the maximum level of viral RNA accumulation was determined, from which the fraction of cells infected could be measured, assuming full yields of RNA from cells infected at any multiplicity. The rate of accumulation of viral RNA was measured in each culture and adjusted according to the fraction of cells infected. It was found that the maximum rate of viral RNA synthesis was identical in cells infected with 30- or with 1 PFU. Since it is known that a large

---

* PFU, plaque forming units.

number of virus particles can function simultaneously in infected cells (e.g., both Type I and Type II virus are represented in yields from single cells even though the relative multiplicity of infection is 10 or more to 1; Ledinko and Hirst, 1961), it follows that multiplication of templates must occur in order for the achievement of maximum rate of RNA synthesis by singly infected cells. In addition, the exponential rate of RNA accumulation, seen in multiply infected cells early in infection, is compatible with an increasing number of functioning templates. From these experiments it can be calculated that the doubling time for 35 S viral RNA, as well as for templates, is about 15 min.

After the demonstration that viral RNA multiplication begins soon after infection, the question was asked—"does early formed RNA enter whole virus?" Examination of the distribution of viral RNA after two, and after three hours of infection revealed that, although viral RNA synthesis had been initiated for well over an hour, no viral RNA had entered whole virus by two hours, whereas a large fraction of the viral RNA had entered virions at three hours. Furthermore, it was shown that the entrance of virus RNA into whole particles takes, at most, a few minutes during the height of the maturation stage. Therefore, early viral RNA either never enters particles or, at least, must wait for over an hour before so doing (Baltimore *et al.,* 1966).

Nature of Viral RNA Templates

Since the discovery by Montagnier and Sanders (1963) of an RNase-resistant infectious form of EMC virus RNA, many laboratories have been trying to ascertain whether or not viral RNA is produced from a double-stranded template by a Watson-Crick base pair copying mechanism. The experiments which we will describe that bear on this question have been designed to accommodate an important feature of viral RNA synthesis: Viral RNA synthesis is asymmetric; that is, much more RNA of the type found in whole virus and in polyribosomes ("plus" strand in the terminology of Weissmann *et al.,* this volume) is formed than any other type (Zimmerman *et al.,* 1963).

Thus, in order to observe the nature of any intermediates in the formation of finished single-stranded "plus" chains, it is necessary to limit the incorporation time to very short periods—preferably to a time in which the majority of a newly incorporated radioactive RNA precursor is in unfinished chains. Figure 1 shows that after a 15 min period of incorporation, more than 85% of the labeled viral-specific RNA sediments as 35 S material, while an exposure of infected cells to uridine-^{3}H for only 2.5 min results in the formation of considerable labeled RNA that

sediments at a rate less than 35 S. Also, it can be seen that the proportion of RNA which is RNase-resistant is higher in the shorter labeling
period. Therefore, it is clear that different types of RNA are being observed in the short than in the longer labeling period.

Advantage can be taken of the observations just described to determine the length of time required to make a chain of viral RNA. It was
seen in Fig. 1 that finished chains of viral RNA can be distinguished
from labeled RNA not yet completed which therefore is presumed to

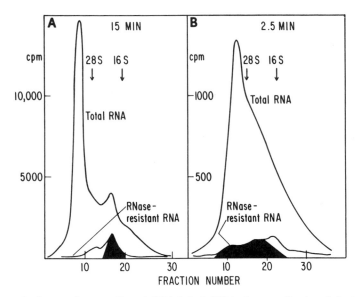

FIG. 1. Sedimentation profile of ^{3}H-labeled RNA from poliovirus-infected HeLa
cells. Infected Actinomycin-treated cells were labeled for 15 (A) or 2.5 (B) min
and RNA extracted by the hot-phenol SDS method (Scherrer and Darnell, 1962).
After sedimentation and fractionation on sucrose gradients, the total acid precipitable
radioactivity and that resistant to RNase treatment was determined in each fraction (Baltimore and Girard, 1966). The OD$_{260}$ markers at 28 S and 16 S are from
the cellular ribosomal RNA.

still be associated with the template (regardless of the nature of the
template). Consider the situation which obtains after the introduction of
^{3}H into the nucleotides of the cell pool. When the average time to synthesize a chain of 35 S viral RNA has elapsed, the amount of label still
associated with the template will equal that incorporated into finished
molecules (Fig. 2). By exposing infected cells to uridine-^{3}H for shorter
and shorter intervals, it becomes possible to determine the time at which
about 50% of the labeled RNA sediments as distinguishable 35 S RNA.
This time is found to be between 2 and 3 min (Fig. 3). However, be-

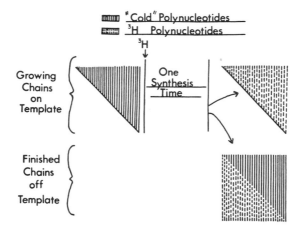

FIG. 2. Distribution of radioactive RNA between template and finished chains of viral RNA. When uridine-³H is introduced into infected HeLa cells, radioactivity will appear first in growing, template-associated RNA chains. After completion of some chains, radioactivity will appear in 35 S RNA *not* attached to template. When a period of time has elapsed that corresponds to the average synthesis time of one viral RNA molecule, there will be an equal amount of radioactivity on and off the template.

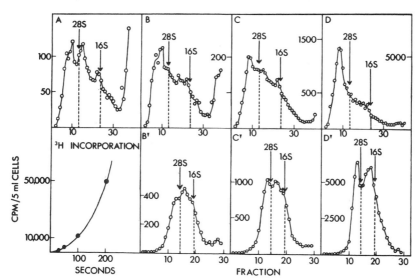

FIG. 3. Synthesis time of individual chains of poliovirus RNA—two separate experiments designed to utilize the principle described in Fig. 2. Infected cells were labeled for 30 (A), 50 (B,B'), 100 (C,C'), or 200 (D,D') sec. RNA was extracted and analyzed to measure the proportion sedimenting as 35 S RNA. Between 100 and 200 sec, over 50% of the label sediments as 35 S RNA. The curve in the lower left panel shows the total radioactive viral RNA recovered from cultures in A-D.

cause the radioactivity in the cell pool is not raised immediately to a maximum, but is increasing rapidly just after exposure to uridine-^{3}H, a correction must be made. For example, in Fig. 3, lower left panel, it can be seen that almost 80% of the total incorporated radioactivity entered the RNA in the last 100 seconds of incorporation. It thus seems likely that it requires no more than 1.5 min for infected HeLa cells to synthesize a chain of poliovirus RNA.

The nature of the labeled RNA taken from cells exposed to uridine-^{3}H for 200 sec or less (pulse-labeled RNA) was examined for evidence of physicochemical differences from 35 S viral RNA or the 18–20 S ribonuclease-resistant double-stranded form of viral RNA, which has been studied in several laboratories (Montagnier and Sanders, 1963; Baltimore *et al.,* 1964; Bishop *et al.,* 1965; Katz and Penman, 1966). The pulse-labeled material (see Fig. 1) was found to contain ribonuclease-resistant RNA not only at 18–20 S, the sedimentation position of double-stranded RNA, but also sizeable amounts of RNase-resistant material sedimenting from 20–40 S. When this more rapidly sedimenting RNase-resistant material was separated and examined, it was. found to be insoluble in LiCl, a property of single-stranded RNA but not of double-stranded RNA (Baltimore, 1966). The character of the material was changed, however, by treatment with RNase (Baltimore and Girard, 1966). A portion of the radioactivity was degraded and that which remained was now LiCl soluble 18–20 S RNA—that is, indistinguishable from authentic double-stranded RNA. The "pulse labeled" RNA was also shown to have a density close to that of single-stranded material until after RNase treatment, when the residual radioactivity assumes the density of double strands. Thus, after very brief labeling periods, it is possible to isolate a type of RNA that is a complex between newly formed single-stranded RNA and double-stranded RNA. This type of RNA has been termed the *replicative intermediate,* a term first used by Fenwick *et al.* (1964), to describe briefly labeled RNA from RNA bacteriophage-infected cells.

If this material is related to viral RNA replication, it should increase exponentially in amount early in infection just as the 35 S viral RNA does. Figure 4 shows an experiment in which the amounts of (1) 35 S single-stranded viral RNA, (2) LiCl-soluble double-stranded RNA, and (3) the double-stranded RNA, recoverable after RNase digestion of the LiCl-precipitable material, were all determined as a function of time after infection. It can be seen that the chains of double-stranded RNA form exponentially early in infection, and, that during the exponential phase of viral RNA synthesis, there is more of the LiCl-precipitable double-strand present than of the free 20 S LiCl-soluble form. This is

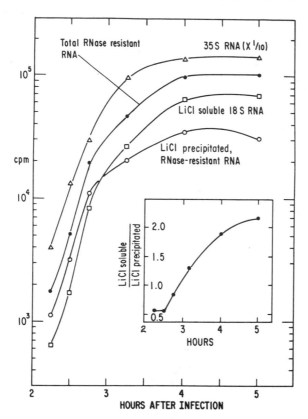

Fig. 4. Accumulation of various types of virus-specific RNA during the replicative cycle of poliovirus. All RNA was derived at indicated times from an actinomycin-treated, infected HeLa cell culture to which uridine-³H had been added at the outset of infection. The various types of RNA were measured as follows: (1) 35 S RNA was measured on untreated aliquots of extracted RNA as in Fig. 1; (2) LiCl-soluble 18 S RNA was measured after the total RNA was treated with 2 *M* LiCl overnight, the pellet separated and the soluble fraction alcohol precipitated, re-suspended in 2 × SSC (0.3 *M* NaCl, 0.03 sodium citrate) and treated with 100 γ/ml RNase at 22°C for 30 min; (3) LiCl-precipitable, RNase-resistant RNA was determined on RNA in pellet after LiCl precipitation. All measurements were made after fractions were analyzed on sucrose gradients.

consistent with the idea that the LiCl-precipitable double-stranded material is the template for viral RNA synthesis.

It is possible, from the data given in Fig. 4, to approximate the number of chains being synthesized on each template at the time of maximal viral RNA synthesis. Figure 4 shows that at 2.75 hours after infection there are approximately 5×10^4 35 S viral RNA molecules per cell (calculated from the fact that $\frac{1}{5}$ of the total RNA that will eventually be made has been made and 2.5×10^5 molecules per cell is the eventual yield), and approximately 2% as much radioactive LiCl-precipitable double-stranded RNA as 35 S RNA. Since the LiCl-precipitable RNA after RNase digestion presumably contains twice as much labeled RNA per molecule as does the 35 S viral RNA, it can be calculated that there are about 500 templates per cell at 2.75 hours after infection ($5 \times 10^4 \times 2\%/2$). During the last 100 min of virus RNA growth, about 2×10^5 35 S viral RNA molecules per cell are synthesized (2000 molecules per min), and the time for the synthesis of a single chain may be calculated to be one minute. Thus (2000 chains per cell)/(500 templates per cell) = 4 chains per template per minute. Each template should therefore contain approximately 4 enzyme molecules.

Site of RNA Synthesis: The Replicative Complex

Once it had been determined that a large proportion of the viral-specific RNA obtained from infected cells exposed for 2–3 min to uridine-^{3}H was in incomplete, newly-synthesized chains, experiments were begun to locate this RNA in cytoplasmic extracts of infected cells. If HeLa cells are swelled in a hypotonic buffer for 5–10 min, their cytoplasm can be very effectively removed by gentle homogenization (Penman *et al.,* 1963). Extracts from infected cells were prepared, treated with desoxycholate and examined by zonal centrifugation through sucrose. A large proportion of the "pulse-labeled" RNA was found in a broad band with a peak at 250 S (Fig. 5). There was almost no labeled RNA in smaller structures. Upon extraction of the 250 S structures, which have been named "replicative complexes," RNA was obtained which had the same characteristics as that recovered from whole cells after brief labeling periods, i.e., heterogeneous material with RNase resistant RNA sedimenting between 20–40 S (Fig. 1).

Because of the size of the replicative complex, it was at first thought that it might represent RNA associated with ribosomes. A number of lines of evidence indicate that this is not the case. (1) Puromycin treatment of infected cells, which causes breakdown of polyribosomes (viral or host cell) has no effect on the replicative complex. (2) Ribosomes are released from polysomes by the action of EDTA, while the replicative complex is unaffected by EDTA. (3) Cells were prelabeled with uridine-^{3}H and then infected with poliovirus. The replicative complex was prepared from this culture and from a parallel unlabeled, infected culture by two zonal centrifugations through EDTA-containing buffer. At least 50% of the replicative complex was recovered while less than

0.02% of the ribosomes (as estimated from recovery of label) were recovered (Girard et al., 1966).

It is concluded, therefore, that the rapid sedimentation behavior of the replication complex is not due to association of nascent RNA chains with ribosomes.

Positive information about the nature of the replicative complex was obtained by exposing it to the proteolytic enzyme, pronase. This enzyme destroyed the replicative complex without affecting the size distribution of the RNA within it (Fig. 6). It was concluded, therefore, that a protein(s) was an integral part of the complex.

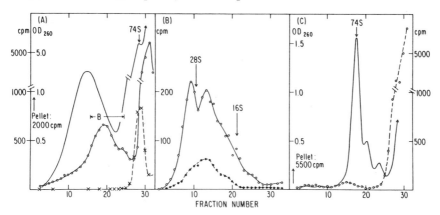

FIG. 5. Demonstration of poliovirus RNA "replicative complex." Cytoplasmic extract from infected, actinomycin-treated cells, labeled for 2–5 min with uridine-³H was prepared (Penman *et al.*, 1963), treated with 0.5% desoxycholate and sedimented through a sucrose gradient (A). The contents of the gradient were analyzed by a flow recorder for OD_{260}, and in a portion of each fraction, acid-precipitable radioactivity was determined (—○—). (B) The remainder of each fraction was treated with phenol-SDS to release RNA which was, in turn, analyzed for total (○), and RNase-resistant (●), acid-precipitable radioactivity, as in Fig. 1. Part C demonstrates the lack of any incorporated ³H in material larger than 4 S.

In addition, it was shown that, after a brief labeling period, the radioactivity in the replicative complex sedimented in parallel with the RNA polymerase activity of a desoxycholate-treated cytoplasmic extract (Fig. 7). Without the detergent treatment, the pulse-labeled RNA and the RNA polymerase activity are bound in large structures which have been termed virus-synthesizing bodies (Penman *et al.,* 1964; Baltimore, 1964).

The final experiment which was done to characterize the role of the replicative complex in viral RNA synthesis involved an examination of

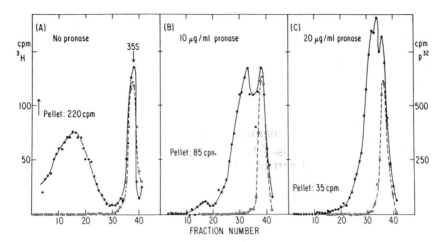

FIG. 6. Effect of pronase on "replicative complex." An extract of infected, ³H-labeled cells, prepared as described in the legend to Fig. 5, was sedimented in a sucrose gradient containing EDTA (Girard *et al.,* 1966). Pronase treatment is seen to destroy the complex.

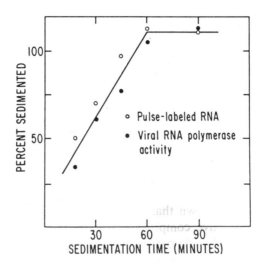

FIG. 7. Cosedimentation of RNA polymerase activity and pulse-labeled RNA. Extracts of pulse-labeled, and unlabeled, infected cells were prepared and treated with desoxycholate, and were then subjected to sedimentation through 3.5 ml of 15% sucrose solution for the times indicated. The pellets were then analyzed for the proportion of pulse-labeled RNA (from labeled extract) or RNA polymerase activity (from the unlabeled extract) which had sedimented.

the accumulation of radioactivity in this structure relative to the remainder of the cytoplasmic structures. Measurement of the radioactivity in the replicative complex was facilitated by its rapid sedimentation behavior and stability in EDTA. Viral RNA in whole virus, in polyribosomes, or in lighter structures could all be recovered in sucrose gradients as less rapidly sedimenting material. The distribution of radioactivity studied in this way revealed that, for the first few minutes of incorporation, almost all the labeled viral RNA was recovered in the replicative complex. The total counts in this structure increased progressively for about 10 min after exposure to label, and this was followed by a rise in radioactivity in other structures. This pattern of accumulation is consistent with the idea that the replicative complex is the site of viral RNA synthesis, and only after it becomes saturated is there an increase in the RNA in other fractions.

Summary of Viral RNA Synthesis

Viral RNA synthesis in poliovirus-infected cells begins within 30 min after infection, and continues exponentially, with a doubling time of 15 min, for 2 to 2½ hours, at which time about 10% of the total viral RNA has been synthesized. Viral RNA does not enter whole virus during most of this phase. A linear rate of RNA production then ensues, which continues for an additional 1½ hours. The time required to make a chain of RNA during this latter interval is about one minute, therefore each infected cell initiates and completes the synthesis of about 2000 chains per minute (about 2×10^5 molecules per cell are made between 2.5 and 4 hours post-infection) during this phase of replication.

These growing chains can be identified, by very brief pulses of uridine-^{3}H, as being in the form of single strands attached to a double-stranded template which is called the *replicative intermediate.*

The RNA which is in the form of the *replicative intermediate* is recoverable after desoxycholate treatment of cytoplasmic extracts in a large structure termed the *replicative complex.* No ribosomes can be detected in this structure, and its integrity is dependent on a protein or proteins which are sensitive to pronase. Since the *in vitro* RNA polymerase activity is also found in the replicative complex, it is tempting to speculate that the structure is composed of polymerase which actively transcribes RNA in the cell.

The nature of any viral-specific proteins in the replication complex and the movement of finished chains from the replicative complex into polyribosomes and whole virus are now under investigation.

Effects of Poliovirus on Host Cell Metabolism

Having considered a number of new experiments dealing with viral RNA replication by the viral RNA polymerase, we would now like to consider the interruption of host cell RNA and protein synthesis which occur during the course of infection. We will first consider the effect on host cell RNA synthesis. HeLa cell RNA metabolism has been the object of intensive research in the past several years (Scherrer and Darnell, 1962; Scherrer *et al.,* 1963; Penman *et al.,* 1963; Girard *et al.,* 1964, 1965) and the following general picture has emerged (Fig. 8). All RNA

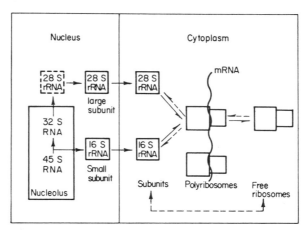

Fig. 8. Diagram of flow, in HeLa cells, of ribosomal RNA from transcription through ribosomal maturation and ultimate appearance in cytoplasm.

synthesis takes place in the cell nucleus. The nucleus can be isolated free of contaminating cytoplasm, and contains no finished ribosomes (Penman, 1966). A technique of nuclear fractionation involving nuclear lysis in a high salt buffer, followed by destruction of the DNA with pancreatic DNase allows the separation by low speed centrifugation of nucleoli (nuclear pellet fraction), and a fraction we refer to as nuclear supernatant (Penman *et al.,* 1966). The 45- and 32 S ribosomal precursor RNA molecules are quantitatively recovered with the nucleoli. A heterogeneous RNA fraction (S values 15–80 S), which has a base composition very much like that of HeLa cell DNA (42–44% G and C), is divided between the nuclear supernatant and nuclear pellet fractions (Warner *et al.,* 1966a; Soeiro *et al.,* 1966). Also present in the nuclear supernatant fraction are subribosomal particles, which are precursors to cytoplasmic subribosomal particles (Vaughan *et al.,* 1967). The cytoplasmic subunits have been shown to be the first cytoplasmic site of appearance of newly formed ribosomal RNA. The ribosomal subunits initially enter polyribosomes preferentially, and eventually equilibrate with the pool of single ribosomes (Girard *et al.*, 1965). A summary of the flow of ribosomal RNA is illustrated in Fig. 8.

The effect of poliovirus on these various processes has been studied, and a specific interruption of ribosomal RNA formation and maturation has been observed.

Figure 9 shows, first, that by 1½ hours post-infection with high multiplicities of poliovirus, the synthesis of the 45 S ribosomal precursor RNA has been substantially depressed, whereas the synthesis of the nuclear supernatant RNA is not seriously affected, and may even be stimulated.

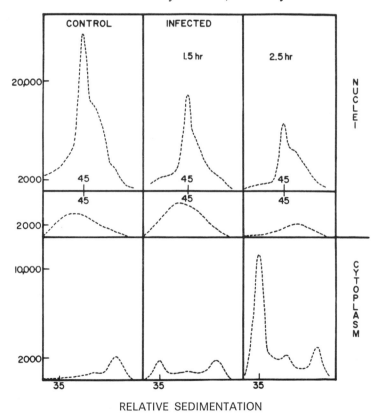

RELATIVE SEDIMENTATION

FIG. 9. Effect of poliovirus infection on host cell RNA synthesis. Cells infected with approximately 100 PFU adsorbed/cell were labeled with uridine-³H for 20 min beginning at 70 and at 130 min after infection. RNA was extracted from various cellular fractions (Penman *et al.*, 1966; Soeiro *et al.*, 1966) and compared to normal cell RNA. The two nuclear fractions are nucleoli, upper section, and nuclear supernatant, middle section.

Quantitation of the appearance of mRNA in the cytoplasm is difficult after brief periods of labeling, but Willems and Penman (1966) have described experiments which indicate that the cytoplasmic appearance of this type of RNA is probably not depressed in infected cells for at least 1–2 hours after infection.

If the time of exposure to label is sufficiently long (e.g. 45 min), radioactivity from the nuclear pellet fraction is seen not only in 45 S ribosomal precursor RNA but also in a second precursor, 32 S RNA. It is believed that the 45 S molecule is divided into one 32 S and one 16 S molecule. This conversion is inhibited or slowed in the virus-infected cell (Fig. 10).

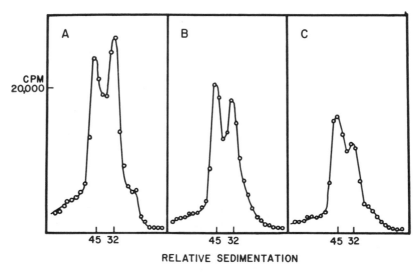

FIG. 10. Effect of poliovirus infection on appearance of 32 S RNA. Uridine-³H labeled RNA from nucleoli of uninfected cells (A), or cells infected for 25 (B), or 45 min (C) was prepared after 45 min of labeling time. Analysis was by sucrose gradient centrifugation for a relatively short period so that 45 S RNA is about 60% of the way down the centrifuge tube (Warner *et al.*, 1966a).

A second dramatic effect of virus infection is seen by examining the nuclear supernatant fraction for the formation of nuclear subribosomal particles. After a 45 min labeling period the larger of these particles, which has been shown to contain 28 S RNA (larger ribosomal RNA molecule), is labeled in the control cells but not in the cells which have been infected for at least 25 min prior to labeling (Fig. 11). This is true in spite of the fact that OD_{260} in the region of the 50 S ribosomal subunit appears to increase progressively during infection.

If cells are labeled 20 min before infection, and the nuclear supernatant examined 90 min after infection, radioactivity as well as optical density is seen to pile up in the nuclear 50 S subunit (Fig. 12). The smaller ribosomal subunit is relatively unaffected, and no accumulation takes place. This is reflected in the cytoplasm, where in the control culture (Fig. 12A') approximately equal amounts of 28 S and 16 S ribosomal

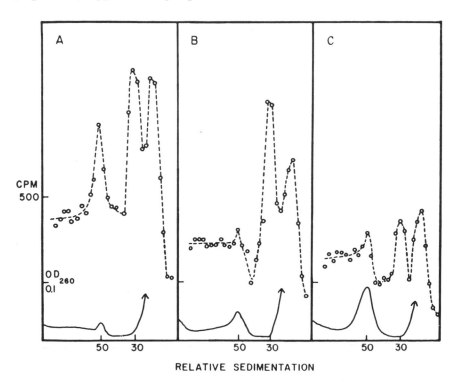

Fig. 11. Interruption of formation of the larger ribosomal subunit by poliovirus infection. Cells were labeled with uridine-³H for 50 min, the nuclear supernatant fraction prepared (Penman *et al.*, 1966; Soeiro *et al.*, 1966) and analyzed to display nuclear subribosomal particles which are precursors of cytoplasmic particles (Vaughan *et al.*, 1966) (A) control cells; (B) infected cells, labeled 20–70 min after infection; (C) infected cells, labeled 40–90 min after infection.

RNA have appeared within the time of the experiment, whereas in the infected cells the emergence of ribosomal subunits containing 28 S RNA has been greatly impaired (Fig. 12B').

This unequal effect (in virus-infected cells) on the completion of or release into the cytoplasm of the larger ribosomal subunit is not due

J. E. Darnell et al.

simply to an inhibition of protein synthesis. Cycloheximide, which immediately stops all protein synthesis, still allows the emergence of approximately equal amounts of 28 and 16 S RNA into the cytoplasm, and no accumulation of 50 S nuclear ribosomal subunits occurs in its presence (Figs. 12C and 12C'). The fact that more total counts are present in both the control and virus-infected cultures than in the cor-

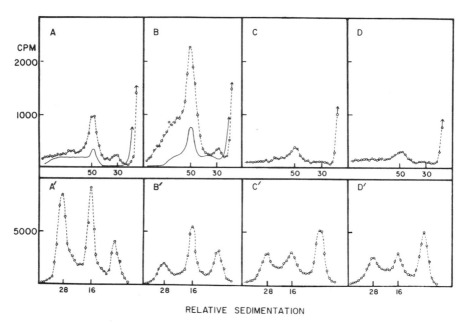

FIG. 12. Analysis of the effect of poliovirus infection on the synthesis of the larger ribosomal subunit. Nuclear supernatants were examined for ribosomal precursor particles as in Fig. 11, and cytoplasmic extracts were examined after the release of RNA by SDS (Girard *et al.,* 1965). Nuclear particles: (A) control, labeled for 90 min; (B) labeled for 15 min, infected for 75 min; (C) labeled for 15 min, treated with cycloheximide (150 γ/ml) for 75 min; (D) labeled for 15 min, infected and treated with cycloheximide for 75 min. A' B' C' D' illustrate the sedimentation behavior of RNA of the cytoplasmic fractions from cell samples A–D.

responding cycloheximide-treated culture is due to the almost immediate depression of overall incorporation of uridine-³H after this drug is added, while virus infection alone does not depress incorporation for some time. The distribution patterns of RNA and nuclear particles in the virus-infected, cycloheximide-treated cultures, and in the uninfected, cycloheximide-treated cultures, are virtually identical (compare Figs. 12C and 12C' with 12D and 12D', respectively). The data obtained from this final culture, (D), indicate that for the virus to exert the observed effect on maturation or release of the large ribosomal subunit, protein synthesis must occur in the infected cell.

The foregoing experiments indicate that some virus product(s) has the capacity to interfere specifically with ribosome maturation. In an effort to determine whether the entering virus genome can induce this suppressing activity, cells were infected in the presence of guanidine, an agent which completely blocks viral RNA synthesis (Crowther and Melnick, 1961; Summers *et al.,* 1965). Both the accumulation of the larger ribosomal subunit, and the inhibition of conversion of 45- to 32 S ribosomal precursor RNA occurred in the guanidine-treated infected culture (Fig. 13), indicating that the entering strand of viral RNA is sufficient to cause the effects on RNA synthesis, albeit at a slower rate than that observed in the absence of guanidine.

One possible mechanism of action of a viral product in blocking normal ribosomal maturation could be that it inhibits methylation of ribo-

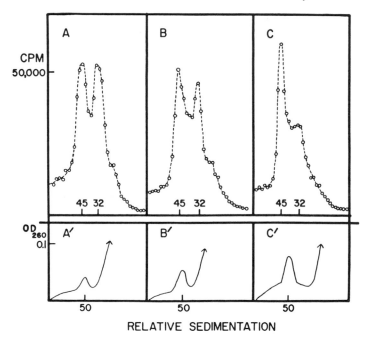

RELATIVE SEDIMENTATION

Fig. 13. Poliovirus suppression of the appearance of 32 S nuclear RNA in absence of viral RNA replication. Top section: RNA was prepared from nucleoli of cells labeled with uridine-³H for 45 min and subjected to sedimentation analysis. All cells were incubated in presence of 3 micromoles/ml of guanidine. (A) control; (B) infected, labeled 55–100 min; (C) infected, labeled 90–135 min. Bottom section: OD₂₆₀ tracings of the nuclear supernatants of samples A–C, showing the increase in 50 S particles.

somal RNA. Cells infected for 25 min, and normal cells, were therefore exposed for 50 min to both uridine-³H and ¹⁴C-methyl-labeled methionine, the known precursor (via S-adenosyl methionine) of —CH₃ groups of HeLa cell RNA (Brown and Attardi, 1965). The incorporation of the ³H label showed that the 45- to 32 S conversion was inhibited in the infected culture, while the incorporation of the ¹⁴C label showed that no inhibition of methylation of the 45 S RNA had occurred (Fig. 14). Several laboratories have described the methylation of HeLa cell 45 S ribosomal precursor RNA (Zimmerman and Holler, 1966; Goodman and Penman, 1966) and, because the —CH₃ groups appear sequentially in 45-, 32- and 16 S RNA just as does uridine-³H, it appears that all or almost all of the methylation takes place at the 45 S stage.

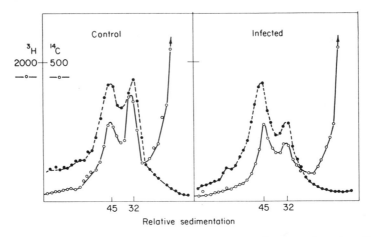

Fig. 14. Methylation of ribosomal precursor RNA in normal and poliovirus-infected cells. Sedimentation analysis of nucleolar RNA was performed on cells that had been labeled for 50 min with both uridine-³H and ¹⁴C-(methyl)methionine. Infected cells were labeled between 40 and 90 min after infection.

Since the methylation does not appear to be inhibited by poliovirus, it seems unlikely that the virus is exerting its action at this level. Of course, improper methylation might be just as damaging to proper maturation of ribosomal RNA as incomplete methylation, and this possibility was not tested by this experiment.

The main point to be made about the effect of poliovirus on ribosome maturation is that it is very different from the effects of two other agents which inhibit protein synthesis, namely, cycloheximide, in the presence of which particles mature normally for some time (Warner *et al.,* 1966b), and puromycin, which completely blocks particle formation (Latham and Darnell, 1965). Further study of this virus-induced abnormality should provide information relevant to the normal steps in ribosome maturation.

Protein Synthesis in Poliovirus Infected Cells

The site of protein synthesis in HeLa cells, as in all other cells that have been examined, is the polyribosome, a group of ribosomes which are simultaneously and independently translating a single molecule of mRNA (Penman *et al.,* 1963). A striking set of events in the control of protein synthesis occurs in HeLa cells infected with poliovirus. Host cell protein synthesis declines (Darnell and Levintow, 1960; Zimmerman *et al.,* 1963), and the polyribosomes become disaggregated. The same ribosomes then are recruited by association with viral RNA into functioning viral polyribosomes (Penman *et al.,* 1963). It was clear from the original experiments on poliovirus-infected cells that this could not be due simply to an interruption of mRNA synthesis, since actinomycin D alone caused a much less dramatic fall in host protein synthesis. It appeared that perhaps a synergistic effect was obtained when virus infection was carried out in the presence of actinomycin. Willems and Penman (1966) have analyzed this situation in detail and have found that rapidly labeled RNA, with the sedimentation characteristics of mRNA, continues to enter the cytoplasm of infected cells and to become associated with structures sedimenting like polyribosomes. If further synthesis of cellular RNA is blocked by actinomycin, a portion of this newly formed RNA disappears very quickly from the rapidly sedimenting structures. They concluded that perhaps the polyribosomes are being maintained at a constantly diminishing level in the infected culture not treated with actinomycin by an enhanced production of mRNA.

In addition, they have reinforced the original observations that the polyribosomes which remain functional in the virus-infected cell are of normal size. The data in Table I indicate that they also are functioning normally, as tested by the amount of nascent protein per ribosome in the remaining polysomes. Thus, it appears, that a host cell polysome is either completely removed by virus infection, or is functioning normally.

The mechanism by which cellular polyribosomes are broken down in the face of poliovirus infection is unknown, and a number of experiments have failed to shed any light on the subject. (1) No new nuclease activity has been found in infected cells (Willems and Penman, unpublished results). (2) Formyl methionine is used as the initiator of protein synthesis in bacteria (Capecchi, 1966). The interruption of its formation by virus infection, plus the substitution in the virus-infected cell of another sRNA as chain initiator might account for the viral inhibition of

TABLE I

NASCENT PROTEIN ON RIBOSOMES[a]

Declining HeLa polysomes		Polio polysomes	
Hours infected	cpm/OD	Hours infected	cpm/OD
Control	2450	2.5	12,500
1.5	3400	4.0	2270
2.0	2300	4.75	700

[a] Data are from two experiments performed similarly to the one illustrated in Fig. 15, except that the polysome region was purified by a second sedimentation through a sucrose gradient before the measurement of nascent protein and polysomal OD_{260} (Summers *et al.*, 1967). Both infected and control cells were labeled for 5 min. The two experiments were not done simultaneously and more ^{14}C-amino acid was used for the polio polysome determination than for the declining HeLa cell polysomes, hence the greater initial specific activity. Declining HeLa polysomes were derived from an infected culture in the continuous presence of guanidine and polio polysomes from an infected culture maintained in guanidine for one hour and removed at time 0.

protein synthesis. However, no formyl methionine was found on sRNA in either infected or normal HeLa cells under conditions in which it was found in *E. coli* sRNA (Table II). (3) On the assumption that a normal cellular RNA might be changed in its methyl groups subsequent to infection, the sRNA was examined in actinomycin-treated normal and virus-infected cultures after exposure to methyl-^{14}C labeled methionine. No differences in the radioactivity incorporated into sRNA in the two cultures could be found, and very little incorporation occured in either culture.

TABLE II

ABSENCE OF *N*-FORMYL METHIONINE ON HELA CELL sRNA[a]

	CPM in Chromatograph in Region of	
sRNA	Methionyl adenosine ester	*N*-formyl methionyl adenosine ester
Control HeLa sRNA	4000	83
Infected HeLa sRNA	7450	160
E. coli sRNA	2100	9700

[a] Soluble RNA was prepared from HeLa cell cytoplasmic extracts of cells exposed to methionine-^{35}S for 5 min by SDS-phenol extraction at 25°C, pH 5.1; *E. coli* sRNA was isolated from a methionine-requiring strain of *E. coli* exposed, in mineral medium which lacked methionine, to methionine-^{35}S for one minute by treating the whole cells with phenol at room temperature. The terminus of sRNA was released by RNase digestion (Marcker and Sanger, 1964) and the hydrolyzate separated by electrophoresis at pH 3.5.

It seems an outstandingly important problem to determine how poliovirus disrupts host protein synthesis, and work will continue on it. At the moment all we have are negative answers as to the possible mechanism.

Virus-Specific Protein Synthesis

Viral polyribosomes function until about 4 to 5 hours after infection, after which there occurs a loss of viral polyribosomes and a consequent depression of virus protein synthesis (Fig. 15). The properties of these declining polyribosomes have also been studied. It has been found that the amount of labeled, nascent protein on the ribosomes from these polysomes is depressed after a brief (5 min) exposure to ^{14}C-amino acid. If the culture is labeled for 20 min, however, the amount of nascent

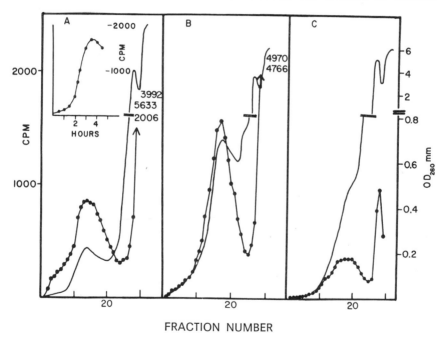

FIG. 15. The decline of poliovirus polysomes late in the infectious cycle. Cytoplasmic extracts of infected cells were prepared after a 5 min exposure to labeled amino acids, and examined for OD_{260} and nascent protein in polyribosomes (Penman *et al.*, 1963; Summers *et al.*, 1965). All cells were treated with guanidine for one hour after infection and then the guanidine was removed to allow virus RNA replication. (A) 2.5 hours after reversal; (B) 4 hours, after reversal; (C) 5 hours after reversal. Inset records Actinomycin-resistant uridine-^{14}C incorporation in a portion of the infected culture.

protein per ribosome increases toward the normal value (Table III). Thus there is evidence for a general slowdown in the rate of protein synthesis late in infection. This finding might be expected as a result of the general disorganization of virus-infected cells by this time. For example, the mitochondria are severely affected (Dales *et al.*, 1965) and the energy supply necessary for protein synthesis might become limiting; on the other hand, the cessation of virus protein synthesis could be a specific control function of the virus genome. At the moment, no decision can be made on this question.

TABLE III

RATE OF NASCENT CHAIN SYNTHESIS IN POLIOVIRUS POLYSOMES[a]

Label time (min)	Early polysomes (cpm/OD)	Late polysomes (cpm/OD)
5	8500	2109
20	—	5300

[a] Amount of nascent protein per ribosomal OD_{260} unit, determined as in the experiment shown in Fig. 15 and Table I, except that pulse time varied as indicated. Only one pulse time was used for the early poliovirus polysomes since the maximum amount of nascent protein is accumulated in 3–5 min (Penman *et al.*, 1964). Early polysomes were taken when 50% of viral RNA had been synthesized, and late polysomes when viral RNA synthesis had been complete for 15 min (see inset graph in Fig. 15).

Further studies of the declining late cycle poliovirus polysomes have revealed several other interesting features of these structures (Summers *et al.*, 1967). By infecting cells which had been prelabeled in their ribosomal RNA with uridine-^{3}H and introducing uridine-^{14}C during the period of viral replication, it was possible to count ribosomes and chains of viral RNA independently in the polysomal structure. It was found that the very large (380–400 S) early cycle polysomes contain about 40 ribosomes, while the smaller (200 S) late cycle polysomes still contain about 20 ribosomes. Since the expected ribosome content of a 200 S polysome (calculated from work with reticulocytes; Gierer, 1963) would be about 6 to 8 ribosomes, it is clear that other factors, such as the spacing between ribosomes, must have an effect on the sedimentation profile of these very large aggregates.

The final point to be made about poliovirus protein synthesis relates to the types of proteins being formed in infected cells. Inhibition of host cell protein synthesis is observed in cells infected in the presence of guanidine (Penman and Summers, 1965), that is, when viral RNA multiplication is completely blocked. After the removal of guanidine to allow viral RNA replication to proceed, it is possible to achieve a situation where all (>95%) of the protein synthesis in infected cells is dependent on the translation of viral RNA (Summers *et al.*, 1965).

With the aid of the new techniques of acrylamide gel electrophoresis and gel fractionation (Maizel, 1966), it is possible to examine the properties of proteins formed in cells infected with poliovirus and then pulse-labeled with ^{14}C-amino acids. Figure 16 shows that purified ^{14}C-

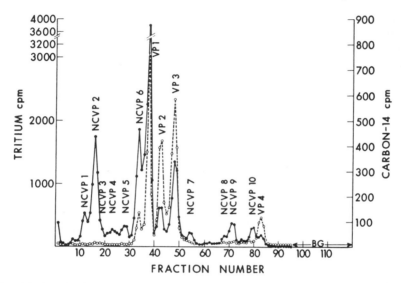

FIG. 16. Acrylamide gel electrophoresis pattern of poliovirus capsid proteins (VP) and of virus-specific noncapsid proteins (NCVP). Purified virus labeled with ^{3}H-amino acids was compared to the total cytoplasmic extract of infected cells labeled with ^{14}C by the techniques of Summers *et al.* (1965).

amino acid-labeled virus can be dissociated into its four constituent chains in the presence of a cytoplasmic extract of infected cells which had been labeled throughout most of the replicative cycle with ^{3}H-amino acids. In addition to the four virus proteins, it can be seen that many polypeptides of differing electrophoretic behavior are found in the cytoplasm. These are believed to represent approximately 10–15 different gene products specified by poliovirus. The peaks are separately designated because in many different gel determinations clear separations of polypeptides in a particular region have been seen. In the case of smaller peaks like NCVP (noncapsid virus protein), 3, 4, and 5, and VP (virus capsid proteins) 2 and 4, the resolution is frequently poor, while in the cases of NCVP 1, 2, 6, 9, and 10, and VP 1 and 3, it is almost always very good.

The polypeptides shown in Fig. 16 have been characterized in a number of ways, in an effort to make certain that each, in fact, represents a different protein (Maizel *et al.,* unpublished results). A gel electrophoresis experiment was performed in which 14 labeled amino acids were used instead of one or two. A number of different peaks were collected, and the distribution of ^{14}C-amino acids in each peak was determined by chromatography. There were significant differences in the relative amounts of various amino acids present in the various peaks. Pending the outcome of partial chain analysis of the polypeptide in each peak (by either tryptic digestion or cyanogen bromide treatment), this is the best available evidence that each peak represents a different peptide chain.

Another type of characterization was made by centrifuging a solution of all the labeled, virus-specific polypeptides through a sucrose solution. Although none of the chains differed in sedimentation size from any of the others by a factor larger than about two, there was a considerable enrichment in the most rapidly sedimenting material for NCVP-1. This shows two things: (1) The spread of sizes is not very great among the various polypeptides, and (2) the slowest moving polypeptide in gel electrophoresis is the fastest sedimenting one.

One of the main questions which can be approached with the use of the gel technique is whether different proteins are made in different amounts and whether the ratio of proteins made in the virus-infected cell changes throughout the replicative cycle. Experiments designed to answer these questions are complicated by various factors: For example, (1) protein turnover (synthesis and degradation) might occur for some proteins but not for others, (2) some of the polypeptides might derive from combinations of others or as scission products of others, (3) partial polypeptides may be distributed irregularly in the gel fractions.

With these reservations in mind, gel analysis was performed on extracts obtained from cells labeled at various times after infection. The labeling interval was relatively short in order to maximize the chance of observing any unstable products, and the cells were "chased" with cold amino acids to ensure completion of all growing peptide chains. Two points can be made about the results of this experiment which are given in Fig. 17. All the discrete viral protein peaks are present at all the times examined, which include times equivalent to from less than 5 to 90% completion of the replication of viral RNA. Also included was a sample from a late time point, at which virus polysomes had declined in size and activity. Thus, no major changes in the manner in which poliovirus RNA is translated take place during the replicative cycle.

It is immediately obvious, however, that many of the protein chains (NCVP 3, 4, 5, possibly 6, 7, 8, 9, 10, and VP 2 and 4) are not made at nearly the same rate as are VP 1 and 3 and NCVP 1 and 2. Since sucrose density gradient centrifugation shows that most of the chains are about the same size, it is clear that different polypeptides are synthesized in different amounts. The mechanism by which this translation control operates on a polycistronic messenger RNA is not known at present. Whether any of the mechanisms (modulation, etc.) proposed for the translation of messenger from whole operons in bacteria are operating here remains to be seen (Ames and Martin, 1964).

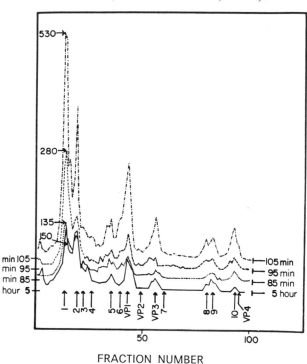

FRACTION NUMBER

Fig. 17. Acrylamide gel patterns of virus proteins at various times in the replicative cycle. Extracts of infected cells labeled with ^{14}C-amino acids for 8 min and chased with ^{12}C-amino acids for 5 min were compared by acrylamide gel electrophoresis. The zero points for the various curves have been displaced as indicated on the sides of the graph in order to decrease the overlap of lines in the graph.

Conclusion

It is clear for this somewhat abbreviated discussion of a very large number of experiments, that the poliovirus-infected HeLa cell provides an excellent system for the study of many problems—viral RNA biosynthesis, virus particle maturation, host cell ribosomal maturation, mechanism of viral interference with host protein metabolism, and the control of translation of polycistronic messenger RNA. The major promise that the small RNA viruses hold is the eventual understanding in molecular terms of the operation of every viral function, and continued work may ultimately bring this goal within reach.

REFERENCES

Ames, B. N., and Martin, B. (1964). *Ann. Rev. Biochem.* **33**, 235.

Baltimore, D. (1964). *Proc. Natl. Acad. Sci. U.S.* **51**, 450.

Baltimore, D. (1966). *J. Mol. Biol.* (in press).

Baltimore, D., and Girard, M. (1966). *Proc. Natl. Acad. Sci. U.S.* (in press).

Baltimore, D., Becker, Y., and Darnell, J. E. (1964). *Science* **143**, 1034.

Baltimore, D., Girard, M., and Darnell, J. E. (1966). *Virology* **29**, 179.

Bishop, J. M., Summers, D. F., and Levintow, L. (1965). *Proc. Natl. Acad. Sci. U.S.* **54**, 1273.

Brown, G. M., and Attardi, G. (1965). *Biochem. Biophys. Res. Commun.* **20**, 298.

Capecchi, M. R. (1966). *Proc. Natl. Acad. Sci. U.S.* **55**, 1517.

Crowther, D. L., and Melnick, J. L. (1961). *Virology* **15**, 65.

Dales, S., Eggers, H. J., Tamm, I., and Palade, G. E. (1965). *Virology* **26**, 379.

Darnell, J. E. (1962). *Cold Spring Harbor Symp. Quant. Biol.* **27**, 149.

Darnell, J. E., and Levintow, L. (1960). *J. Biol. Chem.* **235**, 74.

Darnell, J. E., Levintow, L., Thoren, M. M., and Hooper, J. L. (1961). *Virology* **13**, 271.

Eagle, H. (1959). *Science* **130**, 432.

Fenwick, M. L., Erikson, R., and Franklin, R. M. (1964). *Science* **146**, 527.

Gierer, A. (1963). *J. Mol. Biol.* **6**, 148.

Girard, M., Penman, S., and Darnell, J. E. (1964). *Proc. Natl. Acad. Sci. U.S.* **51**, 205.

Girard, M., Latham, H., Penman, S., and Darnell, J. E. (1965). *J. Mol. Biol.* **11**, 187.

Girard, M., Baltimore, D., and Darnell, J. E. (1966). *J. Mol. Biol.* (in press).

Goodman, H., and Penman, S. (1966). *J. Mol. Biol.* (in press).

Joklik, W. K., and Darnell, J. E. (1961). *Virology* **13**, 439.

Katz, L., and Penman, S. (1966). *Biochem. Biophys. Res. Commun.* **23**, 557.

Latham, H., and Darnell, J. E. (1965). *J. Mol. Biol.* **14**, 13.

Ledinko, N., and Hirst, G. K. (1961). *Virology* **14**, 207.

Levintow, L., and Darnell, J. W. (1960). *J. Biol. Chem.* **235**, 70.

Maizel, J. V. (1966). *Science* **151**, 988.

Marcker, K., and Sanger, F. (1964). *J. Mol. Biol.* **8**, 835.

Montagnier, L., and Sanders, F. L. (1963). *Nature* **199**, 664.

Penman, S. (1966). *J. Mol. Biol.* **17**, 117.

Penman, S., and Summers, D. F. (1965). *Virology* **27**, 614.

Penman, S., Scherrer K., Becker, Y., and Darnell, J. E. (1963). *Proc. Natl. Acad. Sci. U.S.* **49**, 654.

Penman, S., Becker, Y., and Darnell, J. E. (1964). *J. Mol. Biol.* **8**, 541.

Penman, S., Smith, I., and Holtzman, E. (1966). *Science* (in press).

Reich, E., Franklin, R. M., Shatkin, A. J., and Tatum, E. L. (1961). *Science* **134**, 556.

Scherrer, K., and Darnell, J. E. (1962). *Biochem. Biophys. Res. Commun.* **7**, 486.

Scherrer, K., Latham, H., and Darnell, J. E. (1963). *Proc. Natl. Acad. Sci. U.S.* **49,** 240.

Soeiro, R., Birnboim, H. C., and Darnell, J. E. (1966). *J. Mol. Biol.* (in press).

Summers, D. F., Maizel, J. V., and Darnell, J. E. (1965). *Proc. Natl. Acad. Sci. U.S.* **54,** 505.

Summers, D. F., Maizel, J. V., and Darnell, J. E. (1967). *Virology* (in press).

Vaughan, M., Warner, J. R., and Darnell, J. E. (1966). *J. Mol. Biol.* (in press).

Warner, J. R., Soeiro, R., Birnboim, H. C., and Darnell, J. E. (1966a). *J. Mol. Biol.* (in press).

Warner, J. R., Girard, M., and Darnell, J. E. (1966b). *J. Mol. Biol.* (in press).

Willems, M., and Penman, S. (1966). *Virology* (in press).

Zimmerman, E. F., and Holler, B. (1966). *Federation Proc.* **25,** 646.

Zimmerman, E. F., Heeter, M., and Darnell, J. E. (1963). *Virology* **19,** 400.

Genetics and Biochemistry of Arbovirus Temperature-Sensitive Mutants

E. R. Pfefferkorn and Boyce W. Burge

DEPARTMENT OF BACTERIOLOGY AND IMMUNOLOGY, HARVARD MEDICAL SCHOOL,
BOSTON, MASSACHUSETTS

Introduction

Conditional-lethal mutants have been of great value in studying the genetics and physiology of bacteriophages (Epstein *et al.,* 1963). In the last several years, temperature-sensitive, conditional-lethal mutants have also been isolated from a variety of animal viruses: polio virus (Cooper, 1964), Sindbis virus (Burge and Pfefferkorn, 1964), polyoma virus (Fried, 1965), Newcastle disease virus (Kirvaitis and Simon, 1965), and rabbitpox virus (Sambrook *et al.,* 1966). We shall describe some recent experiments that indicate the usefulness of temperature-sensitive conditional-lethal mutants in animal virology.

Our experiments were all done with Sindbis virus, a group A arbovirus, in chick fibroblast tissue cultures. The group A arboviruses contain single-stranded RNA of molecular weight 2×10^6 (Wecker, 1959), and they consist of a nucleoprotein core surrounded by a lipoprotein membrane. Sindbis virus has a rapid growth cycle and yields a large crop of progeny virus. Figure 1 shows a typical growth curve. Release of newly synthesized virus begins about 3 hours after infection; by 4 hours, the virus is released at a maximal rate which is maintained for about 6 hours without gross cellular damage. The rate of viral release then declines, and some hours later the cells begin to show characteristic cytopathic effects. Our biochemical and genetic experiments have concentrated upon an analysis of the latent period and of the period of linear viral release.

Isolation of Temperature-Sensitive Mutants

Since wild-type Sindbis virus is quite heat-labile, we first selected a mutant with a heat-stable virion. This was accomplished through many

403

cycles of heating at 60°C and regrowing the survivors. The result was a genetically stable, heat-resistant (HR)* variant, that was inactivated at 60°C at only one-sixth the rate of the wild-type. The HR strain is presumably a multistep mutant with alterations in all of those structural proteins that determine the heat stability of the virion.

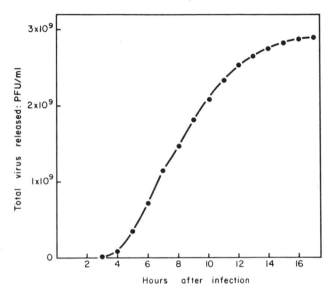

Fig. 1. Growth curve of wild-type Sindbis virus at 37°C. The medium was changed every hour and the newly released virus was titrated. The HR strain has a similar growth curve.

All temperature-sensitive conditional-lethal mutants were derived from this HR virus. The ease with which temperature-sensitive mutants of any organism can be selected and studied is determined in part by the breadth of the temperature range over which it can multiply. We were fortunate that Sindbis virus (wild-type or HR) made plaques over a comparatively broad range, 27° to 41°C. Probably, its usual pattern of alternate growth in mosquitoes and birds has resulted in natural selection for growth at both low and high temperatures. The permissive temperature for all experiments was 28°C. The nonpermissive temperatures were 39° to 41°C.

* Abbreviations: DNase, deoxyribonuclease; HR, a Sindbis virus strain characterized by a heat-stable virion; PFU, plaque forming unit; RNA, ribonucleic acid; *RNA⁺*, the phenotype of temperature-sensitive mutants able to make viral RNA at a nonpermissive temperature; *RNA⁻*, the phenotype of temperature-sensitive mutants unable to make viral RNA at a nonpermissive temperature; RNase, pancreatic ribonuclease.

To isolate mutants, we treated virus with nitrous acid or nitrosoguanidine (Burge and Pfefferkorn, 1966a), and then tested clones prepared at 28°C for the ability to produce plaques at 40°C. Potential mutants (those that failed to produce plaques at 40°C) were cloned twice before use.

The mutants that we isolated showed a wide range of back mutation frequencies. Some had such high frequencies that they had to be discarded, but more than half had frequencies of less than 2×10^{-5}, and were useful for biochemical and genetic tests. This reversion frequency is substantially lower than the 10^{-3} value reported for conditional-lethal mutants of RNA bacteriophages (Zinder and Cooper, 1964). Zinder (1965) has suggested that the high reversion frequency in bacteriophage f2 may be a consequence of the large growth advantage of the wild-type revertants. At 28°C the HR virus has little or no growth advantage over most of our mutants; this may explain the difference in back mutation frequencies.

A few of our mutants yielded stocks with no detectable revertants. However, it was difficult to detect back mutants at a frequency of less than 2×10^{-6} because the large background of mutants interfered with plaque formation.

Twenty-three mutants of low or unmeasurable reversion frequency have been isolated. A complete catalogue of these mutants has been published (Burge and Pfefferkorn, 1966a).

Characterization of Viral Mutants

Characterization of these mutants on the basis of their physiological properties has proved to be difficult. The only reliable distinction, thus far discovered, is the ability of some mutants to make infectious RNA at a nonpermissive temperature. Those mutants able to make infectious RNA (*RNA+* mutants) make nearly as much viral RNA as does the parental HR virus. The *RNA−* mutants make less than 1% as much viral RNA as the HR virus at a nonpermissive temperature. An alternative, simpler way to determine the *RNA* phenotype was to measure the virus-stimulated incorporation of uridine-^{14}C in the presence of 2 μg actinomycin D/ml. There was a perfect correlation between these two results, indicating that no mutant made a defective, noninfectious RNA at the nonpermissive temperature.

Curiously, the great majority (70%) of our mutants are *RNA−*. Since the molecular weight of the viral RNA leads us to expect that it contains about ten cistrons, and since it is unlikely that more than two are involved in viral RNA synthesis, it is surprising that most of our mutations fall into this small portion of the viral genome. At least two explanations may be suggested. First, that portion of the viral genome responsible for the synthesis of viral RNA may be inherently more likely to undergo the sort of mutations that we have selected. Alternatively, only those RNA^+ mutants with an extremely low reversion rate may have been selected. Since the genome of these mutants is extensively replicated at the nonpermissive temperature, it has additional opportunities for reversion. If even one back mutant PFU is released, the infection will not appear to have been initiated by a mutant.

Several other characteristics are strongly correlated with the RNA phenotype of the viral mutants. The heat stability of the mutant virion, the stability of infectious centers at a nonpermissive temperature, and the time of expression of the defect will be considered in turn.

Heat Stability of Mutant Virions

The heat stabilities of a variety of mutants are shown in Table I. Most of the RNA^- mutants are slightly less heat stable than the HR virus from which they were derived, possibly because of other mutations accumulated during the original mutagenesis. More important, with one exception, all of the RNA^+ mutants are more heat labile than any RNA^-

TABLE I

HEAT STABILITY OF THE VIRIONS OF VARIOUS MUTANTS OF SINDBIS VIRUS

Mutant	RNA phenotype	Reduction in titer after 2.5 min. at 60°C ($\log_{10}$)
HR	not a conditional lethal mutant	0.7
ts19	—	0.7
ts20	+	0.8
ts21	—	0.8
ts24	—	1.0
ts17	—	1.0
ts4	—	1.2
ts13	+	1.7
ts15	+	1.8
ts23	+	1.8
ts2	+	2.3
ts10	+	2.8
ts9	+	3.3
Wild	not a conditional lethal mutant	4.2

mutant. This strong correlation suggests that the same mutation that makes the virus a conditional-lethal mutant also yields a very heat-labile virion. That is, the same altered protein that cannot assume or maintain a functional configuration at 40°C is incorporated into virions at 28°C, and renders them heat labile.

It should be noted that the heat lability of the virion is not the cause of the conditional-lethal character; the wild-type virus is more heat-labile than any mutant and yet makes plaques normally at 40°C.

One RNA^+ mutant ($ts20$) fell among the RNA^- mutants with respect to heat lability. Although other explanations are possible, it may be that this mutant makes an altered protein that does not play a role in the heat stability of the virion—perhaps an internal protein. Support for this suggestion comes from the observation that this mutant falls in a different complementation group than do other, heat-labile, RNA^+ mutants (see Table III).

Stability of Virus-Cell Complexes at a Nonpermissive Temperature

The RNA^+ and RNA^- mutants are also clearly different with respect to the stability of the infectivity of cell-virus complexes at a nonpermissive temperature. This property was determined by allowing the mutant viruses to adsorb to monolayer cultures at 4°C, covering the cells with an agar overlay, and incubating the cultures at 39°C. At intervals, cultures were moved to 28°C, and plaques were counted three days later. Thus, the property tested was the ability of the cell-virus complexes to release one PFU after their return to a permissive temperature. Figure 2 shows typical results of such an experiment. The infectious centers due to RNA^+ mutants are initially stable but then are very rapidly lost, despite the fact that they must contain large quantities of infectious RNA. This loss of infectious centers is somewhat similar to the decline in the rate of viral release in a wild-type infection (Fig. 1), and may be a consequence of the production of some cytotoxic viral protein(s). If only one viral protein were involved in cell death, there might exist an RNA^+ temperature-sensitive conditional-lethal mutant that yields stable cell-virus complexes. We have not seen such a mutant.

In contrast, the infectivity of cell-virus complexes of the RNA^- mutants were found to be lost more slowly and with exponential kinetics. Since the RNA of these mutants is unable to replicate at the nonpermissive temperature, this experiment may actually measure the half-life of a foreign messenger RNA quite precisely, because one-percent survival can be detected accurately. The half-life of viral messenger was estimated, from data obtained in this way, to be 7.5 hours at 39°C.

A more conventional method for the estimation of messenger half-life in animal cells employs actinomycin D, an antibiotic that rapidly inhibits cellular RNA synthesis. In this method, the decay of messenger RNA is equated to the decreasing ability of cells to incorporate amino acids into protein after exposure to the antibiotic. In bacteria, this method yields a value for the half-life of messenger RNA activity that is very close to values obtained from more direct measurements (Fan *et al.,* 1964).

The determination of the half-life of messenger RNA in chick fibroblasts by the actinomycin method yielded a value which was dependent on incubation temperature, but was extremely reproducible at any given temperature. At 39°C the half-life was 2.5 hours (Fig. 3).

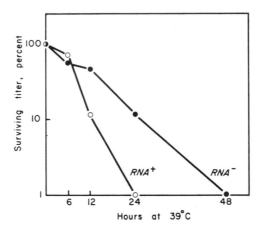

Fig. 2. Decay of infectivity of virus-cell complexes at a nonpermissive temperature. Suitable dilutions of mutant stocks were adsorbed to monolayers at 4°C. The monolayers were covered with agar and incubated at 39°C. At the indicated times, monolayers were shifted to 28°C and incubated 4 more days before plaques were counted. The *RNA+* mutant was *ts*2, the *RNA−* mutant, *ts*4.

We were surprised by this result, since we did not expect that viral RNA, unable to replicate, would have a longer half-life than cellular messenger RNA. If both viral RNA and endogenous messenger RNA were attacked at the same rate by a cellular enzyme, the viral infectivity should be more labile than the messenger function of endogenous RNA, since breaking only one or very few phosphodiester bonds of the viral RNA should eliminate its infectivity.

Several explanations for the stability of viral RNA may be suggested: (1) The viral RNA may be capable of limited replication below our level of detection at the nonpermissive temperature. (2) The viral RNA may be more resistant to enzymatic destruction than cellular messenger RNA. (3) Actinomycin D may not yield a valid estimate of the half-life of messenger RNA in chick fibroblasts because of secondary toxic effects.

Support for the last hypothesis was gained from experiments in which the ability of actinomycin D-treated chick cells to support virus synthesis was determined. Virus production was measured during the fifth through sixth hours of infection, using monolayers that had been exposed to actinomycin D (1 μg/ml) for various lengths of time prior to measurement of virus production. Cells were able to support a normal level of virus production for only 4 to 8 hours after addition of actinomycin—at 17 hours the virus yield was reduced by more than 99%.

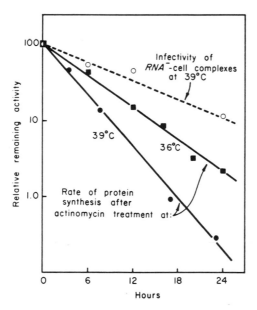

FIG. 3. Comparison of the decay of *RNA⁻* virus-cell complexes at 39°C (from Fig. 2) with the decay of protein synthetic capacity in the presence of actinomycin D (1 μg/ml) at 36° and 39°C. Protein synthetic capacity was measured by incorporation of isoleucine-^{14}C during one hour pulses.

The observed actinomycin-induced decay of the capacity of monolayers to support viral growth may be peculiar to chick cells, since L-cells are capable of supporting normal yields of mengo virus even 24 hours after exposure to actinomycin (Reich *et al.*, 1962). In any event, the above observations suggest that normal cellular activities other than RNA synthesis may be rapidly compromised by actinomycin D (possibly as a consequence of the cessation of messenger RNA synthesis). Therefore, estimation of messenger RNA activity in chick cells by this technique may be unreliable.

"Early" and "Late" Functions

The *RNA* phenotype of the viral mutants was also correlated with the time at which the defect was expressed. RNA^+ mutants do not produce any significant amount of virus at the nonpermissive temperature even when the first 4 hours of the infection are carried out at a permissive temperature (Table II). Thus, these mutants may be said to be blocked in a "late" function.

TABLE II

TIME OF EXPRESSION OF TEMPERATURE-SENSITIVE DEFECT[a]

| Mutant | Phenotype | PFU/cell produced between 4 and 8 hours | |
		8 hours at 41°C	then 4 hours at 28°C 4 hours at 41°C,
*ts*2	RNA^+	0.01	0.5
*ts*4	RNA^-	0.02	100
*ts*6	RNA^-	0.04	3
*ts*11	RNA^-	0.01	90
*ts*15	RNA^-	0.04	200
*ts*16	RNA^-	0.03	30
*ts*17	RNA^-	0.01	30
*ts*19	RNA^-	0.07	45
*ts*21	RNA^-	0.01	2
*ts*24	RNA^-	0.01	3
HR	Not a conditional lethal mutant	150	300

[a] Monolayers were infected with a multiplicity of 10 PFU/cell, rinsed twice after 45 min adsorption, and incubated at either 28° or 41°C. At 4 hours all monolayers were drained, and fresh medium at 41°C was added. The amount of virus produced between 4 and 8 hours was measured.

In contrast, most RNA^- mutants profited from incubation at a permissive temperature during the first four hours of infection. After a shift to a nonpermissive temperature, the majority of the RNA^- mutants go on to produce a nearly normal yield of virus (Table II), and thus their temperature-sensitive defect is an "early" one. However, a few RNA^- mutants (e.g., *ts*6, *ts*21, *ts*24) were found to produce only about 1 to 5% the normal HR yield in such a shift-up experiment, a yield which, though small, is greater than that produced under the same conditions by an *RNA*⁺ mutant. It might be suggested that these *RNA*⁻ mutants are double mutants, blocked in both "early" and "late" functions. Since, however, these 3 mutants complement all *RNA*⁺ mutants, and since 2 of the 3 have the usual reversion frequency, it seems improbable that this explanation can be generally applied.

A more reasonable possibility is that most *RNA*⁻ mutants are defective in the synthesis of viral RNA polymerase at nonpermissive temperatures, but that once the enzyme is formed at a lower temperature, it is stable under conditions of shift-up. However, some *RNA*⁻ mutants might be expected to encode a viral polymerase unstable under shift-up conditions. Such mutants would gain only transient benefit from earlier incubation at permissive temperatures. An RNA phage mutant of this type has been described by Lodish and Zinder (1966). Mutants *ts*6, *ts*21, and *ts*24 may also be examples of this type.

KINETICS OF EXPRESSION OF "EARLY" FUNCTION

We have assumed that *RNA*⁻ mutants are defective in the formation of viral RNA polymerase and that, for the majority of mutants, polymerase formed at a permissive temperature is stable at a nonpermissive temperature. Thus it was possible to carry out an experiment to determine the early kinetics of polymerase formation. In this experiment (Fig. 4), a series of monolayers infected with an *RNA*⁻ mutant were incubated at a permissive temperature. At intervals from one-half to 4 hours after infection, monolayers were rinsed with prewarmed medium and shifted-up to a nonpermissive temperature. Four hours after infection, all monolayers were drained, fresh medium was added, and the virus produced during the next 4 hours at 39°C was measured.

As Fig. 4 shows, incubations of less than 1 hour at the permissive temperature produced no increase in yield over that observed with infected monolayers incubated continuously at 39°C. As the incubation under permissive conditions was increased to 1.5 hours and more, subsequent viral production at the nonpermissive temperature increased in an almost exponential fashion. This result would be expected if the synthesis of polymerase molecules was initiated at 1 to 1.5 hours after infection, and continued to increase in an exponential fashion up to 3 or 4 hours after infection. Presumably, a saturating amount of polymerase, enough to insure adequate synthesis of viral RNA during later incubation at 39°C, was made during the first 3 hours. It is possible, however, that this experiment may actually have measured the accumulation of a double-stranded replicative form of the viral RNA. The results cannot be explained on the basis of an accumulation of a supply of single-stranded viral RNA sufficient for subsequent viral production, because little or no actinomycin D-insensitive RNA synthesis could be detected during the first 3 hours of infection at 28°C.

Results obtained with an RNA^- mutant that benefits only slightly from early incubation at the permissive temperature (ts24) are also included in Fig. 4 for comparison.

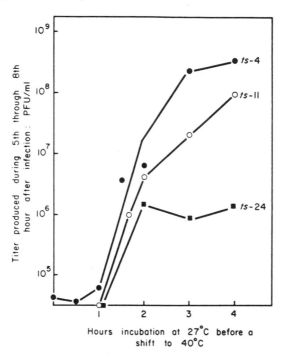

Fig. 4. Virus production by RNA^- mutants at a nonpermissive temperature, after preliminary incubation at a permissive temperature. A series of monolayers were infected with about 10 PFU/cell and incubated at 27°C. After various periods at 27°C, from 0 to 4 hours, monolayers were rinsed twice with 39°C medium and shifted to a 39°C incubator. At 4 hours after infection all monolayers were rinsed and received fresh medium. The virus produced during the subsequent 4 hours incubation at 39°C was determined by plaque titration.

Complementation

We have observed no genetic recombination with Sindbis virus under conditions which would appear to be most favorable for its detection— the use of parents with low reversion frequencies and blocks in different cistrons (RNA^+ and RNA^-). The method used was simply to establish mixed infections at a permissive temperature and to look for recombinant progeny able to make plaques at the nonpermissive temperature. The sensitivity of this test for recombination was determined by the frequency of back mutants in the parental stocks. Since these were approximately 10^{-5}, a recombination frequency of 10^{-4} would have been easily detected.

The absence of genetic recombination is surprising, since recombination has been observed with a variety of RNA-containing animal viruses (reviewed by Fenner and Sambrook, 1964), though not with RNA-containing bacteriophages.

Our genetic analysis was therefore confined to studies of complementation. Since complementation has seldom been used in the study of animal viruses, we shall first describe some features of the phenomenon. In positive tests for complementation (for example, in all crosses between RNA^+ and RNA^- mutants) the yield from mixed infections at a nonpermissive temperature exceeded the sum of the yields, from the two parents grown separately, by a factor of 3 to 300. The magnitude of this factor was determined primarily by the leakiness and the reversion frequency of the parents.

In most cases, the absolute efficiency of complementation was quite low, 1–3% of the yield of the parental HR virus under the same conditions. Because of this low efficiency, one precaution was essential for the demonstration of complementation. A significant fraction of the input virus was adsorbed in such a way that it could not be removed by washing after the adsorption period, but was eluted during the first few hours at the nonpermissive temperature. The similar behavior of WEE virus (a related arbovirus) has been described by Dulbecco and Vogt (1954), who showed that this elution is insensitive to cyanide and unrelated to viral replication. In our routine complementation tests, the infected monolayers were rinsed 4 hours after infection to remove the eluted virus, and true viral synthesis was measured for the next 2 hours. In an experiment designed to illustrate this elution phenomenon, the medium was changed every $1\frac{1}{2}$ hours and the virus was titrated. The release of eluted virus and complementation are both illustrated in Fig. 5. The points before 3 hours represent, primarily, elution of uneclipsed virus; the separate infections showed the same values as the mixed infection. After 3 hours the eluted virus was washed out and viral growth began. Here the mixedly infected cells produced substantially more virus than those infected by either parent alone.

Other conditions do not have to be critically controlled for the demonstration of complementation. For example, complementation occurs between $36°$ and $40°C$, the temperature range in which the growth of mutants is depressed, but in which the parental HR virus grows well.

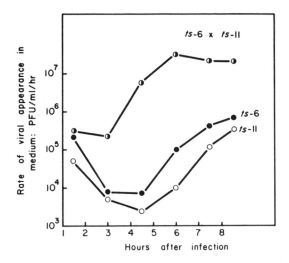

Fig. 5. Elution of uneclipsed virus and increased production of virus due to complementation in a cross between two *RNA⁻* viruses. The media of the mixed and separate infections were changed every 1.5 hours and titrated for virus content. Points in the first 3 hours represent elution of uneclipsed virus while later points represent viral growth. Virus produced after 3 hours by monolayers infected separately with parental viruses was chiefly due to growth of wild-type revertants present in the inoculum.

Effect of Multiplicity on Complementation

Multiplicity of infection has a substantial effect on the efficiency of complementation. This is a factor not often considered in the study of bacteriophages because the host cell is so small. Since our tissue culture cells have about one thousand times the volume of *E. coli,* topography might be expected to be important, especially since there are several cases in which genetic interaction between animal viruses is promoted by using clumps of virus to insure proximity of infection (Abel, 1962; Kirvaitis and Simon, 1965). We therefore examined the effect of multiplicity of infection on the efficiency of complementation. In the first experiment the multiplicities of both parents were altered so as to achieve mixed infections of more than 99% to less than 10% of the cells. Figure 6 shows that the production of virus due to complementation is a linear function of the number of mixedly infected cells when both parents are *RNA⁺*. Since the multiplicities were about 10 PFU of each mutant per cell at the upper end of the scale, and one at the lower end, multiplicity clearly is not important; a cell performs as efficiently in complementation whether it is infected by one PFU of each parent or several. This result is reasonable, for the *RNA⁺* mutants are presumably making essentially

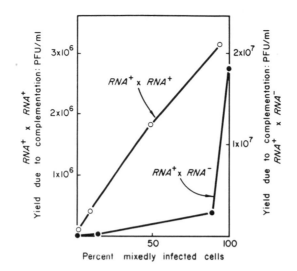

FIG. 6. Yield of virus due to complementation as a function of the fraction of mixedly infected cells. The number of mixedly infected cells (cells infected by at least one PFU of each mutant) was calculated from the number of PFU adsorbed and the number of cells in the monolayer. The yield due to complementation is the yield measured in a mixed infection minus the sum of the yields of the two parents grown separately. In this experiment the multiplicities of both mutants in each complementation pair were varied; the $RNA^+ \times RNA^+$ cross was done with *ts*5 and *ts*10 and the $RNA^+ \times RNA^-$ cross with *ts*10 and *ts*17.

normal amounts of all gene products at the nonpermissive temperature.

In contrast, when one parent was RNA^- (Fig. 6), increasing the multiplicities of both parents, was found to increase the efficiency of complementation. To see which was important, we varied independently the multiplicity of each parent in the $RNA^+ \times RNA^-$ cross. Figure 7 shows that when nearly all of the cells were infected by a constant multiplicity of the RNA^- parent, altering the multiplicity of the RNA^+ parent had little effect. Conversely, a reduction in the multiplicity of the RNA^- parent markedly reduced the efficiency of complementation. Our routine complementation tests were therefore done with adsorbed multiplicities of 10 to 20 PFU per cell of each parent.

Kinetics of Complementation

The kinetics of the release of complementation progeny were determined in order to find the optimal time for complementation analysis. Genetic interactions between animal viruses do not always follow a predictable time course. For example, the release of recombinants of influenza virus does not occur uniformly over the growth cycle of the

virus, but rather is concentrated in the early hours of viral release
(Simpson, 1964).

The kinetics of complementation were determined by changing the
medium at 1½ hour intervals and titrating the virus produced. Figure 8

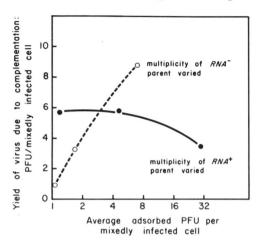

FIG. 7. Yield of virus due to complementation in $RNA^+ \times RNA^-$ crosses in which
the multiplicity of each parent was varied independently. When the multiplicity of
the RNA^+ parent ($ts10$) was altered, that of the RNA^- parent ($ts17$) was constant
at 43 PFU/cell. When the multiplicity of the RNA^- parent was altered, the RNA^+
parent was constant at 4 PFU/cell.

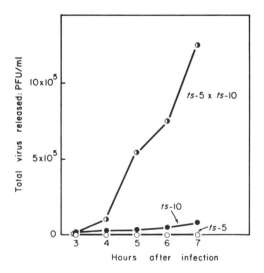

FIG. 8. Yield of virus due to complementation as a function of time. Both mutants
were RNA^+.

shows that the complementation between two *RNA⁺* mutants has exactly the same kinetics as does the growth of wild-type virus (Fig. 1), that is, linear release of virus beginning 4 hours after infection. Thus there is no lag in beginning the complementation. Exactly the same results were obtained with complementation between two *RNA⁻* mutants (Fig. 9). This experiment was done with high input multiplicities of both *RNA⁻* parents. The complementation might have been delayed if low multiplicities had been used.

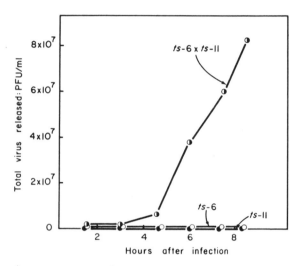

FIG. 9. Yield of virus due to complementation as a function of time. Both mutants were *RNA⁻*.

Genotypes and Phenotypes Produced by Complementation

The genotypes of the progeny that result from complementation were determined by checking the *RNA* phenotype in an *RNA⁺* × *RNA⁻* cross, or by backcrosses with the parents in an *RNA⁺* × *RNA⁺* cross. In each case, both parents were represented, although not necessarily in equal proportions. For example, in complementation between *RNA⁺* and *RNA⁻* mutants, the *RNA⁺* genotype was preferentially represented in the progeny. Presumably the functional viral RNA polymerase was preferentially associated with the genome that determined it.

More can be learned from an analysis of the phenotype of the progeny that result from complementation. All but one of our *RNA⁺* mutants were also characterized by marked heat lability of the virion. We suggested that this heat lability was the result of the incorporation of an altered protein, bearing the temperature-sensitive mutation, into the virion.

The heat-labile RNA^+ mutants fell into two complementation groups (C and D), represented respectively by two and three independently isolated mutants (Table III) The heat stable RNA^+ mutant fell into a third group (E). Let us assume that the two heat-labile RNA^+ complementation groups represent cistrons determining two structural proteins, both of which result in a heat-labile virion. In cells mixedly infected at a nonpermissive temperature, these altered proteins should assume a nonfunctional configuration, and be largely excluded from the complete

TABLE III

COMPLEMENTATION BETWEEN RNA^+ MUTANTS[a][b]

Complementation group	Mutant	Mutant			
		$ts2$	$ts5$	$ts10$	$ts20$
C	$ts2$	2.0×10^4	2.5×10^4 (1.0)	1.0×10^6 (36)	2.7×10^6 (5.2)
C	$ts5$	—	5.0×10^3	4.5×10^6 (345)	4.0×10^6 (8.0)
D	$ts10$	—	—	8.0×10^3	8.5×10^6 (16.7)
E	$ts20$	—	—	—	5.0×10^5

[a] Complementation tests were performed as described by Burge and Pfefferkorn (1966b). Mixed and separate infections with pairs of mutants were established at 39° to 40°C and the yield of virus (PFU/ml) from the fourth to the sixth hour of infection was measured by plaque assay at 28°C. Values on the diagonal are yields produced by each mutant in single infection; values off the diagonal are yields produced in mixed infections with the indicated mutants. In parenthesis is the "complementation level," or the yield from the mixed infection divided by the sum of the yields of the two single infections.

[b] Parental (HR virus) yield in this experiment: 1.4×10^8 PFU/ml.

virions. Lacking both of the altered proteins, the virions produced by complementation should be as heat-stable as is the HR virus. Figure 10 shows that this prediction was, at least in part, fulfilled; the virus produced by complementation was found to be substantially more heat-stable than either parent. Moreover, this was only a phenotypic property, for the progeny of the complementation, once cloned, were all heat-labile.

The virus produced by such a complementation, however, was not identical to the HR virus in heat-stability. Although almost as heat-stable as the HR virus during the first 3 min at 58°C, the complementation progeny were inactivated nearly as rapidly as the less heat-labile parent after 6 min. These kinetics are characteristic of multi-hit killing. We have never seen such kinetics with either wild-type or mutant stocks. The mechanism responsible is obscure; structural heterogeneity within each virion may be involved.

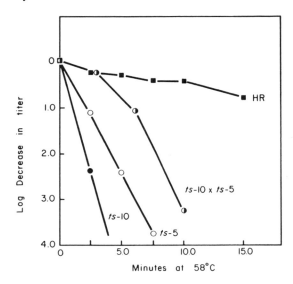

FIG. 10. Heat inactivation kinetics of the virions of HR virus, of two *RNA*⁺ mutants (*ts5* and *ts10*) derived from the HR virus, and of virus produced by complementation between these two mutants.

Complementation between *RNA*⁻ Mutants

The pattern of complementation between *RNA*⁻ mutants is difficult to interpret, because it is hard to distinguish between intracistronic and intercistronic complementation. One *RNA*⁻ mutant, *ts6*, complemented extremely well with all other *RNA*⁻ mutants, producing 5 to 50% of the yield of the parental (HR) infections at the nonpermissive temperature (Table IV).* The remaining *RNA*⁻ mutants exhibited a complex pattern of complementation—some pairs complemented quite well, while others showed no complementation. One explanation for this pattern is that there are two cistrons involved in the *RNA*⁻ phenotype, one represented by *ts6* and the other by all of the remaining *RNA*⁻ mutants, some pairs of which exhibit intracistronic complementation among themselves. It

* The introduction of mutant *ts6* has modified the designation of complementation groups reported earlier (Burge and Pfefferkorn, 1966).

is possible, of course, that this entire pattern is due to intracistronic complementation.

Since the genetic evidence was inconclusive, we sought physiological evidence for two different functions involved in the RNA^- phenotype. A simple, though unsuccessful, experiment examined the stability of the infectivity of cell-virus complexes at the nonpermissive temperature. We hoped that, in one class of RNA^- mutants, the RNA of the infecting virus would remain single-stranded at the nonpermissive temperature, while the RNA of the other class of mutants would be converted to a double-stranded form which would have greater (or lesser) stability. Figure 11

TABLE IV

Complementation between RNA^- Mutants[a][b]

	ts6	ts4	ts11	ts21	ts24
ts6	1.1×10^3	6.5×10^6 (127)	2.5×10^6 (280)	1.9×10^7 (200)	2.0×10^6 (364)
ts4	—	5.0×10^4	5.2×10^5 (9.0)	1.1×10^4 (0.08)	2.0×10^4 (0.36)
ts11	—	—	7.5×10^3	1.8×10^6 (17.6)	5.0×10^4 (4.2)
ts21	—	—	—	9.4×10^4	1.5×10^5 (1.5)
ts24	—	—	—	—	4.5×10^3

[a] For details see footnote to Table III.

[b] Parental (HR virus) yield in this experiment: 5.2×10^7 PFU/ml.

shows that the stabilities of three RNA^- complexes (including those formed by mutant *ts*6) at the nonpermissive temperature were identical.

Thus, if the RNA of any of these mutants were converted to a double-stranded form at the nonpermissive temperature, that form would appear to have the same stability in virus-cell complexes as the single-stranded form. However, if the actual mechanism of RNA replication involves the production of free, complementary, "minus" viral RNA strands, virus-cell complexes of both classes of RNA^- mutants would be expected to have similar stabilities.

Since this approach did not distinguish among our RNA^- mutants, we turned to more direct experiments on the replicative form of Sindbis virus.

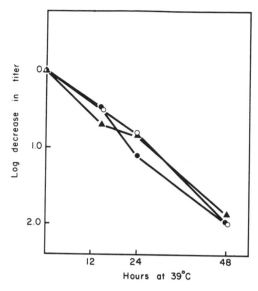

Fɪɢ. 11. Stability of *RNA⁻* mutant-cell complexes at a nonpermissive temperature. Methods described in the legend of Fig. 2. Mutant *ts*4 (▲), mutant *ts*6 (●), mutant *ts*11 (○).

RNase-Resistant Viral RNA in Infected Cells

Montagnier and Sanders (1963) were the first to report the appearance of a double-stranded RNA in virus-infected cells. They described an RNA species present in encephalomyocarditis virus-infected cells that possesses many of the properties expected of a double-stranded molecule active in replication. It is absent immediately after infection and accumulates gradually thereafter; it sediments more slowly than RNA extracted from purified virus but is infectious; it is relatively resistant to RNase, and it shows a sharp thermal transition. Similar RNA species have been observed in cells infected by a variety of RNA animal viruses and RNA bacteriophages. It is possible, however, that this double-stranded RNA plays no essential role in viral RNA replication, but may be merely a by-product. It may even be formed during the extraction procedure through the association of "plus" and "minus" strands of viral RNA.

Lodish and Zinder (1966) have provided some genetic evidence for the involvement of an RNase-resistant species of RNA in the replication of an RNA bacteriophage. They used temperature-sensitive conditional-lethal mutants of bacteriophage f2 that are unable to synthesize viral RNA when infected cells are incubated at high (nonpermissive) temperatures. One of these mutants (*ts*6) ceases production of double-stranded RNA almost immediately after shift-up to a nonpermissive temperature, although it continues to make single-stranded RNA. Lodish and Zinder (1966) postulate that two enzymes (or a bifunctional enzyme) are required for replication of f2 RNA: *Enzyme I,* which produces minus RNA strands complementary to parental plus strands, resulting in double-stranded molecules (this enzyme is presumably rapidly inactivated when *ts*6-infected cells are shifted to a nonpermissive temperature), and *Enzyme II,* which produces new plus strands from the double-stranded template and is presumed to be stable at the nonpermissive temperature.

Some of the techniques developed by Lodish and Zinder in the study of f2 mutants have been applied to the study of RNA synthesis in Sindbis virus-infected cells. Following their terminology we define RNase-sensitive RNA as uridine-^{3}H-labeled material solubilized by exposure to pancreatic RNase (10 μg/ml) in buffer (0.1 M NaCl, 0.05 M Tris, pH 7.5, 0.001 M MgCl$_2$) for 30 min at 25°C. RNase-resistant RNA is resistant to RNase digestion in buffer, but is solubilized by the same concentration of RNase in water.

Phenol-sodium dodecyl sulfate extracts (Montagnier and Sanders, 1963) of actinomycin D-treated cells, infected with Sindbis virus for 4 hours and then labeled with uridine-^{3}H, contained some labeled RNA that was RNase-resistant. Although this species of RNA was defined in subsequent experiments only by its RNase resistance, preliminary experiments showed that it could also be resolved by velocity gradient centrifugation. After sedimentation through a linear sucrose gradient, virus-specific RNA was found in two broad peaks, one centered at 26 S, the other at 40–50 S. After RNase treatment of gradient fractions, a small peak of RNase-resistant material remained in the 20–22 S region. These results are similar to those obtained by Martin (1966) with a related arbovirus, Semliki forest virus. The mere identification of an RNase-resistant species of RNA does not, of course, provide evidence for its involvement in viral RNA replication.

Kinetics of Labeling of RNase-Resistant RNA in Sindbis-Infected Cells

Labeled uridine, added to actinomycin-treated, Sindbis-infected chick fibroblasts became available for incorporation into RNA relatively slowly. The rate of incorporation of uridine-^{3}H into acid-insoluble material reached a maximum only after 20 to 30 min. This delay in equilibration of exogenous uridine with the intracellular pool of UTP prevented full exploitation of the rapid, pulse-chase experiments found useful in studying RNA replication in bacteria infected by bacteriophage f2 (Lodish and Zinder, 1966). It did not, however, prevent observation of the properties of RNA synthesized during relatively short labeling intervals.

In the experiment summarized in Fig. 12, cells infected with HR virus for 4 hours at 39°C were exposed to uridine-³H for intervals of 15 to 60 min, and RNA was extracted. The percentage of labeled material present in a RNase-resistant form was determined. Figure 12 shows that the percentage of RNase-resistant RNA was maximal at the shortest labeling interval, and decreased with longer labeling intervals. This behavior is compatible with a recently proposed model of viral RNA replication, in which RNA precursors are first incorporated into an RNase-resistant, multistranded replicating structure, and only later appear as mature single strands (Fenwick *et al.*, 1964).

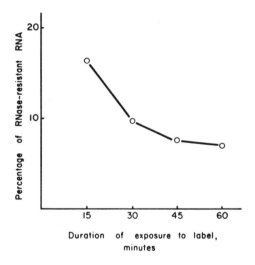

FIG. 12. Influence of labeling interval on percent RNase-resistant RNA. HR-infected, actinomycin D-treated monolayers, incubated at 39°C, were exposed to uridine-³H (0.003 mCi/ml) at 4 hours after infection for the specified interval. RNA was extracted by dissolving washed monolayers in a buffered sodium dodecyl-sulfate solution (Montagnier and Sanders, 1963) and shaking with an equal volume of aqueous phenol at 40–45°C. The phenol phase and interface were extracted a second time with buffer; the aqueous solutions were pooled, made 0.3 M with sodium acetate, and then mixed with 2 volumes of cold ethanol. After several hours at 0°C, the precipitate was collected by centrifugation, dissolved in buffer (0.1 M NaCl, 0.05 M Tris pH 7.5, 0.001 M MgCl₂) and treated with RNase in a 5 ml volume of the same buffer (10 μg/ml RNase, 30 min, 25°C). Carrier protein was added and the acid-insoluble material was precipitated with 0.3 N trichloroacetic acid. The radioactivity of the washed precipitates was determined. RNA not solubilized by RNase in H₂O (10 μg/ml) was subtracted from the value for counts resistant to RNase in buffer. This value was always less than 1% of the total.

Conditional-Lethal Mutants of Sindbis Virus Defective in RNA Synthesis

More evidence for the physiological importance of RNase-resistant RNA comes from experiments with temperature-sensitive mutants of Sindbis virus defective in RNA synthesis (RNA^- mutants). It was found that most such mutants (e.g. *ts*4, *ts*11) produced the same percentage of RNase-resistant RNA as the HR ancestor during incubation at 31°C, 35°C and, under shift-up conditions, at 39°C (Fig. 13). However, one

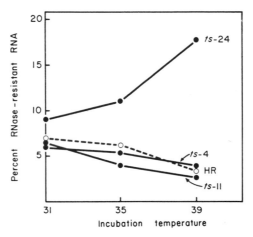

FIG. 13. Percentage RNase-resistant RNA made at 3 different temperatures. Actinomycin D-treated monolayers (1 μg/ml), infected with about 10–50 PFU/cell of the specified virus, were incubated at the indicated temperature and exposed to uridine-³H (.003 mCi/ml) from 3.5 to 5.5 hours after infection. RNase-resistant RNA was measured as described in the legend to Fig. 12. Since RNA^- mutants (*ts*4, *ts*11, *ts*24) did not stimulate uridine uptake when incubated continuously at 39°C, the 39°C determinations were made with monolayers incubated at 31°C during the first 3 hours of infection and then shifted to 39°C for the remainder of the experiment.

mutant, *ts*24, produced a higher percentage of RNase-resistant RNA than the HR virus at all incubation temperatures, particularly under shift-up conditions, at 39°C (Fig. 13). It should be recalled that mutants *ts*4 and *ts*11 are able to produce a nearly normal yield of virus if shifted from a permissive to a nonpermissive temperature a few hours after infection (Fig. 4), and thus it may be argued that enzymes for RNA synthesis produced by these two mutants are stable, once made. Thus, their pattern of RNA synthesis should not differ from that of the wild-type virus following shift-up to a nonpermissive temperature. In contrast, though mutant *ts*24 benefited somewhat from incubation at a permissive temperature during the eclipse period (Fig. 4), it produced only 1%

of the titer achieved by *ts*4 and *ts*11 under the same conditions. It seems probable, therefore, that an enzyme required for viral RNA synthesis in *ts*24-infected cells is labile at 39°C. This enzyme may be the hypothetical *Enzyme II,* discussed above, which synthesizes plus strands from a double-stranded template. If this enzyme were somewhat labile even at temperatures lower than 39°C, then a higher than normal percentage of RNase-resistant RNA might be found at all incubation temperatures, and exactly this behavior was found. Mutant *ts*24, therefore, appears to be a companion to the bacteriophage f2 mutant *ts*6 described by Lodish and Zinder (1966), with *ts*24 (Sindbis) giving rise to a labile *Enzyme II* and *ts*6 (f2), a labile *Enzyme I.*

An objection arises, however: Almost immediately after shift-up, the f2 mutant *ts*6 produces single-stranded RNA exclusively. Though *ts*24 (Sindbis) produced four to five times the normal percentage of RNase-resistant RNA under shift-up conditions, most of the incorporated radioactivity still remained RNase-sensitive. The explanation may lie in the relative labilities of the enzymes synthesized by the two mutants. If the lability of the defective enzyme of *ts*24 (Sindbis) were less extreme than that of the *ts*6 (f2) enzyme, then the balance of RNA synthesis in *ts*24-infected cells would be shifted in favor of RNase-resistant RNA, without exclusion of all synthesis of the RNase-sensitive species. Alternatively, if the real function of *Enzyme I* were to make free minus strands of viral RNA, and the double-stranded form were an artifact resulting from the association of plus and minus strands during the extraction, our mutant *ts*24 might not be expected to make wholly RNase-resistant material even if it synthesized minus strands exclusively after shift-up to a non-permissive temperature.

A Proposed Experiment

If two enzymes *are* required for replication of viral RNA, and if a double-stranded, RNase-resistant molecule *is* a required intermediate in the synthesis of new plus strands, then a strong prediction can be made concerning the behavior of two temperature-sensitive mutants with defects in *Enzyme I* and *Enzyme II,* respectively. Virus defective in *Enzyme I* should be unable to convert the input viral RNA into a double-stranded structure when allowed to infect cells at a nonpermissive temperature, while the virus defective in *Enzyme II* should achieve the synthesis of the double-stranded form but progress no further in RNA synthesis. Virus of sufficient radiochemical purity and specific activity can be prepared to test these possibilities. Since RNase-resistant RNA can be readily detected in arbovirus-infected cells, and since many *RNA*⁻ mutants are available and partially characterized, resolution of this problem may soon be possible.

ACKNOWLEDGMENTS

We wish to thank Mrs. Helen M. Coady for skillful technical assistance. This investigation was supported in whole by United States Public Health Service research grants AI-04531-04 and -05 from the Institute of Allergy and Infectious Diseases. B.W.B. was suported by a predoctoral fellowship from the United States Public Health Service, number 1-F1-GM-31, 037-01.

REFERENCES

Abel, P. (1962). *Virology* **16,** 347.
Burge, B. W., and Pfefferkorn, E. R. (1964). *Virology* **24,** 126.
Burge, B. W., and Pfefferkorn, E. R. (1966a). *Virology* **30,** 204.
Burge, B. W., and Pfefferkorn, E. R. (1966b). *Virology* **30,** 214.
Cooper, P. D. (1964). *Virology* **22,** 186.
Dulbecco, R., and Vogt, M. (1954). *J. Exptl. Med.* **99,** 183.
Epstein, R. H., Bolle, A., Steinberg, C. M. Kellenberger, E., Boy De La Tour, E., Chevalley, R., Edgar, R. S., Susman, M., Denhardt, G. H., and Lielausis, A. (1963). *Cold Spring Harbor Symp. Quant. Biol.* **28,** 375.
Fan, D. P., Higa, A., and Levinthal, C. (1964). *J. Mol. Biol.* **8,** 210.
Fenner, F., and Sambrook, J. F. (1964). *Ann. Rev. Microbiol.* **18,** 47.
Fenwick, M. L., Erikson, R. L., and Franklin, R. M. (1964). *Science* **146,** 527.
Fried, M. (1965). *Virology* **25,** 669.
Kirvaitis, J., and Simon, E. H. (1965). *Virology* **26,** 545.
Lodish, H. F., and Zinder, N. D. (1966). *Science* **152,** 372.
Martin, E. M. (1966). *Proc. Fed. European Biochem. Soc.,* (1966). (in press).
Montagnier, L., and Sanders, F. K. (1963). *Nature,* **199,** 664.
Reich, E., Franklin, R. M., Shatkin, A. J., and Tatum, E. L. (1962). *Proc. Natl. Acad. Sci. U.S.* **48,** 1238
Sambrook, J F., Padgett, B. L., and Tomkins, J. K. N. (1966). *Virology* **28,** 592.
Simpson, R. (1964). *Ciba Found. Symp., Cellular Biol. Myxovirus Infections* pp. 187–206.
Wecker, E. (1959). *Z. Naturforsch.* **14b,** 370.
Zinder, N. D. (1965). *Bacteriol. Rev.* **19,** 455.
Zinder, N. D., and Cooper, S. (1964). *Virology* **23,** 152.

Studies on the Replication of Reovirus

*A. J. Shatkin and B. Rada**

NATIONAL INSTITUTE OF ALLERGY AND INFECTIOUS DISEASES,
LABORATORY OF BIOLOGY OF VIRUSES,
NATIONAL INSTITUTES OF HEALTH,
BETHESDA, MARYLAND

Studies on the replication of RNA viruses have shown that new enzymes, whether RNA replicases or synthetases, are induced following virus infection, and that both double-stranded and single-stranded virus-directed RNA can be isolated from infected cells (Levintow, 1965). In contrast to the single-stranded nucleic acid found in the extensively studied animal and bacterial RNA viruses, the RNA of reovirus is a double-stranded helix (Gomatos and Tamm, 1963a). Its replication presumably proceeds by a mechanism which is different from that described for most RNA viruses, and we have undertaken a study of the synthesis of RNA in reovirus-infected cells.

Preferential Inhibition of Cellular RNA Synthesis by Actinomycin

Analysis of the virus-specific RNA formed in reovirus-infected L cells is complicated because cell RNA synthesis continues after infection and masks viral processes (Gomatos and Tamm, 1963b). Similar difficulties encountered with poliovirus- and mengovirus-infected cells were obviated when it was found that the antibiotic, actinomycin, could be used to completely suppress cellular RNA formation without interfering with virus production (Reich *et al.,* 1962; Shatkin, 1962). It is also possible to preferentially inhibit cellular RNA formation in reovirus-infected cells. Although virus replication is markedly inhibited by 2 μg actinomycin/ml (Gomatos *et al.,* 1962) concentrations of 0.1–0.5 μg/ml permit the synthesis of full yields of infectious reovirus (Shatkin, 1965a; Kudo and

*World Health Organization Fellow. Permanent address: Institute of Virology, Czechoslovak Academy of Sciences, Bratislava 9, Czechoslovakia.

Graham, 1965). This is illustrated in Fig. 1 in which suspension cultures of mouse L-929 fibroblasts were infected with type 3 reovirus at a multiplicity of 100 plaque forming units (PFU) per cell, and the time course of virus formation was compared in the presence and absence of actinomycin. In the absence of the antibiotic, new infectious virus was detected beginning 6–8 hours after infection, and maturation proceeded at a logarithmic rate for the next several hours. The addition of 0.5 μg actinomycin/ml 2½ hours after infection was without effect either on the time course of formation or the final yield of infectious virus. In uninfected cells, this same concentration of actinomycin almost completely inhibited RNA formation (Fig. 2). The specific activity of the RNA in growing cells after a one-hour exposure to uridine-2^{14}C was 180 cpm/μg. This value was reduced by more than 95% following treatment with

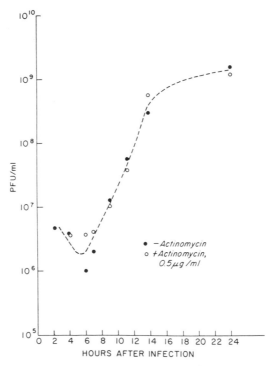

FIG. 1. Time course of reovirus formation. Suspension cultures of L-929 mouse fibroblasts (5 $\times$ 10^6 cells/ml), in Eagle's medium (Eagle, 1959) containing 2% fetal bovine serum were infected with ca. 100 plaque forming units (PFU) type 3 reovirus per cell. After adsorption for 2 hours at 37°C, unadsorbed virus was removed by washing, and the cells were resuspended at a concentration of 5 $\times$ 10^5/ml. Actinomycin was added at 2½ hours after infection, and samples were assayed by a plaque method (Gomatos *et al.*, 1962).

0.5 μg actinomycin/ml for two or more hours. Since virus replication is unaffected under these conditions, it should be possible to measure the rate of synthesis of virus-specific RNA. A series of infected cultures treated with 0.5 μg actinomycin/ml beginning 2½ hours after infection (PI) was exposed to uridine-^{14}C for one-hour intervals during the infectious cycle, and the specific activity of the RNA was determined. As seen in Figs. 1 and 2, there was an increased rate of RNA synthesis in

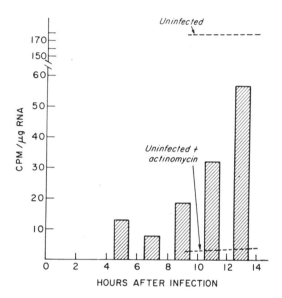

FIG. 2. Stimulation of RNA synthesis in actinomycin-treated cells following reovirus infection. Infected cultures containing 0.5 μg actinomycin/ml were exposed to uridine-2-C^{14} (2 × 10^{-6}M, specific activity = 24.4 μCi/μmole) for one hour. The cells were then chilled, centrifuged, extracted twice with 5% perchloric acid at 4°C, and hydrolyzed with 0.3 N KOH at 37°C for 16 hours. Aliquots were plated for counting; the total RNA was measured by the orcinol method (Mejbaum, 1939). Uninfected cells were treated with actinomycin for two hours or longer before exposure to uridine-2-^{14}C for one hour.

the infected, antibiotic-treated cells which paralleled the time course of virus replication. RNA formation at 12½ to 13½ hours PI proceeded at a fourteenfold greater rate than that observed in an uninfected, inhibited culture and was one-third that of a growing culture. This observation, that an appropriate concentration of actinomycin suppresses cell RNA synthesis without affecting virus replication, provides an experimental basis for a study of the virus-specific RNA species formed in reovirus-infected L cells.

Properties of Newly Formed RNA in Actinomycin-Treated Infected Cells

Double-stranded RNA extracted from reovirus (Krug *et al.*, 1964) or rice dwarf virus (Miura and Muto, 1965) and the double or triple-stranded helices consisting of polyadenylic and polyuridylic acid (Nirenberg and Matthaei, 1961) function poorly, if at all, as messengers for the *in vitro* stimulation of amino acid incorporation into polypeptides. The inactivity of helical polymers is presumed to be due to their inability to bind effectively to ribosomes (Cukier and Nirenberg, 1963; Takanami and Okamoto, 1963). These findings suggest that the formation of a single-stranded RNA which can function as a messenger for viral protein synthesis is a necessary intermediate step in the replication of reovirus. To examine this possibility, infected, actinomycin-treated cells were exposed to uridine-^{14}C for 2 hours at various times during the infectious cycle. The RNA was then purified by extraction with sodium dodecylsulfate and phenol at 60°C and by precipitation with ethanol (Scherrer and Darnell, 1962). Several properties of the newly formed RNA—including size, secondary structure and chemical composition—have been examined.

SIZE AND SECONDARY STRUCTURE

The size of the virus-specific RNA was determined by sedimentation analysis in 15–30% sucrose density gradients (Britten and Roberts, 1960). RNA from both uninfected and infected cells includes the 28 S and 16 S ribosomal species and the 4 S soluble RNA (Fig. 3). The pattern of distribution of newly formed RNA before and after treatment with RNase is also shown. Treatment of uninfected cells with actinomycin completely inhibited ribosomal RNA synthesis, although uridine incorporation into 4 S RNA persisted. Essentially all of the newly synthesized RNA in uninfected cells was degraded to acid-soluble material by RNase. In infected cultures, the RNA formed 6 to 8 hours PI was also largely single-stranded as demonstrated by its susceptibility to degradation by RNase. However, 9% of the newly synthesized RNA was nuclease-resistant (the shaded area). It sedimented at 10–12 S, the position of double-stranded RNA extracted from purified reovirus (Kudo and Graham, 1965). At this time in the infectious cycle, RNA synthesis was increased 30% above that observed in uninfected, inhibited cultures. At 8–10 hours PI, the rate of RNA synthesis increased to fivefold that of uninfected cells. The newly formed RNA was heterogeneous and formed a broad peak at 12–16 S. Of the total RNA synthesized during this interval, 23% was nuclease-resistant. A further stimulation in RNA synthesis to more than ten times the level in uninfected cells was observed at 10–12 hours PI. The rate of synthesis of the single and double-stranded RNA's apparently increased disproportionately during this interval since the double-stranded RNA decreased to 15% of the total formed. Kudo and Graham (1965) also have reported that the major fraction of newly formed RNA in reovirus-infected actinomycin-treated cells is single-stranded and sensitive to degradation by RNase.

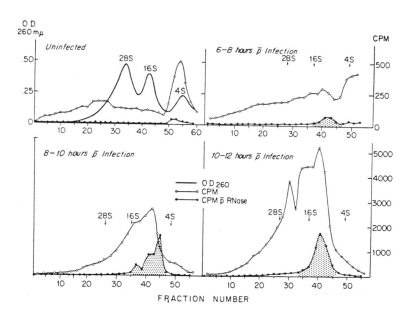

FIG. 3. Sedimentation of newly formed RNA in sucrose density gradients. After exposure to uridine-2-^{14}C, cells were chilled, washed with serum-free media, and resuspended in 0.01 M acetate buffer, pH 5.1 containing 0.1 M NaCl and 0.001 M MgCl$_2$, at a concentration of 10^7 cells/ml. Sodium dodecyl sulfate, 0.35%, was added followed by an equal volume of H$_2$O-saturated phenol. The suspension was extracted at 60°C and rapidly cooled to 0–4°C; then, the extraction was repeated. RNA was precipitated from the aqueous phase with 2 volumes of ethanol, dissolved in a small volume of acetate buffer without MgCl$_2$, and passed through sephadex G-25. It was again precipitated and dissolved, and layered onto 15–30% linear sucrose density gradients containing 0.01 M acetate buffer pH 5.1 and 0.1 M NaCl. After centrifugation for 16 hours, 0.5 ml samples were collected. The even-numbered tubes (○) were precipitated with 5% perchloric acid and 2 mg yeast RNA as carrier and counted. Odd-numbered tubes (●) were brought to 0.15 M NaCl, 0.01 M phosphate buffer pH 7.4 and incubated with 2 μg RNase/ml for 30 min at 37°C before precipitation and counting.

PURIFICATION OF SINGLE- AND DOUBLE-STRANDED RNA ON MAK
COLUMNS

Separation of double-stranded RNase-resistant RNA and single-stranded enzyme-sensitive RNA was achieved by chromatography on methylated albumin-kieselguhr (MAK) columns (Mandell and Hershey, 1960). The elution profiles for RNA isolated from uninfected and infected cells are shown in Fig. 4. As seen in the optical density tracings, soluble RNA eluted at 0.4–0.5 M NaCl and ribosomal RNA eluted at 0.9–1.1 M NaCl. The double-stranded RNase-resistant RNA extracted from infected

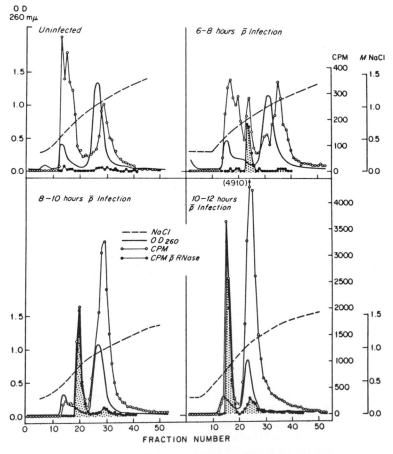

FIG. 4. Chromatography of RNA on MAK's. Phenol-extracted sephadex-treated RNA was dissolved in 0.4 M NaCl, 0.05 M phosphate buffer pH 6.8 and loaded onto the column at this molarity. Aliquots of eluted samples were tested for acid-precipitable radioactivity before (○) and after (●) treatment with RNase.

cells eluted at 0.7–0.8 *M* NaCl (shaded areas) and was well separated from the single-stranded nuclease-sensitive RNA which eluted with and later than the ribosomal RNA at 0.9–1.1 *M* NaCl.

Base Composition Analysis

The finding that actinomycin-inhibited cells synthesize single-stranded RNA in response to reovirus infection suggests that this newly formed RNA is coded for by the double-stranded viral RNA. The results of base composition analysis and annealing experiments support this suggestion. The base composition of ^{32}P-labeled single- and double-stranded RNA extracted from infected cells and separated on a MAK column were compared to RNA-^{32}P extracted from reovirus and purified by equilibrium density gradient centrifugation in Cs_2SO_4 (Shatkin, 1965b). RNA was hydrolyzed with alkali and the distribution of radioactive phosphate among the electrophoretically separated nucleotides was determined (Sebring and Salzman, 1964). As shown in Table I, both the single- and double-stranded RNA's synthesized 9–12 hours after infection resembled purified virus RNA. Similar results were obtained with RNA formed 6–9 hours after infection. In each case the guanylic + cytidylic acid content was 47.4–47.5% and the purine to pyrimidine ratio was close to one. However, the single-stranded RNA consistently had a higher proportion of guanylic than cytidylic acid.

Annealing of Single-Stranded RNA and Reovirus RNA— Evidence for Strand Selection

Base sequence homology between single-stranded RNA isolated from infected cells and double-stranded RNA extracted from purified virus was tested by annealing experiments. RNA was purified from cells exposed to uridine-^{14}C 10–12 hours PI, and the single-stranded RNA was separated on a MAK column. It was heated to 90°C and cooled slowly over a period of several hours in the presence of 0.25 *M* NaCl, 0.005 *M* phosphate buffer, pH 7, and the additions indicated in Table II. Before heating, more than 90% of the RNA was degraded to acid-soluble material by incubation with 2 μg RNase/ml for 30 min at 37°C. After heating and cooling, the RNA remained nuclease-sensitive. However, when heat-denatured reovirus RNA, but not native reovirus RNA or L cell RNA, was present during the heating and cooling, 87% of the single-stranded RNA was converted to an RNase-resistant form. In addition, when this annealed RNA was rechromatographed on a MAK column, it eluted at 0.76 *M* NaCl as expected for double-stranded RNA.

TABLE I

BASE COMPOSITION ANALYSIS[a]

RNA[b]	Cytidylic acid	Adenylic acid	Guanylic acid	Uridylic acid	G + C	$\dfrac{A + G}{C + U}$
RNA from purified reovirus	23.5	26.2	24.0	26.3	47.5	1.01
RNA synthesized 9–12 hours PI						
Double-stranded	24.2	26.1	23.2	26.5	47.4	0.97
Single-stranded	22.2	25.6	25.2	27.0	47.4	1.03

[a] Bases measured in moles/100 moles nucleotides.

[b] The values for virus RNA were obtained with Cs_2SO_4-purified material.

The extent of annealing was dependent upon the concentration of denatured reovirus RNA present (Fig. 5). In the experiment shown, a constant amount of radioactive, single-stranded RNA purified from cells exposed to uridine-[14]C 8–10 hours PI was annealed with increasing quantities of denatured reovirus RNA. At concentrations of 1–3 μg of denatured reovirus RNA/ml, there was a sharp rise in the amount of annealed RNA as measured by RNase resistance. At higher RNA concentrations the annealed fraction continued to increase slowly but did not reach 87%, the value observed with RNA formed 10–12 hours PI and annealed with 54 μg denatured RNA/ml.

TABLE II

ANNEALING OF REOVIRUS-DIRECTED SINGLE-STRANDED RNA[a]

Additions to RNA	Annealed	RNase	Acid-precipitable (cpm/0.5 ml)	% RNase resistant
None	—	—	521	
None	—	+	34	7
None	+	+	64	12
Native reovirus RNA (27 μg)	+	+	45	9
L-cell RNA (16 μg)	+	+	53	10
Denatured reovirus RNA (27 μg)	+	+	451	87

[a] Single-stranded RNA extracted from infected cells exposed to uridine-2-[14]C 10 to 12 hours after infection was separated on a MAK column. The RNA was annealed by slow cooling from 90°C to room temperature in 0.005 M phosphate buffer, pH 7, 0.25 M NaCl. After incubation at 37°C for 30 min with 2 μg RNase/ml, the vessels were chilled, and the RNA was precipitated with 5% perchloric acid and 2-mg-yeast carrier RNA and counted.

It has been demonstrated previously that strand selection occurs when DNA is transcribed by RNA polymerase *in vivo* (Marmur and Greenspan, 1963; Guild and Robison, 1963; Hayashi *et al.*, 1963; Hall *et al.*, 1963; Tocchini-Valentini *et al.*, 1963; Bautz, 1963) and may occur under certain conditions *in vitro* as well (Hayashi *et al.*, 1964). The observation that the single-stranded RNA fails to self-anneal (Table II) indicated that strand selection also occurs during transcription of the double-stranded RNA of reovirus. Since the annealing process is inefficient at low RNA concentrations (Fig. 5), it was necessary to demonstrate that the amount of RNA present during the heating and cooling was adequate to permit renaturation when complementary strands were present. Assuming that the double- and single-stranded species of reovirus-directed RNA are synthesized in infected cells from a common precursor pool, the specific activities of the two species of RNA iso-

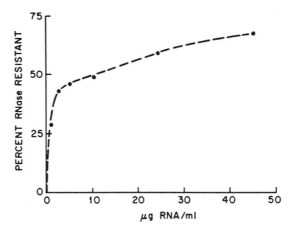

FIG. 5. Annealing of reovirus-directed single-stranded RNA. Actinomycin-treated cells were exposed to uridine-2-^{14}C 8–10 hours after infection. The RNA was purified, and the newly formed single-stranded RNA was separated on a MAK column. The indicated concentration of heat-denatured RNA extracted from purified reovirus was added to a solution of radioactive RNA (402 cpm) in 0.5 ml 0.005 M phosphate buffer, pH 7, 0.25 M NaCl. The mixture was heated to 90°C and slowly cooled to room temperature during a period of several hours before treatment with RNase.

lated from the same cells will be similar, and the radioactivity can be used to estimate the RNA concentration. An infected culture was incubated with uridine-^{14}C 10–12 hours PI, and the single- and double-stranded species were separated on a MAK column. The RNA concentration which would permit renaturation between the separated strands of double-stranded RNA was measured. Double-stranded RNA was heat-denatured and then self-annealed at a concentration of 150 cpm in 0.5 ml of 0.25 M NaCl, 0.005 M phosphate buffer pH 7. As shown in Table III, 52% of the radioactivity became RNase-resistant, whereas the single-

TABLE III

ANNEALING OF DENATURED DOUBLE-STRANDED REOVIRUS RNA[a]

RNA	Acid-precipitable (cpm/0.5 ml)	% RNase resistant
RNA	150	
RNA after RNase	15	10
Annealed RNA after RNase	78	52

[a] Double-stranded RNA formed 10 to 12 hours after infection was separated on a MAK column. After dialysis against 0.03 M NaCl, 0.005 M phosphate buffer, pH 7, the RNA was denatured by heating at 100°C for 10 min and fast-cooled. The solution was then brought to 0.25 M NaCl and annealed.

stranded RNA at a threefold greater concentration did not self-anneal (Table II). These results indicate that the failure of single-stranded RNA to self-anneal is not a concentration effect. Instead, it seems likely that the single-stranded RNA includes copies of only one of the strands of helical reovirus RNA.

POLYRIBOSOME ASSOCIATION OF SINGLE-STRANDED RNA

The basic unit of protein synthesis in a number of diverse organisms is the polyribosome, a cluster of single ribosomes held together by a single-stranded messenger RNA (Gierer, 1963; Warner *et al.,* 1963; Wettstein *et al.,* 1963). Viral protein synthesis also occurs on polyribosomes (Scharff *et al.,* 1963) and, at least in the case of poliovirus replication, the single-stranded viral RNA functions as the messenger RNA (Penman *et al.,* 1963; Summers and Levintow, 1965). We have examined the cytoplasmic fraction of reovirus-infected cells, and observed that the single-stranded, virus-directed RNA is present in polyribosomes. Actinomycin-treated cells which had been exposed to uridine-^{14}C for 1 hour beginning $10\frac{1}{4}$ hours PI were homogenized, and the cytoplasmic fraction was centrifuged for $2\frac{1}{2}$ hours in a 5–30% sucrose density gradient. The acid-precipitable radioactivity formed a broad peak which sedimented more rapidly than the 74 S single ribosomes (Fig. 6, left panel). When the extract was digested with 2 μg RNase/ml for 10 min at 0–4°C before centrifugation, the single-stranded RNA was partially degraded resulting in the disruption of polyribosomes and the formation of single ribosomes with the remaining virus-specific RNA attached (Fig. 6, right panel). Newly formed RNA purified from polyribosomes by phenol extraction was also RNase-sensitive in the presence of 0.25 M NaCl, indicating that it is single-stranded (Table IV). Furthermore, it annealed well with denatured reovirus RNA, but only to a small extent with itself or denatured L cell DNA.

Inactivity of Purified Reovirus RNA as Template *in Vitro* for *E. coli* Polymerases

The enzymatic activities of cells infected with RNA viruses have been studied in plants, animal cells in culture, and more extensively in bacteria (Levintow, 1965). Weissmann and collaborators (1964) have isolated from MS2 bacteriophage-infected *E. coli* an RNA synthetase which is associated with a double-stranded RNA template—the replicative form. Lodish and Zinder (1966) have described the results of experiments with temperature-sensitive mutants of bacteriophage f2 which indicate

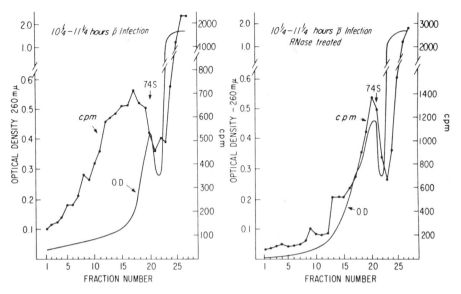

FIG. 6. Sedimentation of cytoplasmic fractions from reovirus infected cells. Actinomycin-treated cultures were exposed to uridine-2-^{14}C 10¼–11¼ hours after infection. The cells were chilled, washed, resuspended at a concentration of 2–5 × 10^7/ml in 0.01 M Tris buffer pH 7.8 containing 0.0015 M MgCl$_2$ and 0.01 M NaCl. After 5 min the cells were ruptured with a Dounce homogenizer. The unbroken cells and nuclei were removed by low-speed centrifugation (2 min–800 rpm) and the extract divided. One-half was incubated for 10 min at 0–4°C with 2 μg RNase/ml (right panel) and both received sodium desoxycholate, 0.25%. The extracts were layered onto 5–30% linear gradients of sucrose dissolved in the above buffer and centrifuged for 2½ hours at 24,000 rpm in the Spinco SW25 rotor. Fractions were collected and optical density and acid-precipitable radioactivity were measured. The sedimented pellets contained 2845 cpm (left panel) and 1052 cpm (right panel).

TABLE IV

ANNEALING OF RNA EXTRACTED FROM POLYRIBOSOMES[a]

Additions to RNA	Annealed	RNase	Acid-precipitable (cpm/0.5 ml)	% RNase resistant
None	—	—	460	
None	—	+	35	8
None	+	+	61	13
Denatured reovirus RNA (27 μg)	+	+	386	84
Denatured L-cell DNA (25 μg)	+	+	77	17

[a] RNA was purified from fractions 1–21 collected from a sucrose density gradient similar to that shown in Fig. 6 (left panel).

that the virus-induced single-stranded and double-stranded RNA's found in infected bacteria are synthesized by two different enzymes. On the other hand, Haruna and Spiegelman (1966) have purified from Qβ bacteriophage-infected *E. coli* an enzyme which replicates infectious viral RNA *in vitro,* apparently without the formation of an intermediate replicative form.

In contrast to the bacterial systems, the enzymes responsible for reovirus replication are not known. It was reported previously by Gomatos *et al.* (1964) that reovirus RNA could function like DNA as a template *in vitro* for *E. coli* DNA-dependent RNA polymerase. This function was inhibited by actinomycin. However, unlike DNA, the double-stranded RNA did not appear to interact significantly with the antibiotic. In view of the observation that 0.5 μg actinomycin/ml inhibits cell RNA synthesis more than 95% without affecting virus replication, it seemed important to examine further the nature of the template activity of double-stranded RNA extracted from purified reovirus.

FAILURE OF REOVIRUS TO BIND ACTINOMYCIN

When actinomycin complexes to helical DNA, there is an increase in the melting temperature (Reich, 1964; Haselkorn, 1964), molecular weight (Cavalieri and Nemchin, 1964), and sedimentation coefficient (Rauen *et al.,* 1960) of the DNA and a decrease in the buoyant density of the DNA (Reich, personal communication). Furthermore, the antibiotic undergoes a shift in its absorption maximum and reduction in its maximum absorbancy (Rauen *et al.,* 1960; Kirk, 1960). Gomatos *et al.* (1964) reported that reovirus RNA did not shift the absorption maximum of actinomycin, although it lowered the absorbancy at the maximum by 2% as compared to 25% for a comparable quantity of salmon sperm DNA. As an additional test of the possible interaction of double-stranded RNA with actinomycin, its effects on the melting temperature and sedimentation coefficient of reovirus RNA were examined. Actinomycin C_1 or X_2 at a concentration twice that of reovirus RNA did not affect the T_m of 81°C measured in 0.05 × standard saline citrate (SSC = 0.15 M NaCl, 0.015 M sodium citrate) (Fig. 7). A number of other compounds which bind to DNA (Ward *et al.,* 1965; Kersten *et al.,* 1966; Cohen and Yielding, 1965) including proflavine, chromomycin A_3, nogalamycin, and chloroquinone were tested and found to be without effect on the melting temperature of reovirus RNA.

Sedimentation of the double-stranded RNA extracted from purified reovirus in the analytical ultracentrifuge revealed the presence of three components. The major component had a sedimentation coefficient of

12.1 S which corresponds to a molecular weight of approximately two million for helical DNA; additional peaks were observed at 14.2 S and 10.7 S. The virus particle contains an amount of RNA equivalent to ten million daltons (Gomatos and Tamm, 1963a), and apparently the nucleic acid is degraded during the extraction (Kleinschmidt *et al.*, 1964; Gomatos and Stoeckenius, 1964). The presence of an equal concentration of actinomycin X_2 or C_1 was without effect on the observed sedimentation coefficients.

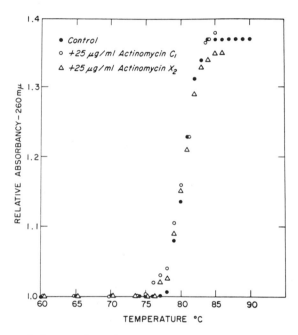

Fig. 7. Thermal denaturation of reovirus RNA. The solution contained 13 μg RNA/ml in $0.05 \times SSC$ (7.5×10^{-3} *M* NaCl, 7.5×10^{-4} *M* sodium citrate, pH 7).

A more direct test of the binding capacity of reovirus RNA was made with radioactive actinomycin. Reovirus RNA and actinomycin C_1-^{3}H were mixed in a small volume and overlaid onto a Cs_2SO_4 solution of average density 1.60 gm/cm^3. After centrifugation to equilibrium, fractions were collected through a hole punctured in the bottom of the tube and the distribution of RNA and actinomycin-^{3}H was determined by absorbancy and radioactivity measurements. As shown in Fig. 8, the buoyant density of the RNA, 1.61 gm/cm^3, was unaffected by the presence of actinomycin, and no binding of the radioactive antibiotic to reovirus RNA was observed. Actinomycin X_2 labeled with ^{14}C also failed to bind to reovirus RNA. The inability of double-stranded reovirus RNA to form complexes with actinomycin is not unexpected in view of the observation that this RNA is similar to the A configuration of DNA (Langridge and Gomatos, 1963), and actinomycin binds to DNA in the B form (Hamilton *et al.*, 1963).

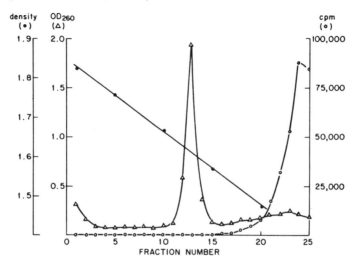

FIG. 8. Failure of reovirus RNA to bind actinomycin C_1-^{3}H. Reovirus RNA, 54 μg, and 11 μg actinomycin were incubated in 0.2 ml SSC for 30 min at 25°C, over-laid onto a Cs_2SO_4 solution of average density 1.60 gm/cm^3, and centrifuged for 5 days at 33,000 rpm in the Spinco SW39 rotor. Fractions of 0.2 ml were collected through a hole punctured in the bottom of the tube. The distribution of RNA was determined from absorbancy measurements at 260 mμ, and actinomycin-^{3}H was counted as described previously (Dingman and Sporn, 1965).

DNA AS THE SOURCE OF TEMPLATE ACTIVITY

Reovirus RNA extracted from purified virus was tested for its template activity with *E. coli* DNA-dependent RNA polymerase. The RNA stimulated the incorporation of nucleotides into an acid-insoluble product, and the reaction was inhibited to an extent similar to the DNA-primed reaction by the addition of 1 μg actinomycin (Table V). Individual preparations stimulated incorporation to an extent which was concentration dependent (Fig. 9). However, template efficiency varied among several RNA preparations suggesting that variable amounts of contaminating DNA might be present. To test this possibility, RNA was further purified by equilibrium density gradient centrifugation in Cs_2SO_4. The purified RNA was tested and found to be inactive as a template (Fig. 9). Exposure of reovirus RNA preparations to Cs_2SO_4 without centrifugation did not per se destroy their template activity; furthermore, Cs_2SO_4-banded

TABLE V

Inhibition of RNA Polymerase Reaction by Actinomycin[a]

Template	1 μg Actinomycin	AMP-³H[b]	GMP-³H[b]	% Inhibition
2.7 μg L-929 DNA	−	1.40	—	—
	+	0.07	—	95
6.7 μg L-929 DNA	−	—	1.19	—
	+	—	0.21	82
2.3 μg reovirus RNA	−	0.06	—	—
	+	0.01	—	83
2.9 μg reovirus RNA	−	—	0.13	—
	+	—	0.03	77
3 μg reovirus RNA (Cs_2SO_4-purified)	−	0.00	0.00	—

[a] Conditions as described for Fig. 9. The specific activities of ATP-³H and GTP-³H were 9.7 and 25 μCi/μmole, respectively.

[b] Measured in mμmoles incorporated.

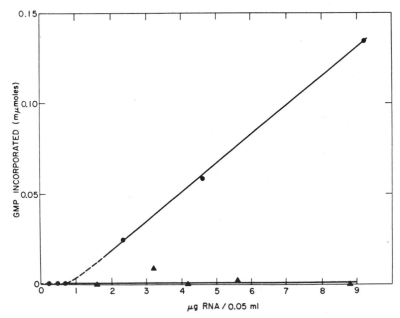

Fig. 9. Template activity of reovirus RNA for *E. coli* RNA polymerase: (●) before, and (▲) after purification by Cs_2SO_4 equilibrium density gradient centrifugation. The assay mixture was that described previously (Chamberlin and Berg, 1962) in a volume of 0.05 ml. After incubation at 37°C for 15 min, the vessels were chilled and 1 ml 5% trichloroacetic acid (TCA) added. The acid-insoluble material was collected on a millipore filter, washed three times with 1 ml 5% TCA, dried, and counted in 15 ml liquifluor-toluene in a scintillation counter. GTP-³H specific activity = 25 μCi/μmole.

TABLE VI

EFFECT OF PURIFIED REOVIRUS RNA ON RNA POLYMERASE REACTION[a]

Additions	(GMP-³H mμmoles incorporated)
Reovirus RNA (22 μg)	0.00
Calf thymus DNA (1 μg)	0.41
Calf thymus DNA (1 μg) + reovirus RNA (5.5 μg)	0.33
Denatured thymus DNA (1 μg)	0.24
Denatured thymus DNA (1 μg) + reovirus RNA (5.5 μg)	0.22
dGdC (0.9 μg)	0.24
dGdC (0.9 μg) + reovirus RNA (5.5 μg)	0.28

[a] The standard assay mixture in a volume of 0.3 ml was supplemented with 7.5 μmoles sodium phosphoenolpyruvate (PEP) and 5 μg PEP kinase. After incubation for 30 min at 37°C the vessels were chilled. Sodium nucleate and bovine serum albumin, 2 mg each, 4 μmoles GTP, and 5 ml 5% TCA were then added. After 15 min the vessels were centrifuged, the pellet was washed three times with 5% TCA, and dissolved in 0.5 ml formic acid. Four ml of ethanol were added followed by 15 ml liquifluor, and the solutions were counted in a scintillation counter. Thymus DNA was denatured by heating at 100°C for 10 min in 0.01 M tris buffer, pH 7.9, 0.01 M NaCl followed by quenching in ice. This experiment was performed with a preparation of RNA polymerase, fraction IV, obtained from A. Cerami and D. Ward.

RNA at a concentration five times that of native or denatured calf thymus DNA or of dGdC did not affect their template activities; this indicates that the purified RNA does not contain an inhibitor of the polymerase reaction (Table VI). On the other hand, RNA not previously purified in Cs_2SO_4, together with DNA at low concentrations, promoted incorporation to an extent equal to the sum of the individual nucleic acids (Table VII).

These results indicate that reovirus RNA does not serve as a template for RNA polymerase. Instead, it seems likely that the RNA preparations contain small amounts of DNA which band at a density of approxi-

TABLE VII

ADDITIVE TEMPLATE ACTIVITY OF DNA AND RNA FOR RNA POLYMERASE[a]

Additions	GMP-³H (mμmoles incorporated)
L-929 DNA (0.034 μg)	0.07
Reovirus RNA (6.87 μg)	0.11
L-929 DNA (0.034 μg) + reovirus RNA (6.87 μg)	0.17

[a] Conditions as for Fig. 9. L-929 DNA prepared by procedure of Marmur (1961). Reovirus RNA was not Cs_2SO_4-purified.

mately 1.4 gm/cm³ in Cs_2SO_4 (Erikson and Szybalski, 1964) and are thus removed by equilibrium centrifugation. The following additional control experiments were performed to test this conclusion (Shatkin, 1965b). (1) Reovirus was purified from HeLa cells which had been grown in the presence of thymidine-³H for 48 hours before virus infection, and the virus RNA was extracted with phenol. Before Cs_2SO_4 centrifugation the RNA solution of 595 μg/ml contained 270 cpm/ml, and 5.9 μg stimulated the incorporation of 0.04 mμmole of CMP. After Cs_2SO_4 purification no radioactivity could be detected in a solution containing 255 μg RNA/ml, and 3.3 μg RNA did not promote the incorporation of CMP. (2) Reovirus RNA is not degraded by DNase I or exonuclease I. However, pretreatment of RNA which had not been Cs_2SO_4-purified with DNase or exonuclease reduced its template activity by 40% and 70%, respectively. (3) Reovirus RNA-³H (43 cpm/μg) was digested to acid-soluble nucleotides with 0.3 *N* NaOH and dialyzed exhaustively. A volume of the dialyzed hydrolysate equivalent to 9.2 μg RNA before digestion was tested with RNA polymerase. It retained 50% of its original template activity, although the RNA had been completely degraded. Treatment with 4% trichloroacetic acid at 90°C for 15 min abolished all template activity. (4) Reovirus was treated exhaustively with micrococcal nuclease and DNase before extraction of the RNA with phenol. This RNA had less than 1% the template activity of a comparable amount of DNA.

Haselkorn has also compared reovirus RNA and T7 bacteriophage DNA as templates for the DNA-dependent RNA polymerase of *Micrococcus lysodeikticus* and found that DNA is at least 30 times more effective. The small RNA template activity which was observed was not diminished by actinomycin (Haselkorn, personal communication). In addition, Miura has found that the double-stranded RNA from rice dwarf virus is inactive as a template for *E. coli* RNA polymerase (Miura, personal communication).

TABLE VIII

Effect of Purified Reovirus RNA on DNA Polymerase[a]

Additions	(dTMP-³H mμmoles incorporated)
Reovirus RNA (14 μg)	0.00
Calf thymus DNA (15 μg)	0.76
Reovirus RNA (14 μg) + thymus DNA (15 μg)	0.72

[a] The assay mixture in a final volume of 0.3 ml was that of Richardson *et al.* (1964). The specific activity of dTTP-³H was 8×10^3 cpm/mμmole. After incubation at 37°C for 15 min, the vessels were chilled and assayed as described in the legend to Table VI.

The priming ability of reovirus RNA for *E. coli* DNA polymerase was also examined (Table VIII). As observed for RNA polymerase, reovirus RNA was inactive after purification in Cs_2SO_4.

Conclusions

These results indicate that purified, double-stranded RNA is not a template for the nucleic acid polymerases of *E. coli in vitro*. Although the possibility remains that the polymerases of L cells in tissue culture can utilize intact reovirus RNA as a template, it seems more likely that the induction of new enzymes is required for reovirus replication. In agreement with this suggestion, it has been found that the inhibition of protein synthesis in infected cells by cycloheximide also prevents the formation of virus-specific RNA (Shatkin and Rada, 1967).

If new enzymes whose synthesis is directed by single-stranded messenger RNA are required for reovirus replication, then it seems reasonable that the cell genome codes for the enzyme which synthesizes the reovirus-specific single-stranded RNA. The single-stranded RNA could then direct the synthesis of both virus coat proteins and an enzyme for the formation of double-stranded virus RNA. This second enzyme might produce a duplex by replicating a complementary strand on a template of newly formed single-stranded RNA or by a semiconservative mechanism with the parental double-stranded RNA as primer. Alternatively, the parental double-stranded RNA may be reduced to a functional single-stranded messenger inside the cell, or the virus particle, in addition to double-stranded RNA, may contain a small undetected amount of single-stranded RNA, adequate to code for a new enzyme.

In reovirus-infected cells in which host RNA synthesis has been suppressed with actinomycin, at least two types of virus-specific RNA are formed: (1) double-stranded, RNase-resistant RNA which represents 10–25% of the total RNA synthesized and is destined for incorporation into mature virus and (2) single-stranded RNase-sensitive RNA which is similar in base sequence to one of the strands of double-stranded RNA extracted from purified virus and presumably is the species which functions as messenger for viral protein synthesis. A detailed understanding of the enzymatic mechanisms by which these RNA's are synthesized will require a study of purified enzymes from reovirus-infected cells.

ACKNOWLEDGMENTS

The authors are grateful to E. Reich for proving the actinomysin X_2-^{14}C and dGdC, H. Weissbach for the actinomycin C_1-H[3] and to R. Basch, A. Cerami, and D. Ward for providing the DNA polymerase and exonuclease I.

REFERENCES

Bautz, E. K. F. (1963). *Cold Spring Harbor Symp. Quant. Biol.* **28**, 205.

Britten, R. J., and Roberts, R. B. (1960). *Science* **131**, 32.

Cavalieri, L. F., and Nemchin, R. G. (1964). *Biochim. Biophys. Acta* **87**, 641.

Chamberlin, M., and Berg, P. (1962). *Proc. Natl. Acad. Sci. U.S.* **48**, 81.

Cohen, S. N., and Yielding, K. L. (1965). *J. Biol Chem.* **240**, 3123.

Cukier, R., and Nirenberg, M. W. (1963). Unpublished results (quoted in Nirenberg *et al.*, 1963).

Dingman, C. W., and Sporn, M. B. (1965). *Science* **149**, 1251.

Eagle, H. (1959). *Science* **130**, 432.

Erikson, R. L., and Szybalski, W. (1964). *Virology* **22**, 111.

Gierer, A. (1963). *J. Mol. Biol.* **6**, 148.

Gomatos, P. J., and Tamm, I. (1963a). *Proc. Natl. Acad. Sci. U.S.* **49**, 707.

Gomatos, P. J., and Tamm, I. (1963b). *Biochim. Biophys. Acta* **72**, 651.

Gomatos, P. J., Tamm, I., Dales, S., and Franklin, R. M. (1962). *Virology* **17**, 441.

Gomatos, P. J., and Stoeckenius, W. (1964). *Proc. Natl. Acad Sci. U.S.* **52**, 1449.

Gomatos, P. J., Krug, R. M., and Tamm, I. (1964). *J. Mol. Biol.* **9**, 193.

Guild, W. R., and Robison, M. (1963). *Proc. Natl. Acad. Sci. U.S.* **50**, 106.

Hall, B. D., Green, M., Nygaard, A. P., and Boezi, J. (1963). *Cold Spring Harbor Symp. Quant. Biol.* **28**, 201.

Hamilton, L. D., Fuller, W., and Reich, E. (1963). *Nature* **198**, 538.

Haruna, I., and Spiegelman, S. (1966). *Proc. Natl. Acad. Sci. U.S.* **55**, 1256.

Haselkorn, R. (1966). Personal communication.

Haselkorn, R. (1964). *Science* **143**, 682.

Hayashi, M., Hayashi, M. N., and Spiegelman, S. (1963). *Proc. Natl. Acad. Sci. U.S.* **50**, 664.

Hayashi, M., Hayashi, M. N., and Spiegelman, S. (1964). *Proc. Natl. Acad. Sci. U.S.* **51**, 351.

Kersten, W., Kersten, H., and Szybalski, W. (1966). *Biochemistry* **5**, 236.

Kirk, J. M. (1960). *Biochim. Biophys. Acta* **42**, 167.

Kleinschmidt, A. K., Dunnebacke, T. H., Spendlove, R. S., Schaffer, F. L. and Whitcomb, R. F. (1964). *J. Mol. Biol.* **10**, 282.

Krug, R. M., Gomatos, P. J., Tamm, I., and Lipmann, F. (1964). Unpublished results (quoted in Gomatos *et al.*, 1964).

Kudo, H., and Graham, A. F. (1965). *J. Bacteriol.* **90**, 936.

Langridge, R., and Gomatos, P. J. (1963). *Science* **141**, 694.

Levintow, L. (1965). *Ann. Rev. Biochem.* **34**, 487.

Lodish, H. F., and Zinder, N. D. (1966). *Science* **152**, 372.

Mandell, J. D., and Hershey, A. D. (1960). *Anal. Biochem.* **1**, 66.

Marmur, J. (1961). *J. Mol. Biol.* **3**, 208.

Marmur, J., and Greenspan, C. M. (1963). *Science* **142**, 387.

Mejbaum, W. (1939). *Z. Physiol. Chem.* **258**, 117.

Miura, K. I. (1966). Personal communication.

Miura, K. I., and Muto, A. (1965). *Biochim. Biophys. Acta* **108**, 707.

Nirenberg, M. W., and Matthaei, J. H. (1961). *Proc. Natl. Acad. Sci. U.S.* **47**, 1588.

Nirenberg, M. W., Jones, O. W., Leder, P., Clark, B. F. C., Sly, W. S., and Pestka, S. (1963). *Cold Spring Harbor Symp. Quant. Biol.* **28**, 549.

Penman, S, Scherrer, K., Becker, Y., and Darnell, J. E. (1963). *Proc. Natl. Acad. Sci. U.S.* **49**, 654.

Rauen, H. M., Kersten, H., and Kersten, W. (1960). *Z. Physiol. Chem.* **321**, 139.

Reich, E. (1964). *Science* **143**, 684.

Reich, E., Franklin, R. M., Shatkin, A. J., and Tatum, E. L. (1962). *Proc. Natl. Acad. Sci. U.S.* **48**, 1238.

Reich, E. (1965). Personal communication.

Richardson, C. C., Schildkraut, C. L., Aposhian, H. V., and Kornberg, A. (1964). *J. Biol. Chem.* **239**, 222.

Scharff, M. D., Shatkin, A. J., and Levintow, L. (1963). *Proc. Natl. Acad. Sci. U.S.* **50**, 686.

Scherrer, K., and Darnell, J. E. (1962). *Biochem. Biophys. Res. Commun.* **7**, 486.

Sebring, E. D., and Salzman, N. P. (1964). *Anal. Biochem.* **8**, 126.

Shatkin, A. J. (1962). *Biochim. Biophys. Acta* **61**, 310.

Shatkin, A. J. (1965a). *Biochem. Biophys. Res. Commun.* **19**, 506.

Shatkin, A. J. (1965b). *Proc. Natl. Acad. Sci. U.S.* **54**, 1721.

Shatkin, A. J., and Rada, B. (1967). *J. Virol.* **1**, 24.

Summers, D. F., and Levintow, L. (1965). *Virology* **27**, 44.

Takanami, M., and Okamoto, T. (1963). *Biochem. Biophys. Res. Commun.* **13**, 297.

Tocchini-Valentini, G. P., Stodolsky, M., Aurisicchio, A., Sarnat, M., Graziosi, F., Weiss, S. B., and Geiduschek, E. P., (1963). *Proc. Natl. Acad. Sci. U.S.* **50**, 935.

Ward, D. C., Reich, E., and Goldberg, I. H. (1965). *Science* **149**, 1259.

Warner, J. R., Knopf, P. M., and Rich, A. (1963). *Proc. Natl. Acad. Sci. U.S.* **49**, 122.

Weissmann, C., Borst, P., Burdon, R. H., Billeter, M. A., and Ochoa, S. (1964). *Proc. Natl. Acad. Sci. U.S.* **51**, 890.

Wettstein, F. O., Staehelin, T., and Noll, H. (1963). *Nature* **197**, 430.

Inhibitory Effect of a Cytidine Analog on the Growth of Rabies Virus: Comparative Studies with Other Metabolic Inhibitors[*]

Roland F. Maes, Martin M. Kaplan,[†] Tadeusz J. Wiktor, James B. Campbell, and Hilary Koprowski

THE WISTAR INSTITUTE OF ANATOMY AND BIOLOGY,
PHILADELPHIA, PENNSYLVANIA

Introduction

In spite of the fact that the disease caused by rabies virus has been known since 2300 BC, proper characterization of the infectious agent, isolated 80 years ago, remains an unfinished task. Rabies virus is referred to as a lipid-containing RNA virus. However, the presence of the RNA component in the viral nucleoprotein has not been determined by direct isolation and characterization of an infectious RNA, but has been deduced from indirect evidence based on reports of the failure of DNA inhibitors, such as bromodeoxyuridine (Hamparian *et al.,* 1963; Kissling and Reese, 1963), fluorodeoxyuridine, actinomycin D, and mitomycin C (Defendi and Wiktor, 1966), to inhibit the growth of rabies virus.

The virus particle resembles a bullet, having a cylindrical shape with one rounded end. When viewed in cross section, the virus may be seen to consist of two membranes surrounding a core filled with convoluted material, which probably consists of nucleoprotein (Hummeler *et al.,* 1967). Thus, it has the same morphologic characteristics as does vesicular stomatitis virus (Howatson and Whitmore, 1962), and other viruses affecting plants, drosophila, mites, trout, and birds (Harrison and Crowley, 1965; Hitchborn *et al.,* 1966; Berkaloff *et al.,* 1965; Ditchfield and

* This work was supported by USPHS Grant No. AI-02954 from the National Institute of Allergy and Infectious Disease, Training Grant No. TI-GM-142 from the National Institute of General Medical Science and a grant from the World Health Organization.

† Present address: World Health Organization, Geneva, Switzerland.

449

Almeida, 1964; Zwillenberg *et al.,* 1965; Murphy *et al.,* 1966). Rabies virus can be partially purified by density gradient centrifugation in CsCl; the majority of the infectious particles have a density of 1.20 gm/ml (Neurath *et al.,* 1966). Apart from these data, however, very little is known about the physical and chemical properties of the virus.

The purpose of the present study was to use various metabolic inhibitors, particularly those which inhibit DNA and protein synthesis, to extend our knowledge concerning the characteristics of rabies virus. The availability of several tissue culture systems susceptible to rabies infection made this study feasible. Although a plaque assay for egg-adapted rabies virus, using chick embryo monolayers, has been reported (Yoshino *et al.,* 1966), we have, as yet, been unsuccessful in developing such an assay using mammalian cells. In the present investigations, quantitative data were obtained by the titration of infectious material in mice, and by counting infected cells by means of the fluorescent antibody technique.

Experimental

Techniques

The PM strain of fixed rabies virus (Wiktor *et al.,* 1964) was grown in BHK/21 cells (Macpherson and Stoker, 1962), maintained as monolayers on coverslips in either plastic or glass petri dishes (60-mm diameter). The cells were maintained in Eagle's in Earle's medium containing 10% inactivated calf serum, aureomycin (50 μg/ml) and mycostatin (25–50 units/ml). The virus inoculum (1 ml), adjusted to a concentration providing an input of 2–10 mouse LD_{50} per cell, was adsorbed for 1 hour at 37°C in presence of DEAE-dextran at a concentration of 25–50 μg/ml (Kaplan *et al.,* 1967). If cultures were pretreated with inhibitors prior to infection, the latter were kept in the virus inoculum during the time of adsorption.

Following adsorption of the virus, the cells were washed with PBS, media containing the inhibitors under investigation were added, and the cultures were incubated at 37°C. The coverslips were removed from the cultures at regular time intervals, and stained with fluorescent antibody to detect the presence of rabies antigen (FA) (Wiktor and Koprowski, 1966). Selected tissue culture media were also titrated in mice (Wiktor and Koprowski, 1966).

Uridine-5-³H (specific activity 7.77 Ci/mmole) and thymidine-6-³H (specific activity 6.70 Ci/mmole) were used as radioactive precursors. Unless otherwise stated, they were left in the medium for the duration of the experiment. Coverslips removed from cultures exposed to these radioactive precursors were rinsed 3 times in PBS, and those exposed to tritiated uridine were fixed by immersion in ethanol. Coverslips exposed to tritiated thymidine were washed twice with 5% TCA, and then again with PBS. Individual coverslips were then placed in glass vials in the presence of scintillation fluid, and assayed for radioactivity in a Tri-Carb scintillation counter.

For control purposes, vesicular stomatitis virus (Indiana strain) and herpes virus were grown on BHK/21 cells under conditions similar to those used for the growth of rabies virus. The infectivity of these cultures was determined by plaque assays using standard agar or agarose overlays.

EFFECT OF METABOLIC INHIBITORS

Metabolic inhibitors were procured from commercial sources, dissolved in tissue culture medium, and tested for toxicity for BHK/21 cells. Based on the results shown in Table I, the concentrations of the various inhibitors used in the experiments with rabies virus were adjusted to the levels shown in Table II. Their effect on the growth of rabies was evaluated by determining the percentage of cells showing the presence of FA, and by observing the type of fluorescing granules in the cytoplasm of individual cells (Kaplan *et al.*, 1967). In addition, the amount of infectious virus present in the various media was determined by titrations in mice.

The results of these experiments (Table II) indicate that inhibitors of DNA synthesis such as actinomycin D, mitomycin C, and 5-fluoro-deoxyuridine (FUdR) have no inhibitory effect on the replication of rabies virus when added to the cultures at any time from 3 hours prior to, to 1 hour after infection. On the contrary, the presence of these three inhibitors was found to enhance the infectious process, as evidenced by the higher percentage of cells showing FA, the increase in the size of fluorescing granules*, and the higher yield of infectious virus in cultures exposed to the inhibitors.

In contrast to these results, 1-β-D-arabinofuranosyl-cytosine (ara-C)

* The classification of the type of fluorescing inclusion body is described in detail by (Kaplan *et al.*, 1967). It is given only in a schematic way below:

Code	Number of granules per cell	Size of granules	Intensity of fluorescence
A	5 or less	Small	Dim
B	10–20	Small and medium	Bright
C	>50	Medium and large	Bright
D	Coalescent	Large when discernible	Bright

TABLE I

Toxicity of Some Metabolic Inhibitors for BHK/21 Cells

Inhibitor	Concentration (μg/ml)	Condition of culture and % detached cells	
		24 hours	48 hours
Actinomycin D	0.1	10	25
	0.2	25	25
	0.4	50	100
	0.8	80	100
Mitomycin C	0.5	0	0
	1.0	10	505
	5.0	50	7
Fluorodeoxyuridine (FUdR)	5	0	0
	25	0	10
	50	10	20
Fluorophenylalanine (FPA)	5	0	0
	10	0	10
	50	50	100
	100	80	100
Arabinosylcytosine (ara-C)	2.5 ↓ 80.0	10–25 ↓ 10–25	10–25 ↓ 10–25
Cycloheximide	2.5	0	0
	5	0	0
	10	10	20

was found to exercise a pronounced inhibitory effect on the replication of rabies virus, as evidenced by the marked decrease in the percentage of FA-positive cells, and the correspondingly lower yield of the virus. Two inhibitors of protein biosynthesis, *p*-fluorophenylalanine (FPA) and cycloheximide, also effectively inhibited the growth of rabies virus, and the effect of FPA could be reversed by the simultaneous addition of phenylalanine (PA) to a phenylalanine-free medium (Table II).

The activity of the DNA inhibitors, at the concentrations used in these experiments, was checked against a DNA virus (herpes), and against vesicular stomatitis virus (VSV)—another RNA virus, morphologically identical to rabies virus. The results indicated that all three inhibitors decreased markedly the growth of herpes virus. Neither actinomycin D nor mitomycin C had any significant effect on VSV multiplication, although a two- to twenty-fold reduction in titer was consistently found following treatment with ara-C (Table III).

TABLE II

EFFECT OF METABOLIC INHIBITORS ON INFECTION OF BHK/21 CELLS WITH RABIES VIRUS

Experimental series	Inhibitor		Exposure of cultures in relation to time of infection	Cells showing presence of FA rabies antigen		Infectivity of the medium for mice $-\log (LD_{50})$
	Type	Dose (µg/ml)		%	Type of granules	
A	Actinomycin D	0.1	+1	60	C-D	4.5
	Control	—	+1	60	B-C	3.6
B	Mitomycin C	2.0	+1	80	C-D	4.6
	ara-C	50.0	−3	10	A	2.5
	FUdR	10.0	−3	90	C-D	—
			+1	55	C-D	—
	FPA	20.0	+1	0	—	1.6
	FPA + PA	20 + 20	+1	50	B-C	3.6
	Cycloheximide	5.0	−3	0	—	<2.0
	Control	—	—	55	B-C	3.4
C	Actinomycin D	0.1	−3	90	C-D	4.5
	Mitomycin C	5.0	−3	75	B-C	3.6
	FUdR	10.0	−3	75	B	4.4
	ara-C	50.0	−3	2	A	2.2
	FPA	25.0	−3	0	—	2.2
	Cycloheximide	5.0	−3	0	—	2.5
	Control	—	—	90	B-C	3.5

TABLE III

COMPARATIVE EFFECT OF DNA INHIBITORS ON RABIES, VESICULAR STOMATITIS VIRUS (VSV) AND HERPESVIRUS (HV)

		Effect[b] of Inhibitors on:			
	Dosage[a]	Rabies		VSV	HV
Inhibitor	(μg/ml)	FA	$-\log LD_{50}$	(PFU/ml)	(PFU/ml)
Actinomycin D	0.1	90 (C)	>4.5	7.5×10^7	$<10^3$
Mitomycin C	2.0	90 (C)	>4.5	4.0×10^7	1.5×10^4
Ara-C	50.0	10 (A)	2.5	1.4×10^7	$<10^3$
None	—	80 (B)	3.8	1.5×10^8	3×10^6

[a] Added to the cultures one hour prior to infection.

[b] Determined 24 hours after exposure of tissue culture to viral infection. VSV was titrated in BHK/21 cells and HV in secondary mouse embryo fibroblasts. (A), (B) and (C) refer to the size of fluorescing granules (see text). FA = percentage of cells showing presence of FA rabies antigen and type of granules. PFU = plaque forming units.

A MORE DETAILED INQUIRY INTO THE INHIBITORY ACTION OF ARA-C

Noninfected BHK/21 cultures were exposed to tritiated uridine and tritiated thymidine (at the concentrations shown in the legend to Fig. 1) in the presence and absence of ara-C, and the uptake of radioactive precursors was determined at hourly intervals during the next 5 hours. The uptake of tritiated thymidine and *eo ipso* synthesis of cellular DNA was markedly depressed by ara-C, whereas the same concentration of this inhibitor had no effect on the uptake of tritiated uridine.

The same concentration of the inhibitor (50 μg/ml) was then added to BHK/21 cultures at various times before and after infection with rabies virus. The results indicated that the inhibitory effect is most pronounced when the cultures are exposed to ara-C for 3 hours prior to infection (Fig. 2). The later the inhibitor was added, the less pronounced was its effect found to be. When ara-C was added 12 hours after infection, and the cultures were examined 12 hours later, no difference in the percentage of FA-positive cells was observed between the control and the treated culture. However, during the subsequent 24 hours, the number of infected cells increased significantly in the control cultures, whereas the number of FA-positive cells in cultures containing ara-C remained almost constant.

The addition of ara-C to already infected cultures never completely suppressed the growth of the virus. Even when cultures were exposed to a concentration of ara-C as high as 2 mg/ml one hour after infection, 5–10% of the cells showed the presence of FA 48 hours later. In contrast, 200 μg/ml of ara-C, added to cultures 2–3 hours before infection, reduced the number of FA-positive cells to zero.

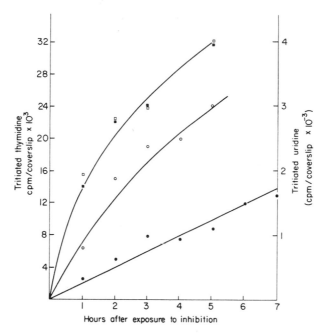

FIG. 1. Uptake of tritiated thymidine (dT-³H; 0.5 μCi/ml) and tritiated uridine* (U-³H 0.125 μCi/ml) by BHK/21 cells in the absence and in the presence of ara-C (50 μg/ml). ☐ Incorporation of U-³H in the absence of ara-C; ■ Incorporation of U-³H in the presence of ara-C; ○ Incorporation of dT-³H in the absence of ara-C; ● Incorporation of dT-³H in the presence of ara-C.

The inhibitory effect of ara-C on the replication of rabies virus was completely reversed by the addition to the cultures of deoxyribosylcytosine (dC) at a concentration of 200 μg/ml. The reversal could be shown when this compound was added at any time up to 8 hours after the addition of ara-C (Fig. 3). When added 11 hours after ara-C (12 hours after infection), dC was ineffective in reversing the inhibition, as judged by the percentage of FA-positive cells present in the culture 24 hours post-infection (Fig. 3). However, during the subsequent 24 hours, the number of FA-positive cells rose rapidly in cultures to which dC had been added as compared to cultures treated with ara-C alone. It is also interesting to note that dC not only reverses the inhibitory effect of ara-C, but if added 3 hours after infection, the ara-C-dC combination has the same enhancing effect on viral growth as do the DNA inhibitors listed in Tables II and III. The reversal of the inhibitory effect of ara-C by dC could also be demonstrated by the higher yields of infectious virus (Table IV) produced in dC-treated cultures.

* Followed by nonlabeled thymidine of 200 times higher concentration.

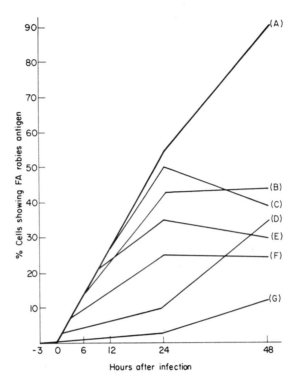

FIG. 2. Inhibitory effect of ara-C (50 μg/ml) on the multiplication of rabies virus. A = control; G = ara-C added to the culture 3 hours prior to infection; D, F, B, E and C = ara-C added to the cultures 1, 3, 6, 9 and 12 hours post-infection, respectively.

TABLE IV

INFECTIVITY OF CULTURES TREATED WITH dC, ARA-C (50 μg/ml) AND THEN WITH dC
(200 μg/ml)

Time of infection (hours)		$-\log \text{LD}_{50}$ Titer in Mice[a] (Per 0.03 ml of Medium)
ara-C	dC	
+1	None	2.4
↓	+1 → +48	4.4
	+6 → +48	>4.5
+48	+12 → +48	<3.5

[a] Determined 48 hours after infection.

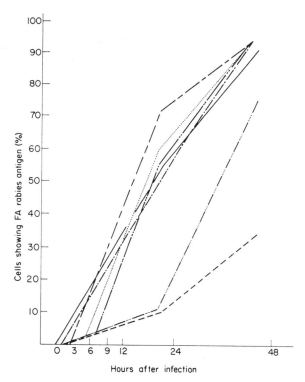

Fig. 3. Reversal of the inhibitory effect of ara-C (50 μg/ml) on the multiplication of rabies virus by deoxyribosylcytosine (200 μg/ml). Ara-C was kept in medium from 1 to 48 hours post-infection. ——————— control, no ara-C, dC; ——— — ———— dC added 3 hours post-infection; · · · · · · · · · · dC added 6 hours post-infection; —· ·—— · ·—— dC added 9 hours post-infection; ——·——·—— dC added 1 hour post-infection; ——· · · · ·——· · · · dC added 12 hours post-infection; — — — — — — ara-C, no dC.

Since it is doubtful that the inhibitory effect of ara-C is mediated through a repression of DNA synthesis, it is equally improbable that the effect of dC can be explained on the basis of a restoration of DNA synthesis. However, in order to rule out this possibility completely, another set of experiments was designed in which dC was used at concentrations which were expected to have an inhibitory effect on DNA synthesis. In addition, cytidine was investigated to see if it would reverse the action of ara-C, and the concentration of the latter was increased to 200 μg/ml in order to suppress DNA synthesis completely.

As shown in Fig. 4, the uptake of tritiated thymidine was not affected by cytidine alone at a concentration of 800 μg/ml, whereas it was completely suppressed by ara-C at a concentration of 200 μg/ml. The inhibitory effect of ara-C on thymidine uptake was not reversed by the addition of cytidine, nor was it reversed by dC when the latter was added to the cultures 3 hours after the addition of ara-C. At the high concentration used (800 μg/ml), dC alone markedly inhibited the uptake of tritiated thymidine (Fig. 4).

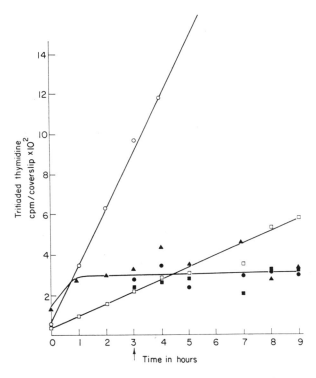

Fig. 4. Uptake of tritiated thymidine (0.3 μCi/ml added at time 0) by BHK/21 cells in the presence of ara-C (200 μg/ml), dC (800 μg/ml) or cytidine (800 μg/ml). ○ Cytidine added at 0 hours; □ dC added at 0 hours; ▲ ara-C added 0 hours; ● ara-C added at 0 hours and cytidine added 3 hours later; ■ ara-C added 0 hours and dC added 3 hours later.

The effect of the same compounds, in the same combinations, on the growth of rabies virus is shown in Table V. Unfortunately, in the first experiment, the input virus multiplicity was lower than that employed in other experiments described herein, and, as a result, only 16% of untreated BHK/21 cells showed the presence of FA 36 hours after infection. However, the infectious process was completely suppressed by ara-C, and dC and cytidine *reversed* this inhibition at concentrations which, as indicated in Fig. 4, *did not reverse* the inhibition of DNA synthesis imposed by ara-C. In the second experiment, the input virus multiplicity was adjusted so that 100% of the cells were FA-positive 20 hours after infection. Again, the suppressive effect of ara-C was reversed by dC at two dosage levels, and by cytidine at a concentration of 200 $\mu g/ml$. Thus, cytidine, at a concentration one-fourth of that used in the first experiment, and which itself had no effect on DNA synthesis, reversed the inhibitory action of ara-C on the replication of rabies virus.

TABLE V

REVERSAL OF INHIBITORY EFFECT OF ARA-C ON RABIES VIRUS BY dC AND CYTIDINE

| | | | Cells showing FA at ara-C[a] concentration in medium | | | |
| | Compound | | 0 | | 200 $\mu g/ml$ | |
Experiment	Type	Amount ($\mu g/ml$)	%	Type of granules	%	Type of granules
1	None	—	16	C-D	0	—
	dC[b]	800	27	B-C.	22	C-D
	Cytidine[b]	800	13	B-C	26	B-C
2	None	—	100	C-D	20	A
	dC[b]	200	100	B-C	100	C-D
		800	100	B-C	100	C-D
	Cytidine[b]	200	100	B-C	100	B-C

[a] ara-C added at −2 hours.
[b] dC and cytidine added at +1 hour.

Discussion

The results described in the preceding section show that the growth of rabies virus in BHK/21 cells is inhibited by FPA, cycloheximide and by ara-C. Although it it was not surprising to find that two inhibitors of the protein biosynthesis inhibit the growth of rabies virus, the results obtained with ara-C, an inhibitor of DNA synthesis, were unexpected. Rabies is apparently an RNA virus, and ara-C is not known to suppress the growth of RNA viruses, with the exception of Rous sarcoma virus (Bader, 1965). However, no parallel can be drawn between Rous sarcoma and rabies viruses, since the replication of the former is blocked by other inhibitors of DNA biosynthesis, such as actinomycin D and BUdR (Bader, 1964), whereas the growth of rabies virus is enhanced by the presence of actinomycin D, mitomycin C and FUdR in the culture medium.

Further proof that inhibition of DNA synthesis by ara-C cannot account for its suppressive effect on the growth of rabies virus, is furnished by the observation that its inhibitory effect can be reversed by cytidine, a compound which is inactive in reestablishing DNA synthesis which has been suppressed by ara-C. It is thus necessary to postulate a different mechanism* of action of ara-C.

Cytidine diphosphate esters are essential co-factors in the synthesis of phospholipids, through the formation of intermediary compounds such as cytidine diphosphate choline for the synthesis of lecithin, and cytidine diphosphate ethanolamine for the synthesis of phosphatidyl ethanolamine, etc. (Kennedy and Weiss, 1956). Synthesis of these compounds may perhaps be inhibited by ara-C, and since the ether-sensitive rabies virus probably contains phospholipids in its coat, formation of these may also be inhibited. With this hypothesis, it is necessary, however, to account for the low effectiveness of ara-C against the multiplication of VSV, another lipid-containing RNA virus. A possible basis for differentiating between the two agents may be obtained by considering briefly their intracellular sites of replication. VSV is assembled at the cell surface, through which it "buds out," ultimately destroying its host cell. In contrast, rabies virus emerges as a morphological unit from an intracytoplasmic matrix (stained by fluorescing antibody), which is formed within 9 hours after penetration of the cell by the virus (Hummeler *et al.,* 1967). The formed particles, for the most part, occur free in the cytoplasm, and only a very small fraction is found "budding out" at the cell surface.

This difference between the two viruses with respect to their sites of intracellular replication, may be the result of a difference in the way in which the two agents incorporate phospholipids into their coats. VSV may obtain preformed cell phospholipids, available from the cell membrane, whereas rabies may have to direct the formation of phospholipids, de novo, in the cytoplasm of the infected cell. If this is the case, these induced phospholipids are probably incorporated into the intracytoplasmic matrix (FA antigen) from which virus particles emerge. Formation of the induced phospholipids in rabies-infected cells would then depend, according to this hypothesis, on the availability of cytidine monophosphate groups acting as co-factors for the various cytidyl transferases. Ara-C, which hypothetically interferes with the formation of such phospholipids, would have no effect on preformed cellular phospholipids, and if the latter are used in the final maturation of VSV, then the multiplication of this agent should be little affected by ara-C. Within the framework of

* The authors are indebted to Dr. Seymour S. Cohen of the University of Pennsylvania for his kind help in suggesting other possible mechanisms of action of ara-C.

this hypothesis, the "enhancing" effect of inhibitors of DNA synthesis on rabies virus replication could be explained on the grounds that the small intracellular cytidine pool becomes available for increased synthesis of phospholipids, rather than being utilized for the synthesis of DNA and RNA.

If this hypothesis is valid, rabies would become the first virus known to require the synthesis of specific phospholipids for its assembly. Up to now, the lipid-containing viruses studied from this point of view have been found to utilize the preformed phospholipids of the cell (Wecker, 1957; Pfefferkorn and Hunter, 1963). If, in the course of current studies, evidence is obtained that rabies virus does indeed direct the formation of virus-specific phospholipids, other facets of the mechanism of rabies virus infection may become clearer. Predilection of the virus for the neurons may be due to the high phospholipid content of these cells. A slow assembly of virus particles—an assembly dependent on the availability of phospholipids—may account for the prolonged incubation period which is characteristic of this virus. Animals exhibiting symptoms of rabies infection often show no loss of structural integrity in their infected tissues. Illness in these cases, then, may be related to the formation of pathological phospholipids, or to the abnormally high utilization of cell lipids.

If all this is true, the universal susceptibility of all warm-blooded animals to rabies virus infection indicates that substrates for the formation of rabies-specific phospholipids are available in the tissues of every species of homoiothermic animals.

REFERENCES

Bader, J. P. (1964). *Virology* **22**, 462.
Bader, J. P. (1965). *Virology* **26**, 253.
Berkaloff, A., Bregliano, J. C., and Ohanessian, A. (1965). *Compt. Rend.* **260**, 5956.
Defendi, V., and Wiktor, T. J. (1966). *Intern. Rabies Symp. Talloires, 1965; Symp. Ser. Immunobiol. Standard.* **1**, 119. (Karger, Basel and New York).
Ditchfield, J., and Almeida, J. (1964). *Virology* **24**, 232.
Hamparian, V. V., Hilleman, M. R., and Ketler, A. (1963). *Proc. Soc. Exptl. Biol. Med.* **112**, 1040.
Harrison, B. D., and Crowley, N. C. (1965). *Virology* **26**, 297.
Hitchborn, J. H., Hills, G. J., and Hull, R. (1966). *Virology* **28**, 768.
Howatson, A. F., and Whitmore, G. F. (1962). *Virology* **16**, 466.
Hummeler, K., Koprowski, H., and Wiktor, T. J. (1967). *J. Virol.* **1**, 152.
Kaplan, M. M., Wiktor, T. J., Maes, R. F., Campbell, J. B., and Koprowski, H. (1967). *J. Virol.* **1**, 145.
Kennedy, E. P., and Weiss, S. B. (1956). *J. Biol. Chem.* **222**, 193.
Kissling, R. E., and Reese, D. R. (1963). *J. Immunol.* **91**, 362.
Macpherson, I., and Stoker, M. (1962). *Virology* **16**, 147.

Murphy, F. A., Coleman, P. H., and Whitfield, S. G. (1966). *Virology* **30,** 314.
Neurath, R., Wiktor, T. J., and Koprowski, H. (1966). *J. Bacteriol.* **92,** 102.
Pfefferkorn, E. R., and Hunter, H. S. (1963). *Virology* **20,** 446.
Wecker, E. (1957). *Z. Naturforsch.* **12b,** 208.
Wiktor, T. J., and Koprowski, H. (1966). *World Health Organ., Monograph Ser.* **23,** 173. 2nd Ed.
Wiktor, T. J., Fernandes, M. V., and Koprowski, H. (1964). *J. Immunol.* **93,** 353.
Yoshino, K., Taniguchi, S., and Arai, K. (1966). *Arch. Ges. Virusforsch.* **18,** 370.
Zwillenberg, L. O., Jenson, M. H., and Zwillenberg, H. H. L. (1965). *Arch. Ges. Virusforsch.* **17,** 1.

Properties of RNA from Vesicular Stomatitis Virus

Joseph Huppert(a), Marta Rosenbergova(a),
Luce Gresland(b), and Louise Harel(b)

(a) INSTITUTE OF VIROLOGY, CZECHOSLOVAK ACADEMY OF SCIENCES, MLYNSKA DOLINA, BRATISLAVA, CZECHOSLOVAKIA; AND (b) RESEARCH INSTITUTE FOR THE NORMAL AND CANCEROUS CELL, VILLEJUIF, FRANCE

RNA, extracted by phenol from ^{32}P-labeled vesicular stomatitis virus (VSV, Indiana strain) shows 3 components on glycerol density gradient centrifugation (Fig. 1). They are a fast-moving fraction with an average sedimentation constant of 45 S, a sharp peak in the 18 S region, and a fraction having a sedimentation velocity of less than 7 S.

Since base analyses of these fractions (Table I) indicate that none of them has a base composition corresponding to that of cellular RNA, we have concluded that they must all represent viral material. These data also show that the three fractions differ significantly from one another with respect to base composition. Various hypotheses may be formulated to explain the existence of these three fractions of RNA, but the one best supported by our results is that the RNA in the VSV virions exists as a single molecule, which is broken, at specific weak points, during the extraction procedure.

In the absence of precise chemical determinations of the RNA content of VSV, the following rough estimate of its molecular weight can be made. Electron micrographs (Howatson and Whitmore, 1962; Hackett, 1966) have shown that VSV particles have an inner core in the form of a helix of 35 turns, and a diameter of about 50 mμ. If this core represents the viral nucleoprotein, the length of the nucleic acid therein would be 50 m$\mu \times \pi \times 35 = 5.5$ μ, which corresponds to a molecular weight (for single-stranded RNA) of about 6×10^6 [the RNA of avian myeloblastosis virus has a length of 8.7 μ and a molecular weight of 10^7 (Granboulan

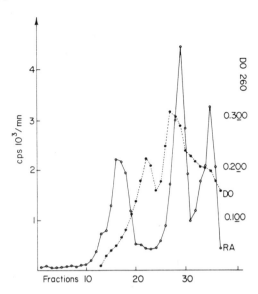

FIG. 1. Sedimentation analysis of RNA extracted from ^{32}P-labeled VSV by the phenol-SDS method. Centrifugation was at 39,000 rpm for 210 min (Spinco SW 39 rotor) in a preformed linear gradient of 5–30% glycerol in 0.02 M phosphate buffer, pH 7.3, containing 0.001 M EDTA. Solid line: radioactivity; broken line: OD_{260} of marker RNA (mouse ascites tumor cell).

et al., 1966)]. The molecular weight of the 45 S component, as calculated by Spirin's formula, is 4.6×10^6.

In order to test the premise that the three components seen in VSV-RNA preparations are due to breakage of a single RNA molecule, the following experiments were done.

TABLE I

Base Composition of RNA from VSV

$C + A + G + U = 100$

Fraction	C	A	G	U	$\dfrac{A + U}{C + G}$
45 S[a]	21.8 ± 0.6	27.2 ± 0.9	20.3 ± 1.0	30.7 ± 0.9	1.37
18 SA[b]	21.2 ± 0.5	29.6 ± 0.3	20.3 ± 0.15	28.9 ± 0.25	1.40
18 SB[b]	18.5 ± 0.5	27.3 ± 1.0	19.6 ± 1.0	34.6 ± 1.0	1.62
7 S[b]	25.4 ± 0.8	29.2 ± 1.1	24.3 ± 1.2	21.1 ± 1.0	1.01
Unfraction-ated RNA	22.5 ± 0.2	28.7 ± 0.6	20.0 ± 0.0	28.5 ± 1.0	1.34

[a] Means of 5 determinations.

[b] Means of 4 determinations.

A ^{32}P-labeled VSV preparation was divided into three aliquots, and RNA was extracted from them by the following procedures:

a. The virus, suspended in 0.1 *M* NaCl–0.01 *M* sodium citrate, was shaken 3 times with phenol in the presence of bentonite and 1% sodium dodecyl sulfate (SDS). Viral RNA (plus some added carrier RNA) was precipated from the aqueous phase with ethanol. The pellet was dissolved in standard buffer (0.1 *M* NaCl–0.01 *M* Tris–0.001 *M* EDTA), and was dialyzed against the same buffer until free of all dialyzable ^{32}P-containing material.

b. The virus was suspended in 30% sucrose containing bentonite, and SDS was added stepwise from an initial concentration of 0.001% to a final concentration of 0.05%. The RNA was then extracted with phenol. It was hoped that this procedure would avoid an explosive opening of the virion.

c. The virus, suspended in 30% sucrose, was introduced into a $ZnCl_2$-treated dialysis bag (Massie and Zimm, 1965). Pronase and bentonite (1 mg/ml) were added, and the preparation was dialyzed for 1 hour at 37°C against 20% sucrose containing 0.05% SDS, then for 1 hour against standard buffer containing 0.05% SDS, and finally for 16 hours in the cold against standard buffer. The preparation was then extracted 8 times with ether, after which the RNA was precipitated with ethanol, redissolved in standard buffer, and examined without any further deproteinization.

Figure 2 shows the sedimentation patterns of these 3 RNA preparations, run simultaneously in preformed 5–20% linear saccharose gradients. It can be seen that each sample contained the three RNA fractions described above, but that the amount of radioactivity in each peak varied from sample to sample. The RNA prepared by the normal phenol method [(a) above] gave a pattern similar to the one shown in Fig. 1. The RNA extracted by phenol in 30% sucrose had only a small proportion of its total radioactivity in the slow (7 S) component, but had a sizeable 18 S peak [in another experiment, the third (7 S) peak was completely absent]. The preparation made without phenol had roughly equal amounts of ^{32}P in the 45 S and 18 S peaks, and contained very little of the slowest component.

In the case of the preparation isolated by method (c), a good deal of the radioactivity appeared in a pellet at the bottom of the tube. When the pellet was resuspended, precipitated with ethanol, redissolved, and rerun on the gradient, 80% of the radioactivity again sedimented into a pellet, and 20% appeared in a peak corresponding to a sedimentation velocity of 45–50 S. When the sediment was extracted with phenol, the RNA so obtained consisted of the usual 3 components.

A control experiment was carried out in which VSV, labeled with ^{14}C-amino acids, was treated with pronase. After the enzyme treatment, the preparation still had a high protein content, but, after density gradient centrifugation, all the radioactivity was found in the pellet.

RNA was recovered from the fractions corresponding to the peaks in Fig. 2 by ethanol precipitation in the presence of carrier RNA, the pellets were dissolved in 0.15 *M* NaCl–0.015 *M* sodium citrate, and then $MgCl_2$ was added to a final concentration of 0.2 *M*. The precipitates were collected by centrifugation, and the radioactivity in both supernatants and

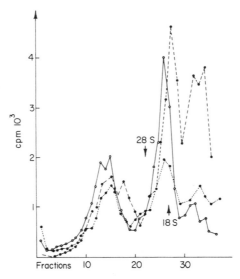

Fig. 2. Sedimentation analysis of RNA extracted from ^{32}P-labeled VSV by a variety of methods. Centrifugation was at 39,000 rpm for 210 min (Spinco SW 39 rotor) in a preformed linear gradient of 5–20% saccharose in 0.01 *M* Tris buffer, pH 7.3, containing 0.001 *M* EDTA. Broken line: RNA isolated by method a (see text); solid line: RNA isolated by method b (see text); dotted line: RNA isolated by method c (see text).

pellets was counted. Finally, both supernatants and redissolved pellets were incubated with 5 µg of RNase for 30 min, after which perchloric acid was added, and acid-soluble and acid-precipitable radioactivity was counted.

The results of this experiment are summarized in Table II, where one can see that the 18 S and 7 S fractions contain a relatively high proportion of $MgCl_2$-soluble, RNase-resistant material. This would indicate that at least part of the RNA has a double-stranded structure.

In the light of these results, we would propose the following model for

VSV-RNA—namely that the RNA molecule contained in the virion is a single, long molecule, of molecular weight about $5-6 \times 10^6$, primarily single-stranded, but containing double-stranded sequences at certain points. We would suggest further, that the junction between single- and double-stranded regions constitutes the points of preferential breakage.

The complete molecule, we believe, would correspond, in the density gradient centrifugations, to the shoulder at tube 13 in Fig. 1, and to the first part of the heavy peak—also at tube 13—in Fig. 2 (the bimodal character of the 45 S peak was found in 5 of 7 experiments). The 18 S fraction likely is made up of fragments of approximately the same

TABLE II

RNA from VSV

Measured portion (%)	Fractions		
	45 S	18 S	7 S
Radioactivity	28.2	37.7	24.0
MgCl₂ soluble	4.3	9.1	75.3
Portion RNase resistant	17.5	41.0	30.4
MgCl₂ Precipitable	95.7	90.9	24.7
Portion RNase resistant	7.5	18.8	26.7
Total RNase resistant	7.6	21.3	33.7

size, but of different composition. The base analyses shown in Table I indicate that the composition of the 45 S and 7 S fractions are constant from experiment to experiment, but that the 18 S component is of variable composition. We believe that this reflects the presence of two types of molecules in this fraction, and we have designated them A and B in Table I.

Preliminary electron microscopy studies, carried out in collaboration with Dr. A. Niveleau (Laboratory of Electron Microscopy, Institute for Cancer Research, Villejuif, France), have provided morphological evidence which tends to support our proposed model. When VSV particles are disrupted with concentrated urea, the nucleic acid is extruded from the virion, and can be seen as filaments of various lengths. The longest of these was found to be about 6 μ, although the population is quite heterogeneous. The electron micrographs also suggest that there are regions sensitive to breakage, which appear as minute displacements of portions of the otherwise linear molecule.

REFERENCES

Granboulan, N., Huppert, J., and Lacour, F. (1966). *J. Mol. Biol.* **16,** 571.
Hackett, A. J. (1966). *Virology* **24,** 51.
Howatson, A. F., and Whitmore, G. F. (1962). *Virology* **16,** 466.
Massie, H. R., and Zimm, B. H. (1965). *Proc. Natl. Acad. Sci. U.S.* **54,** 1636.

Discussion—Part V

Dr. G. A. Gentry: (Question directed to Dr. L. Levintow.) Have you determined the effect of any cross-linking agents on the infectivity of the double-stranded RNA? This might have some bearing on the fate of the infecting molecule.

Answer: Unfortunately, this approach has been inconclusive because the only effective cross-linking agent in our hands (nitrogen mustard) also abolishes the infectivity of double-stranded RNA by mechanisms other than the production of cross links.

Dr. W. Dove: (Question directed to Dr. J. Darnell.) You have presented two lines of argument to demonstrate that there is multiplication of templates in a poliovirus-infected cell; there is something I do not understand about each.

1. At a multiplicity of infection of 30, the eclipse phase is shorter than at a multiplicity of infection of 0.2; but, the burst size is independent of multiplicity of infection. This would provide evidence for multiplication of templates only, if you know that the genetic participation number is greater than 1 in these infected cells.

2. You show that the net synthesis of viral RNA in polio-virus-infected cells is an exponential function of time. Even if one grants the assumption that the synthesis time per genome is constant during growth of the virus, this result can reflect only an exponential distribution of starting times for linear synthesis in the population of infective centers.

Answer: Your second point is valid; thus, we concentrate on showing only that templates must multiply in singly infected cells. This conclusion is based on the fact that (1) many genomes have been shown capable of independent function in single cells and that (2) the final rate of viral RNA synthesis attained is independent of multiplicity. Therefore, we concluded that template multiplication must have occurred.

Since we also find an exponential in areas in the type of RNA, believed to be the template molecule at the proper time in the infectious cycle, then we also believe the exponential accumulation curves are not due to parental virus which initiate infection at various times after adsorption.

Dr. H. S. Ginsberg: (Question directed to Dr. A. J. Shatkin.) What was the size of the RNA that Gomatos isolated from reovirus? Did your reovirus RNA of molecular weight 2×10^6 possess any priming activity?

Answer: (1) The RNA isolated from reovirus by Gomatos was heterogeneous and had an estimated molecular weight of 1–2×10^6 daltons. (2) No. I might add that the nearest neighbor determinations of RNA polymerase product, reported by Gomatos and collaborators, were essentially the same as those reported by Kornberg and collaborators for mouse DNA. Furthermore, the ratio $A + U/G + C$ derived from the data of Gomatos *et al.* is 1.43 as compared to the chemically determined ratios of 1.27 for reovirus RNA and 1.43 for mouse DNA.

PART VI

MAMMALIAN DNA VIRUSES

Early and Late Functions during the Vaccinia Virus Multiplication Cycle*

Wolfgang K. Joklik, C. Jungwirth,† K. Oda,**
and B. Woodson††

DEPARTMENT OF CELL BIOLOGY, ALBERT EINSTEIN COLLEGE OF MEDICINE,
BRONX, NEW YORK

One of the most rapidly moving fields in biology concerns itself with the control and regulation of the expression of genetic information. Virus-infected cells form favorable systems for study; since it is possible to observe the functioning of relatively small pieces of genetic material in them.

The genome of vaccinia virus, which is in all probability one molecule of DNA, has a molecular weight of about 160 million. This represents a tremendous amount of genetic information; it should be able to code for about 500 protein molecules of average size. It seems quite clear that not all these proteins are made throughout the viral replication cycle. For instance, DNA polymerase, which appears in the cytoplasm of infected cells and is almost certainly coded for by the viral genome, is made between 1 and 4 hours after infection (Jungwirth and Joklik, 1965), while over half of the structural viral proteins, which are destined to be incorporated into mature viral progeny, are *not* synthesized before 4 hours, but only between 4 and about 12 hours after infection (Joklik and Becker, 1964). This sort of situation is found for many virus infections, and the proteins coded from viral genomes have been subdivided accordingly into two categories, early and late proteins. There is presumably

* This work was supported by Grants number AI 04913, AI 04153 and GM 876 from the National Institutes of Health, U.S.P.H.S. W. K. J. is a Research Career Awardee of the U.S.P.H.S. (1-K6-AI-22,554).

† Present address: Department of Virology, University of Würzburg, Germany.

** Present address: the Salk Institute of Biological Studies, San Diego, California.

†† Present address: Department of Microbiology, San Francisco Medical Center, University of California, San Francisco, California.

some mechanism in the infected cell which controls which units of genetic information in the viral genome are expressed at any particular time. It would appear that this could take place by either of two mechanisms. On the one hand, utilization of genetic information could be regulated entirely by the availability of the corresponding messenger RNA's— that is, control could operate exclusively at the transcription level. On the other hand, all genetic information might be transcribed at all times, but regulation might operate exclusively at the level of translation. This is difficult to envisage at the moment, although there is evidence from the phage field that a particular messenger RNA, transcribed during the early part of the infection cycle, only begins to express itself much later, whereas other messenger RNA's, also transcribed early, function immediately (Bautz et al., 1966). As in so many other situations, the truth is probably that regulation operates by a *combination* of transcription and translation control.

Here, I will try to review the evidence concerning the regulation of transcription and translation of vaccinia messenger RNA's that we have gathered from the study of three different phenomena: the control of the synthesis of "early" enzymes, the inhibition of vaccinia virus multiplication by isatin-β-thiosemicarbazone (IBT*), and the inhibition of vaccinia virus multiplication by interferon.

The Synthesis of Virus-Induced DNA Polymerase

Under normal conditions, synthesis of DNA polymerase starts at about 1 to $1\frac{1}{2}$ hours after infection and continues until about 4 hours after infection (Jungwirth and Joklik, 1965) (Fig. 1). The enzyme is stable within the cells as can be shown by adding puromycin at any time during the period when the amount of enzyme is increasing. Under these conditions, when protein synthesis is completely arrested, the level of the enzyme activity immediately stabilizes and then remains constant for at least 4 to 6 hours. When synthesis of progeny DNA is inhibited by dFU, synthesis of DNA polymerase is initiated in the normal manner, proceeds at the same rate as in cells infected under normal conditions, but instead of being switched off at 4 hours after infection, continues in an uncontrolled manner. Two points emerge from this experiment: (1) The enzyme which is being made in the presence of dFU is clearly translated from messenger RNA transcribed from the parental viral genome, since no progeny genomes are synthesized in these cells. (2) Whatever causes

* The following abbreviations are used: IBT, isatin-β-thiosemicarbazone; dFU, 5-fluorodeoxyuridine; PFU, plaque forming unit.

the switch-off under normal conditions is not present in cells infected in the presence of dFU, and therefore the switch-off must be due to information stored in progeny genomes, since such information is evidently not available to be read from the parental genomes. It has, in fact, been shown that switch-off does require the synthesis of both messenger RNA and protein, and that it is in all probability due to protein (McAuslan, 1963a,b). We may conclude that DNA polymerase is an early protein and that the switch-off protein is a late protein.

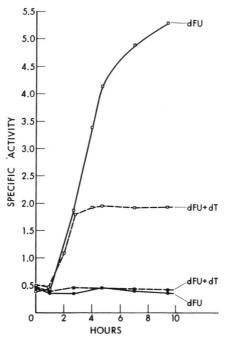

FIG. 1. Effect of inhibiting viral DNA replication on the time course of increase of DNA polymerase activity. Concentration of dFU, 10^{-5} M; of dT, 10^{-4} M; both added at the time of infection. Enzyme activity is expressed as mμmoles ^{3}H-dTTP rendered acid-insoluble/mg protein/30 min/37°C. Closed symbols: uninfected cells. Open symbols: infected cells (reproduced from Jungwirth and Joklik, 1965).

There are other early proteins. In cells infected under conditions precluding synthesis of progeny DNA, or translation of viral messenger RNA transcribed from progeny DNA molecules, from 4 to 6 proteins which will react with vaccinia antiserum can be demonstrated by immunological techniques (principally by the Ochterlony gel diffusion technique), whereas in normally infected cells about twenty such proteins are synthesized (Appleyard and Westwood, 1964). I refer, in particular, to cells infected in the presence of sodium azide (Appleyard *et al.,* 1962) in which no replication of viral DNA proceeds (dFU has unfortunately not yet been tested), or to cells infected in the presence of IBT (Appleyard *et al.,* 1965), in which replication of vaccinia DNA proceeds normally, but in which messenger RNA transcribed from progeny DNA molecules appears to be unable to express itself (Woodson and Joklik, 1965). It seems then that the molecular basis for the division into early and late proteins, and thus into early and late functions, is that the early proteins are those translated from messenger RNA molecules transcribed from parental DNA strands, while late proteins are those which are translated from messenger RNA molecules transcribed from progeny genomes. Alternative hypotheses postulating that messenger RNA molecules coding both for DNA polymerase and for the switch-off protein are transcribed both from parental and progeny genomes are difficult to devise without invoking the existence of further regulatory messenger RNA's and proteins which are themselves regulated at the transcription level. When ways around such restrictions are devised, the required systems are so complex as to be, at this time, only of theoretical significance, and not amenable to direct experimentation. Although they may, in fact, correspond to the true state of affairs, until it is experimentally disproved, it appears more profitable at this stage to accept the above hypothesis, which postulates control at the transcription level.

The switch-off phenomenon also provides evidence for control at the translation level. It can be shown by the use of actinomycin D that the effective life of the messenger RNA molecule coding for DNA polymerase is very long (Fig. 2) (Jungwirth and Joklik, 1965). It is likely, therefore, that switch-off is due to a newly formed factor which abolishes the ability of these messenger RNA molecules to function, and I have mentioned above that there is evidence that this "something" is a protein. It might be argued that the fact that messenger RNA, in the presence of dFU or actinomycin D, continues to code for DNA polymerase, whereas in the absence of either of these inhibitors a switch-off occurs between 3 and 4 hours after infection, is not conclusive proof that protein coded for by messenger RNA transcribed from progeny genomes directly abolishes the ability of the DNA polymerase messenger RNA to function. It could be argued that messenger RNA molecules, which would not appear in cells infected in the presence of either of the above inhibitors, may be synthesized from progeny DNA. This messenger RNA might have a higher affinity for ribosomes than early messenger RNA—and the reason for the switch-off phenomenon may be a simple competition for ribosomes in which early messenger RNA loses. The results obtained with IBT make this explanation unlikely (see below).

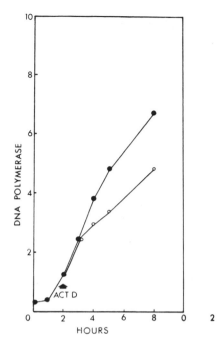

FIG. 2. Effect of inhibiting messenger RNA formation with actinomycin D on the continued synthesis of DNA polymerase. Enzyme activity is expressed as in Fig. 1. Closed circles, no actinomycin; open circles, actinomycin D added at 2 hours after infection (reproduced from Jungwirth and Joklik, 1965).

The Inhibition of Vaccinia Virus Multiplication by IBT

IBT (Fig. 3) is a relatively small molecule which effectively inhibits the multiplication of vaccinia virus. Figure 4 shows the dose response curve obtained with this compound. At a concentration of 15 μmolar this compound decreases the yield of vaccinia virus to 1–3% of normal.

IBT does not prevent the synthesis of the early enzyme DNA polymerase, and in its presence 4 or 5 structural viral proteins are synthe-

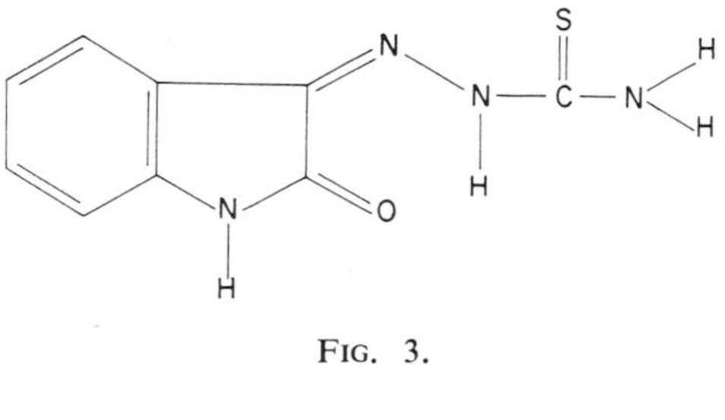

FIG. 3.

sized, as revealed by the Ochterlony gel diffusion technique (Appleyard *et al.*, 1965). These are "early" structural viral proteins. IBT does not prevent the normal replication of vaccinia DNA, nor the transcription of vaccinia messenger RNA from either parental or progeny genomes (Woodson and Joklik, 1965). In fact, during the first 2 or 3 hours IBT has no effect on the course of viral infection. However after this time a drastic change occurs. Figure 5 shows the rate of protein synthesis in normally infected cells and in cells infected in the presence of 15 μmolar IBT. IBT drastically inhibits protein synthesis after 3 to 4 hours post

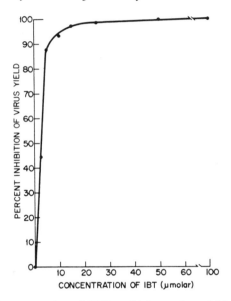

FIG. 4. Effect of concentration of IBT on 24 hour virus yields. Cells were infected at a multiplicity of 4 PFU per cell (reproduced from Woodson and Joklik, 1965).

infection. This inhibition of protein synthesis is due to the breakdown of polyribosomes. Figure 6 shows the polyribosome patterns in normally infected cells at 2 hours after infection; at 2 hours after infection in the presence of IBT, when the pattern is essentially identical; at 4 hours after normal infection, when polyribosomes are still plentiful; and at 4 hours after infection in the presence of IBT, when cells contain practically no polyribosomes. Thus, in cells infected in the presence of 15 μmolar IBT, polyribosomes commence to breakdown after about 3 hours and this breakdown is clearly responsible for the inhibition of protein synthesis. This breakdown of polyribosomes can be followed conveniently by the use of higher concentrations of IBT (Fig. 7). If one exposes cells infected under normal conditions to various concentrations of IBT at 4 hours after

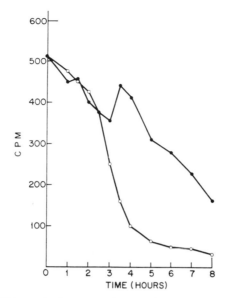

FIG. 5. Effect of IBT on the incorporation of ¹⁴C-amino acids into cells infected in the absence (closed symbols) and presence (open symbols) of 15 μM IBT. Cells were pulsed for 5 min at each time point and incorporation into total cellular protein was determined (reproduced from Woodson and Joklik, 1965).

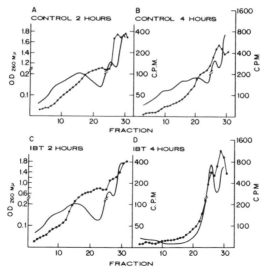

FIG. 6. Optical density and radioactivity profiles in 15–30% sucrose density gradients (centrifuged 2 hours at 25,000 rpm) of cytoplasmic fractions of HeLa cells infected for 2 and 4 hours in the absence (A and B) and presence (C and D) of 15 μM IBT. For each gradient the cytoplasmic fraction of 4×10^7 cells pulse-labeled with uridine-¹⁴C for 10 min was used (reproduced from Woodson and Joklik, 1965).

infection, polyribosomes dissociate completely within 1 hour when the concentration of IBT exceeds 50 μmolar. If one pulse-labels for 3 minutes with radioactively labeled amino acids during this period, and compares the optical density and radioactivity distribution profiles after sucrose density gradient centrifugation, one finds that the decrease in optical density exactly parallels the decrease in the amount of incorporated radioactivity, demonstrating that the physical and functional ribosomes disappearing from the polyribosome region are identical. If one carries out this experiment with cells infected for only 1 hour, IBT has no effect. Thus the polyribosomes during the first 3 hours of infection may be said to be IBT-resistant, and the polyribosomes after the first 3 hours of infection may be said to be IBT-sensitive.

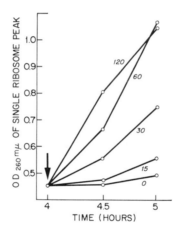

FIG. 7. Effect of IBT concentration on the release of single 74 S ribosomes from polyribosomes. At 4 hours after infection a cell culture was divided into five aliquots. One sample served as control, and IBT was added to the other four to give final concentrations of 15, 30, 60, and 120 μM. Samples were removed for analysis 30 and 60 min later. All samples contained 4×10^7 cells. The cytoplasmic fractions were placed onto 15–30% sucrose density gradients, centrifuged for 2 hours at 25,000 rpm, and the optical density of the 74 S single ribosome peak was measured (reproduced from Woodson and Joklik, 1965).

We may inquire as to the basis of this sensitivity to IBT after the first 3 hours post-infection. Let us examine the fate of nascent vaccinia messenger RNA molecules transcribed at 2 hours after infection, when the polyribosomes are resistant to IBT, and at 4 hours after infection, when they are sensitive. At both times, messenger RNA of normal size is transcribed (this is determined by examining the size of the vaccinia messenger RNA synthesized in the presence and absence of IBT after very short (3–4 min) pulses of uridine-^{14}C). Furthermore, at both 2 and

4 hours after infection, both in the presence and absence of IBT, these messenger RNA molecules combine with ribosomes to form polyribosomes. After a very short (4 min) pulse, the same proportion of newly formed vaccinia messenger RNA is present in the polyribosome fraction in the presence or absence of IBT, and all of this messenger RNA is the same size as when it was transcribed (about 16 S). When the duration of the pulse is increased to ten min, so that one examines not only messenger RNA that is nascent, but also messenger RNA molecules that are 8–10 min old, then the situation is the same in samples taken 2 or 4 hours after infection in the absence of IBT, or in the presence of IBT at 2 hours after infection—that is the size of the messenger RNA is still about 16 S and a large proportion of it is found associated with polyribosomes. However, the situation is different at 4 hours after infection if IBT is present. The median size of the messenger RNA molecules after a pulse of ten min is much smaller than that after a pulse of four minutes (8 S rather than 16 S) (Fig. 8), and most of it is then found not in polyribosomes, but in the free state. Since the results are quite different for four and ten min pulses, changes must have occurred during the extra six-min period. These changes are a halving of the sedimentation coefficient, which corresponds roughly to a two-thirds reduction in the molecular weight, and an efflux of messenger RNA from the polyribo-

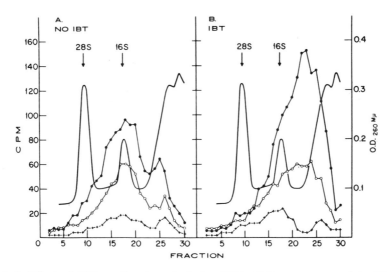

FIG. 8. Effect of IBT on the size of viral mRNA. Cells were exposed to 30 μM IBT at 2 hours after infection, and, 2 hours later, were pulse-labeled with uridine-^{3}H for 4 ($\times$), 10 ($\bigcirc$), and 15 ($\bullet$) min. The cytoplasmic fractions derived from 10^7 cells were treated with SDS and centrifuged for 16 hours at 24,000 rpm in 15–30% sucrose-SDS density gradients (reproduced from Woodson and Joklik, 1965).

some fraction. It means that during the extra six-min period a large proportion of the messenger RNA molecules were broken in two places, and the messenger RNA dissociated from ribosomes. We do not know at the moment whether the messenger RNA is first reduced in size and, as a result, ribosomes dissociate from it, or whether ribosomes are caused to dissociate from the messenger RNA and the resulting free messenger RNA molecules are then reduced in size. Whatever the true answer is, the net result is a lowering of the effective half-life of the messenger RNA from a value significantly greater than thirty min (and possibly more than ninety minutes) to less than five minutes. This dissociation of messenger RNA molecules from ribosomes is the reason for the absence of polyribosomes and the lack of protein synthesis in infected IBT-treated cells after 3 hours post-infection. There is thus a profound lesion at the translational level. As a result there is a gross deficiency in the synthesis of the structural viral proteins necessary for maturation, and therefore inhibition of viral multiplication.

Let us consider now the significance of the time point of roughly 3 hours after infection—before 3 hours polyribosomes are resistant to IBT, after 3 hours they are sensitive. In the system used, 3 hours after infection is the time at which appreciable quantities of gene products transcribed from progeny DNA molecules accumulate in the infected cell. Polyribosomes are resistant to IBT before appreciable transcription from progeny genomes has taken place, and are sensitive after this time. There are two demonstrations of the fact that IBT does not exert its effect under conditions when no progeny DNA molecules arise within the cells.

1. In the presence of dFU, IBT does not switch off or inhibit the synthesis of DNA polymerase after 3 hours. In the presence of dFU, the synthesis of the early enzyme DNA polymerase continues in an uncontrolled manner no matter whether IBT is present in the cell or not (Fig. 9). The 3 hour period after which IBT begins to act is, therefore, not an expression of a clock concerned with IBT activity as such, but is rather an expression of a clock related to DNA replication.

2. Polyribosome patterns in cells infected in the presence of dFU with and without IBT are identical for periods of at least 6 hours. These polyribosomes are those formed both by residual host messenger RNA and by viral messenger RNA transcribed from parental genomes. Again, IBT is without effect under conditions when no progeny genomes arise within the cell. The conclusion from these experiments is that there is synthesized from progeny genomes, and only from progeny genomes, a protein which, provided IBT is present in the cell, disrupts all polyribosomes—not only those held together by viral messenger RNA transcribed from progeny DNA molecules, but those formed by host cell

messenger RNA molecules and viral messenger RNA molecules transcribed from parental genomes as well. This follows from the fact that in the presence of IBT at 4 hours after infection, there are fewer polyribosomes than in the presence of both dFU and IBT.

A most attractive way of connecting the switch-off phenomenon with the mechanism of action of IBT is to postulate that, in the presence of IBT, the protein, which under conditions of normal infection selectively abolishes the ability of early messenger RNA to function, is altered allosterically in such a way as to prevent *all* messenger RNA's from functioning.

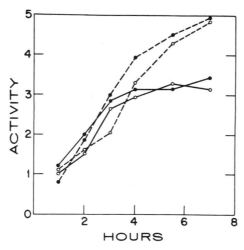

Fig. 9. Effect of IBT and dFU on the synthesis of virus-induced DNA polymerase. IBT: 15 μmolar; dFU: 10^{-5} M; both added at time of infection. Enzyme activity is expressed as in Fig. 1. Open symbols, no IBT; closed symbols, IBT added; continuous curves, no dFU; broken curves, dFU added.

The Inhibition of the Multiplication of Vaccinia Virus by Interferon

Further evidence for control at the translation level comes from experiments which were designed to establish the molecular basis for the action of interferon. There have been many attempts to define the mechanism of action of this substance, but none have yielded particularly convincing results. It seems quite clear now that interferon is a protein with absolute species specificity (Merigan *et al.,* 1965), and that it does not inhibit virus multiplication as such, but that it elicits the formation of a protein which is itself the inhibitor (Friedman and Sonnabend, 1964; Levine, 1964; Taylor, 1964). Furthermore, since interferon inhibits both RNA and DNA viruses, it is clear that unless the multiplication of these two types of viruses is affected by different mechanisms (a most unlikely situation) we must look for the locus of action at some point of the viral replication cycle which is common to both RNA and DNA viruses (Joklik, 1965). Such a point is clearly the translation of information encoded in the viral genome. In the case of viruses whose genome is RNA, viral RNA itself functions as messenger RNA; in the case of viruses whose genome is DNA, messenger RNA is transcribed and then translated. Whatever the nature of the nucleic acid, at some stage of the replication cycle virus specific message must be translated, and we may predict that it would be there that the interferon-induced inhibitor would be most likely to exert its action.

We have carried out a number of experiments to pinpoint the mechanism of action of interferon in L cells infected with vaccinia virus, and have found that, in the interferon-treated infected cell, viral messenger RNA is incapable of forming polyribosomes, so that the genetic information encoded in it is not translated (Joklik and Merigan, 1966). The interferon used in these studies was made in mouse L cells by Dr. Thomas Merigan of Stanford University. At a dilution of 1 in 100 it inhibited the growth of vaccinia virus in these cells by 95 to 98%. L cells were always incubated with interferon for 16 to 20 hours before challenge with vaccinia virus and subsequent biochemical analysis. Interferon has no detectable effect on macromolecular synthesis in uninfected cells—cells multiply perfectly well in its presence. Further, in line with what has been said above, this interferon had no effect whatsoever on the replication of vaccinia virus in human HeLa cells. In mouse L cells, however, induction of the synthesis of virus-induced DNA polymerase is inhibited by at least 95%, and as a consequence viral DNA replication is almost completely inhibited (Figs. 10 and 11). Viral messenger RNA formation, however, is not inhibited; on the contrary, it is often markedly enhanced. Figure 12 illustrates the time course of viral messenger RNA synthesis in L cells infected with vaccinia virus. At a multiplicity of 750, viral messenger RNA synthesis occurs in a very rapid burst soon after infection and tapers off thereafter. In interferon-treated cells infected with such a high multiplicity the initial burst of vaccinia messenger RNA synthesis is even more pronounced, but its rate of synthesis decreases more rapidly than in untreated cells, until it is practically zero by 4–5 hours after infection. As the multiplicity of infection is decreased, the rate of messenger RNA synthesis decreases more rapidly in untreated than in interferon-treated cells, so that at a multiplicity of 175, the rate of messenger RNA synthesis may be as much as five times higher in interferon-treated than in untreated infected cells. However, here too the rate of messenger RNA formation tapers off much

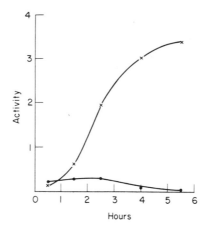

FIG. 10. Synthesis of DNA polymerase in normal and interferon-treated L cells infected with vaccinia virus. Multiplicity: 500. Activity is expressed as in Fig. 1. (×) no interferon; (●) interferon-treated (reproduced from Joklik and Merigan, 1966). Multiplicity in this and all subsequent figures is expressed in terms of the number of elementary bodies adsorbed per cell.

more rapidly with time in interferon-treated than in normal cells. We do not know why the rate of vaccinia messenger RNA synthesis should be enhanced in cells pretreated with interferon. Explanations which we are investigating are (1) that interferon treatment affects uncoating of the infecting virus, (2) that interferon treatment influences the availability of RNA polymerase; and (3) that regulation of transcription from vaccinia DNA is relaxed in some way.

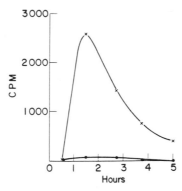

FIG. 11. Replication of vaccinia DNA in normal and interferon-treated L cells. Multiplicity: 500. At each time interval a cell sample (10^7 cells) was pulse-labeled for 10 min with 2 μCidT-^{14}C. The amount of radioactivity in the cytoplasmic fraction was then measured. (×) normal cells; (●) interferon-treated cells.

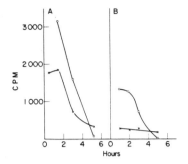

FIG. 12. Rate of formation of vaccinia messenger RNA in normal and interferon-treated cells. Each point was derived from the summation of the appropriate area of an SDS-sucrose density gradient profile obtained as described in Fig. 8. (A) Multiplicity of infection, 750; (B) Multiplicity of infection, 175. (×) no interferon; (○) interferon-treated (reproduced from Joklik and Merigan, 1966).

In spite of this absence of inhibition of viral messenger RNA synthesis, protein synthesis is markedly inhibited very soon after infection of interferon-treated cells (Fig. 13). This inhibition is due to a rapid breakdown of polyribosomes. Figure 14 shows the polyribosome profiles at 0, 1, 2, and 4 hours after infection of normal and interferon-treated L cells with vaccinia virus. Polyribosomes very rapidly disappear in infected interferon-treated cells and are never reformed, as experiments carried out for as long as 8 hours after infection have shown. These profiles were obtained using a multiplicity of about 500 vaccinia virus

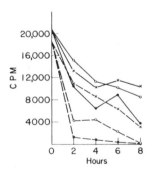

FIG. 13. Rate of protein synthesis in untreated and interferon-treated L cells infected with vaccinia virus. Each point represents the amount of radioactivity incorporated into the cytoplasmic fraction of 1.35×10^7 cells in Eagle's medium containing one-tenth of the usual concentration of amino acids, pulsed for 4 min with 15 μCi of a uniformly ^{14}C-labeled amino acid mixture (1.5 mCi/mg). Continuous curves: no interferon; broken curves: interferon-treated. (×) multiplicity of infection 50; (○) 150; (●) 500 (reproduced from Joklik and Merigan, 1966).

particles per cell (that is, about 10 PFU per cell). Similar results were obtained with a multiplicity of infection as low as 50.

It might appear from these experiments that infection of interferon-treated cells with vaccinia virus leads to a dramatic breakdown of host cell polyribosomes, which does not occur under normal conditions of infection. However, this is probably not the true state of affairs. We currently believe that rapid breakdown of host cell polyribosomes is a normal consequence of infection of L cells with vaccinia virus, but is a phenomenon which is not observable under conditions of normal infec-

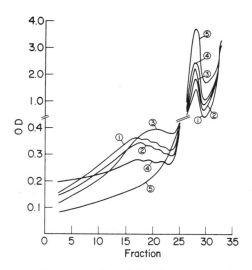

FIG. 14. Polyribosomes in untreated and interferon-treated L cells infected with vaccinia virus. Each profile was derived from a 15–30% sucrose density gradient charged with the cytoplasmic fraction of 5×10^7 L cells; 2 hours centrifugation at 25,000 rpm at 2°C. Multiplicity of infection, 500. Curve 1 = uninfected cells, normal or interferon-treated; curves 2, 3, and 4 = untreated cells infected for 1, 2, and 4 hours respectively; curve 5 = interferon-treated cells infected for 1, 2, or 4 hours (reproduced from Joklik and Merigan, 1966).

tion because most of the ribosomes liberated in this process combine with the large number of viral messenger RNA molecules which are transcribed at this time. However, in interferon-treated cells, the viral messenger RNA does not associate with ribosomes to form polyribosomes and the very rapid breakdown of host cell polyribosomes is therefore unmasked.

Direct demonstration that vaccinia messenger RNA cannot combine with ribosomes to form polyribosomes in the presence of interferon was provided by experiments in which the cytoplasmic fractions of infected cells pulse-labeled for various times were analyzed by sucrose density gradient centrifugation under conditions allowing observation of the fate of newly formed viral messenger RNA molecules (Table I). Whereas in normally infected cells, from 40 to 65% of newly formed viral messenger RNA molecules are found in the polyribosome fraction, less than 20% is found in these structures in extracts of infected, interferon-treated cells. Thus both lines of evidence, (1) normal or even enhanced viral messenger RNA synthesis under conditions of complete polyribosome breakdown and (2) the demonstration of a greatly diminished association of viral messenger RNA molecules with ribosomes, suggest that the primary locus of action of interferon (or rather of the protein, the synthesis of which is induced by interferon treatment) is to prevent the attachment of ribosomes to viral messenger RNA molecules.

TABLE I

PERCENTAGE OF VACCINIA MESSENGER RNA ASSOCIATED WITH POLYRIBOSOMES IN NORMAL AND INTERFERON-TREATED CELLS

Duration of pulse (min)	No interferon	Interferon-treated
4	41	16
9	57	17
15	66	18

The possibility that interferon induces the synthesis of a protein which accelerates the destruction of host cell polyribosomes following infection with *any* virus was ruled out by experiments using Mengovirus. Host cell polyribosomes disappeared at the *same* rate during the first 2 hours in untreated and interferon-treated cells. Very few polyribosomes were then present. After this time there occurred a marked reformation of polyribosomes in cells infected under normal conditions. These polyribosomes were heavier than host cell polyribosomes and were formed by Mengovirus RNA. This reformation was completely absent in interferon-treated cells (Joklik and Merigan, 1966).

A further consequence of infection of interferon-treated cells is the complete disintegration of cells by about 4 to 5 hours after infection (Fig. 15). Only ghosts are then visible under the microscope, and these contain only about 15% of the protein and less than 10% of the ribosomes originally present in the cell.

The net result of viral infection of an interferon-treated cell is, therefore, an abortive infection cycle which yields no mature viral progeny because the translation of viral messages is blocked. Therefore, interferon

does not protect an infected cell. The only cells protected are the uninfected members of a population which do not in turn become infected
because there is no virus yield from the initially infected cells. It may be
said then that those cells in an interferon-treated population which are
unfortunate enough to be infected act as a sponge for the invading virus
particles and effectively neutralize them, sacrificing themselves in the
process.

Let us now recapitulate what these experiments with interferon have
demonstrated. (1) In the normally infected cell, infection with vaccinia
virus leads to a rapid destruction of host cell polyribosomes, while at the
same time polyribosomes are being reformed equally rapidly on vaccinia
messenger RNA. (2) In the interferon-treated cell viral messenger RNA
cannot attach to ribosomes and expression of genetic information is therefore inhibited. Both phenomena indicate highly specific regulation of
translation.

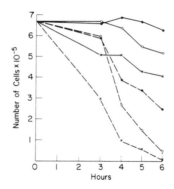

FIG. 15. Number of "intact" cells in cultures of untreated and interferon-treated
L cells infected with vaccinia virus. Continuous curves = untreated cells. Broken
curves = interferon-treated cells. ($\times$) multiplicity of infection of 500; ($\bigcirc$) 200;
($\bullet$) 100. Cell destruction was somewhat less when the initial cell concentration
was 2.5×10^5 cells/ml (reproduced from Joklik and Merigan, 1966).

A further point arising both from the work on IBT and on interferon
is that there must exist fundamental differences between the various types
of messenger RNA. Host cell messenger RNA must be different from
early viral messenger RNA because in vaccinia infected L cells breakdown of polyribosomes held together by host cell messenger RNA proceeds concurrently with reformation of polyribosomes formed by early
viral messenger RNA. Early viral messenger RNA must be different from
late viral messenger RNA since the functioning of early viral messenger
RNA, (that coding for DNA polymerase, for example) is switched off
under conditions when late viral messenger RNA is fully functional. It is

likely that these three types of messenger RNA differ in some basic way, most probably in chemical structure, possibly at the beginning of the polynucleotide chain. It is conceivable for instance, as Noll (1966) has suggested, that the initiating codons differ among these various classes of messenger RNA.

In conclusion, we may say that the orderly synthesis in time of the large number of proteins coded for by the vaccinia viral genome is controlled both at the transcription and the translation levels. There seem to be certain sections of the viral genome which are only transcribed at appropriate times. Whether inability to be transcribed is caused by ionic or steric or other effects is not known. Furthermore, there is also control at the translation level—that is, the ability of messenger RNAs to function is also controlled.

Intrinsic Stability of Messenger RNA

Finally, there is built into the information transfer system one further control, which we have not yet discussed at all—the intrinsic stability of the messenger RNA. There is some evidence that messenger RNA populations in most cells are composed of molecules with a wide spectrum of stability. In bacteria, for instance, it appears that messenger RNA's, the transcription of which is under the control of an operator, are more labile than those which are not (Moses and Calvin, 1965). Nothing is known concerning the molecular basis of the phenomenon that some messenger RNAs are short-lived while others are long-lived. Nor is anything known concerning the mechanism by which messenger RNA's are degraded. Neither ribonuclease, nor nucleotide phosphorylase nor phosphodiesterase appear to be involved (Gesteland, 1966; Sarkar and Dürwald, 1966). Few studies have been carried out on the half-life of messenger RNA in animal cells. The estimates range all the way from 50 min for rat liver messenger RNA, as determined by the rate of reformation of polyribosomes after reversal of the inhibition of RNA synthesis caused by ethionine (Villa-Trevino et al., 1964), to about 3 hours for HeLa cell messenger (Penman et al., 1963), and to over 40 hours for rat liver messenger RNA, as determined by the rate of decrease in the ability to incorporate ^{14}C-labeled amino acids into protein after the administration of actinomycin D (Revel and Hiatt, 1964).

The situation regarding the half-life of vaccinia messenger RNA is equally complex. Three methods have been used for estimating its half-life, all depending on arresting further messenger RNA synthesis with actinomycin D.

1. The ability of messenger RNA to function in the translation of its

specific protein has been measured for certain early messenger RNA's, in particular those coding for DNA polymerase and thymidine kinase (should this prove to be a protein coded by the viral genome; see Joklik, 1966). The enzymes continue to be synthesized at virtually undiminished rates for at least 5 hours after the complete arrest of further messenger RNA synthesis. The half-life of the corresponding messenger RNA's is thus very long. On the other hand, messenger RNA molecules coding for structural viral proteins later on during the infection cycle (at about 5 hours after infection) are said to function for no more than 1 hour (Shatkin, 1963). There appear then to be large differences between the functional stability of various vaccinia messenger RNA's.

2. The functional half-life of messenger RNA has also been estimated by measuring the release of ribosomal monomers from polyribosome form. At 4 hours after infection the kinetics of release of ribosomes is biphasic—about 50% of the total number of ribosomes in polyribosome form are released within 30 min, but further release is slow (Woodson and Joklik, unpublished data). This suggests that, at that time, both stable and unstable messenger RNA molecules are present in the cell.

3. A similar conclusion is reached by measuring the stability of labeled vaccinia messenger RNA molecules. This type of analysis is uniquely possible in cells infected with vaccinia virus, since from about 2 hours after infection, no RNA labeled in the nucleus, either messenger or ribosomal, is transported to the cytoplasm. It is thus possible to label viral messenger RNA, to block further synthesis with actinomycin D, and then to measure as a function of time the amount of label remaining in messenger RNA molecules. If this experiment is carried out with HeLa cells infected for 4 hours, there is a rapid initial loss of label during the first 30 min, which, however, does not affect more than about 60% of the viral messenger RNA molecules. During the next three hours the amount of label in messenger RNA decreases only very slowly (Oda and Joklik, 1967). Thus, there is evidence for a relatively large unstable messenger RNA fraction, although some vaccinia messenger RNA's seem to be fairly long-lived.

In conclusion, the intrinsic half-life of a messenger RNA molecule is another factor which controls the information transfer from viral genome to polypeptide chain. Some vaccinia messenger RNA's appear to be very stable, others labile.

Hybridization of Messenger RNA with Viral DNA

Hybridization provides a direct technique for examining different viral messenger RNA's. The approach, in this case, is to isolate messenger

RNA transcribed at various times after infection, and to determine whether hybridization with viral DNA is inhibited by either "early" or "late" messenger RNA. Of course, this only separates the viral messenger RNA's into two classes (early ones and late ones), and the actual position is undoubtedly far more complicated. But if we at least achieve a separation of messenger RNA's into those coded for by parental genomes and those coded for by progeny genomes, we will have taken an initial step towards elucidating the control mechanisms which operate during viral infection.

We have prepared early vaccinia messenger RNA from polyribosomes isolated at a time before viral DNA replication has commenced, and from the cytoplasmic fraction of HeLa cells infected in the presence of cytosine arabinoside (Oda and Joklik, 1967). These two types of viral messenger RNA compete completely with each other. Vaccinia messenger RNA present in polyribosomes at five hours after infection, which is two hours after the peak of viral DNA replication (Joklik and Becker, 1964), also competes completely with early messenger RNA. All the molecular species of messenger RNA transcribed from parental genomes are therefore present in polyribosomes at a late stage of the infection cycle. Messenger RNA transcribed at late times of infection is competed with *partially* by early messenger RNA, which implies that early messenger RNA molecules are still being transcribed at late stages of infection, either from parental genomes still present, or, more likely, from progeny DNA molecules. Exact quantitation is impossible for a number of reasons. However, one can say that between 60 and 80% of the genetic information present in the vaccinia genome is transcribed from parental genomes and that in all likelihood all information is transcribed from progeny genomes. On this basis, the fate of the viral messenger RNA molecules coding for proteins, the synthesis of which is switched off at three to four hours after infection (see above), is a puzzle, since, as far as one can tell by the hybridization-competition technique, all species of early messenger RNA molecules are present in polyribosomes at five hours after infection.

Examination of Viral Structural Proteins

Finally, one can examine much more closely than has been done heretofore the viral gene products synthesized in the vaccinia virus infected cell. As pointed out above, the vaccinia virus genome is so large that it could code for about 500 proteins. Of these we know only very few—a handful of early enzymes and about 20 structural viral proteins. These structural viral proteins have been studied to a certain extent because of their ability to combine with antibodies to vaccinia virus. The new technique of polyacrylamide gel electrophoresis opens up new horizons; it enables one to follow specifically which structural viral proteins are synthesized at any time during the multiplication cycle (Holowczak and Joklik, unpublished results). As I have already indicated, above, there is every likelihood that about 4 of these proteins are to be classed as early proteins. It will be interesting to see whether the synthesis of these is also subject to the switch-off control mechanism that applies to the early enzymes.

Conclusion

Expression of genetic information in virus-infected cells is subject to 3 control mechanisms: (1) control at the transcription level; (2) control at the translation level; and (3) the intrinsic instability of the individual viral messenger RNA molecules. The various experimental approaches that can be applied to these problems at this time have been discussed: namely, analysis of the rate of synthesis of messenger RNAs, analysis of their half-life, analysis of their ability to function as judged by their incorporation into polyribosomes, analysis of their base sequence identity by means of the hybridization-competition technique, and finally analysis of their functioning by direct estimation of their gene products. The elucidation of these control mechanisms will no doubt provide a fascinating chapter in the unravelling of the mysteries of information transfer in living cells.

REFERENCES

Appleyard, G., and Westwood, J. C. N. (1964). *J. Gen. Microbiol.* **37**, 391.
Appleyard, G., Westwood, J. C. N., and Zwartouw, H. T. (1962). *Virology* **18**, 159.
Appleyard, G., Hume, V. B. M., and Westwood, J. C. N. (1965). *Ann. N.Y. Acad. Sci.* **130**, 92.
Bautz, E. K. F., Kasai, T., Reilly, E., and Bautz, F. A. (1966). *Proc. Natl. Acad. Sci. U.S.* **55**, 1081.
Friedman, R. M., and Sonnabend, J. A. (1964). *Nature* **203**, 366.
Gesteland, R. E. (1966). *J. Mol. Biol.* **16**, 67.
Holowczak, J., and Joklik, W. K. (1967). Unpublished results.
Joklik, W. K. (1965). *Progr. Med. Virol.* **7**, 45.
Joklik, W. K. (1966). *Bacteriol. Rev.* **30**, 33.
Joklik, W. K., and Becker, Y. (1964). *J. Mol. Biol.* **10**, 452.
Joklik, W. K., and Merigan, T. C. (1966). *Proc. Natl. Acad. Sci. U.S.* **56**, 558.
Jungwirth, C., and Joklik, W. K. (1965). *Virology* **27**, 80.
Levine, S. (1964). *Virology* **24**, 586.
McAuslan, B. R. (1963a). *Virology* **20**, 162.

McAuslan, B. R. (1963b). *Virology* **21,** 383.
Merigan, T. C., Winget, C. A., and Dixon, C. B. (1965). *J. Mol. Biol.* **13,** 679.
Moses, V., and Calvin, M. (1965). *J. Bacteriol.* **90,** 1205.
Noll, H. (1966). *Science* **151,** 1241.
Oda, K., and Joklik, W. K. (1967). In press.
Penman, S., Scherrer, K., Becker, Y., and Darnell, J. E. (1963). *Proc. Natl. Acad. Sci. U.S.* **49,** 654.
Revel, M., and Hiatt, H. H. (1964). *Proc. Natl. Acad. Sci. U.S.* **51,** 810.
Sarkar, N. K., and Dürwald, H. (1966). *Biochim. Biophys. Acta* **119,** 204.
Shatkin, A. J. (1963). *Nature* **199,** 357.
Taylor, J. (1964). *Biochem. Biophys. Res. Commun.* **15,** 447.
Villa-Trevino, S., Farber, E., Staehelin, T., Wettstein, F. O., and Noll, H. (1964). *J. Biol. Chem.* **239,** 3826.
Woodson, B. A., and Joklik, W. K. (1965). *Proc. Natl. Acad. Sci. U.S.* **54,** 946.
Woodson, B. A., and Joklik, W. K. (1966). Unpublished data.

Enzyme Inductions in Cell Cultures during Productive and Abortive Infections by Papovavirus SV40[*]

Saul Kit

DIVISION OF BIOCHEMICAL VIROLOGY,
BAYLOR UNIVERSITY COLLEGE OF MEDICINE,
HOUSTON, TEXAS

Introduction

Simian papovavirus SV40 has a number of interesting biological properties. The virus produces tumors when inoculated into neonatal hamsters and multimammate rats (Ashkenazi and Melnick, 1963; Eddy *et al.,* 1962; Girardi *et al.,* 1962; Rabson *et al.,* 1962) and causes proliferative changes ("transformations") in cell cultures of human, hamster, mouse, rat, guinea pig, porcine, or bovine origin (Black and Rowe, 1963a,b; Diderholm *et al.,* 1965, 1966; Koprowski *et al.,* 1962; Shein and Enders, 1962). SV40 replicates with a cytocidal interaction in monkey kidney cell cultures (Sweet and Hilleman, 1960) and also enhances the replication of human adenoviruses in simian cells (Beardmore *et al.,* 1965; Feldman *et al.,* 1965; Rabson *et al.,* 1964; Schell *et al.,* 1966; Wertz *et al.,* 1965). In addition, certain strains of SV40 interact with adenoviruses during mixed infections to produce virus particles in which SV40 genes are incorporated into adenovirus capsids (Boeyè *et al.,* 1966; Butel and Rapp, 1966; Easton and Hiatt, 1965; Huebner *et al.,* 1964; Rapp *et al.,* 1964, 1965b; Reich *et al.,* 1966; Rowe, 1965; Rowe and Baum, 1964, 1965; Schell *et al.,* 1966).

During productive infection of monkey kidney cell cultures by SV40,

* This investigation was aided by grants from the National Science Foundation (GB 3126), the American Cancer Society (E 291), the Robert A. Welch Foundation (Q 163), and Public Health Service Research Grants (CA 06656 and 1–K6–AI–2352).

the virus growth cycle is long compared with those of other classes of DNA-containing animal viruses. The SV40-eclipse period lasts for 20 to 24 hours. Total infectious virus then increases during the ensuing 40 to 48 hours (Fig. 1). Vacuolation is not detected until about 60 hours after infection, but by 72 hours, virtually all the cells in the culture display typical cytoplasmic vacuolation (Kit *et al.,* 1966d). The eclipse period for human adenoviruses in KB and HeLa cells and for simian adenovirus SV15 in CV-1, an established green monkey kidney (GMK) cell line, lasts about 16 hours and infectious virus formation is complete

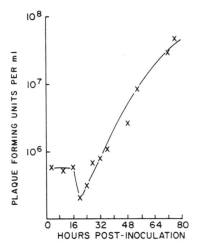

FIG. 1. Growth of SV40 in the CV-1 line of green monkey kidney cells. Cultures containing 4×10^6 cells were inoculated with 277 PFU/cell of SV40. Virus was allowed to adsorb for 2 hours at 37°C, after which the extracellular virus was removed by rinsing the cultures 3 times with 20 ml aliquots of saline-glucose solution. The cultures were treated with SV40 anti-serum, rinsed 3 more times with saline-glucose and 20 ml of medium was added. Cells and supernatant fluid were harvested at the times indicated in the figure and cells were disrupted by ultrasonic treatment. Virus was assayed on CV-1 monolayers and virus yields are expressed as PFU/ml harvest.

by about 36 hours after infection. The replication of vaccinia and herpes simplex viruses in many cell lines is even more rapid, with the eclipse period lasting about 6 hours and maximal titers of infectious virus being obtained at 12 to 16 hours after infection.

As mentioned previously, SV40 is capable of "transforming" primary cultures of mouse kidney cells. In most of these cells, SV40 appears to undergo an abortive infection (Kit *et al.,* 1966f). The virus is adsorbed by mouse kidney cells, though more slowly than to the CV-1 line of GMK cells. In 2 hours, about 60% as much virus is adsorbed by mouse kidney cells as by CV-1 cells. SV40 enters an eclipse phase in mouse kidney cell cultures which lasts about 24 to 32 hours. Thereafter, virus titers increase until about 40 hours and then decline (Fig. 2). Even when high input multiplicities are used, no cytopathic changes are observed, although the virus persists for at least 7 days.

When infected cultures are trypsinized and individual cells plated on noninfected CV-1 monolayers, less than 1% of the mouse kidney cells register as infectious-centers, whereas 60–80% of CV-1 cells do so.

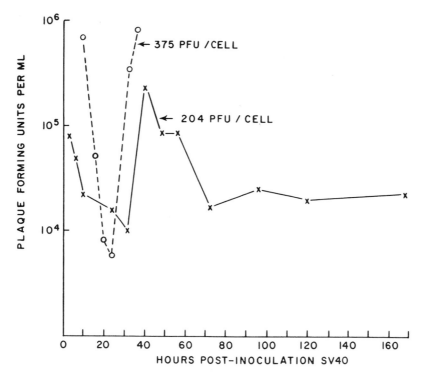

Fig. 2. Replication of SV40 in mouse kidney cell cultures inoculated with high input multiplicities of SV40. Cultures contained 3 to 5×10^6 cells in 20 ml medium. See legend to Fig. 1 for methods.

At 36 to 54 hours, the mouse kidney cultures yield about 1 to 6 plaque forming units (PFU) of SV40 per cell, while CV-1 cultures yield about 100 to 300 PFU per cell. However, the yield of SV40 per infectious center is of the same order of magnitude for the two cell types. At this time, the number of SV40 particles found in supernatant fluids used for analysis of complement-fixing (CF) antigens is about 10^8 per ml for the mouse cells and 10^{10} per ml for the monkey cells. If one assumes that the virus particles are coming from the cells which plate as infectious centers, then again there is no difference in yield between an infected monkey kidney cell and a competently infected mouse cell. These results suggest that infectious SV40 is replicated in only a small percentage of mouse kidney cells.

A study of the enzymology of cell cultures productively infected with SV40 has revealed that several enzymes functioning in DNA metabolism increase appreciably in activity during the latter part of the eclipse period (Kit *et al.,* 1966b,d). The data to be presented here will show that the activities of some of the same enzymes are stimulated during an abortive infection and remain at an elevated level in cell lines transformed by SV40 (Kit *et al.,* 1966f). These observations have raised the question of whether the SV40 genome directs the synthesis of these enzymes.

One of the enzymes, thymidine kinase, has been partially purified. The properties of the partially purified thymidine kinase will be described in detail and discussed in relation to the problem of whether SV40 genes control the synthesis of this enzyme.

RIBONUCLEIC ACID (RNA) AND PROTEIN SYNTHESIS IN CELL CULTURES INFECTED WITH PAPOVAVIRUSES

In a number of cell-virus interactions which lead to cell lysis, an early effect of infection is a reduction of cellular RNA and protein synthesis (Baltimore and Franklin, 1962; Fenwick, 1963; Kit and Dubbs, 1962; Bello and Ginsberg, 1966). In contrast, acute infection of monkey kidney cells with SV40 does not lead to an early arrest of RNA synthesis. Figure 3 shows an experiment in which CV-1 cell cultures were pulse

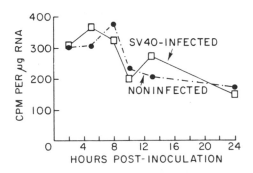

FIG. 3. Incorporation of uridine-³H into RNA of noninfected or SV40-infected CV-1 cells. Seven-day-old cultures (5.4×10^6 cells/culture) were inoculated with 180 PFU/cell of SV40 at time zero. At the times indicated in the figure, uridine-³H 2 µCi and 0.059 µg/ml) was added to each culture, and incubation was continued for an additional 30 min.

labeled for 30 min with uridine-^{3}H at various times after SV40 infection. The incorporation of uridine-^{3}H into RNA of infected cultures was about the same as that of noninfected cultures between 2 and 24 hours post-infection (PI). Carp and Gilden (1966) observed that RNA synthesis was not inhibited after SV40 infection of either stationary or replicating phase GMK cells or in SV40-infected human diploid (strain WI-38) cells, and Dulbecco *et al.* (1965) found that there was no large difference in the rates of incorporation of uridine-^{3}H into the total cell RNA of non-infected and polyoma virus-infected mouse kidney cells. Furthermore, no appreciable differences were found in the incorporation of radioactive amino acids into proteins of control and SV40-infected GMK or WI-38 cells or noninfected and polyoma virus-infected mouse kidney cells (Carp and Gilden, 1966; Dulbecco *et al.*, 1965).

Deoxyribonucleic Acid (DNA) Synthesis in Cell Cultures Infected with Papovaviruses

It has been demonstrated by several laboratories that DNA biosynthesis is appreciably stimulated in murine cell cultures infected with polyoma virus (Dulbecco *et al.*, 1965; Gershon *et al.*, 1965; Kit *et al.*, 1966c; Molteni *et al.*, 1966; Weil *et al.*, 1965; Winocour *et al.*, 1965). It appears that both cellular DNA and viral DNA are made. More recently, in a study in which the techniques of radioautography and immunofluorescence were combined, it was shown that the induction of cellular-DNA synthesis occurs in productively infected mouse kidney cells (Vogt *et al.*, 1966).

Combined radioautographic and biochemical experiments have shown that DNA synthesis is stimulated in confluent monolayer cultures of GMK cells productively infected with SV40 (Fig. 4). A stimulated incorporation of thymidine-^{3}H into DNA was first detected at about 16 hours after SV40 infection. The SV40-infected cultures continued to incorporate thymidine-^{3}H into DNA at a high rate for at least 50 hours PI. At 32–34 hours, the rate of thymidine-^{3}H incorporation into DNA of infected-GMK cultures was 3 to 4 times as great as that into DNA of noninfected cultures. Similar results were obtained with SV40-infected CV-1 cell cultures, pulse labeled with either thymidine-^{3}H or deoxyadenosine-^{3}H (Kit *et al.*, 1966d).

In the noninfected cultures of either GMK or CV-1 cells, less than 10% of the cell nuclei were labeled with thymidine-^{3}H after a 2-hour pulse. This value increased to about 20%, at about 16 hours after the medium was changed, and subsequently declined. In SV40-infected cultures, the percentage of cells with ^{3}H-labeled nuclei increased sharply at

about 12 to 16 hours. By 32 to 34 hours after infection, 70 to 80% of
the cells exhibited nuclear labeling. The radioautographic experiments
are consistent with the infectious center assays which indicate that 60 to
80% of the monkey kidney cells are infected under these conditions. The
radioautographic experiments further demonstrate that many of the cells
not synthesizing DNA, initiate DNA synthesis after SV40 infection.
Colorimetric assay of the total DNA synthesized support the biochemical
and radioautographic findings. At about 30 and 48 hours PI, the total
DNA of SV40-infected cultures exceeded that of noninfected cultures by
43 to 90% respectively (Kit *et al.*, 1966d).

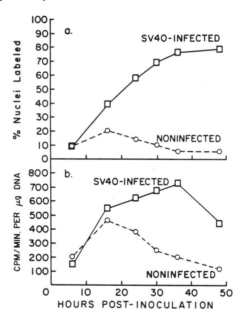

FIG. 4. Radioautographic (a) and biochemical (b) studies on the uptake of thymidine-³H by SV40-infected GMK cells. Replicate 9-day-old cultures (14.7 × 10⁶
cells/culture) were inoculated with approximately 10 PFU/cell of SV40. At the
indicated times, thymidine-³H (0.4 μCi and 3.6 mμg/ml) was added and the cultures
were incubated for an additional 2 hours at 37°C.

The incorporation of thymidine-³H into DNA is stimulated not only in
cell cultures productively infected with SV40 but also in cultures abortively infected with the virus. Figure 5 presents thymidine-³H pulse-labeling and radioautographic experiments on SV40-infected mouse kidney
cell cultures. The incorporation of thymidine-³H into DNA was found to
be stimulated two- to threefold during the interval 16 to 48 hours PI.
In noninfected mouse kidney cultures, the nuclei of only 2 to 5% of the

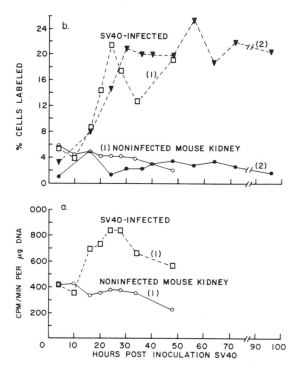

Fig. 5. Radioautographic (b) and biochemical (a) studies on the incorporation of thymidine-³H into DNA of SV40-infected mouse kidney cell cultures. Cultures were infected at input multiplicities of 230 and 620 PFU/cell for experiments 1 and 2, respectively, and were pulse-labeled for 2 hours with thymidine-³H.

cells were found to be labeled after a thymidine-³H pulse, whereas after SV40 infection, approximately 20 to 25% of the nuclei were labeled (Kit *et al.*, 1966f).

Stimulated Enzyme Synthesis in Cell Cultures Productively Infected with Papovaviruses

The finding that DNA synthesis is initiated in cell cultures infected with polyoma virus and SV40 suggests that enzymes of DNA metabolism might be induced by these viruses. This has proved to be the case.

Following polyoma virus infection of murine cell cultures, increases have been observed in the activities of deoxycytidylate deaminase, thymidine kinase, dihydrofolate reductase, thymidylate synthetase, thymidylate kinase and DNA polymerase (Dulbecco *et al.*, 1965; Frearson *et al.*, 1965, 1966; Hartwell *et al.*, 1965; Kit *et al.*, 1966c; Sheinin, 1966). These enzymes catalyze six of the seven final reactions in the pathway of DNA biosynthesis. The increases in the enzyme activities begin during the eclipse period at about the same time that the stimulations in DNA synthesis are observed.

Increases in thymidine kinase, DNA polymerase, dihydrofolate reductase, and thymidylate synthetase activities also occur at the time that DNA synthesis is stimulated in monkey kidney cell cultures productively infected with SV40 (Frearson *et al.,* 1966; Kit *et al.,* 1966b,d,f). However, the activities of deoxycytidylate deaminase and thymidylate kinase do not increase in these cells.

Several additional enzyme activities have been studied in cultures infected with papovaviruses. The activities of uridine kinase, thymidylate phosphatase, deoxyadenylate kinase, and deoxycytidylate kinase have been found not to change appreciably after papovavirus infection (Dulbecco *et al.,* 1965; Kit *et al.,* 1966c,d).

Enzyme Changes in Mouse Kidney Cell Cultures Abortively Infected with SV40

The activities of four of the enzymes of DNA metabolism (deoxycytidylate deaminase, thymidine kinase, thymidylate kinase, and DNA polymerase) have been studied in mouse kidney cell cultures inoculated with SV40 (Kit *et al.,* 1966f), and all were found to be increased considerably. The kinetics of the increases in thymidine kinase and DNA polymerase are illustrated in Fig. 6. The enhanced enzymatic activities were first detected about 16 to 24 hours after infection. The DNA polymerase activity of infected cultures was about 8 times greater than that of noninfected cultures at 48 hours and remained elevated for at least 72 hours. Thymidine kinase activity in infected cultures was eightfold greater than in noninfected cultures at 30 and 40 hours after infection. This activity subsequently declined, but at 72 hours it was still about fourfold higher than the activity found in noninfected mouse kidney cells.

Increases in thymidylate kinase and dCMP deaminase activities occur at about the same time as do the increases in DNA polymerase and thymidine kinase. At 44 to 45 hours after SV40 infection of mouse kidney cell cultures, thymidylate kinase and deoxycytidylate deaminase activities were two- to threefold and two- to fivefold greater, respectively, than were the corresponding activities of noninfected, control cells (Kit *et al.,* 1966f).

Experiments with puromycin and cycloheximide have shown that *de novo* protein synthesis is required if the papovavirus-induced enzyme increases are to occur. Addition of puromycin at the time of virus infection prevents the increase in thymidine kinase, deoxycytidylate deaminase, and DNA polymerase in polyoma-infected mouse kidney cell cultures (Hartwell *et al.,* 1965; Kit *et al.,* 1966c). Cycloheximide or puromycin inhibits the stimulation of thymidine kinase, DNA polymerase, and di-hydrofolate reductase activities in SV40-infected monkey kidney cell cultures (Frearson *et al.,* 1966; Kit *et al.,* 1966d, 1967). If inhibitors of protein synthesis are added to cultures after the increases in the activities of the enzymes have been induced, further increases are blocked. Moreover, removal of the drugs permits a renewal of enzyme synthesis after a lag period.

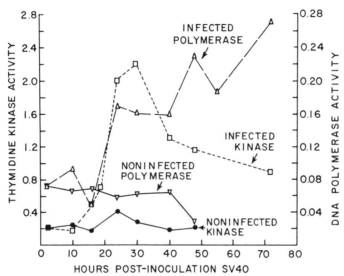

Fig. 6. Kinetics of thymidine kinase and DNA polymerase formation in confluent monolayer cultures of mouse kidney cells inoculated with SV40 at an input multiplicity of 170 PFU/cell. Thymidine kinase activity: $\mu\mu$moles dUMP formed per μg protein in 10 min at 38°C. DNA polymerase activity: $\mu\mu$moles ³H-TTP incorporated into DNA per μg protein in 30 min at 38°C.

Data demonstrating that puromycin inhibits the induction of thymidine kinase and thymidylate kinase by SV40 in mouse kidney cell cultures are shown in Tables I and II.

Effect of Actinomycin D on SV40-Induced Enzyme Synthesis

Low concentrations of actinomycin D inhibit DNA-dependent RNA synthesis. In order to learn whether RNA synthesis is required for the induction of thymidine kinase by SV40, CV-1 cells were treated with 1–5 μg/ml actinomycin D at 2 hours after SV40 infection. This inhibition

TABLE I

EFFECT OF PUROMYCIN (10^{-5} M) ON THE INDUCTION OF THYMIDINE KINASE
IN SV40-INFECTED MOUSE KIDNEY CELLS

Time of puromycin addition (hours PI[a])	Time of enzyme assay (hours PI[a])	Thymidine kinase activity[b]	
		Noninfected	Infected[c]
Not added	2	0.34	—
Not added	30	0.13	0.98
2	30	0.16	0.17
9	30	0.20	0.13
16	30	0.09	0.18

[a] PI, post-infection.
[b] $\mu\mu$moles dUMP formed per μg protein in 10 min at 38°C.
[c] SV40 input multiplicity: 87 PFU/cell.

was found to inhibit completely the SV40-induced increase normally observed at 26 hours PI (Kit *et al.*, 1966b). It may be seen from the data in Table III that if actinomycin D (3.8 μg/ml) addition is delayed until 10 to 14 hours after SV40 infection, a partial induction of thymidine kinase takes place, and that if the drug is added 17 to 21 hours PI, almost normal levels of thymidine kinase are induced. The results suggest that most of the messenger RNA required for thymidine kinase formation is made between 2 and 17 hours PI (Kit *et al.*, 1966e).

TABLE II

EFFECT OF PUROMYCIN (10^{-5} M) ON THE INDUCTION OF THYMIDYLATE (TMP)
KINASE IN SV40-INFECTED MOUSE KIDNEY CELLS

Treatment of cells	Time of enzyme assay (hours PI)	TMP kinase activity[a]	
		Noninfected	Infected[b]
None	2	8.6	
None	16	8.7	10.0
None	32	7.3	17.8
Puromycin, 16–32 hour PI	32	8.6	10.2
Puromycin, 16–32 hour PI	48	8.1	14.7
None	48	10.0	20.7

[a] Enzyme extracts were prepared in buffer containing 0.1 mM TMP. Activity measured in $\mu\mu$mole TDP + TTP formed per μg protein in 10 min at 38°C.
[b] SV40 input multiplicity: 185 PFU/cell.

It was also shown that actinomycin D (1.0–2.5 μg/ml), added at 2 hours after SV40 infection, completely inhibits the increase in dihydrofolate reductase activity normally observed 41 hours after infection of CV-1 cells. If actinomycin D is added 12 hours after SV40 infection, the infected cells show a significant increase in dihydrofolate reductase activity at 41 hours PI, although the increase is less than that observed in infected, untreated cells. Addition at 19 hours PI has little or no effect on the induction of dihydrofolate reductase (Frearson *et al.*, 1966).

TABLE III

EFFECT OF ACTINOMYCIN D (3.8 μG/ML) ON THE INDUCTION OF THYMIDINE KINASE IN SV40-INFECTED GMK CELLS[a]

Experiment	Time of addition of actinomycin D (hours PI)	Time of enzyme assay (hours PI)	Thymidine kinase activity[b]	
			Noninfected cells	Infected cells
A	none added	2	0.4	0.4
	none added	30	1.0	7.7
	2	30	0.3	0.5
	10	30	0.5	1.1
	14	30	0.9	2.3
B	none added	2	0.6	0.4
	none added	30	0.8	2.9
	2	30	0.5	0.6
	17	30	1.1	2.0
	21	30	0.9	2.4

[a] Stationary phase cultures of GMK cells (29 × 10⁶ cells/culture) were inoculated with 90 PFU/cell and 55 PFU/cell of SV40 in experiments (A) and (B), respectively.

[b] Activity measured in μμmoles dUMP formed per μg protein in 10 min at 38°C.

EFFECT OF BROMODEOXYURIDINE (BUdR) ON SV40-INDUCED ENZYME SYNTHESIS

Halogen containing analogs of thymidine have been shown to inhibit cell growth and to prevent the development of infectious SV40 and other DNA-containing animal viruses (Buthala, 1964; Dubbs and Kit, 1965; Rapp *et al.*, 1965a). However, both "early" and "late" proteins (i.e., tumor antigen and virus-coat protein antigen) are synthesized in SV40-infected GMK cell cultures treated with thymidine analogs.

Under the conditions of the present experiments, BUdR inhibited SV40 growth by at least 99.9%. In order to learn whether cell growth or infectious-SV40 formation are required for thymidine kinase synthesis,

SV40-infected and noninfected monkey kidney cell cultures were incubated in media containing BUdR (25 μg/ml). Figure 7 shows the kinetics of thymidine kinase induction in BUdR-treated cultures. BUdR does not affect appreciably the rate at which thymidine kinase is induced by SV40. In infected cultures not treated with BUdR, thymidine kinase activity remains elevated for about 48 hours and then declines sharply at the time that cytopathology becomes pronounced. BUdR treatment of SV40-infected cultures delays for several hours this decline in thymidine kinase activity. At 96 hours after infection of BUdR-treated cultures, the thymidine kinase activity was found to be still sixfold greater than the activity of noninfected cultures.

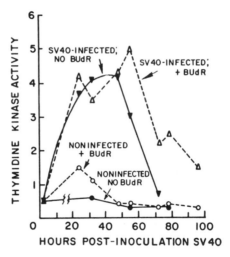

FIG. 7. Kinetics of thymidine kinase induction in stationary phase GMK cell cultures after the addition, at 2 hours PI, of 25 μg/ml bromodeoxyuridine (BUdR). The SV40 input multiplicity was 120 PFU/cell. Thymidine kinase activity: μμmoles dUMP per μg protein in 10 min at 38°C.

Treatment with BUdR does not inhibit the induction of thymidine kinase by polyoma virus in murine cell cultures (Kit *et al.,* 1966c) nor the induction of dihydrofolate reductase by either SV40 or polyoma viruses (Frearson *et al.,* 1966).

EFFECT OF MITOMYCIN C ON SV40-INDUCED THYMIDINE KINASE FORMATION

Mitomycin C is a potent inhibitor of cell growth and DNA synthesis. Treatment of bacterial or mammalian cell cultures with this drug leads to an induction of DNase activity and the breakdown of DNA (E. Reich *et al.,* 1961). At low concentrations, mitomycin C inhibits DNA synthesis without concomitant arrest of RNA and protein synthesis.

Table IV shows the inhibition of DNA synthesis in CV-1 cells by 25 to 75 μg/ml mitomycin C. At 2 to 4 hours after the addition of the drug, the incorporation of thymidine-^{3}H into DNA is inhibited by 40 to 60%, and at 22 to 24 hours inhibition is 80 to 85% complete.

Addition of 15 to 50 μg/ml mitomycin C at 2 hours after SV40 infection did not appreciably inhibit the increase of thymidine kinase activity observed at 30 hours PI (Kit *et al.,* 1966e). Enzyme induction is completely blocked by 100 μg/ml of mitomycin C (Table V).

TABLE IV

Effect of Mitomycin C (MC) on the Incorporation of Thymidine-^{3}H into DNA of SV40-Infected CV-1 Cells[a]

Mitomycin C added (μg/ml. at 2 hours PI)	Thymidine-^{3}H Incorporated into DNA (cpm/μg DNA)			
	4–6 Hours PI		24–26 Hours PI	
	Noninfected	SV40-infected	Noninfected	SV40-infected
0	74	104	229	507
25	44	40	45	75
50	26	40	14	40
75	24	32	2	5

[a] CV-1 cells (8 × 10^6 cells/culture) infected with 190 PFU/cell of SV40. To each culture was added 0.1 ml of thymidine-^{3}H (20 μc and 20.3 μg in a total volume of 20 ml).

It was of interest to learn whether pretreatment of cell cultures with mitomycin C would interfere with the SV40-induced synthesis of thymidine kinase. GMK cells were incubated for 16 hours with 25 μg/ml mitomycin C prior to infection with SV40 (Table VI). In some cultures, mitomycin C was removed at this time and in other cultures, mitomycin C was present throughout the postinfection period. Despite the pretreatment of GMK cell cultures with mitomycin C for 16 hours, thymidine kinase activity was 24 times higher in SV40-infected than in control cultures at 30 hours PI. Moreover, in cultures in which mitomycin C was present both in the preinfection and post-infection periods, the thymidine kinase activity of infected cultures was 9.5 times as great as was that of noninfected cultures.

It is probable that prolonged treatment of cells with high concentrations of mitomycin C not only inhibits DNA synthesis but causes damage

TABLE V

EFFECT OF MITOMYCIN C (MC[a]) ON THE INDUCTION OF THYMIDINE KINASE
ACTIVITY FOLLOWING SV40 INFECTION OF CV-1 CELLS

MC (μg/ml)	Time of enzyme assay (hours PI)	Thymidine Kinase Activity[b]	
		Noninfected Cells	SV40-Infected Cells
0	2	1.7	1.7
0	30	3.6	24.1
15	30	14.9	26.8
25	30	15.1	22.5
50	30	3.7	15.9
100	30	2.5	3.2

[a] MC added to cultures (5.4×10^6 cells/culture) at 2 hours PI. SV40 input multiplicity: 178 PFU/cell.

[b] Measured in $\mu\mu$moles dUMP/μg protein in 10 min at 38°C.

TABLE VI

SV40-INDUCED STIMULATION OF THYMIDINE KINASE ACTIVITY IN GMK CELLS
PRETREATED FOR 16 HR WITH 25 μG/ML MITOMYCIN C (MC)

MC pretreatment	MC present 2 to 30 hours PI	Time of enzyme assay[a] (hours)	Thymidine Kinase Activity[b]	
			Noninfected cells	SV40-Infected cells
0	No	−16	1.0	—
+	No	+ 2	4.2	4.3
+	Yes	+30	0.4	3.8
+	No	+30	0.3	7.2

[a] Zero time defined as the time of SV40 addition to the cultures. GMK cell cultures (30.5×10^6 cells/culture) were inoculated with 52 PFU/cell of SV40. At 16 hours prior to infection, there were 25.1×10^6 cells/culture.

[b] Measured in $\mu\mu$moles dUMP formed per μg protein in 10 min at 38°C.

to preexisting host cell DNA (Iyer and Szybalski, 1964). The failure of mitomycin C to prevent the increase in thymidine kinase activity suggests, therefore, that normal undamaged GMK cell DNA is not required for SV40-induced enzyme synthesis.

EFFECT OF ARABINOFURANOSYLCYTOSINE ON VIRUS-INDUCED
ENZYME SYNTHESIS

The drug, 1-β-D-arabinofuranosylcytosine (ara-C), is a potent inhibitor of DNA synthesis and cell growth (Kit *et al.*, 1966a,d). It has been shown to inhibit the growth of several DNA-containing viruses (Buthala,

1964; Rapp *et al.,* 1965a) and the synthesis of SV40-coat protein antigen, but not the synthesis of SV40-induced tumor antigen (Butel and Rapp, 1965, 1966; Rapp *et al.,* 1965a). The mechanism of action of ara-C is different from that of either mitomycin C or BUdR. It is thought that ara-C curtails deoxycytidine triphosphate synthesis by preventing the reduction of ribonucleotides to deoxyribonucleotides (Chu and Fischer, 1962).

The addition of ara-C (10 μg/ml) 2 hours after either SV40 infection of CV-1 cells or polyoma virus-infection of mouse kidney cells had no inhibitory effect on the induction of thymidine kinase or dihydrofolate reductase by these viruses. This concentration of ara-C suppresses the incorporation of thymidine-^{3}H into the DNA of these cells (Frearson *et al.,* 1966; Kit *et al.,* 1966c,d). The effect of ara-C on the induction of DNA polymerase by SV40 in CV-1 cells is shown in Fig. 8. Ara-C does

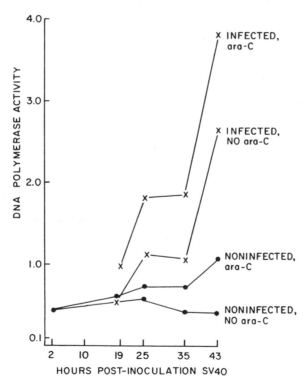

FIG. 8. Effect of ara-C (10 μg/ml) on the induction of DNA polymerase by SV40 in CV-1 cells. Confluent monolayer cultures of CV-1 cells were inoculated with 100 PFU/cell of SV40. Ara-C was added to infected and to noninfected cultures 2 hours after the addition of SV40. Cells were harvested for enzyme assays at the times indicated in the figure. DNA polymerase activity: μμmoles ^{3}H-TTP incorporated into DNA per μg protein in 30 min at 38°C.

not inhibit the induction of this enzyme. In fact, enzyme levels were found to be higher in SV40-infected cultures after ara-C treatment than in untreated cultures.

The effect of ara-C on the induction of TMP kinase in mouse kidney cells abortively infected with SV40 was also studied, and it was shown that the addition of this compound at 2 hours after SV40 infection has little effect on the increase in TMP kinase activity observed at 30 and 47 hours after infection (Kit *et al.,* 1966f).

From all these experiments, it is clear that DNA synthesis, cell growth, and infectious-virus formation are not required for the SV40-induced enzyme increases either in monkey kidney or in mouse kidney cell cultures.

FORMATION OF T-ANTIGEN AND VIRAL-CAPSID ANTIGEN IN SV40-INFECTED CULTURES

In confirmation of the results of others, the SV40-specific tumor (T)-antigen is induced in monkey kidney cells acutely infected with SV40 (Black and Rowe, 1963a,b; Huebner *et al.,* 1964; Hoggan *et al.,* 1965; Rapp *et al.,* 1964; Sabin and Koch, 1964). Figure 9 shows the kinetics of T-antigen formation in CV-1 cells (Kit *et al.,* 1966f). An increase in T-antigen formation may be detected at 10 hours and it attains a maximum at about 30 hours. Thus, T-antigen formation takes place at approximately the same time as does early enzyme formation. Viral-capsid antigen formation occurs somewhat later than T-antigen formation.

Figure 9 also shows that the kinetics of T-antigen formation in mouse kidney cell cultures is similar to that of T-antigen formation in CV-1 cells. However, the average number of CF antigen units produced per cell in the mouse kidney cultures is about one-tenth of the number pro-

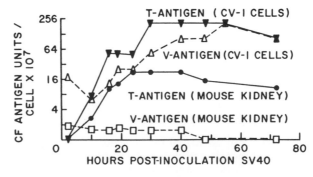

FIG. 9. Kinetics of T-antigen and viral-capsid antigen formation in mouse kidney and CV-1 cell cultures inoculated with SV40.

duced in CV-1 cell cultures. In contrast, the SV40-capsid antigen titer in mouse kidney cell cultures at 40 hours PI is less than 1% of that observed in CV-1 cell cultures.

THE EFFECT OF ULTRAVIOLET LIGHT ON INFECTIVITY AND THYMIDINE KINASE-INDUCING CAPACITY OF SV40

It has recently been shown that the ability of SV40 to induce the synthesis of the SV40-tumor antigen is reduced by exposing the virus to ultraviolet (UV) light, but that the loss of antigen-inducing capacity occurs at a slower rate than the loss of virus infectivity (Carp and Gilden, 1965). The capacity of SV40 to induce an increased thymidine kinase activity is also reduced following UV irradiation (Carp *et al.*, 1966). Doses of UV radiation which reduced infectivity to 0.12% survivors were shown to cause the total elimination of enzyme-inducing capacity. However, UV radiation which had a less pronounced effect upon infectivity permitted appreciable enzyme induction. UV radiation which reduced virus infectivity to 1.7–2.5% survival permitted from 44–73% of the amount of thymidine kinase induction observed with nonirradiated virus when measured at 41 and 48 hours PI. This suggests that the loss of virus infectivity occurs 2.2 to 6.0 times faster than does loss of enzyme-inducing capacity.

It is probable that virus DNA is the primary target of UV radiation. The demonstration that increasing doses of UV radiation progressively inactivate the thymidine kinase-inducing capacity of SV40 suggests that thymidine kinase synthesis in infected cells is controlled by the virus DNA. The observation that the rates of inactivation of enzyme-inducing capacity are consistently slower than the rates at which infectivity is lost suggest that the target size of the genetic area coding for enzyme-inducing capacity is smaller than that concerned with infectivity. However, the UV-radiation experiments do not elucidate the mechanism by which the SV40 nucleic acid controls the formation of thymidine kinase. The UV effects could be due either to a direct action on the coding properties of the virus genome, or to effects on the DNA of the virus which then are translated to the control mechanisms of the cell.

STIMULATION OF THYMIDINE KINASE ACTIVITY BY ARA-C OR MITOMYCIN C IN NONINFECTED MONKEY KIDNEY CELL CULTURES

It is apparent from the preceding discussion that a method for stimulating the thymidine kinase activity of *noninfected* cells would be useful for clarifying the mechanism by which SV40 induces enzyme formation. During studies of the effects of drug treatment on virus-induced thymidine kinase formation, a simple procedure for inducing high levels of thymidine kinase activity in noninfected cells was discovered (Kit *et al.,* 1966a). It was found that the addition, to either GMK or CV-1 cell cultures, of ara-C (10 to 20 μg/ml) caused a significant increase in thymidine kinase activity. Figure 10 illustrates this finding.

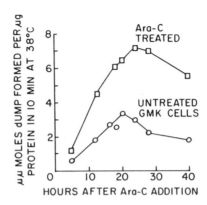

Fig. 10. Kinetics of ara-C induced stimulation of thymidine kinase activity in 9-day-old GMK cell cultures. Ara-C concentration: 10 μg/ml.

The ara-C induced stimulation of thymidine kinase activity takes place in HeLa cell cultures as well as in monkey kidney cell cultures, but does not occur after ara-C treatment of LM and LM(TK$^-$) mouse fibroblast cells, primary mouse kidney cells, or HeLa (BU-100) cells.

Moderate increases in deoxycytidylate deaminase activity were also observed in CV-1 cell cultures treated with ara-C (Kit *et al.,* 1966a). Moreover, as shown in Fig. 9, DNA polymerase activity increases two- to threefold in noninfected CV-1 cell cultures at 35 to 43 hours after ara-C addition.

It was also found that 16 to 30 hours after the addition of mitomycin C (15–25 μg/ml) to monkey kidney cell cultures, there is an appreciable increase in thymidine kinase activity (Kit *et al.,* 1966a). This latter effect is illustrated in Tables V and VI.

Since drug treatment is effective in "inducing" an increase in host cell thymidine kinase activity, it was of interest to compare the properties of the drug-induced enzyme with that of the virus-induced enzyme and the enzyme from normal cells.

Properties of Partially Purified Thymidine Kinase

Using a relatively simple procedure consisting of ammonium sulfate fractionation and negative phosphate gel absorption, thymidine kinase was purified approximately twenty- to fortyfold (Kit *et al.,* 1966c). The enzyme catalyzes the phosphorylation of thymidine, deoxyuridine, or of uracil deoxyribonucleoside analogs in which the constituent on carbon 5 of the pyrimidine is any of several halogens (Kit *et al.,* 1965).

The thymidine analog, trifluorothymidine (F_3TdR) also appears to be a substrate for thymidine kinase. Not only does the 5′-monophosphate derivative of this interesting drug inhibit the formation of thymidylate from deoxyuridylate (Reyes and Heidelberger, 1965), but F_3TdR, it-self, competitively inhibits the phosphorylation of deoxyuridine-³H. Line-weaver-Burk plots illustrating the inhibitory effects of F_3TdR are shown in Figs. 11 and 12.

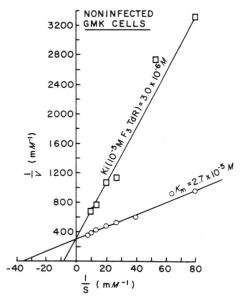

FIG. 11. Lineweaver-Burk plot illustrating the effect of 0.01 mM trifluorothymidine (F_3TdR) on the phosphorylation of deoxyuridine-³H by partially purified thymidine kinase from 7-day-old GMK cells. The enzyme was purified twenty-fold compared with the crude cell extract. The inhibitor constant, K_i, was calculated from the equation $K_p = K_m (1 + i/K_i)$, where i equals the inhibitor concentration, and $-1/K_m$ and $-1/K_p$, respectively, are the intercepts at the abscissa for the uninhibited and the inhibited reactions.

In order to learn whether arabinosylthymine, arabinosyluracil, ara-binosylcytosine, or ribosylthymine are inhibitors of the phosphorylation of deoxyuridine-³H, the arabinosyl and ribosyl compounds were added at concentrations ranging from 0.1 to 0.3 mM to enzyme reaction mix-tures containing 0.075 mM deoxyuridine-³H. The arabinosyl compounds had no effect on the phosphorylation of deoxyuridine-³H. Ribosylthymine, at the highest concentrations used, reduced the phosphorylation of deoxyuridine-³H by only 19 to 27%. These results suggest that the arabinosyl compounds and ribosylthymine are not substrates of simian thymidine kinase.

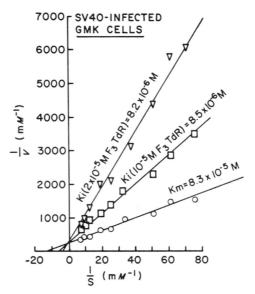

Fɪɢ. 12. Lineweaver-Burk plot illustrating the effects of 0.01 m*M* and 0.02 m*M* trifluorothymidine (F₃TdR) on the phosphorylation of deoxyuridine-³H by partially purified thymidine kinase from GMK cells infected for 41 hours with SV40 at an input multiplicity of 81 PFU/cell. The enzyme used was twentyfold purified with respect to the crude cell extracts.

From Table VII, it may be seen that the Michaelis constant (K_m), with deoxyuridine as substrate, for thymidine kinase from SV40-infected monkey kidney cells is about 3 times as large as that for the enzyme from noninfected cells. As previously stated, the thymidine kinase activity of noninfected monkey kidney cells may be increased appreciably by treating cultures with ara-C or mitomycin C. It was therefore of interest to learn whether ara-C or mitomycin C treatment would also change the K_m of the thymidine kinase of these cells. It was anticipated that ara-C or mitomycin C treatment would alter the intracellular nucleotide pools and that this might perhaps affect the K_m value of thymidine kinase. Table VII shows that this is not the case. The K_m values of the enzymes from ara-C or mitomycin C treated cells were found to be similar to those of the enzymes from noninfected CV-1 or GMK cells and to differ from the K_m of the enzyme from SV40-infected cells.

TABLE VII

MICHAELIS CONSTANTS (K_m)[a] AND INHIBITOR CONSTANTS (K_i)[b] OF PARTIALLY PURIFIED THYMIDINE KINASE FROM NONINFECTED OR VIRUS-INFECTED MONKEY KIDNEY CELLS[c]

Cells	Additions to culture medium	K_m ($\times 10^{-5} M$) deoxyuridine	K_i ($\times 10^{-6} M$) F$_3$TdR
Noninfected GMK	None	2.8 ± 0.2 (3)	4.3 ± 0.7 (3)
	TdR (10 µg/ml) + CdR (10 µg/ml), 16 hours	2.8 ± 0.1 (2)	2.9 ± 0.5 (2)
	TdR (100 µg/ml), 41 hours	3.3 ± 0.1 (2)	4.0 ± 0.6 (6)
	Ara-C (10 µg/ml), 41 hours	3.2 ± 0.3 (2)	2.8 ± 0.3 (5)
	Mitomycin C (20 µg/ml), 16 hours	3.9 ± 0.3 (2)	5.1 ± 0.4 (5)
Noninfected CV-1	None	2.8 ± 0.2 (4)	—
	Ara-C (10 µg/ml), 28 hours	2.8 ± 0.1 (2)	—
GMK-SV40 (41 hours PI)	None	8.4 ± 0.9 (8)	15.0 ± 2.9 (7)
GMK-SV40 (41 hours PI)	TdR (100 µg/ml), 41 hours	11.0 (1)	—
GMK-Vaccinia (7 hours PI)	None	9.0 ± 0.7 (4)	20.0 ± 4.0 (2)
CV1-SV40 (41 hours PI)	None	8.1 ± 0.6 (4)	

[a] With deoxyuridine-^{3}H as substrate.

[b] For trifluorothymidine.

[c] Abbreviations: TdR, thymidine; CdR, deoxycytidine; ara-C, arabinosylcytosine; F$_3$TdR, trifluorothymidine. Values shown are the mean ± standard error of the mean. The numbers in parentheses indicate the number of determinations.

In a further effort to manipulate the intracellular nucleotide pools, non-infected GMK cells were incubated with a high concentration of thymidine (100 μg/ml) or with lower concentrations of thymidine plus deoxycytidine (10 μg/ml). The purpose of the administration of high concentrations of thymidine was to cause an accumulation of thymidine triphosphate, the feedback inhibitor of thymidine kinase. The addition of thymidine plus deoxycytidine to the medium, however, was expected to enhance cell growth. Table VII shows that neither addition of excess thymidine nor addition of thymidine plus deoxycytidine appreciably altered the K_m of partially purified thymidine kinase.

It has been shown (Kit and Dubbs, 1965) that the K_m of thymidine kinase prepared from vaccinia-infected mouse fibroblast cells [strain LM(TK$^-$)] is greater than that of the enzyme from noninfected mouse fibroblast cells. The K_m of the thymidine kinase induced by vaccinia virus in monkey kidney cells is similar to that of the enzyme induced in LM(TK$^-$) cells and differs from the K_m of noninfected monkey kidney cells (Tables VII & X).

F_3TdR-inhibition constants (K_i) for partially purified thymidine kinases are also shown in Table VII. The K_i values for the enzymes prepared from SV40-infected or vaccinia-infected GMK cells are larger than the K_i values for the enzymes from noninfected GMK cells. The preceding experiments demonstrate that thymidine kinases with altered kinetic properties are induced in monkey kidney cells after either SV40 or vaccinia virus infections.

The thermal stability of thymidine kinase induced in LM(TK$^-$) cells by vaccinia is greatly increased compared with the normal mouse fibroblast cell enzyme (Kit and Dubbs, 1965). To learn whether the thymidine kinase induced by SV40 is also more stable than the host cell enzyme, these enzymes were preincubated at 65°C, 70°C, or 75°C for various periods of time prior to assay. The thymidine kinases are all gradually inactivated by heating at these temperatures. However, the kinetics of thermal inactivation at any of the temperatures studied are about the same for the partially purified enzymes from either SV40-infected and noninfected monkey kidney cells. Data illustrating the kinetics of inactivation of thymidine kinase at 70°C are shown in Table VIII.

A number of hypotheses may be advanced to account for the differences in K_m and K_i values between the thymidine kinases from SV40-infected and noninfected monkey kidney cells: (1) SV40 infection causes a gross change in thymidine kinase conformation; (2) thymidine kinase consists of subunits which are either dissociated or aggregated after virus infection; (3) unknown products of SV40 infection combine with monkey kidney cell thymidine kinase, altering the affinity of the enzyme for deoxyuridine and F_3TdR without grossly changing the molecular weight or conformation of the enzyme; and (4) SV40 induces a new, virus-specific thymidine kinase.

If, as a result of SV40 infection, the conformation of a host cell enzyme were to change significantly, one might predict that the sensitivity of the enzyme might be altered with respect to sulfhydryl reagents, heavy metals, or to reagents which denature the protein. Also, sensitivity to thymidine-5'-triphosphate, the feedback inhibitor, might be changed. To investigate

TABLE VIII

KINETICS OF THERMAL INACTIVATION AT 70°C OF PARTIALLY PURIFIED THYMIDINE KINASE FROM SV40-INFECTED AND ARA-C OR MITOMYCIN-C TREATED GMK CELLS

Time of incubation at 70°C (mins)	Activity of thymidine kinase[a] prepared from:		
	SV40 infected cells (41 hours PI)	Ara-C treated cells[b]	Mitomycin C treated cells[c]
0	100 (83)[d]	100 (17)	100 (13)
5	44	68	38
10	42	72	40
15	37	61	35
20	35	68	32
30	32	50	28
45	23	45	12
60	15	12	—

[a] Expressed as % of the activity of the unheated preparations.

[b] 10 μg/ml for 24 hours.

[c] 20 μg/ml for 16 hours.

[d] Values in parentheses represent activity of thymidine kinase (μμmoles dUMP formed per μg protein in 10 min at 38°C) prior to heating.

these possibilities, partially purified thymidine kinase from SV40-infected GMK cells and the enzymes isolated from cells treated with ara-C or mitomycin C, were assayed in the presence of various concentrations of *p*-chloromercuribenzoate, *p*-hydroxy-mercuribenzoate, sodium dodecylsulfate, cobalt chloride, or thymidine-5'-triphosphate (Kit *et al.*, 1967). All of these compounds inhibit the activity of thymidine kinase, and the inhibitor-dose response curves were shown to be about the same for the enzymes from infected and noninfected cells.

To learn whether there were gross differences in the molecular weights of thymidine kinases from SV40-infected and noninfected monkey kidney cells, Sephadex G150 gel filtration studies (Andrewes, 1965) were carried out. Figure 13 shows the relation between elution volume from a Sephadex G150 column and the molecular weights of some known proteins. In determining the molecular weight of thymidine kinase by this procedure, centrifuged cell extracts (S3 fraction) as well as partially purified thymidine kinase preparations were studied. The SV40-induced thymidine kinase was eluted from Sephadex G150 columns after aldolase but prior to liver alcohol dehydrogenase. The molecular weight of this enzyme was estimated to be about 125,000. Thymidine kinase preparations from noninfected CV-1 cells and from ara-C treated CV-1 cells were also studied. The molecular weights of these enzyme preparations were estimated to be 115,000 and 125,000, respectively. It would appear,

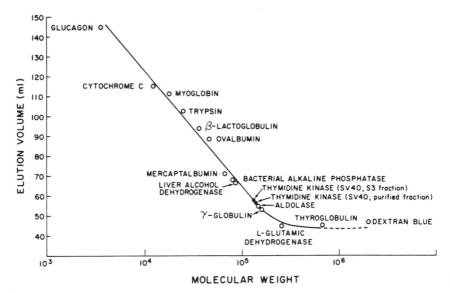

FIG. 13. Estimation of the molecular weight of thymidine kinase by the method of Andrewes (1965) using Sephadex G150 gel filtration. The column (32 × 2.5 cm) was equilibrated with 0.01 *M* Tris-HCl buffer, pH 8.0, containing 0.15 *M* potassium chloride and this buffer solution was employed for eluting the proteins. Recovery from the columns was 81 to 100% of the proteins applied and 75% or more of the thymidine kinase activity. For calibration, the following proteins (2 to 5 mg/sample) were added to the column: crystalline glucagon, Eli Lilley and Co., Indianapolis, Indiana; horse heart cytochrome C and bovine liver L-glutamic dehydrogenase (type II), Sigma Chemical Co., St. Louis, Missouri; myoglobin (IX crystallized), trypsin (3X crystallized); β-lactoglobulin (3X crystallized), ovalbumin (3X crystallized); mercaptalbumin (2X crystallized), aldolase (5X crystallized), human gamma globulin (Fraction II), and thyroglobulin, all from Nutritional Biochemical Corp., Cleveland, Ohio; bacterial alkaline phosphatase and horse liver alcohol dehydrogenase (1X crystallized), Worthington Biochemical Co., Freehold, New Jersey.

therefore, that SV40 infection does not induce a gross change in the molecular weight of monkey kidney thymidine kinase.

The results of these experiments do not support either of the first two hypotheses mentioned above.

THYMIDINE KINASE ACTIVITY AND T-ANTIGEN TITERS OF TRANSFORMED CELL LINES

Cytopathic effects were not seen after infection of mouse kidney cells with SV40 in spite of the use of high input multiplicities. However, 14 to 21 days after infection, colonies of "transformed" cells were noticeable. Some of these "transformed" cells have been subcultured for 40 passages as of this writing.

Primary monolayer cultures of mouse kidney cells show contact inhibition and attain a population density of approximately 3 to 5 million/ 55 cm^2 in 5 to 7 days. The cell population then gradually declines and very little cell growth occurs after subculture. The transformed cultures, however, grow to populations of 10 to 15 million/55 cm^2 and may be subcultured regularly at 3- to 4-day intervals. Mouse kidney cell cultures transformed by SV40 exhibit high levels of thymidine kinase activity. The activity exceeds by a factor of 33 to 60 that of secondary mouse kidney cultures which had been planted at the same time as the transformed lines (Table IX). Moreover, the activities of DNA polymerase and deoxycytidylate deaminase of the transformed cell lines are equal to or greater than the levels of enzymes induced in mouse kidney cells by SV40 infection. The transformed cell lines all contain titers of SV40-T antigen comparable to those of either SV40-transformed hamster tumor cells (H-50) or of primary cultures of mouse kidney cells infected for 48 to 72 hours with SV40 (Fig. 9). However, the transformed cell lines do not contain detectable SV40-capsid antigen or virus particles (Table IX). Attempts to extract infectious virus from the transformed cell lines or to initiate infectious center formation by plating transformed cells on CV-1 cell monolayers have so far been unsuccessful.

MICHAELIS CONSTANTS OF THYMIDINE KINASE FROM VIRUS-INFECTED-MOUSE KIDNEY AND TRANSFORMED CELL LINES

Since the thymidine kinase induced in monkey kidney cells by SV40 has an altered K_m, it was of interest to learn whether the enzymes induced by SV40 or polyoma virus in mouse kidney cells and the enzymes present in cells transformed by SV40 also exhibit changed kinetic properties. The results of these experiments are shown in Table X. It

TABLE IX

THYMIDINE KINASE ACTIVITY AND SV40-T-ANTIGEN TITERS OF SV40-TRANSFORMED MOUSE KIDNEY CELL CULTURES

Cells	Days after SV40 infection when first passage was made[a]	Passage number[b]	Total days in culture	Thymidine kinase[c]	CF antigen units/cell $\times 10^7$	
					T-Antigen	Capsid antigen
Noninfected mouse kidney[a]		2	52	0.1		0
Transformed	1	5	53	3.3	5.4	0
	4	4	53	3.6	8.1	0
	11	4	53	3.4	7.7	0
	14	4	53	3.6	10.2	0
	17	4	53	5.1	7.0	0
	22	7	62	6.0	6.5	0
	30	6	62	4.2	7.3	0

[a] Confluent 5-day-old monolayer cultures were inoculated with SV40 at an input multiplicity of about 150 PFU/cell.

[b] Transformed cells were harvested for enzyme and antigen assays at 3 days after the passage shown in the Table.

[c] Measured in μμmoles dUMP formed per μg protein in 10 min at 38°C.

[d] Replicate primary cultures of noninfected mouse kidney cells were subcultured at 6, 9, 16, 19, and 22 days, respectively. Cultures were fed 2 to 3 times weekly. These sparse monolayers were harvested 52 days after planting, pooled, and assayed for thymidine kinase activity.

may be seen that the K_m values of thymidine kinase from normal mouse kidney, SV40-infected and polyoma-infected mouse kidney, and SV40-transformed mouse kidney are all very similar. Also, the K_m value for thymidine kinase prepared from SV40-transformed hamster cells (Strain H-50) was about the same as that of the enzyme from BHK21 hamster cells. The K_m value of the thymidine kinase from LM mouse fibroblast cells was comparable to that of normal mouse kidney cells and about one-half that of the vaccinia-induced thymidine kinase.

TABLE X

MICHAELIS CONSTANTS (K_m) OF PARTIALLY PURIFIED THYMIDINE KINASE[a] FROM VIRUS-INFECTED AND SV40-TRANSFORMED CELL LINES AND NONINFECTED MURINE AND HAMSTER CELL LINES

Cell line	Treatment of cells	K_m $\times 10^{-5}$ M
Mouse kidney	None	2.4, 2.5, 3.6
Mouse kidney	SV40-infected, 41 hours	2.8, 3.8
Transformed mouse kidney (SV40)	Passage 6	3.2, 2.9, 3.3
Mouse kidney	Polyoma-infected, 28 hours	2.0, 2.5
Mouse fibroblast (LM)	None	3.2, 3.3
Mouse fibroblast [LM(TK⁻)]	Vaccinia-infected, 5 hours	7.2 ± 0.5 (12)[b]
Hamster (BHK21)	None	3.5, 5.6
Hamster tumor (H50)[c] induced by SV40	None	4.2, 6.1, 6.6

[a] Deoxyuridine-³H used as nucleoside substrate.

[b] Kit and Dubbs (1965).

[c] Ashkenazi and Melnick (1963).

Discussion

After either productive infection of monkey kidney cells or abortive infection of mouse kidney cells by SV40, DNA synthesis is appreciably stimulated, as are the activities of enzymes catalyzing the terminal pathway of DNA biosynthesis. These events occur at about the same time after infection of monkey kidney or mouse kidney cells and concomitantly with the appearance of SV40-specific tumor (T)-antigen. After productive infection of mouse kidney cells by polyoma virus, the same enzymes are induced and DNA synthesis is also stimulated. *De novo* protein and RNA synthesis are required for the enzyme inductions and SV40-T-antigen formation, but DNA synthesis, cell growth, and infectious virus formation are not obligatory (Gilden and Carp, 1966; Kit *et al.*, 1966b–f; Rapp *et al.*, 1965a).

The DNA's of SV40 and polyoma viruses consist of polynucleotides having molecular weights of 3 to 5×10^6 daltons. Thus, these viruses have only enough genetic information to code for about 10 proteins of 200 amino acid residues each. At least one of these proteins must be the virus-capsid protein. It seems unlikely that the remaining viral genes directly control synthesis of all six of the papovavirus-induced enzymes. Perhaps one or two of the enzymes are directly determined by viral genes while the others are indirectly regulated by virus infection.

Thymidine kinase, one of the induced enzymes, has been purified and its properties have been studied in detail. The UV-irradiation experiments provide evidence that induction of thymidine kinase is a function of the SV40 genome. However, they do not show whether the SV40 genome derepresses host cell thymidine kinase synthesis or induces the formation of a new enzyme. The Michaelis constant (K_m) for deoxyuridine and the inhibitor constant (K_i) for F_3TdR of the SV40-induced enzyme both differ considerably from the K_m and K_i of the enzyme from noninfected, and drug or nucleoside-treated monkey kidney cells. The differences in K_m and K_i values cannot be ascribed to the aggregation from, or dissociation to subunits of the host cell enzyme after virus infection, since the molecular weights, as estimated by Sephadex G150 gel filtration, are about the same. The experiments on heat inactivation, and inhibition of the enzymes by sodium dodecylsulfate, cobalt chloride, *p*-chloromercuribenzoate, *p*-hydroxymercuribenzoate, and TTP also indicate that gross conformational changes do not provide an explanation for the kinetic differences between the SV40-induced enzyme and the normal monkey kidney cell thymidine kinase. However, these studies do not provide rigorous evidence that the SV40-induced thymidine kinase is a new enzyme and the possibility remains that the K_m and K_i differences are the result of virus-induced changes operating through yet unknown mechanisms.

The view that SV40 infection may stimulate host cell enzyme synthesis gains support from studies on abortive infection of mouse kidney cells. It has been established that the SV40-specific T-antigen is formed in SV40-infected mouse kidney cells and that four enzymes functioning in DNA biosynthesis are increased in activity. Cell lines transformed by SV40 also exhibit high levels of T-antigen and enzyme activities. The thymidine kinase induced by SV40 in mouse kidney cells and the enzyme present in transformed cells do not resemble the enzyme induced by SV40 in monkey kidney cells with respect to their Michaelis constants. Instead, they resemble the normal mouse kidney cell enzyme. Moreover, the thymidine kinase formed after productive infection of mouse kidney cells by polyoma virus has a "normal" Michaelis constant as does the enzyme prepared from SV40-transformed hamster kidney cells. It would appear that these cell lines are "derepressed" with respect to thymidine kinase formation and capacity for DNA biosynthesis. If papovavirus-specific enzymes occur in the transformed or infected mouse kidney cells, more sophisticated techniques are required to detect them.

An inhibition in the expression of SV40 genes occurs in mouse kidney cells. However, the locus of this inhibition is as yet unclear. The inhibition occurs after the synthesis of T-antigen but before the formation of capsid antigens. Although total DNA synthesis is increased in SV40-infected mouse kidney cells, it has not been established whether the newly synthesized DNA is mouse cell or virus DNA. Perhaps, virus-DNA synthesis is also blocked during abortive infection of mouse kidney cells.

It is intriguing to speculate that the failure of mouse kidney cells to make SV40 capsid antigen is due to the phenomenon of intercistronic gene suppression. It is possible that codons in the viral genome for SV40-capsid protein are recognized as genetic sense by permissive, monkey kidney cells, whereas these same codons represent nonsense in nonpermissive mouse kidney cells. Based upon this hypothesis, experiments have recently been initiated in our laboratory to isolate SV40 strains virulent for mouse kidney cells. Attempts are being made to enhance productive SV40 infection of mouse kidney cells by fluorouracil treatment. Studies on highly purified enzyme and capsid proteins may shed light on the differences between productive and abortive infections and the relation between abortive infection and transformation.

REFERENCES

Andrewes, P. (1965). *Biochem. J.* **96,** 595.
Ashkenazi, A., and Melnick, J. L. (1963). *J. Natl. Cancer Inst.* **30,** 1227.
Baltimore, D., and Franklin, R. M. (1962). *Proc. Natl. Acad. Sci. U.S.* **48,** 1383.
Beardmore, W. B., Havlick, M. J., Serafini, A., and McLean, I. W., Jr. (1965). *J. Immunol.* **95,** 422.
Bello, L. J., and Ginsberg, H. S. (1966). *Federation Proc.* **25,** 652.
Black, P. H., and Rowe, W. P. (1963a). *Virology* **19,** 107.
Black, P. H., and Rowe, W. P. (1963b). *Proc. Soc. Exptl. Biol. Med.* **114,** 721.
Boeyè, A., Melnick, J. L., and Rapp, F. (1966). *Virology* **28,** 56.
Butel, J. S., and Rapp, F. (1965). *Virology* **27,** 490.
Butel, J. S., and Rapp, F. (1966). *J. Bacteriol.* **91,** 278.
Buthala, D. A. (1964). *Proc. Soc. Exptl. Biol. Med.* **115,** 69.
Carp, R. I., and Gilden, R. V. (1965). *Virology* **27,** 639.
Carp, R. I., and Gilden, R. V. (1966). *Virology* **28,** 150.
Carp, R. I., Kit, S., and Melnick, J. L. (1966). *Virology* **29,** 503.
Chu, M. Y., and Fischer, G. A. (1962). *Biochem. Pharmacol.* **11,** 423.

Diderholm, H., Stenkvist, B., Ponten, J., and Wesslen, T. (1965). *Exptl. Cell Res.* **37**, 452.

Diderholm, H., Berg, R., and Wesslen, T. (1966). *Intern. J. Cancer* **1**, 139.

Dubbs, D. R., and Kit, S. (1965). *Virology* **25**, 256.

Dulbecco, R., Hartwell, L. H., and Vogt, M. (1965). *Proc. Natl. Acad. Sci. U.S.* **53**, 403.

Easton, J. M., and Hiatt, C. W. (1965). *Proc. Natl. Acad. Sci. U.S.* **54**, 1100.

Eddy, B. E., Borman, G. S., Grubbs, G. E., and Young, R. D. (1962). *Virology* **17**, 65.

Feldman, L. A., Melnick, J. L., and Rapp, F. (1965). *J. Bacteriol.* **90**, 778.

Fenwick, M. L. (1963). *Virology* **19**, 241.

Frearson, P. M., Kit, S., and Dubbs, D. R. (1965). *Cancer Res.* **25**, 737.

Frearson, P. M., Kit, S., and Dubbs, D. R. (1966). *Cancer Res.* **26**, 1653.

Gershon, D., Hausen, P., Sachs, L., and Winocour, E. (1965). *Proc. Natl. Acad. Sci. U.S.* **54**, 1584.

Gilden, R. V., and Carp, R. I. (1966). *J. Bacteriol.* **91**, 1295.

Girardi, A. G., Sweet, B. H., Slotnick, V. B., and Hilleman, M. R. (1962). *Proc. Soc. Exptl. Biol. Med.* **109**, 649.

Hartwell, L. H., Vogt, M., and Dulbecco, R. (1965). *Virology* **27**, 262.

Hoggan, M. D., Rowe, W. P., Black, P. H., and Huebner, R. J. (1965). *Proc. Natl. Acad. Sci. U.S.* **53**, 12.

Huebner, R. J., Chanock, R. M., Rubin, B. A., and Casey, M. J. (1964). *Proc. Natl. Acad. Sci. U.S.* **52**, 1333.

Iyer, V. N., and Szybalski, W. (1964). *Science* **145**, 55.

Kit, S., and Dubbs, D. R. (1962). *Virology* **18**, 274.

Kit, S., and Dubbs, D. R. (1965). *Virology* **26**, 16.

Kit, S., Dubbs, D. R., and Frearson, P. M. (1965). *J. Biol. Chem.* **240**, 2565.

Kit, S., de Torres, R. A., and Dubbs, D. R. (1966a). *Cancer Res.* **26**, 1859.

Kit, S., Dubbs, D. R. de Torres, R. A., and Melnick, J. L. (1966b). *Virology* **27**, 453.

Kit, S., Dubbs, D. R., Frearson, P. M. (1966c). *Cancer Res.* **26**, 638.

Kit, S., Dubbs, D. R., Frearson, P. M., and Melnick, J. L. (1966d). *Virology* **29**, 69.

Kit, S., Dubbs, D. R., and Melnick, J. L. (1966e). *Federation Proc.* **25**, 777.

Kit, S., Dubbs, D. R., Piekarski, L. J., de Torres, R. A., and Melnick, J. L. (1966f). *Proc. Natl. Acad. Sci. U.S.* **56**, 463.

Kit, S., Melnick, J. L., Dubbs, D. R., Piekarski, L. J., and de Torres, R. A. (1967). *In* "Virus-Directed Host Response" (M. Pollard, ed.). Academic Press, New York (in press).

Koprowski, H., Ponten, J. A., Jensen, F., Ravdin, R. G., Moorhead, P., and Saksela, E. (1962). *J. Cellular Comp. Physiol.* **59**, 281.

Molteni, P., de Simone, V., Grosso, E., Bianchi, P., and Polli, E. (1966). *Biochem. J.* **98**, 78.

Rabson, A. S., O'Conor, G. T., Kirschstein, R. L., and Branigan, W. J. (1962). *J. Natl. Cancer Inst.* **29**, 765.

Rabson, A. S., O'Conor, G. T., Berezesky, I. K., and Paul, F. J. (1964). *Proc. Soc. Exptl. Biol. Med.* **116**, 187.

Rapp, F., Melnick, J. L., Butel, J. S., and Kitahara, T. (1964). *Proc. Natl. Acad. Sci. U.S.* **52**, 1348.

Rapp, F., Butel, J. S., Feldman, L. A., Kitahara, T., and Melnick, J. L. (1965a). *J. Exptl. Med.* **121**, 935.

Rapp, F., Butel, J. S., and Melnick, J. L. (1965b). *Proc. Natl. Acad. Sci. U.S.* **54,** 717.

Reich, E., Shatkin, A. J., and Tatum, E. L. (1961). *Biochim. Biophys. Acta* **53,** 132.

Reich, P. R., Baum, S. G., Rose, J. A., Rowe, W. P., and Weissman, S. M. (1966). *Proc. Natl. Acad. Sci. U.S.* **55,** 336.

Reyes, P., and Heidelberger, C. (1965). *Mol. Pharmacol.* **1,** 14.

Rowe, W. P. (1965). *Proc. Natl. Acad. Sci. U.S.* **54,** 711.

Rowe, W. P., and Baum, S. G. (1964). *Proc. Natl. Acad. Sci. U.S.* **52,** 1340.

Rowe, W. P., and Baum, S. G. (1965). *J. Exptl. Med.* **122,** 955.

Sabin, A. B., and Koch, M. A. (1964). *Proc. Natl. Acad. Sci. U.S.* **52,** 1131.

Schell, K., Lane, W. T., Casey, M. J., and Huebner, R. J. (1966). *Proc. Natl. Acad. Sci. U.S.* **55,** 81.

Shein, H. M., and Enders, J. F. (1962). *Proc. Natl. Acad. Sci. U.S.* **48,** 1164.

Sheinin, R. (1966). *Virology* **28,** 47.

Sweet, B. H., and Hilleman, M. R. (1960). *Proc. Soc. Exptl. Biol. Med.* **105,** 420.

Vogt, M., Dulbecco, R., and Smith, B. (1966). *Proc. Natl. Acad. Sci. U.S.* **55,** 956.

Weil, R., Michel, M. R., and Ruschmann, G. K (1965). *Proc. Natl. Acad. Sci. U.S.* **53,** 1468.

Wertz, R. K., O'Conor, C. C., Rabson, A. S., and O'Conor, G. T. (1965). *Nature* **209,** 1350.

Winocour, E., Kaye, A. M., and Stollar, V. (1965). *Virology* **27,** 156.

Studies on the Control of the Infective Process in Cells Infected with Pseudorabies Virus[*]

Albert S. Kaplan, Tamar Ben-Porat, and Celia Coto[†]

DEPARTMENT OF MICROBIOLOGY, RESEARCH LABORATORIES,
ALBERT EINSTEIN MEDICAL CENTER,
PHILADELPHIA, PENNSYLVANIA

Many of the events that follow infection of rabbit kidney (RK) cells with pseudorabies (Pr) virus are similar to those that occur in bacterial cells infected with the T-even bacteriophages: Infection of cells with Pr virus causes a gradual inhibition of the synthesis of cellular DNA, an inhibition that is mediated by a protein (Ben-Porat and Kaplan, 1965), as well as a suppression of the synthesis of cellular proteins (Hamada and Kaplan, 1965). The replication of Pr virus DNA is semiconservative and is geometric in principle (Kaplan and Ben-Porat, 1964); viral progeny DNA accumulates within the infected cells and forms a precursor pool from which it is withdrawn at random to be enclosed in viral particles (Ben-Porat and Kaplan, 1963). Infection of cells with Pr virus also causes an increase in the level of activity of some of the enzymes involved in the synthesis of DNA (Hamada *et al.*, 1966). Increases in the levels of some of these enzymes have been reported for a variety of virus-cell systems (Hanafusa, 1961; Kit *et al.*, 1962, 1965; McAuslan and Joklik, 1962; Green and Pina, 1962; Newton and McWilliam, 1962; Keir and Gold, 1963; Dulbecco *et al.*, 1965; Ledinko, 1966) and, in a number of studies, it has been observed that the enzymes found in virus-infected cells have properties different from the proteins performing the same function in

[*] This investigation was supported by grants from the National Institutes of Health (AI–02432 and AI–03362), and from the National Science Foundation (GB–1386), and by a U.S. Public Health Service Research Career Program Award 5–K3–AI–19335) from the National Institute of Allergy and Infectious Diseases.

[†] Fellow of the Consejo Nacional de Investigaciones Científicas y Técnicas, Argentina.

noninfected cells (Kit and Dubbs, 1965; Keir, 1965; McAuslan, 1963; Nohara and Kaplan, 1963; Sheinin, 1966). These observations have led to the notion that new enzyme proteins are synthesized in the infected cells under the direction of the infecting viral genome.

In view of the many similarities in behavior between Pr virus and the T-even phages, this hypothesis, that has had as its model some of the findings made with phage-infected bacteria (see review by Cohen, 1961), is attractive, especially since Pr virus DNA has a molecular weight of 70×10^6 (Kaplan and Ben-Porat, 1964) and can therefore code for a large number of proteins. However, viruses such as polyoma and SV40, which contain relatively small molecules of DNA, also induce an increase in activity of a number of enzymes (Kit, this volume). The role of the viral genome in the enzymology of cells infected with animal viruses has therefore yet to be established unequivocally.

The increase in the level of activity of the enzymes could be due to the activation of enzyme protein present in the cells at the time of infection or to the induction of the synthesis of cellular enzymes. The changes in some of the characteristics of the enzymes present in the infected cells may be the result of changes in configuration of the enzymes normally present in the cell (Gerhart and Pardee, 1963; Freundlich and Umbarger, 1963; Changeux, 1963; Tomkins *et al.,* 1963). In adenovirus-infected cells, changes in the properties of aspartic acid transcarbamylase have been ascribed to an alteration in the configuration of this enzyme (Consigli and Ginsberg, 1964).

The experiments to be discussed here were designed in an attempt to answer some of these questions in the Pr virus-RK cell system. In addition, the role of the increased activity of thymidine kinase in the infected cells is examined.

Level of Activity of Some of the Enzymes Involved in the Synthesis of DNA in Pr Virus-Infected Cells

The levels of activity of the kinases involved in the phosphorylation of the deoxynucleosides and deoxynucleotides in RK cells at various times after infection with Pr virus are illustrated in Fig. 1. Of the enzymes tested, only the level of activity of thymidine kinase and of (TMP*) kinase increased after infection; the level of activity of the other kinases was not affected by infection.

The increase in activity of the thymidine and TMP kinases was inhibited by cycloheximide or by puromycin. Furthermore, both thymidine

---

* TMP, thymidine monophosphate.

kinase and TMP kinase present in Pr virus-infected cells appear to differ in certain characteristics from the enzymes performing the same function in noninfected cells. Thus, the thymidine kinases in infected and noninfected cells differ antigenically (Hamada *et al.,* 1966), and TMP kinase from infected cells appears to be more thermostable than the enzyme from noninfected cells (Nohara and Kaplan, 1963). These observations suggested that the increase in enzymatic activity in the infected cells may be due to the *de novo* synthesis of proteins under the control either of the viral genome or of part of the cellular genome that is normally not expressed in the noninfected cells. However, the next series of experiments will show that for TMP kinase, at least, this is not the case.

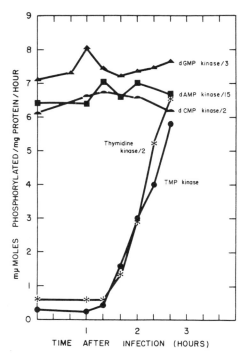

Fɪɢ. 1. The effect of infection with Pr virus on the level of activity in RK cells of some of the kinases involved in the synthesis of DNA. RK cells in stationary phase were infected (adsorbed multiplicity = 10). At various times after infection, the cells were harvested, sonicated, and centrifuged; the activity of the enzymes in the extracts was assayed as described previously (Hamada *et al.,* 1966). The values on the ordinate are for TMP kinase. To obtain the correct values for the other kinases, multiply the value read from the ordinate by the factor given next to each kinase. For the assay of TMP kinase, approximately 1 mg of protein was used per sample; for thymidine kinase, 0.5 mg; for dCMP kinase, 0.5 mg; for dGMP kinase, 150 μg; and for dAMP kinase, 75 μg (from Hamada *et al.,* 1966).

Level of TMP Kinase Activity in Extracts of Infected Cells Which Have Been Stabilized by Substrate

Since TMP kinase is unstable at 0°C and is stabilized by the addition of either thymidine or TMP (Bojarski and Hiatt, 1960), we determined the level of activity of this enzyme in Pr virus-infected cells under conditions in which the inactivation of the enzyme is prevented. The results of this experiment (Fig. 2) show the following: When the enzyme present in the cells at the time of harvest was not stabilized by TMP, the extracts contained increasing levels of enzyme activity as infection proceeded, whereas the samples to which TMP was added at the time of harvest contained a high level of enzyme activity, and no change in the level of enzyme activity was detected after infection. The difference in enzyme activity between extracts of noninfected cells prepared in the presence or absence of TMP suggested that the enzyme present in these extracts is unstable in the

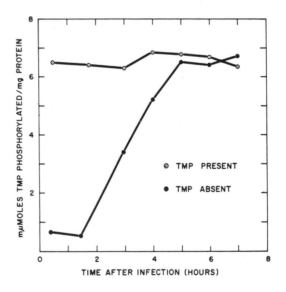

FIG. 2. Level of activity of TMP kinase in cell extracts prepared at various times after infection in the presence or absence of TMP. Cells were infected and were harvested at various times after infection. To part of the samples, TMP (20 μg/ml) was added at the time of harvest and was maintained in the samples throughout the experiment; to another part, no TMP was added. Extracts were prepared as described previously (Hamada *et al.*, 1966). TMP kinase was assayed as follows: To 0.2 ml of cell extract containing approximately 1 mg of protein, 0.55 ml of reaction mixture containing 5 μmoles MgCl₂, 2.5 mμmoles ATP, 25 μmoles Tris pH 7.5, 25 mμmoles TMP-^{14}C at a specific activity of 4×10^2 cpm/mμmole was added. The samples were incubated at 37°C for 15 min. The reaction was stopped and TMP was separated from TDP and TTP on Dowex columns.

absence of TMP. The enzyme present in the extracts of infected cells, on the other hand, appeared to be stable even when TMP was not added.

That this is indeed the case is shown by the fact that immediately after disruption of the cells, the level of TMP kinase in the infected and non-infected samples was the same (Fig. 3). However, whereas the activity of the enzyme in sonicates of infected cells was stable at 0°C, it was lost rapidly in sonicates of noninfected cells. It is clear, therefore, since the

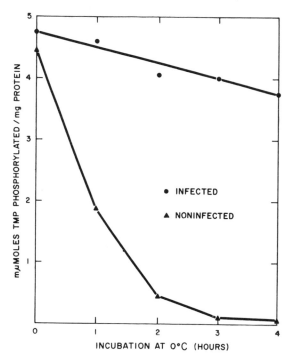

Fig. 3. Stability at 0°C of TMP kinase present in extracts of infected and non-infected cells. Noninfected cells, as well as cells which had been infected 6 hours previously, were harvested and sonicated. The activity of TMP kinase in the cell sonicates was assayed immediately thereafter, and after various periods of incubation at 0°C. Assay conditions as in the legend to Fig. 2.

activity of TMP kinase is assayed usually after partial purification, that the enzyme present in noninfected cells is inactivated during this procedure. By contrast, the TMP kinase in the extracts of infected cells is not affected, and the impression is thus created that infection has caused an increase in enzyme activity. In reality, however, the difference in enzyme activity between the extracts of infected and noninfected cells is due to a difference in stability at 0°C. The observed increase in the activity of

TMP kinase after infection of cells with Pr virus is thus an artifact stemming from an experimental procedure.

Stability of TMP Kinase *in Vivo*

As shown in the preceding section, the activity of TMP kinases in noninfected and in infected cells is approximately the same, but the enzymes differ in stability. This observation suggested that the enzyme present in noninfected cells has been stabilized as a result of infection. However, the same effect would be observed if a steady state existed, in which the TMP kinase normally present in the cells is inactivated at the same rate that new, stable, virus-induced enzyme is synthesized. To test this possibility, the stability of TMP kinase *in vivo* was examined in cells in which the synthesis of new enzyme protein was inhibited. Inactivation of the enzyme *in vitro* was prevented by including TMP in the suspension medium at the time of harvest. Table I shows that TMP kinase is stable *in vivo* in noninfected cells, since no decrease in activity occurred in cells treated with puromycin. Furthermore, the enzyme is not stabilized *in vivo* by thymidine derivatives, since the inhibition of the synthesis of these compounds by 5-fluorodeoxyuridine (FUDR) did not affect the stability of the enzyme in the cells.

That the enzyme present in the cells at the time of infection is also stable in the infected cells was demonstrated in a similar experiment, the results of which are summarized in Fig. 4. In this experiment, puromycin

TABLE I

ACTIVITY OF TMP KINASE IN EXTRACTS OF CELLS INCUBATED
FOR VARIOUS TIMES WITH PUROMYCIN AND FUDR[a]

	Time of incubation of the cultures with inhibitors (hours)			
Incubation medium	0	4	8	12
Eagles[b]	4.7[c]	4.7	4.6	4.2
Eagles + puromycin	—	3.7	3.2	4.0
Eagles + FUDR	—	4.0	4.6	4.8
Eagles + FUDR + puromycin	—	4.0	3.8	3.8

[a] Cultures of RK cells were incubated at 37°C for various times with puromycin (20 µg/ml) or FUDR (2 µg/ml). At the time of harvest, the cells were suspended in a sucrose-Tris-KCl solution containing TMP (20 µg/ml). Extracts were prepared and the activity of TMP kinase in the extracts was determined as described in the legend to Fig. 2.

[b] Eagle's medium (1959) + 3% dialyzed bovine serum.

[c] mµmoles TMP phosphorylated/mg of protein.

was added to part of the cultures 3 hours after infection. To one set of cultures, TMP was added at the time of harvest to stabilize the enzyme normally present in noninfected cells. Extracts of these cells gave the total level of enzyme activity in the infected cells. To a second group of cultures, no TMP was added and the enzymatic activity present in the extract therefore reflected the intracellular level of only stable virus-induced

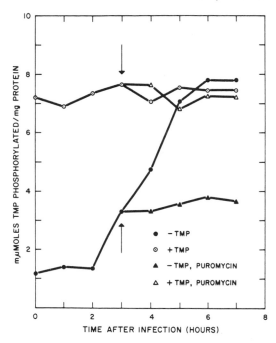

FIG. 4. Stability *in vivo* of TMP kinase present in the cells at the time of infection. Cells in stationary phase were infected. To part of the cultures, puromycin (20 μg/ml) was added 3 hours after inoculation (see arrows in figure). At various times thereafter, cells were harvested. To part of the cells, TMP (20 μg/ml) was added at the time of harvest; to part, no TMP was added. Cell extracts were prepared and the activity of TMP kinase assayed. The assay conditions were as described in the legend to Fig. 2.

enzyme. If the enzyme present in the noninfected cells had been inactivated within the infected cells, as incubation of the cells with puromycin proceeded, the activity of the enzyme in the TMP-stabilized extracts prepared from these cells should decrease to the level of activity present in the extracts which have not been stabilized with TMP. This was not the case and no change in the level of activity of the enzyme could be detected either in the TMP-stabilized or unstabilized extracts prepared from the cells after various periods of incubation with puromycin.

Thus, the TMP kinase present in the cell at the time of infection remains stable *in vivo* in the infected cells. The appearance after infection of enzyme stable *in vitro* must, therefore, represent a stabilization of the enzyme present in noninfected cells, not the synthesis of new enzyme protein.

We have shown that the appearance of TMP kinase activity which is stable *in vitro* requires protein synthesis (see Fig. 4). Therefore, in this case, we are dealing with an enzyme, whose stabilization requires protein synthesis. Hence, the fact that protein synthesis is required if an increase in enzyme activity is to occur in virus-infected cells does not necessarily imply that the enzyme protein is formed *de novo*.

Mechanism of Stabilization of TMP Kinase in Infected Cells

It is simplest to assume that the stabilization of TMP kinase in infected cells is due to the accumulation of thymidine derivatives within these cells. However, the evidence presented below shows that this is not the case. Figure 5 shows that stabilization of TMP kinase occurred at the same rate in infected cells that have been treated with FUDR as in untreated, infected cells. That thymidine was limiting under the experimen-

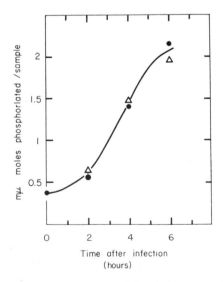

FIG. 5. Level of activity of stable TMP kinase in infected FUDR-treated and untreated cells. Cells were infected and incubated in Eagles medium in the presence or absence of FUDR (2 μg/ml). At various times, cells were harvested, extracts were prepared, and the activity of TMP kinase was assayed (see legend to Fig. 2). ●, FUDR; △, no FUDR.

tal conditions used was shown by the fact that the rate of DNA synthesis in the FUDR-treated cells was decreased by 90%. These results show that thymidine derivatives cannot be responsible for the stabilization of TMP kinase in infected cell extracts.

The activity of TMP kinase in infected cell extracts is also relatively stable to dialysis (Fig. 6), and although some activity is lost upon dialysis, this decrease in activity is not due to a loss of TMP. This can be concluded from the rate of inactivation, upon dialysis, of the enzyme present in extracts of infected cells and in extracts of noninfected cells to which

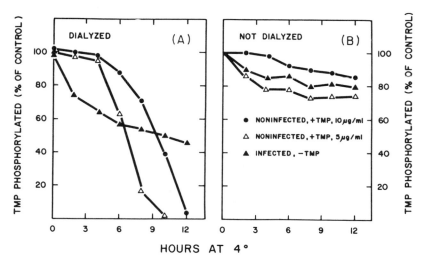

FIG. 6. Effect of dialysis on activity of TMP kinase present in infected cells. Extracts of infected cells (harvested 6 hours post-infection), as well as extracts of noninfected cells to which TMP (5 μg or 10 μg/ml) was added, were prepared. Part of each sample was dialyzed (A) at 4°C against a sucrose-KCl-Tris-solution containing no TMP, while the other part was not dialyzed (B) but was incubated at 4°C. At various times, samples were taken and assayed for TMP kinase as described in the legend to Fig. 2.

TMP has been added. The rapid loss of activity of the enzyme present in the noninfected cell extracts is presumably due to the removal, by dialysis, of TMP. Since the enzyme in the infected cell extract is much more stable to dialysis, the stabilizing agent cannot therefore be TMP.

We have attempted to create conditions in which the TMP kinase obtained from noninfected cell extracts would be stable *in vitro,* thus simulating the conditions prevailing in infected cell extracts. A number of conditions have been employed with varying degrees of success. The enzyme can be stabilized (Fig. 7) if mercaptoethanol is added at least 60 min

prior to the disruption of the cells. However, the addition of mercaptoethanol at the time the cells are disrupted fails to stabilize the enzyme.
Whether sulfhydryl groups are involved in the stabilization of the enzyme
in infected cells remains to be established.

Attempts to stabilize the enzyme of noninfected cells with extracts of
infected cells have yielded inconsistent results. No stabilization was obtained when extracts of infected cells were mixed with noninfected cells

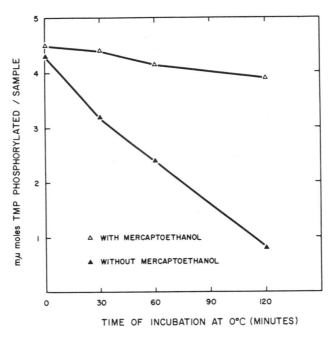

Fig. 7. Stability of TMP kinase of noninfected cells in the presence of mercaptoethanol. RK cells were harvested in a sucrose-Tris-KCl solution containing mercaptoethanol (200 μmoles/ml) and incubated at 0°C for 1 hour. They were then
sonicated and the activity of TMP kinase assayed immediately thereafter, as well
as after various times of incubation at 0°C. Mercaptoethanol has little effect on the
stability of the enzyme, if added immediately prior to sonication. TMP kinase was
assayed as described in the legend to Fig. 2.

and the samples were then disrupted (Table II). That the enzyme present
in extracts of infected cells remains stable despite dilution is shown by the
experiment summarized in Table III. These results show also that the inactivation of the enzyme present in noninfected cell extracts is not due to
an inhibitor, since extracts of noninfected cells did not affect the activity
of the enzyme present in the extracts of infected cells.

A variable amount of stabilization of the enzyme present in noninfected

TABLE II

LACK OF STABILIZATION OF TMP KINASE OF NONINFECTED CELLS
BY EXTRACTS OF INFECTED CELLS[a]

Sample	Time of incubation at 0°C	
	0 hours	2 hours
Noninfected cells (N)	1.45[b]	0.03
Infected cell extract (I)	1.42	1.32
N + I, mixed at 0 hours	1.48	0.86
N + I, mixed at 2 hours	—	0.77

[a] Cells were infected with Pr virus (adsorbed multiplicity = 10). Six hours after infection, the cells were harvested and extracts were prepared. An equal volume of infected cell extract was mixed with noninfected cells either prior to their sonication or after the noninfected cells had been sonicated and incubated at 0°C for 2 hours. The enzyme activity in the samples was assayed at the times indicated.

[b] mμmoles TMP phosphorylated/sample.

cells could, however, be obtained when infected and noninfected cells were mixed prior to sonication. The results obtained from this type of experiment have not always been reproducible and we have been unable to determine and to control the factors involved in enzyme stabilization. Therefore, we are not able to state unequivocally that a factor present in the infected cells can stabilize the enzyme present in noninfected cells.

It is clear from the mixing experiment that the factor(s) leading to the stabilization of TMP kinase in infected cells is itself not stable. The stabilization of the TMP kinase may be due to changes in the intracellular milieu and the establishment of conditions which favor a change in configuration of the enzyme, with this change conferring new properties upon

TABLE III

STABILITY OF TMP KINASE IN EXTRACTS OF INFECTED CELLS[a]

Mixtures[b]	TMP Kinase activity[c] after incubation at 0°C	
	0 hours	5 hours
Infected cell extract + sucrose[d]	1.64	1.58
Noninfected cells + sucrose	1.30	0.20
Infected cell extract + noninfected cells	2.70	1.68

[a] This experiment was performed as described in the legend to Table II.

[b] 0.1 ml aliquots of the components of each mixture were mixed prior to sonication.

[c] mμ moles TMP phosphorylated/sample.

[d] Sucrose = sucrose-KCl-Tris solution.

the enzyme. It is also possible, however, that the stabilization is direct and due to a specific protein. However, since a similar phenomenon has not yet, as far as we are aware, been described for any other system, we think that this possibility is unlikely.

We favor the following interpretation of our results: Infection results in the formation of some factor that causes a change in the configuration of TMP kinase and confers stability upon the enzyme. This factor is either unstable or is diluted and lost upon disruption of the cells. The reverse reaction (that is, a return of TMP kinase to its original configuration with a consequent loss of stability) is slow.

Comparison of the Properties of the TMP Kinases Present in Infected and Noninfected Cells

We have tried to detect a difference in configuration between the enzymes present in infected and noninfected cells. In view of the instability of the enzyme present in noninfected cell extracts, these experiments were performed in the presence of TMP. Under these conditions, the enzymes from both sources behave in an identical fashion upon filtration on Sephadex (G-100) columns. Furthermore, the sensitivity of both enzymes to varying concentrations of parachloromercuribenzoate is identical (Eisenstein and Kaplan, unpublished results). We have also been unable to find any differences between the enzymes in their affinity for substrate and thus, in the presence of TMP, the enzymes appear to behave in an identical fashion. However, the addition of TMP to the enzyme may obliterate the differences in configuration between the enzymes present in infected and noninfected cells. We are now attempting to find conditions in which partial purification of the enzyme in the absence of substrate stabilization can be achieved.

Control of the Cut-Off in the Increase of Enzyme Activity in Virus infected Cells

The increases in the activities of certain enzymes that follow virus infection have been shown in a variety of virus-cell systems to continue for a limited period of time only, unless there is interference with a mechanism which normally regulates the level of these enzymes. This interference can be achieved by UV irradiation of the infecting virus (Dirksen *et al.,* 1960; Delihas, 1961; McAuslan, 1963), by inhibiting viral DNA synthesis with amethopterin (McAuslan, 1963), by the use of mutants that do not allow the synthesis of viral DNA to proceed at a normal rate (Wiberg *et al.,* 1962), or by the inhibition of RNA synthesis during certain stages of the infective process (McAuslan, 1963). We have shown that the cut-off in the increase of enzyme activity does not occur in Pr virus-infected cells when 5-bromodeoxyuridine is substituted for thymidine in progeny viral DNA (Kamiya *et al.,* 1965). If we take these results to mean that uncontrolled synthesis of thymidine kinase or polymerase occurs in the infected cells under these conditions, and if, in contrast to thymidine kinase, TMP kinase is not synthesized *de novo,* higher levels of activity of the latter enzyme should not be observed in the uncontrolled systems. The following experiments, in which an uncontrolled increase in thymidine kinase activity was obtained by incubating the cells either with iododeoxyuridine (IUDR) or actinomycin, showed that this was indeed the case.

Figure 8 illustrates the levels of activity of thymidine kinase and TMP kinase in cells treated with IUDR. The presence of high concentrations of thymidine or IUDR in the medium stabilizes TMP kinase in the cell extracts, since the level of activity of the enzyme is high in noninfected cells and does not change during the infective process. However, although the level of activity of TMP kinase was the same in the IUDR- and the thymidine-treated cells, the activity of thymidine kinase reached a higher level in the IUDR-treated cells. Since a high intracellular con-

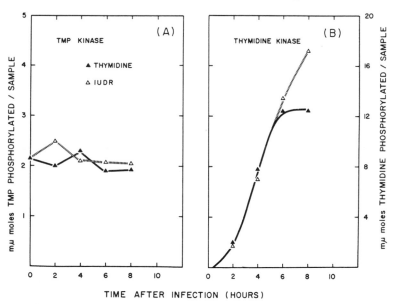

FIG. 8. Effect of IUDR on the level of activity of TMP (A) and thymidine kinases (B). Stationary phase cells were infected and incubated with Eagle's medium containing FU (10 μg/ml) and either IUDR (100 μg/ml) or thymidine (100 μg/ml). (FU was added to inhibit the synthesis by the cells of thymidine derivatives.) At various times, cells were harvested, extracts were prepared, and the activities of thymidine kinase and TMP kinase were determined.

centration of thymidine derivatives may influence the pattern of behavior of the TMP kinase, we determined the behavior of this enzyme when actinomycin was added to the cultures 4 hours after infection. In this case, an uncontrolled increase in the level of thymidine kinase activity occurred (Fig. 9). TMP kinase activity, however, did not increase much above the level found in the infected, untreated cells.

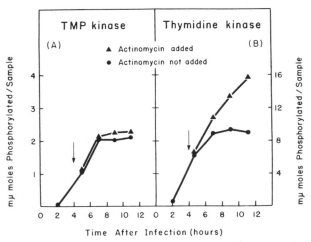

FIG. 9. Effect of actinomycin on the level of activity of TMP (A) and thymidine kinases (B). Stationary phase cells were infected and incubated at 37°C in Eagle's medium. Four hours after infection, actinomycin (2 μg/ml) was added to part of the cultures. At various times thereafter, cells were harvested, and extracts were prepared. TMP (20 μg/ml) was added to the extracts 30 min prior to the determination of their level of enzyme activity. The activities of thymidine kinase and TMP kinase were determined as described previously.

These results are a further corroboration of the premise that the increase in the level of activity of TMP kinase in the infected cells is due to the stabilization of enzyme present in the cells at the time of infection and is not due to the *de novo* synthesis of enzyme protein. Since TMP kinase is stable *in vivo* in the noninfected cells, one wonders at the possible biological significance of a change which results in a stabilization of the enzyme *in vitro*. It seems possible that this stabilization is incidental to changes in the properties of the enzyme which endow it with increased activity *in vivo*.

Speculations on the Significance of Increased Activities of TMP and Thymidine Kinases in Infected Cells

The fact that the activity of the enzymes leading to the formation of thymidine triphosphate may be a key step in the synthesis of DNA has been discussed extensively. The increase in activity of these enzymes after infection of cells in stationary phase, in which the level of activity of these enzymes is low and which, consequently, would otherwise be unable to synthesize DNA, is therefore essential to the survival of the infecting virus. However, a considerable increase in enzyme activity also occurs in cells infected in logarithmic phase, in which the overall rate of DNA synthesis after infection is not changed appreciably. This is shown by the experiments illustrated in Fig. 10 and Table IV. From these data, it would appear that infected log-phase cells contain almost ten times as much enzyme activity as is necessary to allow a given rate of DNA synthesis. This would appear to be an inefficient process, and the possible survival value for the virus of this high level of activity is puzzling.

The possible role of thymidine kinase in growing cells has been the sub-

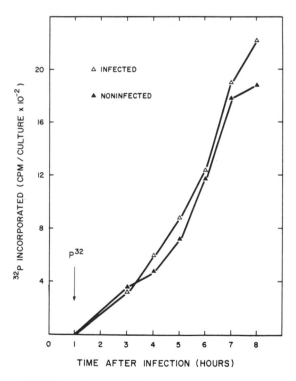

Fig. 10. Rate of DNA synthesis in infected and noninfected log-phase RK cells. Five-day-old RK cultures were infected and incubated in Eagle's medium containing ^{32}P (1μCi/ml). Mock-infected cultures were treated identically. The amount of ^{32}P incorporated into DNA was determined as described previously (Kaplan and Ben-Porat, 1963).

ject of much speculation. The function of thymidine kinase may be explained in terms of competition between anabolic and catabolic enzymatic pathways as suggested by Potter (1958, 1960) and Canellakis *et al.* (1959). An increase in the level of activity of this enzyme provides an alternative pathway for TMP synthesis. However, the results of Dubbs and Kit (1964) seem to indicate that the presence of thymidine kinase in a virus-infected cell has no survival value for the virus, since virus mutants lacking the ability to induce the formation of this enzyme grow as well as wild-type virus grows.

TABLE IV

LEVEL OF ACTIVITY OF KINASES IN INFECTED AND NONINFECTED LOG-PHASE CELLS[a]

Mixtures	Thymidine kinase, Thymidine phosphorylated (mμmoles/sample)	TMP kinase, TMP phosphorylated (mμmoles/sample)
Infected	9.6	2.8
Noninfected	1.2	0.1

[a] Noninfected and infected log-phase cells (6 hours after infection) were suspended in sucrose-Tris-KCl at an approximate density of 10^7 cells/ml. Extracts were prepared and thymidine kinase and TMP kinase were assayed as described previously (Hamada *et al.*, 1966).

It has been established with a variety of tissues that actively growing cells contain lower phosphatase levels than do nongrowing cells (Maley and Maley, 1961; Beltz, 1962; Fiala *et al.,* 1962; Eker, 1965), and it is conceivable that while thymidine kinase may have no survival value when infection occurs in growing cells, it may be of importance if viral DNA synthesis is to occur in resting cells. In order to overcome the effect of the catabolic enzymes present in the nongrowing cells, the infecting virus must establish a mechanism to ensure the accumulation of the deoxynucleotide triphosphates necessary for the synthesis of viral DNA. This could be achieved by an inactivation of the catabolic enzymes. On the other hand, the same end could be attained by an increased level of activity of deoxynucleoside and deoxynucleotide kinases, thereby counteracting the effects of the phosphatases.

The data in Table V illustrate what may be the relative importance of thymidine kinase in infected and noninfected cells. The level of phosphatase activity is indeed much greater in stationary-phase cells than in log-phase cells, and infection does not change the level of activity of the phosphatases in either case (see Table V A). Since the dephosphorylation of TMP occurs at a rapid rate in stationary-phase cells, the accumula-

TABLE V

CYCLIC CONVERSION OF TMP IN INFECTED STATIONARY-PHASE CELLS[a]

	TMP kinase	Phosphatase
	TMP phosphorylated (mμmoles/sample)	TMP dephosphorylated (mμmoles/sample)
A. No ATP present:		
1. Log-phase cells		
Infected	0	0.8
Noninfected	0	0.9
2. Stationary-phase cells		
Infected	0	4.7
Noninfected	0	4.5
B. ATP present:		
1. Log-phase cells		
Infected	2.2	0.3
Noninfected	0.1	0.7
2. Stationary-phase cells		
Infected	3.8	0.1
Noninfected	0.1	4.5

[a] In this experiment, primary monolayer cultures of RK cells were in logarithmic phase 5 days and in stationary phase 8 days after seeding. The cells were infected (adsorbed multiplicity = 10) and were harvested 6 hours after infection. Extracts were prepared and the activity of TMP kinase was assayed as described in the legend to Fig. 2. The activity of the phosphatases able to dephosphorylate TMP was assayed by omitting ATP from a reaction mixture similar to that used for the assay of TMP kinase.

tion of TMP in these cells would presumably be prevented. It is obvious, however, (see Table V, Part B) that thymidine is rapidly rephosphorylated in the extracts of infected stationary-phase cells and that the high phosphatase activity in these extracts is thus completely overcome. If the results of these *in vitro* experiments may be assumed to reflect the conditions within the cell, the important role of thymidine kinase in infected stationary-phase cells is obvious. The relatively high activity of thymidine kinase induced in cells by infection has survival value only if the virus infects cells in stationary phase in which the level of activity of phosphatases is high.

Summary and Conclusions

After infection with Pr virus, extracts of rabbit kidney cells exhibit an increased level of activity of TMP kinase and thymidine kinase. The increase in the level of activity of TMP kinase, although prevented by inhibitors of protein synthesis, is not due to the synthesis of new enzyme protein. Instead, as a result of infection, the enzyme present in noninfected cells, which is unstable *in vitro,* seems to acquire new properties which confer upon it stability *in vitro.* The difference in the level of activity of TMP kinase present in extracts prepared from infected and noninfected cells is thus an artifact resulting from the difference in stability *in vitro* of the enzyme present in these extracts.

TMP kinase is stable *in vivo* in noninfected, as well as infected cells, and since an increase in the stability of the enzyme *in vitro* presents no obvious advantage to the growth of the virus, we assume that this virus-induced change may also affect other properties of the enzyme that cause it to possess greater activity *in vivo.* However, aside from the difference in stability, we have been unable to detect any other differences in behavior between the enzymes present in infected and noninfected cells. These experiments were done with cell extracts in which the enzymes were stabilized with relatively high concentrations of TMP, and thus it is conceivable that these conditions may have obliterated differences between the enzymes.

Experiments, in which the cells were exposed to conditions in which normal cut-off of the increase in enzyme synthesis does not occur, revealed that whereas the cut-off of thymidine kinase activity is subject to a control mechanism, the cut-off of TMP kinase activity is not. This finding lends further support to the premise that the mechanisms by which the levels of activity of these two enzymes are increased in infected cells differ from one another.

The reason for the increase in the activity of thymidine kinase, in the infected cells was examined. Although the large increase in the level of thymidine kinase activity seems to have little survival value for Pr virus when the infective process occurs in growing cells, it may be essential when the virus infects stationary phase cells in which the level of activity of phosphatases is high. The same state of affairs would also be true for TMP kinase, if the configuration of TMP kinase present in the infected cells is changed in such a way as to confer upon the enzyme higher activity *in vivo.*

REFERENCES

Beltz, R. E. (1962). *Arch. Biochem. Biophys.* **99,** 304.
Ben-Porat, T., and Kaplan, A. S. (1963). *Virology* **20,** 310.
Ben-Porat, T., and Kaplan, A. S. (1965). *Virology* **25,** 22.
Bojarski, T. B., and Hiatt, H. H. (1960). *Nature* **188,** 1112.
Canellakis, E. S., Jaffee, J. J., Mantsavinos, R., and Krakow, J. S. (1959). *J. Biol. Chem.* **234,** 2096.

Changeux, J-P. (1963). *Cold Spring · Harbor Symp. Quant. Biol.* **28,** 497.

Cohen, S. S. (1961). *Federation Proc.* **20,** 641.

Consigli, R. A., and Ginsberg, H. S. (1964). *J. Bacteriol.* **87,** 1034.

Delihas, N. (1961). *Virology* **13,** 242.

Dirksen, M. L., Wiberg, J. S., Koerner, J. F., and Buchanan, J. M. (1960). *Proc. Natl. Acad. Sci. U.S.* **46,** 1425.

Dubbs, D. R., and Kit, S. (1964). *Virology* **22,** 214.

Dulbecco, R., Hartwell, L. H., and Vogt, M. (1965). *Proc. Natl. Acad. Sci. U.S.* **53,** 403.

Eagle, H. (1959). *Science* **130,** 432.

Eisenstein, S., and Kaplan, A. S. (1966). Unpublished results.

Eker, P. (1965). *J. Biol. Chem.* **240,** 419.

Fiala, S., Fiala, A., Tobar, G., and McQuilla, H. (1962). *J. Natl. Cancer Inst.* **28,** 1269.

Freundlich, M., and Umbarger, H. E. (1963). *Cold Spring Harbor Symp. Quant. Biol.* **28,** 505.

Gerhart, J. C., and Pardee, A. B. (1963). *Cold Spring Harbor Symp. Quant. Biol.* **28,** 491.

Green, M., and Pina, M. (1962). *Virology* **17,** 603.

Hamada, C., and Kaplan, A. S. (1965). *J. Bacteriol.* **89,** 1328.

Hamada, C., Kamiya, T., and Kaplan, A. S. (1966). *Virology* **28,** 271.

Hanafusa, T. (1961). *Biken's J.* **4,** 97.

Kamiya, T., Ben-Porat, T., and Kaplan, A. S. (1965). *Virology* **26,** 577.

Kaplan, A. S., and Ben-Porat, T. (1963). *Virology* **19,** 205.

Kaplan, A. S., and Ben-Porat, T. (1964). *Virology* **23,** 90.

Keir, H. M. (1965). *Biochem. J.* **94,** 3P.

Keir, H. M., and Gold, E. (1963). *Biochim. Biophys. Acta* **72,** 263.

Kit, S., and Dubbs, D. R. (1965). *Virology* **26,** 16.

Kit, S., Dubbs, D. R., and Piekarski, L. J. (1962). *Biochem. Biophys. Res. Commun.* **8,** 72.

Kit, S., Frearson, P. M., and Dubbs, D. R. (1965). *Federation Proc.* **24,** 596.

Ledinko, N. (1966). *Virology* **28,** 679.

McAuslan, B. R. (1963). *Virology* **21,** 383.

McAuslan, B. R., and Joklik, W. K. (1962). *Biochem. Biophys. Res. Commun.* **8,** 486.

Maley, F., and Maley, G. F. (1961). *Cancer Res.* **21,** 1421.

Newton, A. A., and McWilliam, S. (1962). *Biochem. J.* **84,** 112P.

Nohara, H., and Kaplan, A. S. (1963). *Biochem. Biophys. Res. Commun.* **12,** 189.

Potter, V. R. (1960). "Nucleic Acid Outlines," p. 219. Burgess, Minneapolis, Minnesota.

Potter, V. R. (1958). *Federation Proc.* **17,** 691.

Sheinin, R. (1966). *Virology* **28,** 47.

Tomkins, G. M., Yielding, K. L., Talal, N., and Curran, J. F. (1963). *Cold Spring Harbor Symp. Quant. Biol.* **28,** 461.

Wiberg, J. S., Dirksen, M. L., Epstein, R. H., Luria, S. E., and Buchanan, J. M. (1962). *Proc. Natl. Acad. Sci. U.S.* **48,** 293.

Control of Biosynthesis of Host Macromolecules in Cells Infected with Adenoviruses

Harold S. Ginsberg,* Leonard J. Bello,†
and Arnold J. Levine**

DEPARTMENT OF MICROBIOLOGY, UNIVERSITY OF PENNSYLVANIA
SCHOOL OF MEDICINE, PHILADELPHIA, PENNSYLVANIA

Introduction

The study of the regulation and control of cellular processes offers an approach to an understanding of the basis for differentiation and development, and even oncogenesis. It is abundantly clear, from the preceding chapters in this volume, that at least this selected population of virologists considers the control of host biosynthetic processes to be a fundamental biological problem, and virus-infected cells to be an excellent tool with which to investigate it. We have already learned that there are repressors, suppressors, depressors, and "impressors," but the nature of and the mechanism by which any of these factors act, except for the latter, are ambiguous.

From the earliest studies of Cohen, describing the biochemical events in T-even phage-infected cells (Cohen, 1948, 1949), it was clear that viral infection could effect inhibition of biosynthesis of host DNA (Hershey *et al.,* 1953; Nomura *et al.,* 1962). The effect of phage infection on host protein synthesis was demonstrated in a series of papers beginning with that of Monod and Wollman, showing that β-galactosidase could not be induced in the phage-infected cell (Monod and Wollman, 1947; Benzer, 1953; Levin and Burton, 1961). Subsequent investigations of the biochemistry of mammalian cells infected with poliovirus (Penman and Summers, 1965; Zimmerman *et al.,* 1963), Mengo virus (Frank-

* Present address: Institute of Biochemistry, University of Lausanne, Lausanne, Switzerland.

† Present address: Laboratory of Microbiology, School of Veterinary Medicine, University of Pennsylvania, Philadelphia.

** Present address: Division of Biology, California Institute of Technology, Pasadena, California.

lin and Baltimore, 1962), pseudorabies virus (Kaplan and Ben-Porat, 1963), and vaccinia virus (Becker and Joklik, 1964; Joklik and Becker, 1965; Salzman *et al.,* 1964; Shatkin, 1963) indicated that the synthesis of host proteins as well as of nucleic acids may be inhibited by viral infection. The finding that host cell biosynthetic processes could be restricted while viral macromolecules were being made, suggested that some selective control mechanisms were responsible for the effects observed. Hence, the virus-infected cell seemed to hold out an opportunity for an exploration of cellular regulatory processes.

The use of viruses has additional advantages for investigating control mechanisms in cells. The time at which the controlled events occur are regular and predictable, and the effects on the biosynthesis of host macromolecules can be correlated with the production of viral subunits and the sequential control of their synthesis.

Type 5 adenovirus, the agent employed in the experiments to be described here, appeared to be a particularly suitable agent with which to investigate this complex problem. It is readily propagated in cell cultures, it can be quantitatively assayed by the plaque technique, and the viral yield is high,—average 8,000–10,000 PFU/cell. The virus is stable in crude suspensions, but can be easily purified by isopycnic centrifugation in a linear density gradient of CsCl (Wilcox and Ginsberg, 1963a). Hence, the chemical, physical, and immunological characteristics of the virus were known, and the structure of the virion and its subunits had been determined (Allison *et al.,* 1960; Klemperer and Pereira, 1959; Valentine and Pereira, 1965; Wilcox and Ginsberg, 1961, 1963a).

The virions are naked icosahedrons approximately 700 Å in diameter (Horne *et al.,* 1959). The capsid consists of 252 capsomers, of which 240 are on the faces of the virion (Horne *et al.,* 1959; Valentine and Pereira, 1965; Wilcox and Ginsberg, 1963a). These are polygonal-like structures 70 Å in diameter, with a central hole (Valentine and Pereira, 1965; Wilcox and Ginsberg, 1963a). Each has 6 neighbors and, therefore, is termed a hexon (Ginsberg *et al.,* 1966). At the 12 corners of the icosahedral virion are fibers attached to a base (Valentine and Pereira, 1965). This complex structure is at the center of the fivefold axis of symmetry, and is termed a penton because each is surrounded by 5 neighboring capsomers (Ginsberg *et al.,* 1966) (Fig. 1). The capsomers are joined in the capsid by noncovalent bonds which are readily disrupted by high pH and detergents (Lawrence and Ginsberg, 1966; Wilcox and Ginsberg, 1963a).

Adenoviruses contain DNA, and like type 2 (Green, 1962), type 5 virus contains about 13% DNA. The DNA is a typical double-stranded molecule, having a molecular weight of about 23×10^6 daltons, and a

ADENOVIRUS MORPHOLOGY

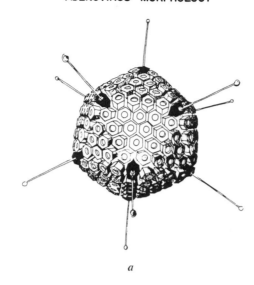

a

ADENOVIRUS MORPHOLOGY

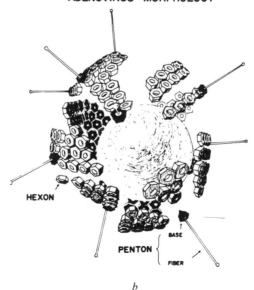

HEXON

PENTON { BASE

FIBER

b

FIG. 1. Diagrammatic model of type 5 adenovirus particle: (*a*) the intact virion, (*b*) partially disrupted particle demonstrating the structure and position of the known capsid units and the viral core.

guanine + cytosine content of approximately 58%. The DNA has a sharp melting profile with a T_m of 92.4°C, and a bouyant density in CsCl of 1.718 gm/cm^3. The cells which are hosts for adenoviruses possess a DNA which is distinctly different from that of the virus, having a guanine + cytosine content of about 42%. As a consequence, the viral DNA can be readily separated from host cell DNA by physical means.

The biosynthetic events leading to the formation of infectious viral particles follows an orderly sequence, commencing with the required synthesis of RNA about 9 hours after virus is added to susceptible cells. The stepwise production of viral DNA, viral capsid proteins, and virions follows (Flanagan and Ginsberg, 1962, 1964; Wilcox and Ginsberg, 1962, 1963b). The reason for the long delay before viral replication begins is still unclear, but it may be related to a tardiness in the entry of the viral genome into the nucleus, or to a requirement for a second stage in uncoating the viral DNA rather than to the initiation of "uncoating" (Lawrence and Ginsberg, 1965). Despite the orderliness of the biosynthetic process involved, the production of virions is an inefficient process, since only 5–10% of the viral DNA and protein is actually assembled into mature particles (Ginsberg and Dixon, 1961).

The objective of this communication is to examine the interrelationships between the steps of viral replication and the biosynthesis of host macromolecules, and to explore the factors that might play a role in the control of these processes.

Experimental

Biosynthesis of DNA, RNA, and Protein in Uninfected, and Type 5 Adenovirus-Infected KB Cells

Cells dividing in suspended-cell cultures exhibit a linear increase in the levels of DNA, RNA, and protein. When exponentially growing cells are infected with type 5 adenovirus, there is no difference between infected and uninfected cells with respect to the accumulation of DNA and protein for at least 40 hours after infection (Fig. 2). RNA synthesis, however, appears to cease in infected cells at about 24 hours after infection. Hence, although infected cells do not divide and the synthesis of viral DNA and protein begins 10 and 12 hours, respectively, after infection, there is no difference between uninfected and virus-infected cells in either the amounts of DNA and protein synthesized or the rates at which they are made. Therefore, it follows that production of host DNA and protein must decline at approximately the same time as that at which the synthesis of the viral precursors commences. Although the synthesis of

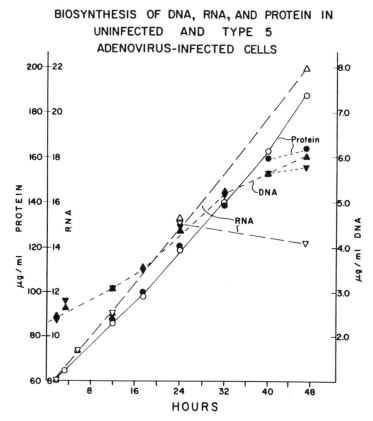

FIG. 2. Synthesis of DNA, RNA, and proteins in uninfected, and type 5 adeno-virus-infected KB cells in spinner cultures. Cells were infected with a multiplicity of 100 PFU/cell.

virus-specific RNA (i.e., mRNA) begins about 9 hours after infection, the amount made is so small relative to the total cellular RNA that it cannot be detected by these measurements. Biosynthesis of host RNA must stop by 24 hours after infection.

Biosynthesis of Host DNA in KB Cells Infected with Type 5 Adenovirus

To determine when the synthesis of host DNA is inhibited in infected cells, it was necessary to separate viral from host DNA. On the basis of differences in the GC content between the DNA's of the virus and host cell, DNA separation was obtained by chromatography on methylated albumin-kieselguhr (MAK) columns (Mandell and Hershey, 1960), and by equilibrium centrifugation in a cesium chloride density gradient. Because larger quantities of DNA could be separated on MAK columns than by CsCl centrifugation (thereby facilitating additional characterization), chromatography was used in the experiments to be described here.

Cells were disrupted with 1.0% desoxycholate, and the homogenate treated successively with 0.25 *M* 2-mercaptoethanol, and pronase (0.5 mg/ml). DNA was then isolated by extracting the suspension at least 3 times with redistilled, water-saturated phenol. The results of a representative experiment, presented in Fig. 3, demonstrate the elution char-

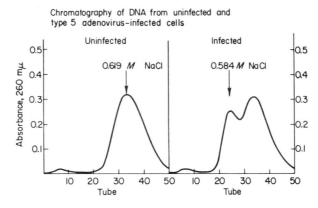

Fig. 3. Chromatography of DNA on methylated albumin-kieselguhr (MAK) columns. The DNA was extracted from uninfected, and type 5 adenovirus-infected KB cells grown in suspended cell cultures. Cells were infected with 50 PFU/cell. DNA was eluted from MAK with a linear gradient of NaCl between 0.50 and 0.85 *M* at pH 6.8.

acteristics of DNA obtained from uninfected and infected cells. There appeared, in the DNA from infected cells, a new species of DNA, which was readily separated from host DNA because it eluted at a lower NaCl concentration.

To identify the new DNA, it was centrifuged to equilibrium in a CsCl density gradient. Its buoyant density was found to be 1.718 gm/cm^3— a value identical to that of viral DNA obtained from purified type 5 adenovirus. To characterize further the new species of DNA obtained from infected cells, it was mixed with DNA extracted from purified virus that had been propagated in the presence of carrier-free, ^{32}P-orthophosphate, and the mixture was chromatographed on a MAK column. The isotopically labeled viral DNA and the new species of DNA which accumulated in adenovirus-infected cells were found to have identical chromatographic characteristics on MAK columns.

Effect of Adenovirus Infection on Synthesis of KB Cell DNA

It was predicted, on the basis of data summarized graphically in Fig. 2, that the synthesis of host cell DNA would be inhibited in cells infected with type 5 adenovirus, and that the inhibition would be evident by the time replication of viral DNA began. To test these predictions, KB cells, in the exponential phase of growth, were infected with an input viral multiplicity of 100 PFU/cell. Thymidine-^{3}H was added to the infected and uninfected cultures for a 2 hour period—between 10 and 12 hours post-infection. The DNA was extracted from both cultures; 50 μg of unlabeled DNA from purified virus was added to each as an OD marker; and, the mixtures were chromatographed on MAK columns. The results of this experiment, summarized in Fig. 4, demonstrated that by 10 hours

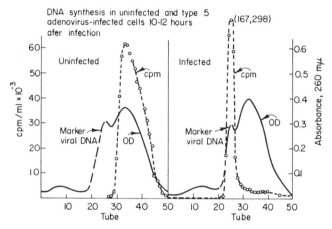

FIG. 4. Inhibition of synthesis of host cell DNA 10 to 12 hours after infection with 100 PFU/cell of type 5 adenovirus. Ten hours after infection thymidine-^{3}H was added to uninfected and infected cultures. DNA was extracted from both cultures 12 hours after infection. To each sample was added 50 μg of unlabeled DNA obtained from purified virus, and each mixture was chromatographed on MAK columns using a linear gradient from 0.50 to 0.85 M NaCl at pH 6.8. The OD$_{260}$ of each sample was determined in a Vanguard recording spectrophotometer, and the ^{3}H content was assayed in a Packard Tricarb scintillation spectrometer.

after infection, synthesis of host cell DNA was markedly inhibited, whereas viral DNA was synthesized at a rapid rate during this period. It could not be determined with certainty that host cell DNA synthesis was completely blocked, but based on the specific activity of the DNA made in uninfected cultures, it could be calculated that a maximum of about 3% of the normal complement of cellular DNA was made in the virus-infected cells.

The time at which the inhibition of synthesis of cellular DNA is first demonstrable was determined by measuring the incorporation of thymidine-^{3}H at various times after infection with 200 PFU/cell. The results of a representative experiment, summarized in Table I, indicate that host cell DNA was synthesized at the usual rate until approximately 6 hours after infection. By 6 to 8 hours after infection, the synthesis of cellular DNA was diminished by about 65% and the biosynthesis of viral DNA was just beginning (probably during the last part of the 2 hour period). During the period 8 to 10 hours after infection, replication of viral DNA was well established, while only approximately 10% of the usual amount of host DNA was synthesized. Only viral DNA was produced by 10 hours after infection.

TABLE I

TIME AT WHICH SYNTHESIS OF KB CELL DNA IS INHIBITED IN CELLS INFECTED WITH
TYPE 5 ADENOVIRUS

Time after infection (hours)[a]	% Inhibition[b]
4–6	6
6–8	66
8–10	91[c]
10–12	96[c]

[a] 200 PFU/cell.

[b] Measured by incorporation of thymidine-^{3}H.

[c] Minimal values for inhibition; no peak of radioactivity corresponding to host DNA was detectable.

Effect of Multiplicity of Infection on Synthesis of Host Cell DNA

To determine whether the multiplicity of viral infection affected the time at which the replication of cellular DNA was blocked, experiments in which the virion: cell ratio varied from 250 : 1 to 10 : 1 were carried out. The results of these experiments, which are summarized in Table II, indicate that when cells were infected with an input viral multiplicity of 10 PFU/cell, only minimal inhibition of host cell DNA synthesis was detectable during the period 8 to 10 hours post-infection. When the multiplicity was 100 or 250 PFU/cell, inhibition of host DNA synthesis commenced during the period 6 to 8 hours post-infection. Moreover, the extent to which host DNA was synthesized was inversely proportional to the input multiplicity of virus.

These data show that the time at which the synthesis of host cell DNA was inhibited was directly dependent upon the number of virions infecting

TABLE II

EFFECT OF MULTIPLICITY OF INFECTION ON SYNTHESIS OF KB CELL DNA

Multiplicity PFU/cell	Time at which synthesis was measured (hours post-infection)	Inhibition of cellular DNA[a] (%)	Viral DNA synthesis[a]
250	6–8	56	++
100	6–8	45	+
10	6–8	0	0
250	8–10	89[b]	++++
100	8–10	77	++++
10	8–10	42	+

[a] Measured by incorporation of thymidine-³H.

[b] Minimum value; i.e., no peak of radioactivity was detectable.

a cell, and suggest that either a component of the infecting viral particles effects the inhibition directly, or that a subunit of the virions induces the synthesis of a substance which in turn blocks the replication of host cell DNA.

The Role of Protein Synthesis in the Inhibition of Synthesis of KB Cell DNA

It was reasoned that if the infecting virions induce the synthesis of a macromolecule which inhibits replication of host DNA, a basic protein would be the most likely candidate. To test this possibility, protein synthesis was inhibited in duplicate cultures of KB cells with cycloheximide (10 μg/ml). One culture was infected with a viral multiplicity of 200 PFU/cell, and DNA synthesis was measured in both by the incorporation of thymidine-³H during the period 8 to 10 hours after infection. Cycloheximide was found to inhibit protein synthesis in both uninfected and infected cells by more than 95%. However, DNA synthesis was similarly affected by cycloheximide. The results of a representative experiment are summarized in Table III. They show that the inhibition of synthesis of cellular DNA in infected, cycloheximide-treated cells was not significantly greater than in the uninfected control cells. Similar results were obtained with puromycin.

Unfortunately, these data do not answer, without ambiguity, the question as to whether protein synthesis is essential for the cessation of host DNA synthesis in type 5 adenovirus-infected cells. The reasons are as follows: (1) The mechanism by which cycloheximide inhibits the biosynthesis of DNA is uncertain; (2) although adenovirus can be "uncoated" in the absence of protein synthesis (Lawrence and Ginsberg, 1965), it is not known whether or not the virus is transported to the nucleus in a cell so markedly affected by these toxic chemicals; and (3) the amount of DNA synthesized in uninfected cells in the presence of cyclohexamide (less than 2% of that synthesized in untreated cells) approximates the maximum amount of cellular DNA synthesized in infected cells; it may represent an irreducible minimum of thymidine incorporation under these conditions. The nature of the material into which the small amount of thymidine is incorporated has not been characterized.

TABLE III

EFFECT OF CYCLOHEXAMIDE ON SYNTHESIS OF HOST DNA EIGHT TO TEN HOURS AFTER
INFECTION WITH TYPE 5 ADENOVIRUS

Virus PFU/cell	Cyclohexamide μg/ml	Specific activity[a] of cellular DNA cpm/OD Unit	Inhibition (%)
0	0	185,531	—
200	0	14,242	92[b]
0	10	3,675	—
200	10	3,116	16

[a] Determinations made on peak 6 tubes from MAK chromatography.

[b] Minimum value; no peak of radioactivity corresponding to cellular DNA was detectable.

Biosynthesis of Host Proteins in KB Cells Infected with Type 5 Adenovirus

Viral structural proteins are made in relatively large amounts beginning approximately 14 hours after infection (Wilcox and Ginsberg, 1963b). Since total protein synthesis in infected cells is not significantly affected at this time, it follows that there must be an inhibition of the synthesis of normal, cellular proteins. The extent to which the synthesis of cellular proteins is inhibited, and the time at which inhibition begins were tested by two different techniques: (1) separation of pulse-labeled host and viral proteins by immunological precipitation of the latter species, and (2) measurement of host protein synthesis by assaying the rate of accumulation of six different normal cell enzymes.

Measurement of Synthesis of Host Proteins by Immunological Precipitation

The synthesis of host and viral proteins was measured by the incorporation of valine-^{14}C for 2 hour periods at various times after infection, followed by the precipitation of viral proteins with specific antiserum. The acid-precipitable radioactivity remaining in the supernatant fluids was considered to represent normal, cellular proteins. More than 95% of the viral proteins could be precipitated by this procedure. The more direct method of precipitating host proteins was abandoned because of our inability to prepare an effective precipitating antiserum.

The results of a representative experiment, presented graphically in Fig. 5, indicate that host protein synthesis began to decline by 15 hours

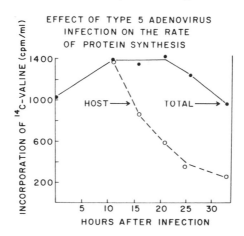

FIG. 5. Synthesis of host and viral structural proteins in KB cells infected with 200 PFU/cell of type 5 adenovirus. Valine-^{14}C was added to suspended cell cultures for 2 hour periods after infection. Viral proteins were precipitated with specific antiserum, and the number of acid-precipitable counts remaining in the supernatant fluid was considered to be the labeled host protein.

after infection, and continued to do so for the remainder of the experiment. These data also confirm earlier studies which demonstrated that the production of viral protein began about 14 hours post-infection (Wilcox and Ginsberg, 1963b). The rate of synthesis of viral structural proteins began to decrease about 20 hours post-infection, but the decrease in the rate of their synthesis was less than that observed for the cellular proteins.

Measurement of Host Protein Synthesis by Assays of Host Enzymes

If one assumes that the synthesis of normal cellular enzymes reflects the synthesis of all cellular proteins, then assays of the activities of such enzymes in exponentially growing cells should provide an index of the formation of host cell proteins. The enzymes selected were acid and alkaline phosphatases, phosphoglucose isomerase, lactic dehydrogenase, fu-

marase, and deoxyribonuclease. Their syntheses were studied in both uninfected and type 5 adenovirus-infected cells. In the infected cells, all the
enzymes, except alkaline phosphatase, were synthesized at the same rate
as in the control cells for 10 to 20 hours after infection. The linear
increase in enzyme activities ceased by 20 hours post-infection. The results obtained with the enzyme fumarase are presented in Fig. 6 by

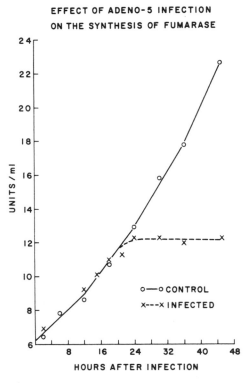

FIG. 6. Increase in fumarase in uninfected and type 5 adenovirus-infected KB cells.
Exponentially growing cells were infected with 200 PFU/cell, and samples of uninfected and infected cultures were disrupted by sonication at the times indicated.

way of illustration. The increase in alkaline phosphatase activity stopped
about 12 hours after infection. Numerous control experiments demonstrated that the cessation of enzyme accumulation in the infected cells
was a true reflection of the inhibition of protein synthesis, and not the
result of the inability to release enzymes from infected cells, the appearance of enzyme inhibitors, the abnormal inactivation or breakdown
of enzymes, or the absence of enzyme activators.

The Effect of DNA Synthesis on Increase in Enzyme Activity

The data described above indicate that the synthesis of viral structural proteins begins prior to the reaction which inhibits the production of most host cell proteins. These findings suggested to us that the shut-off might require the replication of viral DNA—that it might be a "late" function of the viral genome, and that the viral antigens might play a role in the inhibitory reactions. Hence, in an effort to determine whether the synthesis of host proteins would cease in the absence of the formation of the proteins encoded in the viral genome, experiments were devised in which the production of the late viral proteins was inhibited. These experiments were based on the fact that the production of viral structural proteins depends upon the prior replication of viral DNA (Flanagan and Ginsberg, 1962). Viral DNA synthesis was inhibited by $2 \times 10^{-6}\ M$ 5-fluoro-2-deoxyuridine (FUdR). The data presented in Fig. 7 describe an experiment in which fumarase activity was measured in infected cells

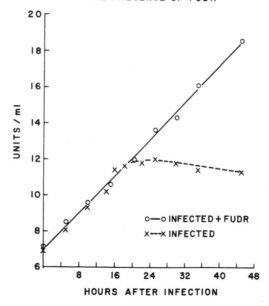

EFFECT OF ADENO-5 INFECTION
ON THE SYNTHESIS OF FUMARASE
IN THE PRESENCE OF FUDR

o—o INFECTED + FUDR
x---x INFECTED

FIG. 7. The effect of 5-fluoro-2-deoxyuridine (FUdR) on synthesis of fumarase in KB cells infected with 200 PFU/cell of type 5 adenovirus. To one culture 2×10^{-6} *M* FUdR was added. At the indicated times after infection, washed cells were disrupted by sonication, and the fumarase activity was determined.

maintained in the presence of FUdR. In contrast to the situation that prevailed in cells infected in the absence of FUdR (see Fig. 6), fumarase activity was found to increase continuously, at a linear rate, during the experimental period of 45 hours, when DNA and viral protein synthesis were blocked by FUdR. Thus, in the absence of synthesis of viral DNA and proteins, it would appear that the synthesis of cellular proteins continued without interruption. These data suggest that the cessation of host protein synthesis requires the replication of viral DNA, and furthermore, they support the suggestion that the inhibition results from the interaction of either viral proteins or virus-specific mRNA with the protein synthesizing system.

Biosynthesis of Host and Viral mRNA's in Adenovirus-Infected Cells

The fact that the inhibition of host protein synthesis depends on the replication of viral DNA raised the question as to whether the block occurred owing to a failure to transcribe or translate host mRNA. A direct approach to this question was made by measuring the synthesis of cellular and virus-specific mRNA by means of the hybridization technique of Nygaard and Hall (1963). Uridine-^{3}H was added to cultures for periods of 1 hour at various times after infection, and RNA was extracted from there with hot phenol. To measure host mRNA, heat-denatured purified DNA from uninfected KB cells was used, while denatured DNA from purified type 5 adenovirus (Wilcox and Ginsberg, 1963a) was employed to detect viral mRNA.

The synthesis of host mRNA was found to proceed at a constant rate until about 15 hours post-infection (Fig. 8), after which time there was a slow decline in its rate of synthesis to a level about 15% of the pre-infection rate (Bello and Ginsberg, 1966). Virus-specific mRNA was first detected 9 to 10 hours after infection. Its rate of synthesis increased sharply until about 16 hours post-infection, at which time the rate of synthesis slowed down, to be followed by a sharp decline at approximately 20 hours. Maximum synthesis of viral mRNA occurred when the synthesis of host mRNA was being inhibited, and, in fact, the rate of synthesis of viral mRNA was at least 5 times greater than that of the host mRNA during this period.

If the synthesis of host protein were inhibited because transcription of host mRNA was blocked, the inhibition of mRNA synthesis should occur earlier, and proceed at a more rapid rate than the inhibition of protein synthesis. A comparison of the rates of syntheses of host protein and

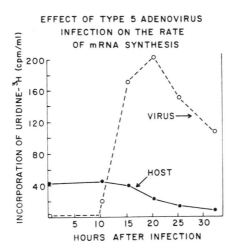

FIG. 8. The rate of synthesis of host and virus-specific mRNA in KB cells infected with type 5 adenovirus. Cells infected at a multiplicity 200 PFU/cell were labeled with uridine-³H for 1 hour periods at the indicated times after infection. RNA was extracted with hot phenol and mixed with either purified, denatured host or viral DNA for hybridization.

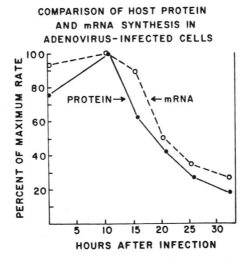

FIG. 9. Comparison of the rates of synthesis of host protein and mRNA in KB cells infected with type 5 adenovirus. Data were obtained from experiments summarized in Fig. 5 and 8.

mRNA presented in Fig. 9 clearly shows that this was not the case. Inhibition of synthesis of both species of macromolecules began at about the same time; and, the rate of decline of host protein synthesis appeared to be somewhat faster than the decrease in synthesis of host mRNA. The inevitable conclusion to be drawn from these data is that host protein synthesis is blocked at some step in the translation of host mRNA. The inhibition of mRNA synthesis may possibly be related, but it is not a causal event.

The Effect of Viral Structural Proteins on Synthesis of DNA, RNA, and Protein

In 1960, Pereira reported that the adenovirus C antigen (now termed the fiber antigen) but not the A (i.e., hexon) antigen could inhibit the multiplication of adenoviruses, poliovirus, and vaccinia virus, without affecting adsorption of the challenge viruses to their host cells (Pereira, 1960). These results suggested that the fiber antigen might have a general inhibitory effect on the biosynthesis of macromolecules in mammalian cells.

In confirmation of Pereira's findings, highly purified fiber, but not hexon antigen was shown to inhibit the multiplication of type 5 adenovirus and type 1 poliovirus. Fifty percent inhibition of viral multiplication was obtained when 35 μg/10^6 cells was added to cultures 6 hours prior to infection. It was also demonstrated that the inhibitory effect was not the result of inhibition of viral attachment, penetration, or uncoating. When the effect of these viral antigens on the syntheses of host cell macromolecules was investigated, it was found that the production of DNA, RNA, and protein was inhibited about 20 hours after the fiber antigen was added to either uninfected or infected cells (Fig. 10). These findings implied that viral multiplication was inhibited by blocking the synthesis of nucleic acids and protein. Furthermore, it suggested that the viral fiber protein could possibly inhibit the biosynthesis of cellular macromolecules in adenovirus-infected cells (Levine and Ginsberg, 1965).

The time interval between the addition of protein and the cessation of synthesis of macromolecules was long—approximately 20 hours. It seemed possible that this long lag period could be accounted for if the fiber attached to and/or formed an irreversible association with KB cells at a slow rate. Purified fiber antigen, extracted from cells infected and incubated in the presence of algal hydrolysate-^{14}C, was used in experiments designed to examine this possibility. Irreversible association was measured as the quantity of fiber that could not be separated from the cells by repeated washing, although these measurements could not distinguish between intracellular· or surface-associated fiber protein. DNA synthesis was also measured in the same cells. The data summarized in Fig. 11 indicate that the association of the fiber with cells has complex kinetics. The initial association of the fiber antigen was found to be slow, but to become more rapid after about 20 hours. A maximum of only 0.3–0.4% of the added fiber became cell-associated. Hexon antigen did not attach to KB cells at all. It was noted that DNA synthesis was in-

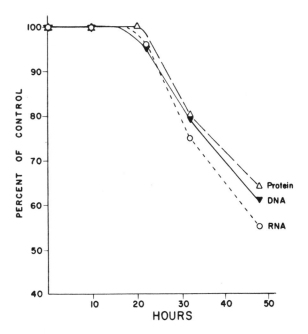

FIG. 10. The effect of fiber antigen on synthesis of DNA, RNA, and proteins in uninfected, exponentially growing KB cells. Purified fiber antigen 250 μg/ml, was added to spinner cultures of dividing KB cells, and measurements made at the times indicated.

hibited at the time at which the rate of association of antigen with cells increased. It is possible that inhibition of DNA replication occurred when the intracellular pool of antigen attained an adequate concentration. However, these data do not distinguish this hypothesis from the possibility that with the cessation of biosynthesis of host macromolecules, there is a marked increase in the permeability of the cells to the protein.

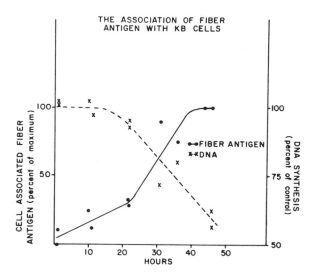

FIG. 11. Adsorption of purified fiber antigen to suspension of KB cells. Fiber antigen was labeled with ^{14}C-amino acids, and adsorption was determined by the number of counts which remained associated with cells after washing twice. For comparison DNA synthesis was measured in the same cells.

The Reaction of Fiber and Hexon Antigens with Viral and KB Cell DNA

If fiber antigen were to inhibit RNA transcription and DNA replication directly, it is likely that it could complex readily with host DNA. To examine this possibility, the reaction of viral protein with DNA was studied, using purified, isotopically labeled DNA. The DNA-protein complexes were precipitated by antibodies directed toward the viral proteins, and the associated DNA was measured by the quantity of precipitated ^{14}C- or ^{3}H-DNA (Levine *et al.*, 1966). The results of experiments using 50 μg of either fiber or hexon antigens, and increasing amounts of viral or KB cell DNA are summarized in Fig. 12. The linear portions of the curves indicate that on a weight basis, the fiber antigen complexed with either cellular or viral DNA about twice as efficiently as did the hexon. At the lower concentrations of DNA used, the source of the DNA did not appear to influence the extent of the binding. At the higher DNA concentrations, however, both proteins were found to combine more extensively with KB cell DNA than with viral DNA, perhaps owing to differences in the sizes of the DNA molecules, or to the differences in their chemical composition.

The association of viral proteins with DNA was found to be maximum at a NaCl concentration of less than 10^{-2} *M*—the capacity of either antigen to bind to DNA was reduced at NaCl concentrations between 10^{-2}

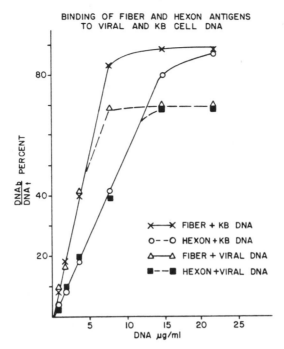

FIG. 12. Assoication of fiber and hexon antigens with viral and KB cell DNA. DNA's were labeled with ^{3}H- or ^{14}C-thymidine. To 50 μg of fiber or hexon antigen, increasing quantities of either DNA was added. Antigen was precipitated with specific antibody; the quantity of unassociated DNA was determined by the number of counts remaining in the supernatant fluid. The quantity of complexed DNA was determined by subtracting unassociated DNA from total input DNA.

and 10^{-1} *M*. At salt concentrations above 0.2 *M* NaCl, only 3 to 5% binding was attained; this suggests that the interaction between DNA and antigen is ionic in nature.

The Effect of the Fiber and Hexon Antigens on DNA-Dependent RNA and DNA Polymerases

Histones bind to DNA, and inhibit the *in vitro* activity of DNA and RNA polymerases (Allfrey *et al.*, 1963; Barr and Butler, 1963; Gurley *et al.*, 1964; Huang and Bonner, 1962; Shalka *et al.*, 1966). Since the fiber and hexon antigens also bind to DNA, the possibility that these viral proteins may play an important biological role in the control of synthesis of host DNA and RNA was investigated. Specifically, experiments were carried out to determine whether or not the purified viral structural proteins can reduce the *in vitro* activities of RNA and DNA polymerases. Enzymes from both uninfected and infected cells were extracted and partially purified (Furth and Ho, 1965). Their activities were assayed both in the absence of viral protein, and after the template DNA had been incubated with one of the purified antigens (Levine and Ginsberg, 1966).

It was found that both fiber and hexon antigens markedly inhibited the DNA-dependent RNA polymerase (Fig. 13), and that the inhibition was not affected by the source of the DNA template (i.e., KB cell or viral), or by the source of the enzyme (i.e., uninfected or infected cells). If the polymerase activity was inhibited by binding of antigen to template DNA, the inhibition should decrease with increasing concentrations of DNA, and the data summarized in Fig. 13 indicate that this was the case. Moreover, if antigen binding to the DNA template was responsible for the diminution of enzyme activity, the inhibition should be reversed when the antigen-DNA complex is dissociated at high salt concentration. In line with this prediction, it was found that when the reaction was carried out at 0.01 M NaCl, fiber antigen inhibited RNA polymerase activity to the extent of 78%, whereas when the reaction was carried out in 0.4 M NaCl, enzyme activity was not reduced.

The activity of DNA polymerase was also inhibited by the hexon and fiber antigens (Table IV). Moreover, the fiber antigen, which on a weight basis complexes with DNA more effectively than does the hexon antigen, was found to inhibit DNA polymerase activity more effectively than

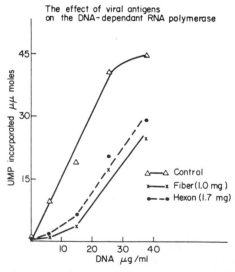

Fig. 13. The effect of fiber and hexon antigen on the activity of DNA-dependent RNA polymerase. The assay was initiated by mixing the partially purified enzyme with a DNA-antigen complex or with DNA alone.

TABLE IV

EFFECT OF VIRAL ANTIGENS ON THE ACTIVITY OF DNA POLYMERASE FROM INFECTED
AND UNINFECTED CELLS[a]

Cell Extract	Antigen added	Incorporation of dGMP $\mu\mu$moles	% Inhibition
Uninfected[b]	None	72	—
	Fiber[d]	26	64
	Hexon[e]	35	52
Infected[c]	None	36	
(17 hours	Fiber[d]	9	75
post-infection)	Hexon[e]	13	64

[a] 11.6 μg/ml of type 5 adenovirus DNA was employed. 14 μM dGTP-[14]C (1060 cpm/mμmole) was used in all cases.

[b] Protein added = 200 μg.

[c] Protein added = 205 μg.

[d] Protein added = 1.0 mg.

[e] Protein added = 1.5 mg.

did the latter. Enzymes from uninfected and infected cells were inhibited to a similar extent, although the specific activity of the enzyme from uninfected cells was approximately twice that of the enzyme extracted from cells 17 hours post-infection. When cell cultures were assayed at intervals after infection, it was found that the activities of the DNA and RNA polymerases began to decrease about 15 hours after infection, and continued to decrease for the next 17 hours.

Discussion

The biochemical events in cells infected with an adenovirus can now be described, and hypotheses may be made concerning the control of syntheses of both host and viral macromolecules. The most complete evidence has been obtained with the type 5 virus, but comparative data, where available, suggest that similar events occur, but at different times, with the other adenovirus types which produce virulent infections.

The sequential biosynthetic steps in the production of the virions' component parts are summarized diagrammatically in Fig. 14. This diagram also suggests a model of the possible interrelationships between the events in the biosynthesis of the virus, and the observed inhibition of the production of host DNA, mRNA, and protein. The mechanisms proposed to explain the cessation of synthesis of host macromolecules in the infected cell attempt to account for the apparent selectivity that permits viral subunits to be made.

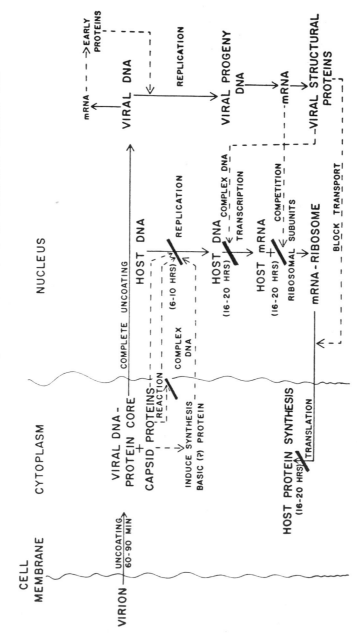

Fig. 14. A model of the possible mechanisms for inhibiting synthesis of host macromolecules in cells infected with type 5 adenovirus. The steps in the production of viral subunits are also summarized.

The initial step in the replicative cycle, dissociation of the viral genome from its protective coat, begins as soon as penetration of the host cell is accomplished, and probably occurs in phagocytic vacuoles (Dales, 1965; Lawrence and Ginsberg, 1965). By 60 to 90 min after penetration of the virion, this stage is completed, and the DNA of more than 85% of the infecting virions is sensitive to DNase (Lawrence and Ginsberg, 1965). These reactions do not require protein synthesis, and probably occur in the cytoplasm. However, the so-called "uncoated" DNA is still associated with a large piece of protein in a complex having a sedimentation velocity of about 400 S (Lawrence and Ginsberg, 1966). The evidence suggests that the viral DNA may be associated with an internal viral protein. This DNA-protein complex then enters the nucleus, where it is possible that a second "uncoating" step is required to free the viral genome for transcription (i.e., the synthesis of mRNA), and for viral DNA replication. Synthesis of viral DNA begins about 7 to 10 hours after infection, depending upon the multiplicity of infection. The rate of synthesis begins to decline 16 to 22 hours post-infection (Flanagan and Ginsberg, 1962; Ginsberg and Dixon, 1961, 1966).

Virus-specific mRNA is first detected 9 to 10 hours after infection. It is likely that this mRNA is that necessary for making the viral structural proteins (i.e., "late proteins"), since, in the presence of FUdR, the appearance of virus-specific mRNA could not be detected. Using the Nygaard and Hall hybridization technique (Nygaard and Hall, 1963), virus-specific mRNA for "early proteins" was not detected, and although early protein synthesis appears to be required for maximal production of viral DNA (Ginsberg and Dixon, 1966; Wilcox and Ginsberg, 1963b), the appearance of *new* enzymes, coded for by the viral DNA, has not been clearly demonstrated. Synthesis of virus-specific mRNA begins to level off at about 16 hours (Bello and Ginsberg, 1966), and its rate of synthesis decreases rapidly after about 20 hours post-infection.

Production of viral structural proteins begins about 2 hours after the synthesis of viral DNA is initiated, and continues at a constant rate for 8 to 12 hours. Their continued production after the inhibition of synthesis of virus-specific mRNA is probably due to the stability of the mRNA.

The cessation of host cell DNA synthesis 6 to 10 hours after viral infection constitutes the first demonstrable effect on the biosynthesis of host macromolecules. There are several possible mechanisms that could account for the observed inhibition: (1) Viral capsid proteins from the infecting virions may participate in the reaction by entering the nucleus and complexing with host DNA, thus inhibiting DNA polymerase activity.

Protein synthesis may be essential for the viral proteins to attain an intranuclear position. (2) Induction of the synthesis of an inhibitory protein (e.g. a histone). Unfortunately, the experiments designed to determine whether protein synthesis was an essential prerequisite to the observed block in DNA synthesis yielded ambiguous data because inhibitors of protein synthesis markedly reduce DNA synthesis. The very small amount of DNA synthesized in cycloheximide-treated cells (less than 2% of that in untreated cells) was considerably less than the maximal synthesis of host cell DNA observed under the usual conditions of infection. Nevertheless, it is possible that the viral proteins, or the viral DNA-protein core, may induce the synthesis of a functional protein which inhibits DNA synthesis. (3) The structural proteins of the infecting virions may react with the nuclear membrane, and inhibit DNA synthesis in a manner analogous to that induced by the reaction of ghosts of T-even phage with *E. coli* (French and Siminovitch, 1955; Herriott, 1951). Although the latter two possibilities are consistent with most of the data, they do not appear to offer the selectivity that would permit viral DNA to be made while host DNA synthesis is blocked.

Inhibition of host mRNA synthesis, 16 to 20 hours after infection, occurs well after the production of viral structural proteins begins. It is possible that the binding of viral proteins to host DNA reduces the activity of the DNA-dependent RNA polymerase *in vivo,* just as it does *in vitro*. If a virus-specific RNA polymerase is not induced, and if the host RNA polymerase has a greater affinity for viral DNA than for host DNA, the rapidly increasing viral DNA could sequester the available enzyme so that synthesis of host mRNA would stop. Preliminary evidence suggests that this is not the case, but detailed studies of the enzyme kinetics must be done using both host and viral DNA's as templates before a definite conclusion can be drawn. The possibility that mRNA synthesis is inhibited, because production of the polymerase stops, was explored. However, the data obtained indicate that the enzyme persists for too long after protein synthesis is inhibited by cycloheximide to explain the inhibition on these grounds.

Synthesis of host proteins begins to decrease at about 16 to 20 hours after infection—at approximately the same time at which the rate of host mRNA synthesis also begins to decline. Host protein synthesis, however, decreases at a faster rate than does the synthesis of host mRNA. Hence, adenovirus infection does not inhibit host protein synthesis simply by stopping the transcription of host mRNA. Moreover, a selective mechanism must be operative to permit the production of viral structural proteins to continue for several hours after the synthesis of host proteins begins to decrease. After the synthesis of virus-specific mRNA begins, its rate of synthesis is at least 5 times that of the host mRNA. These data suggest the possibility that the synthesis of host protein is inhibited because the large pool of viral mRNA competes successfully with the smaller pool of host mRNA for ribosomal subunits. A second hypothetical mechanism, which offers the selectivity required, is that the viral infection alters the nuclear membrane so that the transport into the cytoplasm of the host mRNA-ribosomal subunit complexes to form polyribosomes is blocked. The earlier finding that nuclei of adenovirus-infected cells are more resistant to disruption by either sonication or freezing and thawing supports this notion (Denny and Ginsberg, unpublished observations).

A comparison of the control of the synthesis of host and viral macromolecules indicates that the synthesis of host mRNA and protein and of viral DNA and mRNA are all reduced beginning about 16 hours after infection. Only the production of host DNA and viral structural proteins appears to be affected differently. The degree of selective control provided, which actually may be limited to only two reactions, may result merely from temporal and quantitative effects, rather than from complex biochemical reactions. Thus, the partial uncoating of parental virions permits capsid proteins to detach from the viral core, and to be free to complex with host DNA. This mechanism could limit the synthesis of host DNA, while permitting the production of viral DNA to commence shortly thereafter. The required selectivity for the cessation of host protein synthesis, while the production of viral structural proteins continues, could be provided by the successful competition of viral mRNA for a limited supply of ribosomal subunits, or by a block in the transport of host RNA from the nucleus to the cytoplasm—the preferred site for host protein synthesis.

The mechanisms suggested require only noncovalent complexing, or competitive reactions between macromolecules, and do not imply a requirement for new types of reactions of a high order of complexity. Indeed, the possibilities are sufficiently reasonable that they are subject to experimental test.

Acknowledgments

This investigation was partially conducted under the sponsorship of the Commission on Acute Respiratory Diseases, Armed Forces Epidemiological Board, and was supported partially by the Office of the Surgeon General, Department of the Army. It was also supported by U.S. Public Health Service grants AI–03620, AI–05731, and 2 TI AI–203.

REFERENCES

Allfrey, V. G., Littau, V. C., and Mirsky, A. E. (1963). *Proc. Natl. Acad. Sci. U.S.* **49**, 414.
Allison, A. C., Pereira, H. G., and Farthing, C. P. (1960). *Virology* **10**, 316.

Barr, G. C., and Butler, J. A. V. (1963). *Nature* **199**, 1170.
Becker, Y., and Joklik, W. (1964). *Proc. Natl. Acad. Sci. U.S.* **51**, 577.
Bello, L. J., and Ginsberg, H. S. (1966). *Federation Proc.* **25**, 2615.
Benzer, S. (1953). *Biochim. Biophys. Acta* **11**, 383.
Cohen, S. S. (1948). *J. Biol. Chem.* **174**, 281.
Cohen, S. S. (1949). *Bacteriol. Rev.* **13**, 1.
Dales, S. (1965). *Am. J. Med.* **38**, 699.
Denny, F. W., and Ginsberg, H. S. (1959). Unpublished observations.
Flanagan, J. F., and Ginsberg, H. S. (1962). *J. Exptl. Med.* **116**, 141.
Flanagan, J. F., and Ginsberg, H. S. (1964). *J. Bacteriol.* **87**, 977.
Franklin, R. M., and Baltimore, D. (1962). *Cold Spring Harbor Symp. Quant. Biol.* **27, 175.**
French, R. C., and Siminovitch, L. (1965). *Can. J. Microbiol.* **1**, 757.
Furth, J. J., and Ho, P. L. (1965). *J. Biol. Chem.* **240**, 2601.
Ginsberg, H. S., and Dixon, M. K. (1961). *J. Exptl. Med.* **113**, 283.
Ginsberg, H. S., and Dixon, M. K. (1966). Manuscript in preparation.
Ginsberg, H. S., Pereira, H. G., Valentine, R. C., and Wilcox, W. C. (1966). *Virology* **28**, 782.
Green, M. (1962). *Cold Spring Harbor Symp. Quant. Biol.* **27**, 219.
Gurley, L. R., Logan, I., and Holbrook, D. J. (1964). *Biochem. Biophys. Res. Commun.* **14**, 527.
Herriott, R. M. (1951). *J. Bacteriol.* **61**, 752.
Hershey, A. D., Dixon, J., and Chase, M. (1953). *J. Gen. Phys.* **36**, 777.
Horne, R. W., Brenner, S., Waterson, A. P., and Wildy, P. (1959). *J. Mol. Biol.* **1**, 84.
Huang, R. C., and Bonner, J. (1962). *Proc. Natl. Acad. Sci. U.S.* **48**, 1216.
Joklik, W. K., and Becker, Y. (1965). *J. Mol. Biol.* **10**, 452.
Kaplan, A. S., and Ben-Porat, T. (1963). *Virology* **19**, 205.
Klemperer, H. G., and Pereira, H. G. (1959). *Virology* **9**, 536.
Lawrence, W. C., and Ginsberg, H. S. (1965). *Federation Proc.* **24**, 1383.
Lawrence, W. C., and Ginsberg, H. S. (1966). Manuscript in preparation.
Levin, A. P., and Burton, K. (1961). *J. Gen. Microbiol.* **25**, 307.
Levine, A. J., and Ginsberg, H. S. (1965). *Federation Proc.* **24**, 597.
Levine, A. J., and Ginsberg, H. S. (1966). Manuscript in preparation.
Levine, A. J., Furth, J. J., and Ginsberg, H. S. (1966). *Bacteriol. Proc.* p. 131.
Mandell, J. D., and Hershey, A. D. (1960). *Anal. Biochem.* **1**, 66.
Monod, J., and Wollman, E. (1947). *Ann. Inst. Pasteur* **73**, 937.
Nomura, M., Matsubara, K., Okamoto, K., and Fujimura, R. (1962). *J. Mol. Biol.* **5**, 535.
Nygaard, A. P., and Hall, B. D. (1963). *Biochem. Biophys. Res. Commun.* **12**, 98.
Penman, S., and Summers, O. (1965). *Virology* **27**, 614.
Pereira, H. G. (1960). *Virology* **11**, 590.
Salzman, N. P., Shatkin, A. J., and Sebring, E. D. (1964). *J. Mol. Biol.* **8**, 405.
Shalka, A., Fowler, A. V., and Hurwitz, J. (1966). *J. Biol. Chem.* **241**, 588.
Shatkin, A. J. (1963). *Nature* **199**, 357.
Valentine, R. C., and Pereira, H. G. (1965). *J. Mol. Biol.* **13**, 13.
Wilcox, W. C., and Ginsberg, H. S. (1961). *Proc. Natl. Acad Sci. U.S.* **47**, 512.
Wilcox, W. C., and Ginsberg, H. S. (1962). *J Bacteriol.* **84**, 526.
Wilcox, W. C., and Ginsberg, H. S. (1963a). *J. Exptl. Med.* **118**, 295.
Wilcox, W. C., and Ginsberg, H. S. (1963b). *Virology* **20**, 269.
Zimmerman, E. F., Huter, M., and Darnell, J. E. (1963). *Virology* **19**, 400.

Discussion—Part VI

Dr. S. S. Cohen: (Question directed to Dr. W. K. Joklik.) Can the interferon experiments be interpreted to indicate that translation does not control transcription, at least in the control of mRNA production on viral genomes?

Answer: The experiments are quite clear in demonstrating that vaccina messenger RNA transcription can proceed at very high rates in cells containing negligible amounts of polyribosomes. The control by translation of transcription, which has been most seriously proposed, is that messenger RNA molecules combine with ribosomes even before they have been fully transcribed and that it is this combination which promotes transcription. There is no evidence for this process in mammalian cells; and, it is certainly very unlikely to occur in interferon-treated infected cells. In this sense, the experiments can conceivably be interpreted to indicate that translation does not control transcription.

Dr. G. A. Gentry: (Question directed to Dr. S. Kit.) In the experiments in which you flooded the monolayers with thymidine at a concentration of 100 μg/ml in order to look for the effect on the Michaelis constant, do you have any measurements which give some assurance that the thymidine actually got into the cells at levels which would be expected to influence the Michaelis constant?

Answer: The experiment was based upon previous observations showing that 100 μg/ml thymidine drastically inhibited growth of suspension cultures of LM cells (Dubbs and Kit, S. (1964). *Exper Cell Res.* **33,** 19–28) and that 50 μg/ml administered to 2-day-old (growing) cultures of LM cells prevented the decline in thymidine kinase activity which otherwise took place between 48 and 72 hours after subculture (Kit, S. *et al.* (1965). *J. Biol. Chem.* **240,** 2565–2573). In the experiments with monkey kidney cells, confluent monolayer cultures were employed and there was little cell growth during the period of the experiment. When administered at a concentration of 100 μg/ml, tritiated thymidine was incorporated into the DNA of both noninfected and SV40-infected monkey kidney cell cultures.

Radioactivity was found in acid-soluble extracts made from the infected and the noninfected cultures. However, our data do not permit a precise evaluation of the intracellular dTTP concentrations attained. It should be noted, however, that even if the necessary dTTP concentrations were reached, these concentrations would have been diluted tenfold when the enzyme extracts were prepared; and furthermore, any dTTP not firmly bound to thymidine kinase would have been removed after the ammonium sulfate precipitation step.

PART VII

ONCOGENIC VIRUSES I

Studies on the Basis for the Observed Homology between DNA from Polyoma Virus and DNA from Normal Mouse Cells[*]

Ernest Winocour

SECTION OF GENETICS, WEIZMANN INSTITUTE OF SCIENCE,
REHOVOTH, ISRAEL

The utilization of some nucleic acid homology techniques to demonstrate the presence of viral genes in transformed cells is complicated by the degree of nucleotide sequence homology to normal cell DNA that has been reported for DNA extracted from polyoma virus (Axelrod *et al.,* 1964; Winocour, 1965b), Shope papilloma virus (Winocour, 1965b), and SV40 virus (Reich *et al.,* 1966). Three explanations can be considered to account for this observed homology to normal cell DNA: (1) The homology may be due to the contamination of virus preparations by a cellular DNA-containing component that is not part of the virus particle (virion); (2) it may be due to host-cell DNA that is enclosed within the coats (encapsidated) of some or all virions during the maturation stage of virus synthesis and, hence, is not an integral part of the infectious virus DNA molecule; or (3) it may be due to the fact that a region of the infectious virus DNA molecule does indeed contain sequences homologous to part of normal cellular DNA, as has been suggested in the case of the homology between phage lambda DNA and the DNA of *E. coli* (Cowie and McCarthy, 1963; Green, 1963).

The objective of the studies discussed herein was to determine the basis for the homology to normal mouse cell DNA that is displayed by the DNA

[*] This work was supported, in part, by Research Grant DRG 860-A from the Damon Runyon Memorial Fund. The homology between DNA from polyoma virus and DNA from normal mouse cells refers, in the experiments discussed in this article, to the homology which is observed between DNA from polyoma virus and a synthetic RNA made using normal mouse DNA as a primer for the DNA-dependent RNA polymerase.

extracted from polyoma virus. It will be shown that (1) the component responsible for the homology to normal cell DNA cannot be separated from the virions by any of six independent purification procedures, and (2) that fractionation of the total DNA extracted from purified polyoma virions can result in the separation of an infectious DNA component with reduced or no detectable homology to normal cell DNA. These results, together with supporting evidence from a study of transmethylation in polyoma infected cells, have led to the conclusion that most of the observed homology between the DNA of mouse cells and the DNA from polyoma virus is due to the encapsidation of cellular DNA fragments by polyoma virions.

The Detection of the Apparent Homology Between DNA from Mouse Cells and DNA from Polyoma Virus

The nucleic acid homology test-system employed in these experiments (Winocour, 1965a,b, 1967) is based upon hybridization between DNA from polyoma virus or DNA from normal mouse cells, and synthetic tritium-labeled RNA (^{3}H-RNA) made using either normal mouse cell DNA, or polyoma virus DNA, as primer for the DNA-dependent RNA polymerase enzyme, isolated from *E. coli* (Chamberlin and Berg, 1962). The hybridization reaction and assay was carried out according to the method of Gillespie and Spiegelman (1965), in which the denatured DNA

TABLE I

HYBRIDIZATION BETWEEN DNA FROM POLYOMA VIRUS OR NORMAL MOUSE CELLS
AND POLYOMA VIRUS ^{3}H-RNA OR NORMAL MOUSE ^{3}H-RNA[a]

Synthetic ^{3}H-RNA	Incubated ^{3}H-RNA (cpm)	^{3}H-RNA (cpm) retained by filter containing		
		Polyoma virus DNA (10μg)[b]	Normal mouse DNA (10μg)	None[c] (blank filter)
Normal mouse	338,000	3650	18,400	342
Polyoma virus	450,000	93,000	950	187

[a] Each vial contained 6 ml of the respective ^{3}H-RNA (in 0.3 M NaCl, 0.03 M sodium citrate) and 3 filters; one with 10μg normal mouse cell DNA, one with 10μg polyoma viral DNA, and one blank filter. The incubation period was 15 hours at 65°C. The filters were then washed, treated with RNase (20 μg/ml for 1 hour at 22°C), rewashed, and counted for the amount of ^{3}H-RNA radioactivity retained (procedure of Gillespie and Spiegelman, 1965). The estimated specific activity of the ^{3}H-RNA was 1×1^6 cpm/μg.

[b] The virus was purified by equilibrium centrifugation in a CsCl density gradient, treated with trypsin and DNase, and rebanded in a CsCl density gradient (procedure 1 in Table II).

[c] ^{3}H-RNA (cpm) retained by control filters containing 10μg of phage T4 DNA were the same as blank filters.

is immobilized on nitrocellulose membrane filters. Table I illustrates that the apparent homology between DNA from normal mouse cells and DNA from polyoma virus can be observed experimentally in the hybridization reaction between normal mouse ³H-RNA and polyoma DNA, as well as in the reaction between polyoma virus synthetic ³H-RNA and normal mouse cell DNA. For the present studies, the reaction between mouse ³H-RNA and virus DNA was preferred, since it was technically more feasible to hybridize the large number of polyoma DNA preparations under test with one or two preparations of synthetic mouse ³H-RNA.

An estimate of the fraction of the DNA extracted from virus which is homologous to normal mouse ³H-RNA (and, therefore, homologous to normal mouse cell DNA) can be made from the data shown in Fig. 1. In

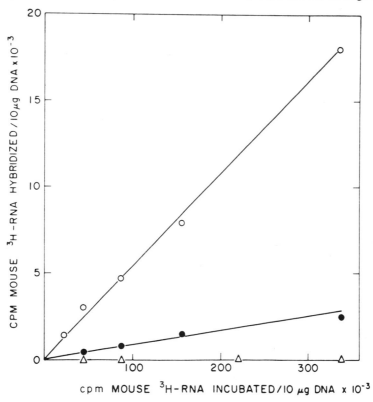

FIG. 1. Hybridization of mouse ³H-RNA with mouse DNA or polyoma virus DNA (Winocour, 1967). Filters loaded with 10 μg of mouse DNA (○————○) polyoma virus DNA (●————●) or phage T4 DNA (△————△) were incubated with the indicated amounts of mouse ³H-RNA, and the amount of ³H-RNA hybridized per 10 μg of DNA determined according to the procedure of Gillespie and Spiegelman (1965). The virus DNA was obtained from virus purified by procedure 1 in Table II.

this experiment, the relationship between the amount of mouse ^{3}H-RNA incubated and the amount of mouse ^{3}H-RNA hybridized by 10 μg of mouse or virus DNA was linear in the range of RNA and DNA concentration used. It will be noted that the *initial* slope of the mouse DNA curve is 6.3 times greater than that of the virus DNA curve. Since complete homology must exist between the mouse DNA and the synthetic mouse ^{3}H-RNA made from it, this result indicates that about 16% of the total DNA extracted from purified polyoma virions is homologous to normal mouse DNA.

The Homology to Mouse ^{3}H-RNA of DNA Obtained from Virus Purified by Different Procedures

To determine if the homology to mouse ^{3}H-RNA arises from a random contamination of the virus preparations by cellular DNA-containing components, polyoma virus was subjected to six different purification procedures (Winocour, 1967). These procedures included enzyme digestions (receptor-destroying enzyme, trypsin, RNase, DNase) to aid in the removal of material superficially attached to the virions; equilibrium centrifugation in CsCl density gradients to remove contaminating particles of different buoyant density (Vinograd and Hearst, 1962) to that of the virion; differential centrifugation and fractionation by sedimentation in sucrose gradients to remove particles of different size or configuration; and column chromatography on hydroxylapatite to remove contaminating particles with different surface properties. At the end of each purification procedure, DNA was extracted by phenol from the virions and tested for its ability to hybridize with mouse ^{3}H-RNA.

The results of these tests are presented in Table II. It can be seen that none of the purification procedures abolished the homology between the extracted virus DNA and mouse ^{3}H-RNA. Although the average percentage of mouse ^{3}H-RNA hybridized varied from 1.2 to 3% with the different DNA preparations tested, no correlation has been found, in repeated experiments, between any particular purification procedure and a particular value in this range. Thus the homology to normal mouse ^{3}H-RNA does not appear to be due to random contamination by a cellular DNA component which can be separated from the polyoma virions.

The Homology to Mouse ^{3}H-RNA of Polyoma DNA Fractionated in Sucrose Gradients and by MAK Column Chromatography

Experiments were carried out to determine if fractionation of the total DNA extracted from purified virions could result in the isolation of an

TABLE II

HYBRIDIZATION BETWEEN MOUSE ³H-RNA AND DNA FROM POLYOMA VIRUS PURIFIED BY DIFFERENT PROCEDURES

	Virus purification procedure[a]			Average percentage of mouse ³H-RNA hybridized[b]
1.	CsCl-DG + (DNase, trypsin) + CsCl-DG			2.0
2.	CsCl-DG + (DNase, trypsin) + CsCl-DG			1.2
3.	CsCl-DG + (DNase, trypsin) + CsCl-DG	+ Hydroxylapatite		2.7
4.	CsCl-DG + Sucrose gradient + CsCl-DG	+ "Crystallization"		1.5
5.	Differential centrifugation + CsCl-DG + (DNase, trypsin)	+ Sucrose gradient		3.0
6.	Deoxycholate + (RNase, DNase, trypsin) + CsCl-DG			2.5

[a] Full details on each purification procedure are given in Winocour (1967) from which the data is taken. "CsCl-DG" means equilibrium centrifugation in a CsCl density gradient; "hydroxylapatite" refers to column chromatography of the purified virus in hydroxylapatite; "sucrose gradient" refers to sedimentation through a preformed 5–20% linear sucrose gradient; "differential centrifugation" refers to centrifugation of the crude lysate at 12,000 g for 15 min followed by sedimentation of the virus, from the 12,000 g supernatant, at 105,000 g for 90 min. Procedure 6 is essentially the same as that reported by Benjamin (1966).

[b] Filters, each containing 10 μg of DNA from virus purified by the procedure listed, were incubated with 3 different concentrations of mouse ³H-RNA. For each DNA, the relationship between cpm ³H-RNA hybridized/10 μg DNA and the total cpm ³H-RNA incubated was linear, and thus the average percentage of ³H-RNA hybridized (average of the values obtained for the 3 different RNA concentrations used) reflects the initial slope of a saturation experiment (see Fig. 1). The amount of ³H-RNA retained by control filters (containing no DNA or 10 μg of phage T4 DNA) was 0.05 to 0.1% of the total ³H-RNA radioactivity incubated. Hybridization procedure was that of Gillespie and Spiegelman (1965).

infectious polyoma DNA component with reduced or no homology to normal mouse ^{3}H-RNA.

Fractionation by Sedimentation Velocity in a Sucrose Gradient

Sedimentation velocity analyses in preformed linear sucrose gradients, or by band-centrifugation in CsCl (Vinograd *et al.*, 1963) have revealed the presence in polyoma DNA preparations of 3 components with sedimentation coefficients of 20 S, 16 S, and 14 S (Dulbecco and Vogt, 1963; Weil and Vinograd, 1963; Vinograd *et al.*, 1965). The fast component (20 S), which accounts for most of the total DNA, is a circular helix, which is infectious (Dulbecco and Vogt, 1963; Bourgaux *et al.*, 1965) and which shows spontaneous renaturation (Weil, 1963). The slow component (16 S), which is believed to arise from a discontinuity in one of the strands of the 20 S component, exhibits a denaturation behavior similar to that of linear double-stranded DNA, and is reported (Bourgaux *et al.*, 1965) to possess a specific infectivity somewhat lower than that of the 20 S component. The origin of the 14 S component, which is not usually detectable in preparative ultracentrifugation, is obscure.

In previous studies (Winocour, 1965b) some homology to mouse ^{3}H-RNA was found with both the fast and slower sedimenting components of polyoma DNA. More recent experiments, however, have indicated that there is a very significant difference in the relative degree of homology shown by these 2 components. In these experiments, the total DNA extracted from purified polyoma virions was sedimented through a preformed 5–20% linear sucrose gradient. The fractionation into fast (FC) and slower (SC) sedimenting components that was obtained is shown in Fig. 2A: The control reconstruction experiment, illustrated in Fig. 2B, confirms that the 2 virus DNA sedimenting components can be separated from the bulk of mouse cellular DNA by this technique (Weil *et al.*, 1965).

The results of the hybridization test between mouse ^{3}H-RNA and polyoma DNA samples, before and after sucrose gradient fractionation, are shown in Table III. The hybridization test with the synthetic complementary polyoma virus ^{3}H-RNA was used to correct for possible differences either in the absorption to filters of the denatured polyoma DNA sample or in the availability of the absorbed DNA for hybridization. From the results in Table III, it is evident that the slow sedimenting component of polyoma DNA hybridizes with mouse ^{3}H-RNA to a much greater extent than does the fast component (the slightly increased hybridization of the slow component with polyoma virus ^{3}H-RNA does not significantly affect this conclusion). It may be concluded, therefore, that sedimentation in sucrose gradients of the total DNA extracted from purified polyoma virions will produce two components which show very different degrees of homol-

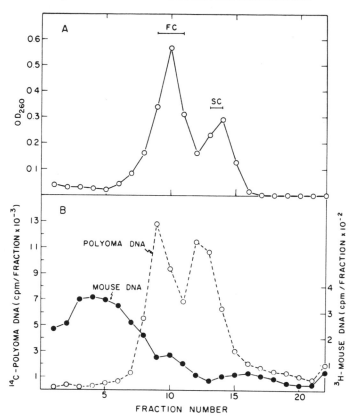

FIG. 2. Sedimentation in sucrose gradients of polyoma DNA and mouse DNA (Winocour, 1967). (A) One hundred and sixty micrograms of polyoma DNA (extracted from virus purified by procedure 6, Table II) was layered on 26 ml of a 5–20% sucrose gradient, and centrifuged at 23,000 rpm in the Spinco SW 25 rotor for 17 hours. Twenty-two fractions of 10 drops each (1.2 ml) were collected from the bottom of the tube and their absorbancy at 260 mμ was measured. The combined fractions 9–11 (FC or fast component) and 13–14 (SC or slow component) were used in the hybridization experiment shown in Table III. (B) A mixture of ^{14}C-polyoma DNA and ^{3}H-mouse kidney cell DNA (phenol extracted by the same procedure as that used for virus) was sedimented in a 5–20% sucrose gradient under the same conditions as in (A) above. The [14]C- and ^{3}H-radioactivity in each of the 22 fractions collected was measured. The proportion of slow components obtained with the ^{14}C-labeled polyoma DNA marker used in this experiment was higher than that obtained with other polyoma DNA preparations.

ogy to mouse ^{3}H-RNA, and that the DNA component with maximum homology to mouse ^{3}H-RNA has a sedimentation coefficient in the range 14 S to 16 S. Further work is required to determine if the component with maximum homology to normal cell DNA is the 14 S component of polyoma DNA.

TABLE III

HYBRIDIZATION BETWEEN POLYOMA VIRUS DNA, BEFORE AND AFTER SUCROSE GRADIENT SEDIMENTATION AND MOUSE OR POLYOMA VIRUS ³H-RNA

	Mouse ³H-RNA		Polyoma virus ³H-RNA	
DNA on filter[a] 203,000[b]	109,000	56,000	562,000	274,000
Polyoma DNA before sucrose gradient sedimentation 4493[c]	2230	1720	80,500	47,140
Polyoma DNA after sucrose gradient sedimentation:				
Fast component 841	601	340	91,882	45,882
Slow component 7243	3487	NT	116,663	NT
Blank (no DNA on filter) 146	114	91	368	136

[a] The polyoma DNA was derived from virus purified by procedure 6 in Table II. Details of the velocity sedimentation in the sucrose gradient are given in the legend to Fig. 2. The fast and slow component fractions are those marked FC and SC in Fig. 2. In each case the filters were loaded with 10 µg of the respective polyoma DNA sample. Hybridization with ³H-RNA was carried out according to Gillespie and Spiegelman (1965).

[b] Total cpm ³H-RNA incubated per vial. Each vial contained 3 to 4 DNA-filters in 8 ml of the respective ³H-RNA solution, and each DNA sample was incubated with the respective ³H-RNA at 2–3 different concentrations.

[c] ³H-RNA (cpm) retained by filter (data from Winocour, 1967).

FRACTIONATION BY METHYLATED-ALBUMIN-KIESELGUHR (MAK)
COLUMN CHROMATOGRAPHY

The technique for the separation of polyoma DNA from mouse cellular DNA by MAK column chromatography has recently been improved by Sheinin (1966). Sheinin observed that after a brief heat-treatment of polyoma virus DNA and mouse cellular DNA, the denatured mouse DNA was retained by the MAK column whereas the virus DNA (presumably due to spontaneous renaturation) was eluted at a NaCl concentration similar to that at which native virus DNA is eluted.

The effect of heat-treatment (5 min boiling in 0.015 *M* NaCl, 0.0015 *M* sodium citrate) on the MAK column chromatography of mouse cell and polyoma virus DNA's is illustrated by the reconstruction experiments shown in Fig. 3. It will be seen that only a partial separation of mouse cell from virus DNA is obtained by direct chromatography on a MAK column (Fig. 3A) because of the considerable overlap between the 2 elution peaks. After the heat-treatment (Fig. 3B) the recovery of mouse cell DNA from the MAK column was only 14% of the added radioactivity whereas that of the virus DNA was 91% of the added radioactivity. In four similar reconstruction experiments, the recovery of the heat-treated virus DNA was found to range from 42 to 91%, and that of the heat-treated mouse cell DNA from 5 to 14%. It was thus found possible to obtain a six- to eightfold purification of virus DNA with respect to mouse DNA in these reconstruction experiments.

The results of hybridization tests with DNA (from purified polyoma virions) that had been subjected to the heat-treatment and MAK column chromatography procedure, are shown in Table IV. It will be noted that the ability of the virus DNA component eluted from the MAK column to hybridize with mouse ^{3}H-RNA was reduced by a factor close to 10, compared with that of the heat-treated sample before MAK chromatography. The virtually identical capacities of the heat-treated viral DNA samples before and after MAK chromatography to hybridize with complementary polyoma virus ^{3}H-RNA indicates that the chromatographic procedure did not affect either the absorption of the DNA to filters, or the availability of the absorbed DNA for hybridization. In a second experiment of this type, the DNA from purified polyoma virions that had been subjected to heat-treatment and MAK column chromatography showed no detectable homology to normal mouse cell ^{3}H-RNA.

To determine if the DNA component eluted from the MAK column is an infectious polyoma DNA molecule, tests were made on the plaque-forming ability (Weil, 1961) of the heat-treated virus DNA before and after MAK chromatography. It was found that the virus DNA eluted

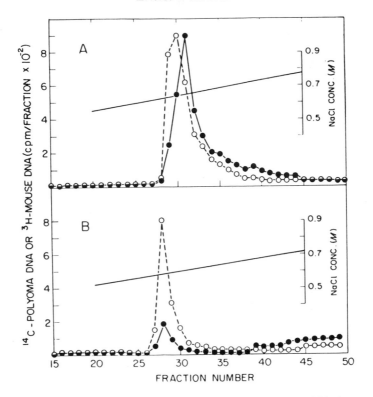

Fig. 3. MAK column chromatography of ^{14}C-labeled polyoma DNA (O——O) and ^{3}H-labeled mouse DNA (●——●) (from Winocour, 1967). (A) (Without heat-treatment): A mixture of ^{14}C-polyoma DNA and ^{3}H-mouse DNA was chromatographed on the MAK column. The recovery of ^{14}C and ^{3}H-radioactivity was 86% and 72% respectively. (B) (With heat-treatment): The mixture of ^{14}C-polyoma DNA and ^{3}H-mouse DNA was boiled for 5 min in 0.015 M NaCl, 0.0015 M sodium-citrate, rapidly cooled in ice water, and chromatographed on the MAK column. The recovery of [14]C- and ^{3}H-radioactivity was 91% and 14%, respectively.

from the MAK column (which showed no homology to mouse ^{3}H-RNA) was fully infectious with respect to plaque formation (Winocour, 1967).

The aforementioned results have shown that the component responsible for the observed homology to normal cellular DNA cannot be separated from the physical virus particle by any of the six purification procedures that were used. On the other hand, fractionation of the total DNA extracted from purified virions, either by sedimentation in sucrose gradients or by the heat-treatment and MAK chromatography procedure, was found to produce an infectious DNA whose homology to mouse ^{3}H-RNA was considerably reduced. With virus DNA subjected to the heat-treatment and MAK chromatography procedure, the homology to normal

TABLE IV

HYBRIDIZATION BETWEEN POLYOMA VIRUS DNA, BEFORE AND AFTER MAK COLUMN CHROMATOGRAPHY, AND MOUSE OR POLYOMA VIRUS ³H-RNA[a]

DNA on filter	Mouse ³H-RNA			Polyoma virus ³H-RNA		
	180,000[b]	90,000	45,000	360,000	156,000	78,000
Polyoma virus DNA before MAK chromatography	4470[c]	2043	1067	49,250	20,148	11,581
Polyoma virus DNA after MAK chromatography	306	233	111	46,281	28,985	12,565
Blank (no DNA on filter)	130	84	59	154	104	105

[a] A sample of polyoma virus DNA (extracted from virus purified by procedure 1 in Table II) was boiled for 5 min in 0.015 M NaCl, 0.0015 M sodium citrate, and rapidly cooled in ice water. A part of this material was retained as the virus DNA sample before MAK chromatography. The remainder of the heat-treated DNA was chromatographed on a MAK column (Sheinin, 1966). After dialyses against 0.0015 M NaCl, 0.00015 M Na-citrate, the samples of polyoma DNA before and after MAK chromatography were denatured by boiling for 15 min, rapidly cooled in ice water, and loaded on to filters (10 μg/filter). Hybridization with ³H-RNA was carried out according to Gillespie and Spiegelman (1965).

[b] Total cpm of ³H-RNA incubated per vial. Each vial contained 3 DNA-filters in 6 ml of the respective ³H-RNA solution, and each DNA sample was incubated with the respective ³H-RNA at 3 different concentrations.

[c] CPM of ³H-RNA retained by filter (data from Winocour, 1967).

mouse ^{3}H-RNA was reduced to a nondetectable level in some experiments. To account for these results it is proposed (Winocour, 1967) that during the maturation stage of polyoma virus replication, a segment of host cell DNA is acquired by some or all virions. It is not presently known if the acquired cellular DNA fragments are merely enmeshed in the virus capsid or are encapsidated in a more internal form. It must be stressed, however, that the above interpretation is advanced to explain the homology displayed by 16% of the total DNA extracted from purified virions. The presence of a small homologous region occupying less than 1% of the infectious DNA molecule (less than 50 nucleotide pairs) would not have been detected in these experiments. It should also be noted that any homology to mouse DNA sequences which were not transcribed in the synthesis of the ^{3}H-RNA would not have been detected.

The 5-Methyl Cytosine Marker System

Supporting evidence for the above proposal, and further information on the nature of the encapsidated cellular DNA fragments, have been obtained from studies on the synthesis and transmethylation of DNA in polyoma infected mouse kidney cultures (Winocour *et al.,* 1965). These studies showed that in polyoma-infected mouse kidney cultures labeled with L-methionine (methyl-^{3}H), the methyl group of this compound is incorporated into cytosine to form 5-methyl cytosine, which is found in both the cellular DNA and the total DNA extracted from the purified polyoma virions. Evidence for a direct transmethylation of the DNA was also provided by the observations that (1) the incorporation into 5-methyl cytosine was increased relative to the other bases in the presence of unlabeled sodium formate, and (2) no radioactivity from sodium formate-^{14}C was incorporated into 5-methyl cytosine. Furthermore, it was calculated that the amount of 5-methyl cytosine found in the DNA extracted from purified polyoma virions was only one-tenth of the amount present in mouse cell DNA.

In more recent experiments (Kaye and Winocour, 1967), the DNA from virus grown in the presence of L-methionine (methyl-^{3}H) was subjected to the same heat-treatment and MAK chromatography procedure that was used to remove the component with homology to cellular DNA. After this treatment of the virus DNA, no 5-methyl cytosine could be detected (the amount present in the control heat-treated sample before MAK chromatography was the same as that in nonheated polyoma DNA). An approximate level of sensitivity of detection in these experiments was less than 0.01%; that is, less than one 5-methyl cytosine per molecule. These results suggest, therefore, that the original observation regarding the presence of a small amount of 5-methyl cytosine in polyoma virus DNA was due to the cellular DNA fragments encapsidated by the virion.

Since contact inhibited, or X-irradiated mouse cells supporting the replication of polyoma virus also show a stimulation of cellular DNA synthesis (Dulbecco *et al.,* 1965; Weil *et al.,* 1965; Winocour *et al.,* 1965; Gershon *et al.,* 1965; Vogt *et al.,* 1966) the question arises as to whether the cellular DNA encapsidated by the virions is derived from the new cellular DNA that is synthesized as a consequence of the virus infection. A partial answer to this question is supplied by the transmethylation studies since they showed that radioactivity was incorporated into the encapsidated cellular DNA fragments from L-methionine (methyl-^{3}H) supplied to the cells after infection. This, together with the fact that no evidence was obtained for a virus-induced "hypermethylation" of the cellular DNA formed prior to infection suggests that most, if not all, of the encapsidated DNA is derived from cellular DNA synthesized after the virus infection.

The incorporation of the radioactive methyl group, from methionine into 5-methyl cytosine thus provides an independent experimental system which reveals the presence of encapsidated cellular DNA fragments in polyoma virions, and, since the infectious polyoma DNA molecule has now been shown to lack 5-methyl cytosine, it also provides an additional marker system to distinguish between cellular and viral DNA in the infected cell complex. Thus, the incorporation (relative to controls) of radioactivity from L-methionine (methyl-^{3}H) into 5-methyl cytosine in the DNA of the infected cell complex, and the possible encapsidation of these marked cellular DNA fragments by the virions, might provide evidence for a stimulation of cellular DNA synthesis in systems where this has not been detected by the methods currently in use—such as in polyoma-infected exponentially growing mouse embryo cells (Sheinin, 1966) and in BS-C-1 cells infected with SV40 virus (Gershon *et al.,* 1966).

Summary and Conclusions

Results have been discussed which suggest that most of the observed homology between DNA from polyoma virus and DNA from normal mouse cells arises as the result of the encapsidation of cellular DNA fragments by virions during the maturation stage of virus development. This conclusion is based upon the fact that 6 different virion purification procedures failed to remove the component responsibile for the homology to cellular DNA, whereas fractionation of the total DNA extracted from the purified virions led to the isolation of an infectious DNA component with no detectable homology to normal cell DNA. Furthermore, the evidence from another experimental system, in which the 5-methyl cytosine in cellular DNA is used as a marker, supports this conclusion, and also suggests that most of the encapsidated cellular DNA is derived from DNA whose synthesis was initiated by the virus infection.

Sucrose gradient analyses, in the preparative ultracentrifuge, of the total DNA extracted from purified polyoma virions has shown that the sedimentation coefficient of the component with homology to cellular DNA is in the range 14–16 S, and MAK column chromatography has demonstrated that its behavior upon denaturation is like that of double-stranded linear DNA. It is tempting to assume an identity between the 14 S polyoma DNA component and the component with homology to cellular DNA, as has been suggested previously by others (Dulbecco, 1964; Weil *et al.,* 1965). However, the final answer to the question of this identity must await the results of analytical ultracentrifugation studies on the homologous segments isolated from the MAK column.

A question of considerable interest is whether the encapsidation of cellular DNA fragments plays a role in any of the known biological activities of polyoma virus. Since the fast component of polyoma DNA has both plaque-forming and transforming activity (Crawford *et al.,* 1964; Bourgaux *et al.,* 1965) and the heat-treated virus DNA eluted from the MAK column has unimpaired plaque-forming ability, it appears that with respect to these two biological activities, the encapsidated cellular DNA fragments play no essential role. Attempts to detect a stimulation of cellular DNA synthesis in mouse cells exposed, under hypertonic salt conditions, to the total DNA extracted from polyoma virions have, so far, been unsuccessful. Thus it is not presently known if the cellular DNA fragments in the virion play a role in the virus-induced stimulation of host cell DNA synthesis.

Although a degree of homology to cellular DNA or synthetic cellular RNA has been observed in the DNA extracted from polyoma, Shope papilloma, and SV40 viruses, Benjamin (1966), in his successful demonstration of virus-specific RNA in polyoma transformed cells, has reported that no homology was detected between purified polyoma DNA and RNA from normal mouse cells. The lack of detectable homology to normal cell DNA in Benjamin's studies can be attributed either to the use of the fast component of purified polyoma DNA, or to the fact that the homologous cellular DNA regions were not transcribed in the *in vivo*-synthesized RNA that was employed. It would be of interest to determine if the phenomenon of encapsidation of cellular DNA fragments is a peculiarity of the tumor-inducing viruses—one report (Rose *et al.,* 1965) has shown that DNA extracted from the nontumorogenic type 2 adenovirus displays no homology to DNA of the host KB cell—and to investigate the possibility that these viruses may effect the transfer of cellular genes in a manner analogous to that of the transducing bacteriophages.

REFERENCES

Axelrod, D., Bolton, E. T., and Habel, K. (1964). *Science* **146,** 1466.

Benjamin, T. L. (1966). *J. Mol. Biol.* **16,** 359.

Bourgaux, P., Bourgaux-Ramoisy, D., and Stoker, M. (1965). *Virology* **25,** 364.

Chamberlin, M., and Berg, P. (1962). *Proc. Natl. Acad. Sci. U.S.* **48,** 81.

Cowie, D. B., and McCarthy, B. J. (1963). *Proc. Natl. Acad. Sci. U.S.* **50,** 537.

Crawford, L., Dulbecco, R., Fried, M., Montagnier, L., and Stoker, M. (1964). *Proc. Natl. Acad. Sci. U.S.* **52,** 148.

Dulbecco, R. (1964). *Proc. Roy. Soc.* **B160,** 423.

Dulbecco, R., and Vogt, M. (1963). *Proc. Natl. Acad. Sci. U.S.* **50,** 236.

Dulbecco, R., Hartwell, L. H., and Vogt, M. (1965). *Proc. Natl. Acad. Sci. U.S.* **53,** 403.

Gershon, D., Hausen, P., Sachs, L., and Winocour, E. (1965). *Proc. Natl. Acad. Sci. U.S.* **54,** 1584.

Gershon, D., Sachs, L., and Winocour, E. (1966). *Proc. Natl. Acad. Sci. U.S.* **56,** 918.

Gillespie, D., and Spiegelman, S. (1965). *J. Mol. Biol.* **12,** 829.

Green, M. H. (1963). *Proc. Natl. Acad. Sci. U.S.* **50,** 1177

Kaye, A. M., and Winocour, E. (1967). *J. Mol. Biol.* (In press.)

Reich, P. R., Black, P. H., and Weissman, S. M. (1966). *Proc. Natl. Acad. Sci. U.S.* **56,** 78.

Rose, J. A., Reich, P. R., and Weissman, S. M. (1965). *Virology* **27,** 571.

Sheinin, R. (1966). *Virology* **28,** 621.

Vinograd, J., and Hearst, J. (1962). *Fortschr. Chem. Org. Naturstoffe* **20,** 372.

Vinograd, J., Bruner, R., Kent, R., and Weigle, J. (1963). *Proc. Natl. Acad. Sci. U.S.* **49,** 902.

Vinograd, J., Lebowitz, J., Radloff, R., Watson, R., and Laipis, P. (1965). *Proc. Natl. Acad. Sci. U.S.* **53,** 1104.

Vogt, M., Dulbecco, R., and Smith, B. (1966). *Proc. Natl. Acad. Sci. U.S.* **55,** 956.

Weil, R. (1961). *Virology* **14,** 46.

Weil, R. (1963). *Proc. Natl. Acad. Sci. U.S.* **49,** 480.

Weil, R., and Vinograd, J. (1963). *Proc. Natl. Acad. Sci. U.S.* **50,** 730.

Weil, R., Michel, M. R., and Ruschmann, G. K. (1965). *Proc. Natl. Acad. Sci. U.S.* **53,** 1468.

Winocour, E. (1965a). *Virology* **25,** 276.

Winocour, E. (1965b). *Virology* **27,** 520.

Winocour, E. (1967). *Virology* **31,** 15.

Winocour, E., Kaye, A. M., and Stollar, V. (1965). *Virology* **27,** 156.

On the Interaction of Polyoma Virus with the Genetic Apparatus of Host Cells[*]

Roger Weil, Gudmundur Pétursson, Jindřich Kára[†] and Heidi Diggelmann[**]

DEPARTMENT OF VIROLOGY, SWISS INSTITUTE FOR EXPERIMENTAL CANCER RESEARCH,
LAUSANNE, SWITZERLAND

Introduction

Polyoma (Py)[††] virus produces a broad spectrum of histologically different tumors in mice, hamsters, and some related rodents (Eddy and Stewart, 1959). Infection of cultured mouse kidney (MK) cells leads to the production of progeny virus followed by cell death (lytic infection) (Winocour, 1963). In contact-inhibited MK cells, replication of Py virus takes place in two distinct phases (Weil *et al.,* 1965). Phase 1, which is similar to the "eclipse" period observed with other viruses, precedes and initiates phase 2, which is characterized by virus-induced syn-

[*] The work described herein was supported by grants No. 3341, 3901, and 3723 from the Swiss National Foundation for Scientific Research.

[†] Permanent address: Institute of Experimental Biology and Genetics, Czechoslovak Academy of Sciences. Prague, Czechoslovakia.

[**] Present address: Department of Biophysics, University of Chicago, Chicago, Illinois.

[††] The following are the abbreviations used throughout this chapter: ATP, adenosine-5′-triphosphate; Ci, curie; CDP, cytidine-5′-diphosphate; CdR, deoxycytidine; dCMP, dCDP, dCTP: deoxycytidine-5′-mono-, di- and triphosphate; cpm, counts per minute; dAMP, deoxyadenosine-5′-monophosphate; dGMP, deoxyguanosine-5′-monophosphate; DNA, deoxyribonucleic acid; dUMP, deoxyuridine-5′-monophosphate; EDTA, ethylene diamine tetraacetate; FUdR, 5-fluorodeoxyuridine; MK, mouse kidney; PFU, plaque forming units; PI, *post infectionem;* Py, polyoma; RNA, ribonucleic acid; S, Svedberg; SDS, sodium dodecylsulfate; TdR, deoxythymidine; dTMP, dTDP. dTTP: deoxythymidine-5′-mono-, di- and triphosphate.

thesis of cellular DNA and the production of progeny virus (Dulbecco *et al.*, 1965; Weil *et al.*, 1965; Winocour *et al.*, 1965). Experimental evidence presented in this chapter and elsewhere is compatible with the assumption that, in contact-inhibited MK cells, viral infection leads to a specific activation (derepression?) of the DNA synthesizing apparatus. This activation is neither preceded by an overall stimulation of RNA synthesis, nor followed by mitosis.

The other type of virus-cell interaction, referred to as nonlytic infection or "transformation," prevails in hamster tissue culture cells and does not lead to the production of progeny virus (Vogt and Dulbecco, 1962; Defendi, 1966). The molecular events associated with this process remain essentially unknown. It probably represents an abortive infection in which viral replication is blocked prior to the induced synthesis of cellular and viral DNA.

Polyoma is a small, icosahedral virus with a diameter of 450 Å (Wildy *et al.*, 1960). It is similar to several other tumor viruses, such as simian virus SV 40, Shope papilloma virus, and the virus of the human wart (*verruca vulgaris*), all of which contain circular DNA (Weil and Vinograd, 1963; Crawford, 1965; Crawford *et al.*, 1966; Kleinschmidt *et al.*, 1965). However, their differences in host range and antigenicity, and the absence of base-sequence homology between their DNA's show that they represent distinct viruses (Weil, 1964; Winocour, 1965a).

Sedimentation velocity analysis of the DNA extracted from highly purified preparations of Py virus reveals 3 discrete components: I, II, and III, with sedimentation coefficients of 20 S, 16 S and 14 S, respectively. Py DNA I accounts, in most preparations, for 80% or more of the total DNA; it consists of a double-stranded, circular helix (made up of two separately continuous strands), which, after thermal denaturation, exhibits spontaneous monomolecular renaturation. It is infective—that is, it contains the genetic information for the replication of intact progeny virus, and for the induction of tumors (Di Mayorca *et al.*, 1959; Eddy and Weil, 1963; Weil, 1961, 1963; Dulbecco and Vogt, 1963; Weil and Vinograd, 1963). The introduction of a single discontinuity into one of the strands of Py DNA I (20 S) converts it into a circular helix with a sedimentation velocity of 16 S (II') (Vinograd *et al.*, 1965). This molecule exhibits a melting behavior similar to that of a linear duplex. The physicochemical properties of the conversion product II' are comparable to those of naturally occurring Py DNA II. Both I and II have molecular weights of approximately 3×10^6, and exhibit, in CsCl solutions, a buoyant density of 1.709 gm cm^{-3}, which corresponds to a base composition of 49 mole% cytosine-guanosine (Weil and Vinograd, 1963). Py DNA undergoes semiconservative replication, as do other double-stranded DNA's. Replication of circular Py DNA, however, must involve a process of strand opening and closing, the nature of which remains unknown (Hirt, 1966). Py DNA III accounts, in most preparations, for 5–10% of the total DNA. It is made up of linear, double-stranded DNA molecules with a molecular weight similar to that of I and II, but with a lower buoyant density (approximately 1.702 gm cm^{-3}). Py DNA III is encased in viral capsids with properties similar to those containing Py DNA I or II. Experimental evidence suggests that at least a fraction of the DNA molecules contained in Py DNA III are derived from cellular DNA (Michel, 1967).

The results reported in this chapter are an extension of earlier work; they represent an attempt to gain a deeper understanding of the interaction of Py virus with the genetic apparatus of host cells.

Material and Methods

A more detailed description of the material and the methods used in this work will be given elsewhere (Kára and Weil, 1967; Pétursson, 1967; Pétursson and Weil, 1967; Diggelmann, 1967; Ruschmann *et al.,* 1967).

Tissue Cultures

Tissue cultures (in 60-mm plastic Petri dishes) were prepared according to the method described by Winocour (1963) from kidneys of 10-day-old inbred mice. In most experiments, CR 1 mice (Wander AG, Bern) were used. In earlier studies (Weil *et al.,* 1965), and in some of the present experiments, cultures prepared from Sn-A mice (originally obtained from Prof. G. Klein, Stockholm) were used. In these cells, infection proceeds more slowly than in cells of cultures from CR 1 mice.

One to three days after the cultures were confluent, they were infected with the same strain of wild-type Py virus used in our previous work. The virus (0.2 ml) was allowed to adsorb for 2 hours at 37°C. Py-infected and mock-infected cultures were then covered with 5 ml of culture medium (Eagle's medium with 10% horse serum) and incubated at 37°C in a water-saturated atmosphere containing 5% CO_2.

In experiments in which 5-fluorodeoxyuridine (FUdR) was used to inhibit dTMP formation and DNA synthesis (Cohen *et al.,* 1958), the cultures were washed with Eagle's medium after adsorption of the virus. Then, they were incubated either in serum-free Eagle's medium or in medium containing dialyzed horse serum. In all these experiments, FUdR was present at a concentration of 6×10^{-5} M (15 μg/ml). Since Flanagan and Ginsberg (1962) had observed that FUdR is inactivated during prolonged incubation, we renewed the FUdR-containing culture medium in all those experiments which continued for longer than 20 hours.

Pulse Labeling of Cellular and Viral DNA

Pulse-labeling of cellular and viral DNA with TdR-^{3}H was performed as described elsewhere (Weil *et al.*, 1965). After removal of the medium, the cells were lysed as in our previous work by adding to each culture 1 ml of a solution made up of 0.6% sodium dodecylsulfate (SDS) and 0.01 M ethylene diamine tetraacetate (EDTA). SDS was precipitated from the lysates by the addition of $\frac{1}{5}$ volume of 5 M NaCl. The extracts were subjected to sedimentation velocity analysis by band-centrifugation as described earlier (Vinograd *et al.*, 1963; Weil *et al.*, 1965); radioactivity was determined in a low-background beta-spectrometer (Nuclear Chicago model 725).

Autoradiography

Coverslip cultures were incubated with medium containing TdR-^{3}H as specified in the section, Results. After the pulse-labeling, they were rinsed and incubated at 37°C for 45 min in nonradioactive medium in order to allow the incorporation of intracellular TdR-^{3}H into DNA. Then the coverslips were washed and fixed in glacial acetic acid-ethanol (1 : 3 vol/vol). The washed and dried coverslips were dipped into NTB-2 Eastman Kodak liquid emulsion and exposed for 4 days.

Immunofluorescence

Coverslip cultures were fixed in 95% ethanol at various times after infection. They were incubated with polyoma capsid antiserum (mouse) at 37°C for 30 min. They were then carefully rinsed and incubated with highly purified fluorescein-labeled γG-immunoglobulins (rabbit) directed against highly purified mouse γG-immunoglobulins. Uninfected control cultures were treated in the same way. The coverslips were mounted in buffered glycerol; the slides were examined in a fluorescence microscope. The fluorescein-labeled antibodies were prepared by Dr. T.-C. Cerottini according to a modification of the method described by Wood *et al.* (1965).

Electron Microscopy

The cultures were fixed in glutaraldehyde, postfixed in osmium tetroxide, dehydrated in acetone, and embedded in Araldite (Durcupan ACM, Fluka) or Vestopal W. Thin sections were cut with a Porter-Blum micro-

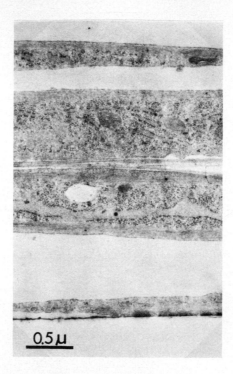

Fig. 1. Electron micrograph of a multilayered colony of spindle-shaped cells from an uninfected, confluent MK culture. The colony is made up of 4 superimposed cell layers which were cut at right angles to the surface of the dish. The culture was fixed 8 days after seeding. The dark line at the bottom corresponds to the carbon layer evaporated onto the surface of the Petri dish. × 24,400.

tome (Sorvall) and mounted on uncoated copper grids. The sections were stained first with aqueous uranyl acetate, and then according to Karnovsky's method A (1961). Observation and photography were done in either an RCA EMU-3 or a Zeiss EM-9 electron microscope. In order to obtain sections of cells cut at right angles to the surface of the Petri dish (see Fig. 1), the cells were grown in carbon-coated plastic dishes, fixed, dehydrated (ethanol), and embedded in situ. After polymerization, the layer of Araldite containing the cells was detached from the Petri dish and cut into elongated strips, which again were embedded in Araldite in the desired position.

Feulgen-Microspectrophotometry

The cultures were fixed at various times after infection (11 to 65 hours PI) and stained by the Feulgen method described by Leuchtenberger

(1958). Absorbance was measured with an automatic scanning and integrating microspectrophotometer (UMSP I, Zeiss).

EXTRACTION AND ANALYSIS OF RNA DERIVED FROM PY-INFECTED AND FROM UNINFECTED MK CULTURES

Total RNA was extracted by the hot phenol method described by Scherrer and Darnell (1962). Pulse-labeling with uridine-^{3}H (2–10 μCi/ml, specific activity 4.4 Ci/mM), for either 30 min or 4 hours, was performed immediately prior to extraction. For hybridization experiments, total RNA was extracted from Py-infected or uninfected MK, HeLa, or hamster cell cultures which had been pulse-labeled with uridine-^{3}H for 30 min. Py DNA was extracted from viral preparations purified by CsCl density gradient centrifugation followed by sedimentation through sucrose gradients. Either total Py DNA (containing components I, II, and III), or Py DNA fractionated by band-centrifugation into a fast (Py DNA I) and a slow (Py DNA II + III) fraction, were used as indicated in the text. In order to obtain irreversible strand separation, Py DNA, which was to be used for hybridization experiments, was either boiled for 30 min, or sonicated prior to boiling for 3–5 min (Weil, 1963). In most experiments, 150 μg of pulse-labeled cellular RNA was hybridized in solution at 68°C for 5 hours with 5 μg Py DNA, according to the method of Nygaard and Hall (1964). The incubation mixture was then filtered through nitrocellulose membranes (Schleicher and Schüll, Bac-T-Flex Coarse B 6). The filters were washed and then incubated for 15 min in a solution (2 × SSC) containing 20 μg/ml ribonuclease. The digested RNA was carefully washed off; the filters were dried and the radioactivity was determined in a low-background beta-spectrometer. In control experiments, the DNA-RNA hybrids were analyzed by CsCl density gradient equilibrium centrifugation.

EXTRACTION AND ASSAY OF ENZYMES

The assay of CDP-reductase was performed under conditions similar to those described by Moore and Reichard (1964). Separation and identification of the reaction products were performed as described elsewhere (Kára and Weil, 1967).

CdR-kinase activity was assayed under conditions comparable to those used for TdR-kinase (see below).

Extraction and enzymatic assay of dTMP- (and dTDP-) kinases were generally performed in the presence of 0.1 mM dTMP added as stabilizer (Grav and Smellie, 1965; Bojarski and Hiatt, 1960).

Activities of dCMP- (and dCDP-) kinases were determined at pH 9.0 under conditions similar to those used for the assay of dTMP-kinase, but without added dTMP.

Extraction of dCMP-deaminase was done in the presence of 2×10^{-4} M dCTP, and its activity was determined according to the method of Kára and Šorm (1964). The activities of dAMP- and of dGMP-kinase were assayed according to Sugino *et al.* (1966).

DETERMINATION OF TdR-KINASE* ACTIVITY

Three cultures containing a total of 10^7 cells were suspended in 0.6 ml of a buffer made up of 0.0025 M mercaptoethanol and 0.02 M Tris-HCl, pH 7.9. In the experiment depicted in Fig. 7 TdR (0.12 mM) was added to the buffer. Cell-free extracts were prepared by freezing and thawing (4 times) or by ultrasonication, followed by centrifugation at 12,000 g for 30 min at 2°C. Protein, in cell-free extracts, was quantitatively determined according to the method of Lowry *et al.* (1951). Enzymatic assays were performed immediately after extraction under conditions very similar to those described by Bollum and Potter (1959). Each assay mixture had a total volume of 0.250 ml and contained 25 μliters 0.1 M ATP; 25 μliters 0.1 M MgCl$_2$; 50 μliters 1.0 M Tris-HCl, pH 7.9; 25 μliters TdR-2-^{14}C (0.05 μCi, specific activity 30 mCi/mmole; NEN); 125 μliters cell free extract (300–400 μg protein). The mixture was incubated for 30 min at 37°C; the reaction was stopped by boiling for 2 minutes. The denatured proteins were sedimented by low-speed centrifugation; samples of the supernatant (100 μliters) were analyzed by descending paper chromatography (Schleicher and Schüll No 2043a). The chromatograms were developed for 16–20 hours in isopropylalcohol-25% NH$_4$OH-water (7 : 1 : 2 vol/vol). Radioactive zones were localized in a low-background scanning beta-spectrometer (Actigraph II, Nuclear Chicago); radioactivity was quantitatively determined on paper segments in a low-background beta-spectrometer.

Specific enzyme activities are expressed as $\mu\mu$moles TdR-5'-phosphate formed at 37°C in 30 min per mg protein.

DETERMINATION OF "DNA-POLYMERASE"† ACTIVITY

The cells from three cultures were suspended in ice-cold distilled water containing 10^{-4} M mercaptoethanol. They were homogenized for 3 min in a glass homogenizer and, then, centrifuged at 12,000 g for 30 min at

---

* ATP: TdR-5'-phosphotransferase.

† Deoxynucleosidetriphosphate: DNA deoxynucleotidyltransferase.

2.°C. Samples of the supernatant were immediately assayed for enzymatic activity under conditions similar to those used by Jungwirth and Joklik (1965). Each assay mixture had a total volume of 0.300 ml and contained 15 μliters 1.0 M Tris-HCl, pH 7.8; 15 μliters 0.1 M MgCl$_2$; 10 μliters 0.01 M mercaptoethanol; 10 μliters 4 mM dATP; 10 μliters 4 mM dCTP; 10 μliters 4 mM dGTP; 10 μliters 3.5 mM dTTP and 25 μ liters (0.125 μCi) dTTP-2-[14]C (specific activity 30 mCi/mM; NEN); 30 μliters (30 μg) heat-denatured calf thymus DNA (primer); 150 μliters cell-free extract (300–400 μg protein) and 15 μliters distilled water.

Assay mixtures were incubated at 37°C for various lengths of time (up to 120 min), and the reactions stopped by the addition of 0.3 ml 10% trichloroacetic acid (TCA) to each. The mixtures were kept on ice for 30 min and, then, filtered through Millipore membrane filters (type HA 0.45 μ; HAW PO 2500). The filters were washed 3 times with 4 ml 5% TCA; on the dried filters, radioactivity was determined in a low-background beta-spectrometer. Assay mixtures which did not contain cell-free extract (blanks) were treated in the same manner.

DETERMINATION OF INTRACELLULAR PHOSPHORYLATION OF TdR-^{3}H BY PULSE-LABELING FOR THREE MIN

Infected and uninfected MK cultures were pulse-labeled with TdR-^{3}H in a specially constructed hood for 3 min, at 37°C, in a water-saturated atmosphere (Hirt, 1966; Kára and Weil, 1967). The pulse was started when the culture medium was replaced with 2 ml of Eagle's medium (37°C) containing 25 μCi/ml TdR-methyl-[3]H (specific activity 13 Ci/mM), and it was stopped by placing the Petri dishes on ice. The radioactive medium was sucked off, and the cultures were washed 3 times with 5 ml of an ice-cold, isotonic buffer, pH 7.4.

The cells were lysed by adding 0.5 ml of a solution made up of 0.6% SDS and 0.01 M EDTA, pH 7.4 to each dish. The viscous lysate was transferred to a glass tube, rapidly mixed with 40 μliters of 70% perchloric acid, and, then, kept on ice for 10 min. The acid-precipitable material was sedimented by low-speed centrifugation, the supernatant neutralized with 5 N KOH, and the precipitated KClO$_4$ removed by low-speed centrifugation. Samples (100 μliters) of the supernatant were analyzed by descending paper chromatography (Schleicher and Schüll No. 2043a). After development for 24 hours in isobutyric acid-0.5 N NH$_4$OH (10 : 6 vol/vol) the chromatograms were dried, cut into segments of 1 or 2 cm, and the radioactivity associated with individual segments was determined in a low-background beta-spectrometer.

Results

EVIDENCE FOR ASYNCHRONOUS REPLICATION OF PY VIRUS

Earlier studies had indicated that in contact-inhibited, confluent MK cultures, replication of Py virus proceeds asynchronously (Weil *et al.,* 1965). For the interpretation of experimental observations such as virus-induced synthesis of cellular DNA and the increase in enzymatic activities (see below), as well as for the analysis of transcription of virus-specific RNA, it was necessary to carry out a detailed study of the time course of infection. In what follows, the relevant results of an investigation which will be reported in detail elsewhere (Weil and Pétursson, 1967) are summarized.

Properties of Uninfected MK Cultures

MK cells, within two days after plating, form confluent sheets which contain approximately 3.5×10^6 cells (in 60-mm plastic Petri dishes). Examination by phase-contrast microscopy, 1–3 days after confluence is attained, reveals two cell types with distinct morphological and biological properties. Epithelial-like, polygonal cells, most of which will produce progeny virus after infection, account for 80–90% of the population. The remainder are fibroblast-like, spindle-shaped cells, few of which are virus producers. It should be noted that the uninfected, spindle-shaped cells share some of the properties characteristic of cells "transformed" by Py virus: They exhibit neither contact-inhibition of multiplication nor of movement. Thus, they develop into multilayered colonies, which continue to increase in size after the MK cultures have reached confluence (see Fig. 1, p. 597). However, it was found that the inoculation of 10^7 spindle-shaped cells into each of more than a hundred newborn, isologous mice failed to give rise to tumors (Leuchtenberger, 1966). This shows that they are not "malignant." Confluent MK cultures (60-mm Petri dishes) contain a total of 15–20 μg DNA, determined chemically. Thus, the DNA content per cell ($2n$) corresponds, on the average, to twice that of mouse spermatozoids (n) (Vendrely, 1955). Measurements with the automatic scanning and integrating microspectrophotometer (UMSP I, Zeiss), performed on nuclei of Feulgen-stained cells, reveal a bimodal distribution of the absorbance values (Fig. 2), which remains essentially unchanged for at least one week after confluence. Most cells (80–90%) exhibit values of $2n$, which shows that their mitotic cycle was arrested prior to the duplication of the DNA. The majority of the remaining cells contain twice that amount of DNA ($4n$) as judged by absorbance, and only very few cells exhibit values which are between $2n$ and $4n$. It is not known whether the

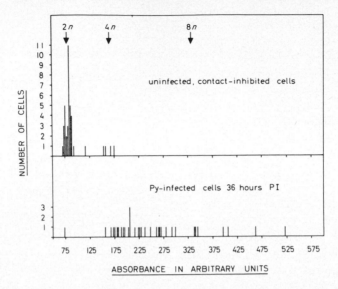

Fig. 2. DNA content of contact-inhibited uninfected and Py-infected mouse kidney cells as judged by Feulgen-microspectrophotometry. Upper: Uninfected MK cultures 4 days after confluence. Individual cells were measured at 563 mμ with the automatic scanning and integrating microspectrophotometer UMSP I (Zeiss). Lower: Parallel cultures infected with Py virus on the second day after confluence. Cultures were fixed and Feulgen-stained at 36 hours PI. It should be noted that Feulgen-absorbance, both in uninfected and in Py-infected cells, is confined to the nucleus.

$4n$ cells are diploid cells arrested in G_2; or, whether they are tetraploid cells arrested prior to the duplication of their DNA.

Time Course of Infection

Under the experimental conditions used (infective titer ca. 10^9 PFU/ml), by 20 hours PI (*post infectionem*), about 50% of the cells exhibit dark and enlarged nucleoli. Around 25 hours PI, the nuclei of a small number of cells become enlarged, and within the next 5–10 hours the number of cells with ballooned nuclei rapidly increases. As early as 25 hours PI, some cells detach from the surface of the Petri dish. At 35 hours PI, the cultures still look "morphologically intact," though they may have lost as many as 30% of the cells present prior to infection. Pulse-labeling experiments show that uninfected MK cultures incorporate TdR-^{3}H at a low rate for one week or longer after confluence. Autoradiography indicates that the residual DNA synthesis proceeds mainly in spindle-shaped cells, which explains the finding that residual incorporation of TdR-^{3}H approximately parallels the relative number of spindle-shaped cells present in confluent

MK cultures. About 14 hours after infection, the rate of TdR-^{3}H incorporation increases and reaches a maximum at 30–35 hours PI. Thereafter, incorporation drops quite rapidly to the level of uninfected controls, or even lower (Weil *et al.,* 1965; Kit *et al.,* 1966). Autoradiography reveals that the number of DNA-synthesizing cells increases after 14 hours PI; the increase proceeds most rapidly between 15 and 25 hours PI. At 25 hours PI, 50–60% of the cells in infected cultures incorporate TdR-^{3}H as compared to the 1–5% in uninfected control cultures. At later stages of infection (>30 hours PI), the relative number of cells which synthesize DNA reaches 60–80%. If relative numbers are indicated, it should be remembered that later than 30 hours PI, a considerable fraction of the cells originally present have become detached from the infected cultures. The apparent paradox, that the number of DNA-synthesizing cells reaches a maximum at a time when incorporation of TdR-^{3}H drops, will be discussed later.

Immunofluorescence tests using fluorescein-labeled antibodies directed against viral capsid protein reveal that the increase in the relative number of positive nuclei is asynchronous. If the number of positive nuclei is plotted as a function of time, a sigmoid curve, which is probably the expression of a population distribution (Fig. 3), is obtained. The increase in the number of immunofluorescence-positive cells follows the increase in the number of DNA synthesizing cells with a delay of about 3–4 hours. The first nuclei which react with fluorescein-labeled antibodies are detected between 15 and 16 hours PI; at 25 hours PI 40–50% of the cells are positive. Around 30 hours PI, the values approach a plateau, with

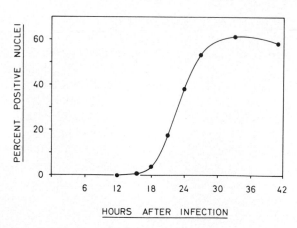

FIG. 3. Asynchronous increase in the relative number of nuclei containing polyoma capsid protein. Coverslip cultures were fixed for immunofluorescence studies at the times indicated in the figure. For each point, 1000 cells were counted.

60–80% of the nuclei being positive. After 35 hours PI, the relative number of positive nuclei often decreases. This can be explained by the fact that, at later stages of infection, a considerable fraction of the virus-producing polygonal cells is lost from the cultures and that as a result the proportion of nonvirus-producing spindle-shaped cells increases.

The comparison of the results obtained by immunofluorescence, by autoradiography and by physicochemical methods suggests that after the start of induced DNA synthesis, 3–4 hours elapse before viral capsid protein becomes detectable by immunofluorescence.

It should be pointed out that the time course of infection depends on several factors which will be discussed in detail elsewhere (Weil and Pétursson, 1967). It varies to some extent with different mouse strains used for preparing the kidney cultures; and, it depends markedly on the input multiplicity of the infecting virus. At the relatively high infective titers used for infection, the number of cells which synthesize progeny virus does not vary linearly with input multiplicity. Preliminary experiments, however, indicate that the average length of phase 1 is shortened if the input multiplicity is increased. This observation supports the working hypothesis (see the section, Discussion) that during phase 1, Py DNA may direct the synthesis of an activator which is required for the initiation of viral and cellular DNA synthesis. The decrease in the average length of phase 1 at higher input multiplicities thus might be the expression of a gene-dosage effect.

EVIDENCE FOR VIRUS-INDUCED SYNTHESIS OF CELLULAR DNA

If MK cultures are gently lysed with SDS, the cellular DNA can be extracted in the form of large complexes, which are separable by ultra-centrifugation from the smaller and more slowly sedimenting viral DNA (Fig. 4). Sedimentation velocity analyses of SDS-extracts from Py-infected and from uninfected MK cultures, pulse-labeled with TdR-^{3}H, do not reveal significant differences up to 14–15 hours PI. Thereafter, in extracts from Py-infected cultures a new, radioactive band, which corresponds to Py DNA I (20 S), appears (Weil et al., 1965). By a modified SDS-procedure, which allows selective extraction of viral DNA, Hirt (1967) found that viral progeny DNA is synthesized by a small number of cells as early as 12 hours PI. As time proceeds, the relative amount of radioactivity associated with viral DNA increases. Surprisingly, at all times after infection, at least 60% of the TdR-^{3}H taken up by infected MK cultures is incorporated into cellular DNA; this conclusion was also reached by Dulbecco et al. (1965). Incorporation of TdR-^{3}H into cellular DNA was thought to be either the expression of virus-induced, *de novo* synthesis, or of a repair mechanism. We have excluded the latter possibility; and, we have obtained direct evidence for virus-induced replication of cellular DNA, from experiments in which infection of confluent MK cultures proceeded in the presence of BUdR-^{14}C and FUdR (Weil *et al.,* 1965). It was found that if FUdR was present at a concentration sufficient to block the formation of dTMP, TdR was replaced by BUdR-^{14}C in the newly synthesized DNA. After substitution of TdR by BUdR in one strand, mouse DNA in Py-infected and also in uninfected cells is unable to undergo a second cycle of replication (Weil *et al.,* 1965; Hirt, 1966). Thus, under the conditions used, cellular DNA which

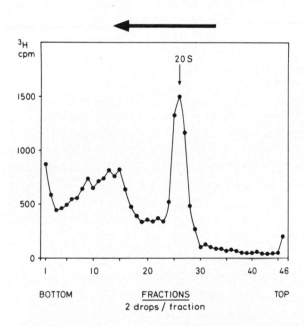

Fɪɢ. 4. Sedimentation velocity analysis by band-centrifugation of TdR-^{3}H-labeled viral and cellular DNA extracted from Py-infected MK cultures. DNA synthesis was inhibited at 2 hours PI by the addition of FUdR (6×10^{-5} M). At 25 hours PI, the cultures were released from the block imposed by FUdR by the addition of TdR (5 μg/ml); 8 hours later (33 hours PI), they were pulse-labeled for 30 min. TdR-[3]H (specific activity 13 Ci/mM) was added to the TdR-containing culture medium to a final concentration of 5 μCi/ml. For sedimentation velocity analysis, 0.25 ml of the SDS-extract was layered on a CsCl solution (3 ml; $\rho_{25°}$ 1.514 gm cm^{-3}, pH 7.5) and was centrifuged in a Spinco model L ultracentrifuge, rotor SW 39, at 35,000 rpm for 3.5 hours at 4°C. Fractions of 2 drops each were collected from the bottom of the tube on filter papers which were dried and assayed for radioactivity in a low-background beta-spectrometer (Nuclear Chicago model 725). Each sample was counted 3 times for 4 min. The values shown in the figure were not corrected for counting efficiency.

has participated in virus-induced replication is present as a "hybrid" consisting of an "old" unsubstituted strand synthesized prior to infection and a "new" BUdR-^{14}C-substituted strand synthesized during infection. Examination by CsCl density-gradient equilibrium centrifugation, of cellular DNA extracted at various times between 30 and 50 hours PI, reveals that 60–70% has undergone one cycle of replication (that is, is present in the form of a "hybrid"). From these results it follows that most, and probably all virus-producing cells (60–80% of the cells in the culture) participate in the induced replication of cellular DNA.

Cultures infected in the presence of BUdR + FUdR exhibited few if any cytopathic effects and the cells had not become detached from the dishes by the time the cultures were lysed with SDS. These results rule out the possibility that replication of cellular DNA would be the consequence of the loss of cells from Py-infected cultures.

Additional evidence for virus-induced synthesis of cellular DNA was obtained by microspectrophotometry performed on infected, Feulgen-stained cells (Fig. 2). The increase in the number of cells which exhibit elevated absorbance ($>2n$) is similar to the increase in the number of DNA-synthesizing cells seen by autoradiography. At later stages of infection (>30 hours PI), a considerable fraction of the cells were found to exhibit an absorbance $>4n$, which suggests that in these cells cellular DNA had undergone more than one cycle of induced replication. The markedly increased absorbance cannot be due to Py DNA, which at all times after infection accounts for only 10% or less of the total DNA (Weil et al., 1965; Hirt, 1967). Total (viral and cellular) DNA present in Py-infected cultures never exceeds that of uninfected controls by more than 50%, as revealed by chemical methods. The estimate obtained in this way on total DNA synthesized in infected cultures is necessarily too low, due to the loss of a considerable number of cells.

Comparative examination of cellular DNA from Py-infected and from uninfected MK cultures, by preparative and analytical CsCl density gradient equilibrium centrifugation and other physicochemical methods, failed to reveal significant differences. These results support the assumption that total cellular DNA participates in induced replication (Weil et al., 1965; Weil, 1966).

Synchronization of DNA Synthesis in Polyoma-Infected Cultures by 5-Fluorodeoxyuridine (FUdR)

In an attempt to synchronize viral replication and to investigate the possible dependence on DNA replication of (1) the transcription of viral messenger RNA, (2) the synthesis of capsid protein, and (3) the virus-induced increase in enzymatic activities, FUdR, an inhibitor of DNA synthesis (Cohen *et al.,* 1958) was used. In the following section, the results of a study which will be published in detail elsewhere are summarized (Pétursson and Weil, 1967).

FUdR at a concentration of 6×10^{-5} M inhibits DNA synthesis in more than 99% of Py-infected MK cells, as judged by autoradiography. The finding that TdR reverses the block imposed by the inhibitor shows that FUdR exerts its effect by the specific inhibition of dTMP formation. In cells in which DNA synthesis is blocked prior to the onset of the replication of viral DNA ($<$12 hours PI), no capsid protein is synthesized as judged by immunofluorescence studies, by hemagglutination, and by electron microscopy (Fig. 5). This observation is in agreement with the finding of Consigli (1967), but it is at variance with the results reported by Sheinin (1964). The metabolic events of phase 1, however, that precede and initiate the activation of the DNA synthesizing apparatus, occur in the presence of FUdR. In the FUdR-inhibited cultures, the activated cells accumulate and start DNA synthesis immediately after the addition of TdR. Autoradiography shows that, after the addition of TdR, the same number of cells synthesize DNA as do so in parallel cultures infected in the absence of FUdR (Table I).

TABLE I

SYNCHRONIZATION OF DNA SYNTHESIS IN POLYOMA-INFECTED CELLS BY FUdR[a]

MK cultures	Autoradiography % labeled nuclei
Uninfected	3.4
Uninfected + FUdR	2.8
Py-infected	52.0
Py-infected + FUdR	50.0

[a] FUdR, at a concentration of 6×10^{-5} M, was present in the medium of coverslip cultures from 2 hours PI until the start of the TdR-^{3}H pulse at 25.5 hours PI. Pulse (60 min; 1 μCi/ml; specific activity, 13 Ci/mM), fixation, and autoradiography were performed as specified in the section on Material and Methods.

Sedimentation velocity analyses of SDS-extracts from MK cultures, infected under standard conditions, and pulse-labeled at 25 hours PI, reveal—as expected—a band of Py DNA I which contains 30–40% of the total radioactivity. However, in SDS-extracts from cultures infected for the same length of time in the presence of FUdR, no radioactive viral DNA can be detected during the first hour after the addition of TdR. Thereafter, a small band of radioactive Py DNA I appears, which in-

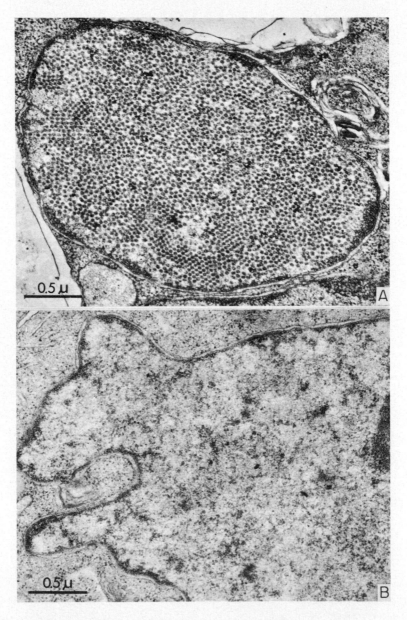

FIG. 5. (A) Electron micrograph of a cell from a MK culture infected with Py virus. The nucleus contains a great number of virus particles. × 34,500. (B) Electron micrograph of a cell from a MK culture infected with Py virus in the presence of FUdR. No viral particles can be detected. FUdR (6×10^{-5} M) and uridine (10^{-4} M) were added to the culture medium at 2 hours PI. The cells were fixed at 33 hours PI. × 34,500.

creases in size with time after release (see Fig. 4). It is likely that replication of viral DNA begins either simultaneously with the synthesis of cellular DNA, or only shortly thereafter, but that the number of replicating viral DNA molecules present during the first hour after release is too small to be detectable by the method used. Viral capsid protein can be visualized by immunofluorescence 3–4 hours after the addition of TdR. At this time, approximately 15% of the nuclei are positive; within the next few hours, this number increases. Electron microscopy, however, fails to reveal virus particles prior to 6 hours after the addition of TdR. This suggests that the synthesis of immunologically reactive capsid protein precedes the assembly of viral capsids by 2–4 hrs. Addition of puromycin (60 μg/ml) 30 min prior to the addition of TdR inhibits the synthesis of capsid protein completely; and, actinomycin D, when added at the same time as TdR, has a comparable effect. If these observations are considered together with the results on the formation of virus-specific RNA (see later), it appears likely that the messenger RNA which directs the synthesis of viral capsid protein is transcribed only after the start of viral and induced cellular DNA synthesis.

ACTIVATION BY POLYOMA VIRUS OF THE ENZYMATIC PATHWAYS INVOLVED IN THE SYNTHESIS OF PYRIMIDINE-DEOXYRIBONUCLEOTIDES AND OF DNA

The observations, that viral infection stimulates cellular DNA synthesis and that the onset of induced DNA replication can be synchronized by FUdR, suggested the hypothesis that during phase 1 viral infection might activate the enzymatic pathways involved in the biosynthesis of pyrimidine-deoxyribonucleotides and of DNA (Fig. 6). The experimental results presented below and in more detail elsewhere (Kára and Weil, 1967) support this assumption. They represent an extension of the work reported by other authors, who have described increased activities of a number of enzymes involved in DNA synthesis in Py-infected mouse tissue cultures (Dulbecco *et al.,* 1965; Frearson *et al.,* 1965; Hartwell *et al.,* 1965; Kit *et al.,* 1966).

CDP-Reductase

The main pathway for the biosynthesis of pyrimidine-deoxyribonucleotides involves an enzymatic reaction that reduces CDP to dCDP (Moore and Reichard, 1964). Specific activity of the reductase system increases in confluent MK cultures after infection with Py virus. Around 30 hours PI, it reaches a plateau of maximal activity, which is maintained during the later stages of infection ($>$30 hours) and which exceeds by three- to fivefold the activity of uninfected cultures (Kára and Weil, 1967).

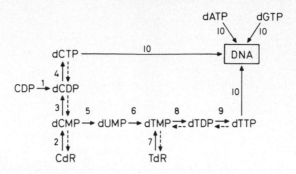

Fig. 6. Scheme of the biosynthesis of pyrimidine-deoxyribonucleotides and of DNA. Drawn out arrows refer to steps catalyzed by: (1) CDP-reductase; (2) CdR-kinase; (3) dCMP-kinase; (4) dCDP-kinase; (5) dCMP-deaminase; (6) dTMP-synthetase; (7) TdR-kinase; (8) dTMP-kinase; (9) dTDP-kinase; (10) DNA-polymerase. Dotted arrows indicate reactions catalyzed by phosphatases. Enzymes 1 (Reichard *et al.*, 1961), 5 (Maley and Maley, 1964), and 7 (Ives *et al.*, 1963) exhibit feedback inhibition by dTTP and enzyme 5 is activated by dCTP (Maley and Maley, 1964).

CdR-Kinase

The activity of this enzyme increases markedly after infection; after 30 hours PI, it reaches and maintains a plateau which is 3–4 times higher than the activity in uninfected controls (Kára and Weil, 1967). The increase in CdR-kinase activity is similar to that observed in the case of TdR-kinase. It should be noted, however, that CdR is not phosphorylated by TdR-kinase.

dCMP- (and dCDP-) Kinase

Py-infected MK cultures regularly show a small but significant increase in dCMP- (and dCDP-) kinase activity, which never exceeds the activity in uninfected controls by more than 25% (Kára and Weil, 1967).

dCMP-Deaminase

Hartwell *et al.* (1965) reported that the activity of this enzyme increases at around 12 hours PI; at 30 hours PI, it reaches a plateau that is maintained thereafter. Our own results are in accord with their findings; furthermore, they show that the three- to ninefold increase in activity is not significantly altered when DNA synthesis is inhibited by the addition of FUdR at 2 hours PI (Kára and Weil, 1967).

dTMP-Synthetase

Frearson *et al.* (1965) performed an extensive comparative study on dTMP-synthetase activity in cell-free extracts from cultures infected with

Py, herpes, and vaccinia viruses. Only infection with Py virus was found to lead to an increase in the activity of this enzyme.

TdR-Kinase

The activity of this enzyme in cell-free extracts increases around 12 hours PI; at 30 hours PI or later, the activity reaches a plateau of maximal activity which is generally five to ten times higher than in extracts from uninfected controls. Under all experimental conditions tested, the specific enzymatic activity was found to remain essentially unchanged at later stages of infection (>30 hours PI) (Fig. 7). When DNA synthesis was

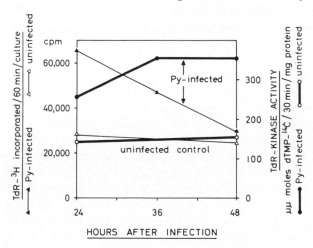

HOURS AFTER INFECTION

FIG. 7. Evidence for stability of TdR-kinase activity at later stages of infection. This experiment shows that decreased incorporation of TdR-³H into DNA at later stages of infection cannot be explained by a decreased activity of TdR-kinase. Parallel experiments show that the activities of dTMP- and dTDP-kinase also remain high. Incorporation of TdR-³H into DNA was determined by pulse-labeling MK cultures for 60 min (5 μCi/ml; specific activity 13 Ci/m*M*. In this experiment, due to the relatively low titer of the virus used for infection (ca. 2 × 10⁸ PFU/ml) the decrease in the rate of TdR-³H incorporation occurs more slowly than at higher input multiplicities. The general pattern, however, remains the same.

blocked with FUdR at 2 hours PI, the enzymatic activity was found to increase in the same way as in cells infected under standard conditions. These observations are in accordance with the results of Kit *et al.* (1966). Hartwell *et al.* (1965), however, noted an increase in activity up to 30 hours PI, followed by a rapid decline.

We compared specific enzymatic activities in extracts from Py-infected MK cultures with activities in various resting or growing mouse tissues. Enzyme activity is relatively low in kidneys of adult mice, in confluent

MK cultures and also in growing (polygonal) MK cells. It is relatively high in cultures of growing mouse embryo fibroblasts, in spindle-shaped cells in confluent MK cultures and in kidneys of 10-day-old mice. These results show that the specific activity of TdR-kinase differs markedly in different cell types and that the activity in Py-infected cultures never significantly exceeds that of growing mouse tissues.

dTMP- (and dTDP-) Kinase

Since dTMP-kinase is very unstable in the absence of substrate (Bojarski and Hiatt, 1960; Grav and Smellie, 1965), extractions and assays were performed in the presence of 0.1 mM dTMP as stabilizer. By 20 hours PI, extracts from infected MK cultures show a small but significant increase in activity that reaches a maximum by approximately 30 hours PI. In the plateau region, this activity is 50% higher than that in extracts from uninfected control cultures. A comparable increase in enzymatic activity is observed in extracts of cultures to which FUdR is added at 2 hours PI (Kára and Weil, 1967). The presence of FUdR, even for prolonged periods of time (20–40 hours), does not significantly alter the enzymatic activity in either Py-infected or in uninfected cells.

If extractions and assays are done without adding dTMP as stabilizer, enzymatic activity is relatively high in extracts from Py-infected cultures. However, it is 5–10 times lower in extracts from cultures infected in the presence of FUdR, or in extracts from uninfected controls (Kára and Weil, 1967). The effect of dTMP on enzymatic activity can possibly be explained by the assumption that Py-infected MK cells contain a sizeable pool of TdR-5'-phosphates, which stabilizes dTMP-kinase during extraction and assay. In uninfected MK cells, this pool is very small and in MK cells infected in the presence of FUdR, it has been exhausted.

The marked instability of dTMP-kinase *in vitro* is in contrast to the relative stability of the enzyme *in vivo* after inhibition of protein synthesis by puromycin. These results suggest that the turnover of dTMP-kinase is relatively low (Kára and Weil, 1967).

"DNA-Polymerase"

When assay mixtures are incubated for a period of one hour or less, cell-free extracts from Py-infected cultures show a markedly higher enzymatic activity than do extracts from uninfected controls (Fig. 8). A similar increase in enzymatic activity was found by Dulbecco *et al.* (1965). Figure 8 also shows that cell-free extracts from cultures infected in the presence of FUdR exhibit a comparable increase in enzymatic activity. When incubation of the assay mixtures is carried on for 2 hours, the quantity of radioactivity present in acid-precipitable material is increased (relative to values obtained after a 1-hour incubation period) in extracts from uninfected controls and from cultures infected in the presence of FUdR. However, in extracts derived from MK cultures, infected under standard conditions, less radioactivity is associated with acid-precipitable material after incubation for 2 hours than after 1 hour.

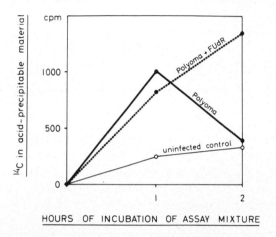

HOURS OF INCUBATION OF ASSAY MIXTURE

FIG. 8. Effect of infection with polyoma virus on "DNA polymerase" activity. After incubation for 1 hour the assay mixture containing a cell-free extract from Py-infected cultures contains more radioactivity in acid-precipitable material ("DNA") than does the mixture which contains an extract from uninfected confluent controls. Inhibition of DNA synthesis by the addition of FUdR (6×10^{-5} M) to the culture medium at 2 hours PI does not influence the increase in enzymatic activity. If the incubation is continued for 2 hours, more radioactivity is found in acid-precipitable material (as compared to the values after 1 hour) in the assay mixture containing extracts from FUdR-treated and from uninfected cultures. In contrast, the quantity of acid-precipitable radioactivity is considerably lower in the assay mixture which contains an extract from cultures infected under standard conditions.

Enzymatic activity is expressed as radioactivity (cpm) in acid-precipitable material per milligram protein in the assay mixture. The values of "blanks" have been deducted.

These data suggest that after replication of cellular and viral DNA has started, an exonuclease, which hydrolyzes the newly synthesized radioactive "double-stranded" DNA, is either synthesized *de novo* or is activated.

The results obtained from these experiments do not allow one to decide whether the activity of cell-free extracts is due to the presence of one or more nucleotidyl-transferases (virus-specific "DNA-polymerase"?).

Phosphorylation of Purine-Deoxyribonucleotides

Specific activities of enzymes involved in the phosphorylation of TdR are markedly higher in extracts from growing- than from resting tissues. In contrast, the activity of enzymes involved in the phosphorylation of purine-deoxyribonucleotides is high in both types of tissue. (Gray *et al.,* 1960). In view of these findings, it was of interest to study the activities of dAMP- and dGMP-kinases after infection of confluent MK cultures with Py virus. Experimental results reported elsewhere show that activities are high in extracts from uninfected cultures and that they do not increase significantly after infection (Kára and Weil, 1967).

EVIDENCE FOR CONTINUED DNA SYNTHESIS AT LATER STAGES OF INFECTION

At times later than 30–35 hours PI, the rate of TdR-^{3}H incorporation drops to, or even below, the level of that of uninfected cultures. Paradoxically, autoradiography and microspectrophotometry indicate that at this time ·(30–40 hours PI) the relative number of· DNA-synthesizing cells has reached a maximum (Weil *et al.,* 1965; Kit *et al.,* 1966). Therefore, decreased uptake of exogenous TdR-^{3}H can not be due to a switch-off of DNA synthesis. Similarly, the fact that, at later stages of infection TdR-, dTMP-, and dTDP-kinases exhibit high specific activities, rules out the possibility that the decreased incorporation of TdR-^{3}H into DNA might be due to a drop in the activity of the phosphorylating enzymes. Earlier observations on the incorporation of BUdR-^{14}C into cellular DNA suggested that Py-infected cells contain an increased pool of TdR-5′-phosphates (Weil *et al.,* 1965). We studied this problem in more detail by determining the intracellular phosphorylation of TdR-^{3}H (Kára and Weil, 1967). If MK cultures are pulse-labeled for 3 min, 90% of the cell-associated radioactivity can be extracted with acid in the form of free TdR-^{3}H and the corresponding 5′-mono-, di-, and trinucleotides, which are separable by paper chromatography. The remaining 10% of the label is found in DNA. The results show that conversion of TdR-^{3}H to dTTP-^{3}H proceeds rapidly and efficiently in uninfected, confluent MK cultures. This is probably due to the fact that in uninfected confluent cultures, DNA synthesis is taking place primarily in spindle-shaped cells, which exhibit a high TdR-kinase activity. After infection, the ability to convert TdR-^{3}H to dTTP-^{3}H increases concomitantly with TdR-kinase activity (Fig. 9).

When parallel cultures are labeled for either 60 or 3 min, the incorporation of TdR-^{3}H into DNA (60 min) parallels the efficiency of dTTP-

³H formation (3 min). In uninfected MK cultures, dTTP-³H formation, as well as incorporation of TdR-³H into DNA, are changed little if at all by FUdR. Similarly, FUdR added at 2 hours PI inhibits neither the increase in the activities of TdR-, dTMP-, and dTDP-kinases, nor the increase in the efficiency of intracellular conversion of TdR-³H to dTTP-³H. Actually, intracellular dTTP-³H formation is more efficient in cultures that are infected in the presence of FUdR (added at 2 hours PI) than in cultures infected under standard conditions. FUdR, when added at later stages of infection (>35 hours PI) leads to a fairly rapid depletion of endogenous TdR-5'-phosphates (Hirt, 1966; Kára and Weil, 1967). Thus,

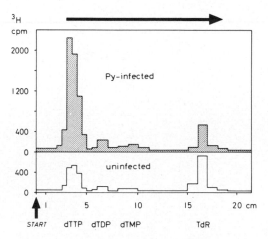

Fig. 9. Intracellular phosphorylation of TdR-³H in uninfected and in Py-infected (28 hours PI) cultures as determined by pulse-labeling for 3 min. Acid-soluble TdR-³H and the corresponding 5'-mono-, di-, and triphosphates were separated by descending paper chromatography as described in the section on Material and Methods.

pretreatment of infected cultures for 3–5 hours with FUdR leads to an increased conversion of exogenous TdR-³H to dTTP-³H, as determined by pulse-labeling for 3 min (Table II); it also leads to a similar increase in the rate at which TdR-³H is incorporated into cellular and viral DNA. These observations are taken as additional evidence for the assumption that dilution of exogenous TdR-³H in an increased pool of endogenous TdR-5'-phosphates plays an important role in the observed decrease in TdR-³H incorporation. It should be pointed out, however, that other factors, such as the loss of cells from infected cultures and a decreased rate of synthesis of cellular DNA, also play a role (Weil and Pétursson, 1967).

These results indicate that the utilization of exogenous TdR-³H for

TABLE II

Intracellular Phosphorylation of TdR-³H in Polyoma-Infected Cultures: Increased Formation of dTTP-³H in Cultures Pretreated with FUdR[a]

Extracts derived from	dTTP-³H cpm	dTDP-³H cpm	dTMP-³H cpm	TdR-³H cpm	Total cpm in acid-soluble TdR and TdR-5'-phosphates
Py-infected cultures	2615	713	238	491	4057
Py-infected cultures as above but pretreated for 5 hours with FUdR	4666	912	207	419	6204

[a] Three min pulses were performed at 43 hours PI.FUdR (6×10^{-5} M) was present from 38–43 hours PI and during the pulse. TdR-³H and the corresponding 5'-phosphates were separated by descending paper chromatography. Pulse and paper chromatography were performed as described in the section on Material and Methods (see also Fig. 9).

DNA synthesis depends on several factors; for this reason, incorporation of TdR-³H into DNA cannot be used as a reliable measurement of the rate of DNA synthesis in Py-infected cells.

EFFECT OF INFECTION ON RNA METABOLISM

Stimuli such as partial hepatectomy and others (Lieberman and Kane, 1965; Lieberman *et al.*, 1963; Bresnick, 1965; Cooper and Rubin, 1966) trigger the mitotic machinery of differentiated cells. The cells respond rapidly to stimulation by a markedly increased synthesis of RNA, which precedes and initiates the duplication of cellular DNA, which, in turn, is followed by mitotic division.

The effect of Py virus on cellular DNA synthesis differs from these systems in several respects. Microspectrophotometric determinations suggest that Py-induced DNA synthesis does not stop after duplication and that cellular DNA may undergo more than one cycle of induced replication. Furthermore, phase-contrast microscopy shows that Py-induced DNA synthesis is not followed by mitosis.

Therefore, it was of interest to study the effect of viral infection on the rates of synthesis of ribosomal-, transfer- and messenger RNA and on RNA content in contact-inhibited MK cultures. The following is a summary of the results of a study to be published in detail (Diggelmann, 1967).

By the method of Scherrer and Darnell (1962), total RNA was extracted with hot phenol from infected MK cultures at various times after infection (6–30 hours PI), and from uninfected controls. No significant differences were found in the amounts of total RNA in Py-infected and in uninfected cultures. Similarly, in experiments in which MK cultures were pulse-labeled with uridine-³H for 4 hours starting at 5, 10, 15, 20, and 25 hours PI, specific activities of total RNA derived from Py-infected and from uninfected cultures did not differ significantly. When orotic acid-³H was used for pulse-labeling instead of uridine-³H, comparable results were obtained. Sedimentation velocity analyses, in sucrose gradients of pulse-labeled RNA (uridine-³H; 30 min or 4 hours) extracted from Py-infected (5–30 hours PI) and from uninfected MK cultures, gave patterns which were indistinguishable with respect to the distribution of optical density and radioactivity.

From these results we conclude that Py-induced replication of cellular DNA is not preceded by an overall stimulation of cellular RNA synthesis. It is to be expected, however, that viral infection leads to qualitative and quantitative changes in the transcription of cellular messenger RNAL-changes, which escape detection by the methods used.

Evidence for Base-Sequence Homology between Py DNA III and RNA from Uninfected MK Cells

Uninfected, confluent MK cultures were pulse-labeled for 30 min with uridine-^{3}H, and RNA was extracted by the hot phenol method. Total cellular RNA was hybridized with Py DNA in solution, and the ribonuclease-resistant DNA-RNA hybrids were determined quantitatively, by the nitrocellulose filter method (Nygaard and Hall, 1964), and qualitatively, by CsCl density gradient equilibrium centrifugation. Py DNA was derived from viral preparations which had first been purified by CsCl density gradient equilibrium centrifugation and thereafter by sedimentation through sucrose gradients. The latter step was necessary to remove small quantities of contaminating, heterogeneous mouse DNA present in viral preparations which had been purified by centrifugation in CsCl density gradients only (Ruschmann *et al.*, 1967).

In experiments where total Py DNA (containing fractions I, II, and III) was incubated with pulse-labeled RNA from uninfected MK cultures, approximately 0.1% of the radioactivity present in the incubation mixture was found in DNA-RNA hybrids. In contrast, the extent to which pulse-labeled RNA from HeLa or hamster tissue culture cells hybridized with Py DNA was very small (0.030%)—the amount of radioactivity retained on the filters ("hybrids") was only slightly greater than the values of blanks, in which radioactive RNA had been incubated in the absence of Py DNA.

Earlier observations had suggested that at least some of the molecules of Py DNA III might be derived from mouse cellular DNA. Therefore, hybridization experiments were performed with fractionated Py DNA I or III. The experimental results showed that essentially only Py DNA III hybridizes with pulse-labeled RNA from uninfected MK cells (Michel, 1967).

The finding that Py DNA I does not hybridize, to a significant extent, with RNA from uninfected MK cultures is in agreement with the results reported by Benjamin (1966). It is at variance, however, with the results of Winocour (1965b), who claimed the existence of base-sequence homology between Py DNA I and mouse RNA. The experimental approach used by the latter author differs, however, from that used by Benjamin and by ourselves. Winocour performed his hybridization experiments with mouse RNA which had been synthesized *in vitro* with *E. coli* RNA polymerase, using mouse DNA as a primer. If further work confirms that Py DNA I hybridizes with "synthetic" but not with "natural" mouse RNA, it may be postulated that mouse DNA contains a region homologous to Py DNA I, which, in uninfected cells, is not transcribed into RNA. The fact that the viral preparations used by

Winocour were purified only by CsCl density gradient centrifugation points to the need to exclude the possibility that the positive results reported by this author might be due to the presence of contaminating cellular DNA in his preparations of Py DNA.

EVIDENCE FOR THE PRESENCE OF VIRUS-SPECIFIC RNA IN PY-INFECTED MK CULTURES

To eliminate hybrid formation due to the presence of Py DNA III, all hybridization experiments, carried out with the objective of detecting virus-specific RNA in extracts from Py-infected cultures, were performed with purified Py DNA I. We designate as "virus-specific" that fraction of pulse-labeled RNA that forms ribonuclease-resistant hybrids with Py DNA I. It remains to be determined whether some of the "virus-specific" RNA might be messenger RNA transcribed (after infection) from cellular DNA homologous to a region of Py DNA I.

Hybridization experiments with Py DNA I demonstrated the presence of a small amount of virus-specific RNA by 6 hours PI. The amount of this material was found to increase rapidly after 12 hours PI—that is, at the time when the relative number of DNA-synthesizing cells increases (Fig. 10). These results are comparable to those reported by Benjamin (1966).

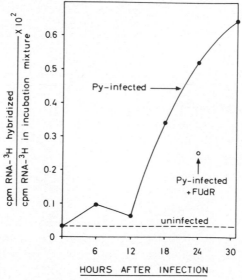

FIG. 10. Time course of the appearance of virus-specific RNA in Py-infected mouse kidney cultures. Formation of "hybrids" between Py DNA I and pulse-labeled RNA (uridine-³H; 30 min) from Py-infected MK cultures. °FUdR (6×10^{-5} M) present from 2 hours PI until extraction of RNA at 24 hours PI.

To investigate the possible dependence of RNA synthesis on DNA replication, experiments similar to those outlined above were performed under conditions in which DNA synthesis had been inhibited with FUdR immediately after the adsorption of the virus (2 hours PI). The experimental results indicate that, in both Py-infected and in uninfected MK cultures, neither RNA content nor the rate of overall synthesis of RNA is significantly altered by FUdR. However, by 24 hours PI, the amount of virus-specific RNA in extracts of FUdR-treated cultures is significantly lower; it corresponds to only about 50% of that present in extracts of cells infected in the absence of the inhibitor (Fig. 10).

The results of the hybridization experiments suggest the possibility that transcription of the genetic information contained in Py DNA I may occur in three steps. "Early" virus-specific RNA may be transcribed during phase 1 (that is, prior to the onset of induced cellular and viral DNA synthesis). The transcription of "late" virus-specific RNA, responsible for the production of viral capsid protein, apparently takes place only after DNA synthesis has started. The virus-specific RNA, present by 24 hours PI in FUdR-treated cultures, possibly represents a messenger RNA, the transcription of which coincides in time with the activation of the DNA-synthesizing apparatus of the host cells, but does not depend on the actual replication of DNA. However, the possibility that it might represent early virus-specific RNA has not been excluded.

Effect of Py Virus on Hamster Cells: Evidence for Abortive Infection

Only a very small proportion (<1%) of hamster tissue culture cells produces Py progeny virus, even when infected with very high input multiplicities (Vogt and Dulbecco, 1962; Fraser and Gharpure, 1962). Several lines of experimental evidence suggest that the relative number of cells that participate in a nonlytic infection with Py virus is probably considerably larger.

Infection by Py virus of hamster embryo fibroblasts is followed by the appearance of "transformed" cells; most of these cells contain new antigens, which are specific and heritable imprints of the transforming virus. The results reported by several laboratories indicate that neither specific morphological properties, nor loss of contact-inhibition or transplantability ("malignancy") are regularly associated with Py-transformed cells. Thus, all attempts to define "transformed" cells other than by the presence of new antigens have so far met with failure (Stanners *et al.,* 1963; Stanners, 1963; Koprowski, 1966; Defendi, 1966; Weil, 1964). Since reliable criteria are not available, it remains impossible at present to estimate, with any degree of accuracy, the relative number of cells which undergo a nonlytic interaction.

Experiments were performed in our laboratory in order to determine whether infection of hamster cells might stimulate cellular DNA synthesis in the absence of viral replication.

Cultures of hamster embryo fibroblasts were infected at different times prior to and after confluence had been attained (Weil *et al.*, 1965). As expected, the infected cultures were found to produce only very small quantities of progeny virus. However, 2–3 days after infection, 20–30% of the cells exhibited severe nuclear fragmentation (as compared to 5% or less in uninfected control cultures). The cultures were pulse-labeled with TdR-^{3}H at various times (up to 4 days) after infection. Chemical analyses and autoradiography revealed only a very slight increase in TdR-^{3}H uptake; this increase corresponds to a stimulation of DNA synthesis in approximately 1% of the infected cells. Examination, by band-centrifugation, of SDS-extracts of cultures pulse-labeled with TdR-^{3}H failed to reveal the presence of radioactive Py DNA. It is likely that the slight stimulation of TdR-^{3}H incorporation is the expression of virus-induced synthesis of cellular DNA in those few cells which produce progeny virus and that the small amount of replicating viral DNA could not be detected by the methods used. In further experiments, we infected cultures of BHK cells (a baby-hamster kidney cell line obtained from Dr. Paul Tournier, Paris) which support the production of little if any progeny virus (Fraser and Gharpure, 1962). In these cultures, infection leads neither to nuclear fragmentation nor to a detectable stimulation of TdR-^{3}H incorporation (Ruschmann and Weil, 1965). From these results, we conclude that in hamster cells which undergo a nonlytic interaction, infection is blocked at a stage prior to the induced replication of cellular and viral DNA.

Discussion

In contact-inhibited MK cells replication of Py virus proceeds in two distinct phases: Phase 1 represents a sequence of metabolic events that precede and initiate phase 2, that is the activation of the DNA-synthesizing apparatus and the production of progeny virus.

Due to the varying length of phase 1 in individual cells, the time course of infection is very asynchronous. The first cells enter phase 2 around 12 hours PI. Thereafter, the number of activated cells increases rapidly and reaches a plateau at 30 hours PI or later. If the relative number of activated cells is plotted as a function of time, a sigmoid curve, which is probably the expression of a population distribution, is obtained.

If DNA synthesis is inhibited immediately after adsorption of the virus (2 hours PI), the metabolic events of phase 1 take place and the DNA-synthesizing apparatus is activated as it is in cultures infected in the absence of the inhibitor. The activated cells accumulate and they are then ready to start DNA synthesis immediately after the addition of TdR, which releases the block imposed by FUdR. In those cells in which DNA synthesis is blocked, no viral capsid protein is synthesized. Experimental evidence presently available suggests that in the absence of induced cellular and viral DNA synthesis, late messenger RNA, which codes for capsid protein, is not transcribed.

The observation that infection activates the DNA synthesizing apparatus and that induced DNA synthesis can be synchronized by FUdR suggested the possibility that infection might activate the enzymatic pathways involved in the synthesis of pyrimidine-deoxyribonucleotides and of DNA. The experimental results reported in this paper and elsewhere (Kára and Weil, 1967) support this assumption, and show, furthermore, that activation of the enzymatic pathways occurs under conditions in which DNA synthesis is blocked by FUdR. This rules out the possibility that increased enzymatic activities are the consequence of induced DNA synthesis.

To interpret the increase in enzymatic activities (Dulbecco et al., 1965; Hartwell et al., 1965; Frearson et al., 1965; Kit et al., 1966; Kára and Weil, 1967) the asynchrony of viral replication must be considered: Both the relative number of activated cells and the enzyme activities increase around 12 hours PI and reach a maximum at 30 hours PI or later. The experimental results available are compatible with the assumption that, in individual cells, the plateau of enzymatic activities is reached rather rapidly during activation, and that the level of activity remains essentially unchanged until the cells lyse. Therefore, the increase of enzymatic activities in cell-free extracts from Py-infected cultures is, to a large extent, the expression of the increase in the number of activated cells with time after infection. It appears likely that the observed increase in enzymatic activities is due to an accelerated de novo synthesis of enzyme molecules, though no direct evidence for this assumption has yet been reported. The finding that puromycin, an inhibitor of protein synthesis, blocks the increase in enzymatic activities (Hartwell et al., 1965; Kit et al., 1966) cannot be accepted as an unequivocal proof for increased rates of de novo synthesis, since puromycin may also exert its effect by inhibiting further activation of cells. Therefore, it remains to be determined whether activation of preexisting enzymes, or shifts in dynamic equilibria between anabolic and catabolic processes may also play a role.

At present, no direct experimental evidence is available as to whether the enzymes exhibiting increased activities are coded by the viral or by the cellular genome. The fact that at least nine enzymes exhibit increased activities excludes the possibility that they all are coded by the small viral DNA with a molecular weight of 3×10^6. Several lines of indirect evidence favor the assumption that most or all of the analyzed enzymes are coded by cellular DNA. A comparison of the properties of enzymes from infected and uninfected cultures has failed to reveal any significant differences (Hartwell *et al.*, 1965; Kit *et al.*, 1966; Kára and Weil, 1967). Furthermore, the level of enzymatic activities in the plateau region (>30 hours PI) depends on the relative number of activated cells rather than on viral input multiplicity (Kára and Weil, 1967). Thus, enzymatic activities do not exhibit the gene-dosage effect expected if the enzymes were coded by viral DNA. This does not, however, exclude the possibility that Py DNA might contain genetic information for the production of one or more virus-specific enzymes.

In differentiated cells, DNA synthesis can be induced by stimuli such as partial hepatectomy (Lieberman and Kane, 1965; Bresnick, 1965), phytohemaglutinin, specific antigens and others (Cooper and Rubin, 1966; Lieberman *et al.*, 1963). In these instances, however, stimulation mobilizes the mitotic machinery of the target cells; stimulation is followed rapidly by a marked increase in RNA synthesis, which precedes and initiates the duplication of cellular DNA. A few hours after duplication of their DNA, the cells undergo mitotic division.

In contrast, Py-induced cellular DNA synthesis is neither preceded by a stimulation of overall synthesis of cellular RNA, nor followed by mitotic division. Thus, in contact-inhibited MK cells, infection specifically activates the DNA synthesizing apparatus.

The results obtained from Feulgen-microspectrophotometry suggest that cellular DNA may undergo more than one cycle of induced replication and experimental evidence supports the assumption that most, or all, of the cellular DNA participates in it.

It is not known whether or not induced synthesis of cellular DNA is a prerequisite for the replication of viral DNA. The existence of a close relationship between the two events is suggested by studies of the non-lytic interaction of Py virus with hamster tissue culture cells, in which infection is blocked prior to induced cellular and viral DNA synthesis.

The mechanism of Py-induced cellular DNA synthesis remains unknown. As a working hypothesis, we consider the possibility that an early viral function leads to the synthesis of an activator, which, after reaching a critical concentration interacts with a cellular regulatory element, and thus activates the DNA synthesizing apparatus. If it is assumed that mitotic activity in contact-inhibited MK cells is blocked by a process analogous to repression, known to operate in microorganisms, then viral infection leads to a derepression of the DNA synthesizing apparatus.

Several lines of experimental evidence suggest that at later stages of infection (>30 hours PI), Py-infected cells contain an increased pool of endogenous TdR-5'-phosphates. This might be the result of the production of larger amounts of TdR-5'-phosphates than are needed for the slowed down synthesis of cellular DNA. The experimental evidence is compatible with the hypothesis that feedback control is lost at the level of the information-transfer system involved in the production of TdR-5'-phosphates, possibly by an irreversible derepression of messenger RNA transcription.

During the lytic infection with Py and related viruses, new, intranuclear, virus-specific ("T") antigens appear prior to synthesis of viral capsid protein; the chemical nature and function of the "T"-antigens remain unknown (Takemoto *et al.,* 1966; Defendi, 1966). During lytic infection, the increase in the number of cells which contain the new antigen is asynchronous, and it is reminiscent of the asynchronous increase in the number of DNA synthesizing MK cells. Both the development of the new antigens and the activation of the DNA synthesizing apparatus remain unaffected if DNA synthesis is blocked by FUdR. This parallelism points to the possibility that the new nuclear antigen may be an early viral product, which functions in contact-inhibited MK cells as an activator of the DNA synthesizing apparatus.

ACKNOWLEDGMENTS

We are indebted to Dr. J.-C. Cerottini for preparing the fluorescein-labeled antibodies; we thank Dr. N. Odartchenko for his help in the preparation of autoradiographs. We express our gratitude to Professor C. Leuchtenberger for help and advice in Feulgen-microspectrophotometry. We are grateful to Professor H. Ginsberg for reading the manuscript; and, we thank Mr. I. Marcovici and Mr. O. Jenni for their help in preparing it.

The electron micrographs were made with the skillful help of Mrs. M. Schreyer at the Center for Electron Microscopy of the University of Lausanne (Director: A. Gautier).

REFERENCES

Benjamin, T. L. (1966). *J. Mol. Biol.* **16,** 359.
Bojarski, T. B., and Hiatt, H. H. (1960). *Nature* **188,** 1112.
Bollum, F. J., and Potter, V. R. (1959). *Cancer Res.* **19,** 561.
Bresnick, E. (1965). *J. Biol. Chem.* **240,** 2550.
Cohen, S. S., Flaks, J. G., Barner, H. D., Loeb, M. R., and Lichtenstein, J. (1958). *Proc. Natl. Acad. Sci. U.S.* **44,** 1004.
Consigli, R. (1967). In preparation.

Cooper, H. L., and Rubin, A. D. (1966). *Science* **152**, 516.

Crawford, L. V. (1965). *J. Mol. Biol.* **13**, 362.

Crawford, L. V., Follett, E. A. C., and Crawford, E. M. (1966). Personal communication.

Defendi, V. (1966). *Progr. Exptl. Tumor Res.* **8**, 125–188.

Diggelmann, H. (1967). In preparation.

Di Mayorca, G. A., Eddy, B. E., Stewart, S. E., Hunter, W. S., Friend, C., and Bendich, A. (1959). *Proc. Natl. Acad. Sci. U.S.* **45**, 1805.

Dulbecco, R., and Vogt, M. (1963). *Proc. Natl. Acad. Sci. U.S.* **50**, 236.

Dulbecco, R., Hartwell, L. H., and Vogt, M. (1965). *Proc. Natl. Acad. Sci. U.S.* **53**, **403**.

Eddy, B. E., and Stewart, S. E. (1959). *Proc. Can. Cancer Res. Conf.* **3**, 307.

Eddy, B. E., and Weil, R. (1963). Unpublished data.

Flanagan, J. F., and Ginsberg, H. S. (1962). *J. Exptl. Med.* **116**, 141.

Fraser, K. B., and Gharpure, M. (1962). *Virology* **18**, 505.

Frearson, P. M., Kit, S., and Dubbs, D. R. (1965). *Cancer Res.* **25**, 737.

Grav, H. J., and Smellie, R. M. S. (1965). *Biochem. J.* **94**, 518.

Gray, E. D., Weissman, S. M., Richards, J., Bell, D., Keir, H. M., Smellie, R. M. S., and Davidson, J. N. (1960). *Biochim. Biophys. Acta* **45**, 111.

Hartwell, L. H., Vogt, M., and Dulbecco, R. (1965). *Virology* **27**, 262.

Hirt, B. (1966). *Proc. Natl. Acad. Sci. U.S.* **55**, 997.

Hirt, B. (1967). In preparation.

Ives, D. H., Morse, P. A., Jr., and Potter, V. R. (1963). *J. Biol. Chem.* **238**, 1467.

Jungwirth, C., and Joklik, W. K. (1965). *Virology* **27**, 80

Kára, J., and Šorm, F. (1964). *Biochim. Biophys. Acta* **80**, 154.

Kára, J., and Weil, R. (1967). *Proc. Natl. Acad. Sci. U.S.* **57**, 63.

Karnovsky, M. J. (1961). *J. Biophys Biochem. Cytol.* **11**, 729.

Kit, S., Dubbs, D. R., and Frearson, P. M. (1966). *Cancer Res.* **26**, 638.

Kleinschmidt, A. K., Kass, S. J., Williams, R. C., and Knight, C. A. (1965). *J. Mol. Biol.* **13**, 749.

Koprowski, H. (1966). *Harvey Lectures* **60**, 173.

Leuchtenberger, C. (1958). *Gen. Cytochem. Methods* **1**, 219.

Leuchtenberger, C. (1967). In preparation.

Lieberman, I., and Kane, P. (1965). *J. Biol. Chem.* **240**, 1737.

Lieberman, I., Abrams, R., and Ove, P. (1963). *J. Biol. Chem.* **238**, 2141.

Lowry, O. H., and Rosebrough, N. J., Farr, L. A., and Randall, R. J. (1951). *J. Biol. Chem.* **193**, 265.

Maley, G. F., and Maley, F. (1964). *J. Biol. Chem.* **239**, 1168.

Michel, M. (1967). In preparation.

Moore, E. C., and Reichard, P. (1964). *J. Biol. Chem.* **239**, 3453.

Nygaard, A. P., and Hall, B. D. (1964). *J. Mol. Biol.* **9**, 125.

Pétursson, G. (1967). In preparation.

Pétursson, G., and Weil, R. (1967). In preparation.

Reichard, P., Canellakis, Z. N., and Canellakis, E. S. (1961). *J. Biol. Chem.* **236**, 2514.

Ruschmann, G. K., Brunner, T., and Weil, R. (1967). In preparation.

Ruschmann, G. K., and Weil, R. (1965). Unpublished results.

Scherrer, K., and Darnell, J. E. (1962). *Biochem. Biophys. Res. Commun.* **7**, 486.

Sheinin, R. (1964). *Virology* **22**, 368.

Stanners, C. P. (1963). *Virology* **21**, 464.

Stanners, C. P., Till, J. E., and Siminovitch, L. (1963). *Virology* **21,** 448.

Sugino, Y., Teraoka, H., and Shimono, H. (1966). *J. Biol. Chem.* **241,** 961.

Takemoto, K. K., Malmgren, R. A., and Habel, K. (1966). *Virology* **28,** 485.

Vendrely, R. (1955). *In* "The Nucleic Acids" (E. Chargaff and J. N. Davidson, eds.), Vol. 2, pp. 155–180. Academic Press, New York.

Vinograd, J., Bruner, R., Kent, R., and Weigle, J. (1963). *Proc. Natl. Acad. Sci. U.S.* **49,** 902.

Vinograd, J., Lebowitz, J., Radloff, R., Watson, R., and Laipis, P. (1965). *Proc. Natl. Acad. Sci. U.S.* **53,** 1104.

Vogt, M., and Dulbecco, R. (1962). *Virology* **16,** 41.

Weil, R. (1961). *Virology* **14,** 46.

Weil, R. (1963). *Proc. Natl. Acad. Sci. U.S.* **49,** 480.

Weil, R. (1964). *In* "Chemotherapy of Cancer," p. 263. Elsevier, Amsterdam.

Weil, R. (1966). Unpublished results.

Weil, R., and Pétursson, G. (1967). In preparation.

Weil, R., and Vinograd, J. (1963). *Proc. Natl. Acad. Sci. U.S.* **50,** 730.

Weil, R., Michel, M., and Ruschmann, G. (1965). *Proc. Natl. Acad. Sci. U.S.* **53,** 1468.

Wildy, P., Stoker, M. G. P., MacPherson, I. A., and Horne, R. W. (1960). *Virology* **11,** 444.

Winocour, E. (1963). *Virology* **19,** 158.

Winocour, E. (1965a). *Virology* **25,** 276.

Winocour, E. (1965b). *Virology* **27,** 520.

Winocour, E., Kaye, A. M., and Stollar, V. (1965). *Virology* **27,** 156.

Wood, B. T., Thompson, S. H., and Goldstein, G. (1965). *J. Immunol.* **95,** 225.

Deoxyribonucleic Acid Synthesis in Cells Infected with Polyoma Virus*

Rose Sheinin

ONTARIO CANCER INSTITUTE AND DEPARTMENT OF MICROBIOLOGY,
UNIVERSITY OF TORONTO, TORONTO, CANADA

Depending upon the nature of the host cell and to some extent on the physiological conditions of cell-virus interaction, polyoma virus may initiate three different responses in animal cells infected *in vitro*. The virus may replicate (Eddy *et al.*, 1958; Stewart, 1960); it may infect abortively (Medina and Sachs, 1961, 1963; Sheinin, 1966c); or it may transform host cells with respect to their morphology, growth characteristics, and neoplastic potential (Vogt and Dulbecco, 1960; Medina and Sachs, 1961; Stoker and Abel, 1962). It is the latter response of infected cells which is undoubtedly the most interesting aspect of the polyoma-cell interaction. It seems probable that clarification of the mechanism of the process of cell transformation will require knowledge of the total biochemical potential of the polyoma genome, and of the factors which control expression of this potential in each cell-virus system.

Because transformation by polyoma virus can be observed to occur in only a small fraction of an infected cell population (Vogt and Dulbecco, 1960; Medina and Sachs, 1961; Stoker and Abel, 1962; Stanners *et al.*, 1963; J. F. Williams and Till, 1964), it is difficult, at the present time, to examine directly by biochemical methods the events underlying this phenomenon. Our investigations, therefore, have been focused on the other cell-virus interactions—virus replication, in which most, if not all, of the polyoma genome is expressed, and abortive infection.

Biochemical analysis of the events which transpire during polyoma replication has been facilitated as a result of the isolation by Stanners (1963)

* The work described herein was supported by the Medical Research Council of Canada, the National Cancer Institute of Canada and the National Institutes of Health, U.S.A.

627

of a polyoma variant (TSP1) which elicits replication in essentially all mouse embryo cells infected with it at high multiplicity (Sheinin and Quinn, 1965). This variant of polyoma T, which produces small plaques, was isolated from polyoma-transformed hamster cells. Abortive infection has been examined using rat embryo cells, since it has been shown that these cells take up polyoma efficiently (Sheinin, 1966c); however, less than 0.1% support viral replication (Sheinin, 1966c) and only 5–12% undergo transformation (Medina and Sachs, 1963; J. F. Williams and Till, 1964).

From the work of Di Mayorca *et al.* (1959) and Crawford *et al.* (1964) it is clear that virus replication and cell transformation are elicited by the DNA of polyoma virus. These observations, and those, which indicate that polyoma multiplies in the nucleus of the cell (Henle *et al.,* 1959; Howatson *et al.,* 1960; M. G. Williams and Sheinin, 1961; Khare and Consigli, 1965), make it probable that both replication and transformation result from the interaction of the virus DNA with the DNA metabolism of the host. Therefore, our investigations have been directed to an examination of the effect of polyoma virus on DNA synthesis in host cells.

Studies of DNA metabolism in cells supporting polyoma virus replication have been carried out with mouse embryo cells infected with polyoma TSP1. It was found by Sheinin and Quinn (1965) that when logarithmically growing cultures of mouse embryo cells are infected with at least 10^3 plaque-forming units (PFU) of virus per cell, essentially 100% of the cells participate in virus formation in a single cycle of virus growth. Autoradiographic analyses of such cultures indicated that the synthesis of cell DNA is inhibited in cells making polyoma virus.

This observation was directly confirmed using techniques in which the procedure of chromatography of DNA on methylated albumen kieselguhr (MAK) columns (Mandell and Hershey, 1960) was adapted so that a complete operational distinction could be made between polyoma and cell DNA. It had been demonstrated initially by Dulbecco *et al.* (1965) and was later confirmed by ourselves (Sheinin, 1966b) and others (Gershon *et al.,* 1965) that polyoma DNA and mouse cell DNA can be separated by chromatography on MAK columns. Polyoma DNA was eluted from such columns with NaCl at concentrations between 0.51 to 0.57 M and mouse cell DNA was eluted with salt at concentrations between 0.54 to 0.68 M (Sheinin, 1966b).

However, variable amounts of virus and cell DNA are eluted together at NaCl concentrations between 0.54 to 0.57 M. Therefore, a more rigorous separation of the two DNA molecules was sought. It was achieved as a result of the application of two observations—first (Sueoka and Cheng, 1962), heat-denatured, single-stranded DNA from animal and bacterial cells was eluted from MAK columns at a much higher salt concentration than that required to remove the native DNA; and second (Weil, 1962, 1963), polyoma DNA behaved differently from cellular DNA, in that it renatured very readily after heating.

These two observations were applied successfully to the complete separation of polyoma DNA from cellular DNA (Sheinin, 1966b). It was found that preparations of mouse embryo cell DNA, which had been heated at 100°C for 10 min and then quick-cooled in ice (Marmur, 1961) before being applied to the MAK column, eluted only if the salt concentration of the eluting solution was increased above that required to elute native cell DNA. Polyoma DNA, heat-treated in the same way, chromatographed on MAK columns in the same way as did the native viral DNA, i.e., it was eluted with 0.51–0.54 M NaCl. It was possible, therefore, to use these characteristics of native and heat-treated DNA to distinguish between cellular and viral DNA.

Using these criteria, analyses were carried out to examine the nature of the DNA made in growing mouse embryo cells replicating polyoma virus (Sheinin, 1966b). It was found that only cellular DNA was synthesized in such cells during the first 12 hours after infection. At about 12 hours post-infection the synthesis of virus DNA was initiated. At the same time inhibition of cell DNA synthesis began, and increased progressively so that by 20 hours after infection no cell DNA formation was detectable in polyoma-infected cultures.

These observations and those obtained from autoradiographic analyses (Sheinin and Quinn, 1965) show clearly that cell DNA synthesis is inhibited in growing mouse embryo cells productively infected at high multiplicity with polyoma TSP1. Inhibition of cell DNA synthesis in mouse cells infected with polyoma virus has also been reported by Birnie and Fox (1965) and by Kramer (1966).

Although it is clear that polyoma virus infection can inhibit the synthesis of host DNA, such inhibition does not always occur. Indeed a number of workers have observed a stimulation of cell DNA synthesis in mouse cell cultures infected with polyoma after cell growth had come to a halt (Dulbecco *et al.*, 1965; Weil *et al.*, 1965; Winocour *et al.*, 1965; Gershon *et al.*, 1965). The recent studies of Vogt *et al.* (1966) leave no doubt that this induction of cell DNA formation occurs in cells destined to replicate polyoma virus.

It is not clear, however, whether the cell DNA which is made in these circumstances is entirely normal. Ben-Porat . *et al.* (1966) and

Murakami (1966) have found that the cell DNA synthesized in polyoma-infected mouse cells is unstable *in vivo* and is degraded to a molecular size similar to that of polyoma DNA. This finding is compatible with the observations (Winocour, 1966, this volume) that a cell DNA component newly synthesized in polyoma-infected mouse cells is enveloped by virus capsid in the course of lytic infection. These observations may indicate that even under conditions in which cell DNA synthesis is initiated in polyoma-infected mouse cells, a normal cycle of cellular DNA synthesis in fact does not occur.

One general conclusion which can be drawn from the evidence just cited is that polyoma virus infection interferes with normal cell DNA metabolism. Under some conditions of productive infection, this results in an absolute block in cell DNA formation; under others, a stimulation of synthesis of perhaps abnormal cellular DNA occurs.

The disruption of the normal cycle of DNA duplication may be closely tied to the inhibition of cell division, which has been found to occur early in the eclipse phase in productively infected cells (Sheinin and Quinn, 1965; Kramer, 1966), since the synthesis of DNA and mitosis are finely interwoven in the growth cycle of cells. This regulation may occur through the direct involvement of the membrane and the DNA of the cell (Jacob *et al.,* 1963; Ganesan and Lederberg, 1965). Thus, the inhibition of DNA synthesis and of cell division which occur in polyoma-infected cells may be of great significance to the phenomenon of cell transformation, which by its very nature is associated with altered patterns of cell multiplication, and with changes in the surface properties of the transformed cells (Defendi and Gasic, 1963; Ambrose, 1966).

The conclusion that inhibition of synthesis of normal cell DNA occurs in association with polyoma replication has derived support from studies of another aspect of the expression of the polyoma genome. It has been found that in cells making polyoma virus, there is formed a thymidine kinase, some of the properties of which differ from those of the enzyme of normal mouse cells (Sheinin, 1966a). The thymidine kinase of polyoma-producing cells exhibits a higher affinity for its substrate than does the normal cell enzyme. The K_m for thymidine, of the enzyme from polyoma-infected cells was found to be 1.8×10^{-7} $M,$ while that for the normal cell enzyme is 3.5×10^{-6} M (see Table I). The respective inhibition coefficients, K_s^1 (Dixon and Webb, 1964), for the two enzymes were found to be 0.01 mM and 0.1 m$M,$ respectively. In addition, the cell enzyme is much more thermolabile than the enzyme from the polyoma-infected cells—the former being inactivated to the extent of 18.6% in one minute at 63°C, and the latter being inactivated to the extent of 5.2% under similar conditions.

TABLE I

PROPERTIES OF THE THYMIDINE KINASE FOUND IN NORMAL
AND POLYOMA-INFECTED CELLS

Source of enzyme	Properties of the enzyme[a]		Thermal stability at 63°C (% inactivation per min)[b]
	K_m for thymidine	K_s^1 for thymidine	
Mouse embryo cells	3.5×10^{-6} M	0.1 mM	18.6
Polyoma-infected mouse embryo cells	1.8×10^{-7} M	0.01 mM	5.2
Rat embryo cells	4.0×10^{-7} M	0.1 mM	5.2
Polyoma-infected rat embryo cells	3.7×10^{-7} M	0.1 mM	4.6

[a] The K_m was calculated by the method of Lineweaver and Burk (1934); the K_s^1 by the procedure described by Dixon and Webb (1964).

[b] The relative thermal stabilities were calculated from the linear portions of the graphs of inactivation of thymidine kinase activity at 63°C (Sheinin, 1966a).

These observations do not, of course, provide unequivocal evidence that the thymidine kinase formed in mouse cells supporting the replication of polyoma virus is an enzyme coded for by the virus genome. However, the observations do indicate that this enzyme has properties which are unlike those of the enzyme present in normal cells. Therefore these are consistent with the observations which indicate that normal cell DNA synthesis is inhibited in cells which replicate polyoma virus replication and may indeed reflect the formation of a virus-specific thymidine kinase.

The mechanism whereby normal cell DNA synthesis is inhibited in mouse cells producing polyoma virus is at present not understood. It is unlikely that it is caused by fully formed virus capsid, since we have found no such inhibition in cells treated with inactive particles derived from the empty capsid fraction of a polyoma preparation (Abel and Crawford, 1963) or with virus inactivated by ultraviolet light (Quinn, 1966). Both preparations have normal hemagglutinating activity, and both are taken up by the treated cells.

The inhibition of cell DNA synthesis which occurs in mouse embryo cells infected with polyoma TSP1 may result from one of at least four processes: (1) degradation of the DNA of the cell genome to small acid-soluble fragments or to larger oligonucleotides; (2) alterations in the DNA of the cell genome which make it unavailable for replication; (3) changes in the enzymes which participate in cell DNA synthesis, and (4) formation of an inhibitor (or inhibitors) which specifically blocks the formation of cellular DNA.

Degradation to acid-soluble components of the DNA present in the cell prior to infection with polyoma TSP1 does not occur. Sheinin and Quinn (1965) prelabeled the cellular DNA with thymidine-^{3}H, and showed that, even 48 hours after infection, none of the labeled DNA could be detected in the acid-soluble pool, but was recovered *in toto* in the DNA fraction of the cells.

These experiments, however, do not exclude the possibility that the preexistent DNA may be degraded to the level of large, acid-precipitable oligonucleotides. Nor could they detect more subtle alteration in the molecule; for example, its partial or complete denaturation, or chemical modification of end groups which might be involved in the initiation of DNA synthesis (Richardson *et al.*, 1964).

The hypothesis that the preformed cellular DNA might be completely denatured to the single-stranded state was tested by means of the technique of chromatography on a MAK column. The DNA of approximately 10^9 TSP1-infected mouse embryo cells was extracted at 20 hours post-infection as previously described (Sheinin, 1966b). For analysis, the unlabeled DNA from infected cells was combined with ^{14}C-labeled DNA from normal mouse embryo cells. A portion of the DNA mixture was chromatographed on a MAK column in the native form. Another portion was first heated at 100°C for 10 min and then plunged into ice prior to chromatography on MAK. The recovery of the marker cell DNA was followed by measuring the radioactivity of the fractions, and that of the DNA from polyoma-infected cells by measuring the OD_{260} mμ. Since the ^{14}C-DNA was added in trace amounts, its presence did not contribute significantly to the optical density of the samples. Nor was the very large optical density of the preexistent cell DNA from the infected cells significantly affected by the presence of the relatively much smaller amount of viral DNA made in the cultures between 8–20 hours.

The results obtained from this study, and shown in Fig. 1, indicate that essentially all of the DNA of the polyoma-infected cells behaves in exactly the same way as does the DNA from normal cells; that is, the native DNA from both sources is eluted from the MAK column with 0.54–0.68 M NaCl, whereas heat-denatured material is not eluted even with 0.8 M NaCl. These results clearly establish that the DNA formed in the cell prior to infection with polyoma virus is not denatured to the single-stranded state in the course of infection.

Preliminary studies have been carried out to test the hypothesis that preexisting DNA might be degraded to large polydeoxynucleotides, or might be altered in such a way as to make it nonfunctional as a template for macromolecule synthesis. DNA was isolated from uninfected cultures and from polyoma-infected cultures 20 hours after infection, when cell

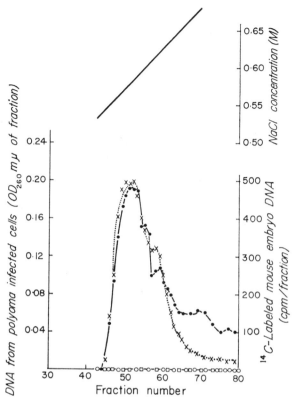

FIG. 1. Chromatography on MAK column of a mixture of ¹⁴C-labeled DNA from mouse embryo cells and unlabeled DNA isolated from polyoma-infected cells 20 hours after infection. The procedures used to isolate and analyze the DNA preparations used have been described (Sheinin, 1966a,b). ¹⁴C-labeled mouse embryo cell DNA was combined with the unlabeled DNA of polyoma-infected cells harvested 20 hours after infection. The DNA's were chromatographed in their native state and after heat treatment (Sheinin, 1966b,c). (●), native ¹⁴C-labeled DNA from normal cells; (○), heat-denatured ¹⁴C-labeled DNA from normal cells; (×), native DNA from virus-infected cells, measured by OD₂₆₀ mμ; (□), heat-treated DNA from virus-infected cells, measured by OD₂₆₀ mμ.

DNA synthesis was totally blocked. The ability of these DNA preparations to serve as primers, *in vitro,* for replication and for transcription by the DNA polymerase (Kornberg, 1959) and RNA polymerase (Hurwitz and August, 1963) respectively, was examined using enzymes isolated from *E. coli*.* Both DNA preparations were found to be equally

* The DNA polymerase (Fraction 3) and the RNA polymerase (Fraction 4), prepared according to the procedure of Chamberlin and Berg (1962), were kindly supplied by Mr. H. Eisen.

effective as templates for *in vitro* replication and transcription. In these respects the DNA isolated from polyoma-infected cells appeared to be similar to that from uninfected cells.

It seems unlikely that the inhibition of cell DNA synthesis which occurs in polyoma-infected cells can be explained by a loss of activity of one or more enzymes associated with DNA synthesis. It has been shown that there is a very large increase in thymidine kinase activity (Hartwell *et al.,* 1965; Sheinin, 1966a; Frearson *et al.,* 1965), and in the activity of other enzymes (Frearson *et al.,* 1965) which provide substrates for DNA formation, as well as in the activity of DNA polymerase itself (Sheinin, 1965; Hartwell *et al.,* 1965). It is possible, of course, that the DNA polymerase formed in polyoma virus-infected cells has a higher affinity for viral DNA than for cell DNA, due to the peculiar configuration of the former. However, even if this were so, it could not explain the observed inhibition since, under some circumstances, cell DNA synthesis of some kind does occur.

One is left with the fourth hypothesis—namely, that there is formed in TSP1-infected cells an inhibitor which specifically blocks cell DNA formation. Evidence that such an inhibitor is produced in cells infected with another DNA-containing virus, pseudorabies, has been obtained (Ben-Porat and Kaplan, 1965).

All the experimental work discussed above suggests that a primary function of polyoma, which is expressed during virus replication, is the disruption of the normal cycle of duplication of cell DNA. In growing cultures, infected at high multiplicity, this disruption is evident in the complete inhibition of cell DNA formation. In nongrowing cultures, in which cell DNA synthesis, in general, is not proceeding, polyoma virus initiates the formation of a cellular DNA, which may be abnormal, however, since it is subsequently degraded *in vivo* to material of low molecular weight, relative to the normal DNA of the cell.

One may now ask what role, if any, the disorganization of cell DNA synthesis plays in the process of cell transformation elicited by polyoma virus. Although not definitively proven, it seems unlikely that transformation proceeds in cells destined to replicate virus, since these eventually undergo lytic death. It is possible, however, that transformation may ensue in abortively infected cells. Accordingly, DNA synthesis has been studied in rat embryo cells, since these are known to be refractory to productive polyoma infection (Medina and Sachs, 1963; Sheinin, 1966c).

Secondary cultures of rat embryo cells (Buffalo strain) were infected with at least 10^3 PFU/cell of polyoma TSP1. It could be shown by immunofluorescent staining that all of the cells took up polyoma virus into their cytoplasm. Nevertheless, fewer than 0.1% of cells were found to make virus (Quinn, 1966; Sheinin, 1966c). Using similar conditions of infection, J. F. Williams and Till (1964) observed that 5–12% of the cells underwent morphological transformation.

The exact nature of the block in polyoma virus multiplication which occurs in rat embryo cells is not known. However, studies using immuno-fluorescent staining and direct plaque assay reveal that the majority of virus neither eclipses nor is uncoated.

Autoradiographic studies carried out by Quinn (1966) suggested that DNA synthesis might be stimulated in rat embryo cells infected with polyoma TSP1. A direct study of the incorporation of tritiated thymidine into the acid-precipitable DNA revealed that such incorporation is indeed stimulated in confluent cultures of rat embryo cells, between 10–16 hours after infection with polyoma virus (Sheinin, 1966c).

Therefore, studies were carried out to determine the kind of DNA that is synthesized in polyoma-infected rat embryo cells. Once again the technique of chromatography on MAK columns was applied. Using ^{32}P-labeled DNA isolated from purified polyoma virus and ^{14}C-labeled DNA isolated from normal rat embryo cells, it was shown (Fig. 2) that the two native DNA's can be grossly separated on MAK columns. If the DNA's were heated at 100°C for 10 min and then quick-cooled in ice prior to being absorbed to the MAK, the polyoma DNA was eluted with 0.52–0.56 M NaCl—the same salt concentration that elutes native polyoma DNA. However, the heat-denatured rat cell DNA were not eluted even with 0.8 M NaCl. It was possible, therefore, to distinguish clearly between heat-denatured rat cell DNA and polyoma DNA, which renatures readily after heating.

To determine the nature of the DNA made in confluent cultures of polyoma-infected rat embryo cells, the latter were incubated in medium containing tritiated-thymidine for 6-hour intervals during a period of 24 hours. The ^{3}H-labeled DNA was isolated and combined with marker ^{14}C-labeled DNA from uninfected cells. The DNA mixtures were chromatographed on MAK columns in the native form, and after heat-treatment. The results obtained revealed that all the DNA made during any 6-hour interval between 0–24 hours post-infection in the polyoma-infected rat cells behaved as does normal rat cell DNA (Sheinin, 1966c).

The observations of Ben-Porat *et al.* (1966) and of Murakami (1966) that the cell DNA synthesized under certain conditions in polyoma-infected mouse embryo cells is subsequently partially degraded, led us to examine the stability of the DNA made in polyoma-infected rat embryo cells. Secondary cultures of rat embryo cells were infected with at least 10^3 PFU/cell of polyoma TSP1. The infected cultures were incubated with ^{3}H-TdR medium (Sheinin, 1966b,c) for 4-hour intervals, as indicated in Table II, and were subsequently incubated in nonradioactive medium for varying periods of time. The cultures were then harvested, the DNA precipitated with cold 5% trichloroacetic acid, collected on Millipore filters and its radioactivity measured (Sheinin, 1966b). The data presented in Table II show that all the radioactive thymidine, which was incorporated into DNA during any 4-hour interval in polyoma-infected

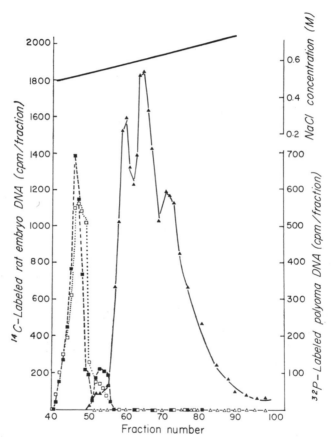

FIG. 2. Chromatography on MAK column of a mixture of DNA from polyoma virus and from normal rat embryo cells. ^{32}P-labeled polyoma DNA and ^{14}C-labeled DNA from normal rat embryo cells were combined and chromatographed on a MAK column, either in the native state, or after heat-treatment. The procedures for the preparation, heat-treatment and chromatography of the DNA have been described (Sheinin, 1966b,c). The recovery of native (■) and heat-treated (□) polyoma DNA was 72.8% and 78.4%, respectively. The recovery of native (▲) and heat-denatured (△) rat cell DNA was 93.6% and 0%, respectively. The salt concentration shown on the line graph was calculated using the formula of Mandell and Hershey (1960).

rat embryo cells, was fully recoverable as acid-insoluble material for as long as 48 hours post-infection. This suggests that at least during this period of time no extensive degradation of this DNA occurred.

These data, however, provide no information concerning the possible partial degradation or denaturation of the DNA in such cultures. To test for such changes, polyoma-infected rat embryo cell cultures, which had been incubated for 6-hour intervals in medium containing 0.1 μCi/ml of tritiated thymidine, were subsequently incubated with nonradioactive medium for up to 48 hours post-infection. The DNA was then isolated and chromatographed on MAK columns as previously described (Sheinin, 1966b,c), with ^{14}C-labeled DNA from uninfected rat embryo cells as a marker.

TABLE II

STABILITY OF DNA SYNTHESIZED IN NORMAL AND POLYOMA-INFECTED RAT EMBRYO CULTURES

Period of cell incubation with ^{3}H-TdR	CPM in DNA of 10[6] cells immediately after labeling	CPM in the DNA of 10[6] cells at		
		28 hours[a]	36 hours	48 hours
a. Control cultures, sham-infected:				
0–4	160	212	286	270
4–8	300	487	424	452
8–12	350	700	723	692
12–16	115	416	450	392
16–20	90	123	117	100
b. TSP1-infected cultures:				
0–4	203	317	342	307
4–8	350	999	1086	1036
8 12	500	1260	1600	1434
12–16	600	890	1000	1000
16–20	100	152	103	112

[a] The increase in radioactivity of the DNA which occurred during the time at which the ^{3}H-TdR medium was removed and 28 hours, is due to the slow equilibration of labeled and unlabeled medium constituents.

The illustrative data presented in Fig. 3 were obtained with ^{3}H-DNA which was labeled between 12–16 hours, and isolated at 48 hours post-infection. They show clearly that the DNA from polyoma-infected cultures and the DNA from normal cells behaved in an identical manner on MAK columns. Both preparations, if absorbed to MAK in the native state, were eluted with NaCl of molarity between 0.52 and 0.58. If, however, these DNA's were heated at 100°C and quick-cooled in ice, and then put onto MAK, neither was eluted even with 0.8 *M* NaCl.

These observations suggest that the DNA made in polyoma-infected rat embryo cultures between 0–24 hours is, in fact, rat cell DNA, normal with respect to its molecular size, molecular configuration and stability *in vivo*.

Having established that the synthesis of normal cell DNA is stimulated in abortively infected rat embryo cells, one is led to ask how this increase in DNA synthesis is initiated. One may also ask whether such synthesis

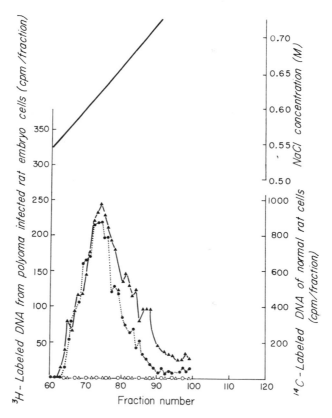

Fig. 3. Cochromatography on MAK column of a mixture of DNA from uninfected and polyoma-infected rat embryo cells. The procedures employed to derive the data shown have been described (see Fig. 3). Secondary cultures of rat embryo cells were infected with polyoma TSP1 (10^3 PFU/cell). These were incubated with ^{3}H-TdR-containing medium for the period 12–16 hours post-infection. The ^{3}H-labeled DNA formed during this time was isolated 48 hours post-infection. This ^{3}H-labeled DNA was combined with ^{14}C-labeled DNA prepared from uninfected rat embryo cells. The combined DNA's were chromatographed on MAK in the native state and after heat treatment. The recovery of native (▲———▲) and heat-treated (△———△) ^{3}H-labeled DNA was 73.1% and 0%, respectively. The recovery of native (●····●) and heat-denatured (○———○) ^{14}C-labeled normal cell DNA was 87.2% and 0%, respectively.

is associated with an increase in the thymidine kinase activity of the cells. It seems likely from studies of DNA synthesis in animal (Bollum and Potter, 1950; Weissman *et al.,* 1960; Beltz, 1962; Bianchi *et al.,* 1962; McAuslan and Joklik, 1962; Behki and Morgan, 1964; Kit *et al.,* 1963; Gentry *et al.,* 1965), plant (Hotta and Stern, 1963, 1965; Wanka *et al.,* 1964), and insect cells (Brookes and Williams, 1965) that the onset of DNA synthesis is preceded by the formation of this enzyme.

The thymidine kinase activities of normal and polyoma-infected rat embryo cells were, therefore, measured. [The infections, enzyme preparations, and enzyme assays were carried out by procedures already described (Sheinin, 1966a).]

The data shown in Fig. 4 indicated that the enzyme activity of control and TSP1-infected rat embryo cultures declines during the first 8–10 hours post-infection. After this time there was a small decrease in the thymidine kinase activity of the noninfected cells. However, the thymidine

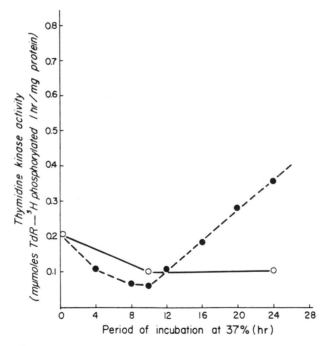

Fig. 4. Formation of thymidine kinase in normal and polyoma TSP1-infected rat embryo cells. Secondary cultures of rat embryo cells were grown to confluency. Forty-eight hours later, control cultures were mock-infected and the test cultures were infected with 1.9×10^3 PFU/cell. At the times noted, extracts were made of approximately 2×10^8 cells and the thymidine kinase activity of each was determined under optimal conditions (using 0.09 mM thymidine as substrate). ($\bigcirc$), control cultures, mock-infected; ($\bullet$), polyoma infected cultures.

kinase activity of the polyoma-infected cells increased after 10 hours and by 28 hours after infection, at which time the rate of enzyme formation was undiminished, the enzyme activity of the virus-infected cultures was approximately 4.5 times that of the control cultures.

The possibility was considered that the stimulation of cell DNA synthesis in the polyoma-infected rat embryo cells is triggered by the introduction into the cells of information coding for the formation of a virus-specific thymidine kinase. If this were so, then one might expect to discover that the thymidine kinase of such cells is different from that of uninfected rat embryo cells, and that it should exhibit properties similar to those shown by the enzyme isolated from polyoma-infected mouse embryo cells.

Accordingly, the K_m, K_s^1 and thermostability of the thymidine kinase of normal, and polyoma-infected rat embryo cells were measured. As is indicated in Table I, no difference was detected between the thymidine kinase activity of normal rat embryo cells and of that of polyoma-infected rat embryo cells. The K_m for thymidine, for both enzyme preparations was about 4×10^{-7} M; the K_s for thymidine was approximately 0.1 mM for both preparations and the rate of inactivation of enzyme activity at 63°C was the same for both preparations. These observations suggest that the thymidine kinase formed in cells nonproductively-infected with polyoma virus is cellular enzyme.

However, the data cannot be unequivocally interpreted at present, since the properties of the enzyme of normal rat embryo cells were found to be similar to those of the enzyme from polyoma-infected mouse embryo cells. That the thymidine kinase of rat embryo cells differs in several of its characteristics from the enzyme of mouse cells is not, in itself, surprising. It is known that similar enzymes from different organisms, indeed from the same tissue or cell population, do exhibit markedly different properties (Kaplan, 1963; Markert, 1963–1964). However, the finding that the thymidine kinase of normal rat embryo cells is similar to the enzyme from polyoma-infected mouse cells makes the results of the rat cell study difficult to interpret. Such interpretation must await studies with purified enzymes.

The studies described herein, as well as those of Gershon *et al.* (1965), clearly show that the synthesis of cell DNA is stimulated in rat embryo cells abortively infected with polyoma virus. The DNA which is made in these cells appears to be normal by all the criteria which have been applied in the present studies.

That the stimulation of DNA synthesis which occurs in polyoma-infected cells may be related to the phenomenon of cell transformation, is suggested by the observations of Gershon *et al.* (1965), who found that the activity of polyoma virus with respect to the stimulation of DNA synthesis (as measured in infected mouse kidney cultures) is inactivated by nitrous acid treatment at the same rate as is its transforming capacity (as measured in hamster embryo cultures). These two functions were found to be much more resistant to nitrous acid treatment than is the function of virus replication.

Studies of mouse cells which support polyoma virus replication, and of rat cells abortively infected with this virus, indicate that a primary action of the virus on the host cell is the disruption of its DNA metabolism. In the nonproductively infected cells, this disruption is manifest in the stimulation of synthesis of normal cellular DNA. In cells in which polyoma is produced, two aspects of aberrant cell DNA formation can be detected. In growing cells, infection with polyoma virus at high multiplicity results in a complete inhibition of cell DNA synthesis. In cells which have entered stationary phase, infection with polyoma virus at low (Dulbecco *et al.,* 1965; Weil *et al.,* 1965; Winocour *et al.,* 1965; Vogt *et al.,* 1966) or high multiplicity (Gershon *et al.,* 1965), results in the initiation of the synthesis of cell DNA. However, the DNA which is made in this situation is degraded *in vivo* to material of a molecular weight similar to that of polyoma DNA (Ben-Porat *et al.,* 1966; Murakami, 1966). This degradation may be due to the initial synthesis of an aberrant DNA which is susceptible to degradation by existing cellular deoxyribonucleases. It may result from the action of a new nucleolytic enzyme, the synthesis of which is stimulated by the virus genome, or indeed it may reflect the activation or derepression of synthesis of cellular nucleolytic enzymes.

An important problem which remains for consideration is the significance, to the phenomenon of cell transformation, of the finding that polyoma virus interferes with the duplication of cell DNA. It is possible that successful transformation is dependent upon the initiation of synthesis of normal cell DNA. Such synthesis can only proceed in cells abortively infected with polyoma virus, since the total expression of the polyoma genome in productively infected cells is clearly associated either with the complete inhibition of cell DNA synthesis, or with the initiation of an aberrant cycle of DNA formation.

The mechanism whereby cell DNA synthesis is induced in polyoma-infected cells remains to be clarified. Its elucidation should throw considerable light on the transforming and oncogenic functions of the virus.

REFERENCES

Abel, P., and Crawford, L. V. (1963). *Virology* **19**, 470.
Ambrose, E. J. (1967). *Proc. Can. Cancer Res. Conf.* (in press).
Behki, R. M., and Morgan, W. S. (1964). *Arch. Biochem. Biophys.* **107**, 427.
Beltz, R. E. (1962). *Biochem. Biophys. Res. Commun.* **9**, 78.
Ben-Porat, T., and Kaplan, A. S. (1965). *Virology* **25**, 22.
Ben-Porat, T., Coto, C., and Kaplan, A. S. (1966). *Virology* **30**, 74.

Bianchi, P. A., Crathorn, A. R., and Shooter, K. V. (1962). *Biochim. Biophys. Acta* **61,** 728.

Birnie, G. D., and Fox, S. M. (1965). *Biochem. J.* **95,** 41P.

Bollum, F. J., and Potter, V. R. (1959). *Cancer Res.* **19,** 561.

Brookes, V. J., and Williams, C. M. (1965). *Proc. Natl. Acad. Sci. U.S.* **53,** 770.

Chamberlin, M., and Berg, P. (1962). *Proc. Natl. Acad. Sci. U.S.* **48,** 81.

Crawford, L., Dulbecco, R., Fried, M., Montagnier, L., and Stoker, M. (1964). *Proc. Natl. Acad. Sci. U.S.* **52,** 148.

Defendi, V., and Gasic, V. (1963). *J. Cellular Comp. Physiol.* **62,** 23.

Di Mayorca, G. A., Eddy, B. E., Stewart, S. E., Hunter, W. S., Friend, C., and Bendich, A. (1959). *Proc. Natl. Acad. Sci. U.S.* **45,** 1805.

Dixon, M., and Webb, E. C. (1964). "Enzymes," 2nd ed., p. 75. Academic Press, New York.

Dulbecco, R., Hartwell, L. H., and Vogt, M. (1965). *Proc. Natl. Acad. Sci. U.S.* **53,** 403.

Eddy, B. E., Stewart, S. E., and Berkeley, W. (1958). *Proc. Soc. Exptl. Biol. Med.* **98,** 848.

Frearson, P. M., Kit, S., and Dubbs, D. R. (1965). *Cancer Res.* **25,** 737.

Ganesan, A. T., and Lederberg, J. (1965). *Biochem. Biophys. Res. Commun.* **18,** 824.

Gentry, G. A., Morse, P. A., Jr., Ives, D. H., Gebert, R., and Potter, V. R. (1965). *Cancer Res.* **25,** 509.

Gershon, D., Hausen, P., Sachs, L., and Winocour, E. (1965). *Proc. Natl. Acad. Sci. U.S.* **54,** 1584.

Hartwell, L. H., Vogt, M., and Dulbecco, R. (1965). *Virology* **27,** 262.

Henle, G., Deinhardt, F., and Rodriguez, J. (1959). *Virology* **8,** 388.

Hotta, Y., and Stern, H. (1963). *Proc. Natl. Acad. Sci. U.S.* **49,** 648.

Hotta, Y., and Stern, H. (1965). *J. Cell Biol.* **25,** Part 2, 99.

Howatson, A. F., McCulloch, E. A., Almeida, J. D., Siminovitch, L., Axelrad, A. A., and Ham, A. W. (1960). *J. Natl. Cancer Inst.* **24,** 1131.

Hurwitz, J., and August, J. T. (1963). *Progr. Nucleic Acid Res.* **1,** 59.

Jacob, F., Brenner, S., and Cuzin, F. (1963). *Cold Spring Harbor Symp. Quant. Biol.* **28,** 329.

Kaplan, N. O. (1963). *Bacteriol. Rev.* **27,** 155.

Khare, G. P., and Consigli, R. A. (1965). *J. Bacteriol.* **90,** 819.

Kit, S., Piekarski, L. J., and Dubbs, D. R. (1963). *J. Mol. Biol.* **6,** 22.

Kornberg, A. (1959). *Rev. Mod. Phys.* **31,** 200.

Kramer, R. (1966). *Intern. J. Cancer* **1,** 149.

Lineweaver, H., and Burk, D. (1934). *J. Am. Chem. Soc.* **56,** 658.

McAuslan, B. R., and Joklik, W. K. (1962). *Biochem. Biophys. Res. Commun.* **8,** 486.

Mandell, J. D., and Hershey, A. D. (1960). *Anal. Biochem.* **1,** 66.

Markert, C. L. (1963–1964). *Harvey Lectures* **59,** 187.

Marmur, J. (1961). *J. Mol. Biol.* **3,** 208.

Medina, D., and Sachs, L. (1961). *Brit. J. Cancer* **15,** 885.

Medina, D., and Sachs, L. (1963). *Virology* **19,** 127.

Murakami, W. T. (1966). Personal communication.

Quinn, P. A. (1966). M. A. Thesis, University of Toronto.

Richardson, C. C., Lehman, I. R., and Kornberg, A. (1964). *J. Biol. Chem.* **239,** 251.

Sheinin, R. (1965). *Federation Proc.* **24,** 309.

Sheinin, R. (1966a). *Virology* **28,** 47.
Sheinin, R. (1966b). *Virology* **28,** 621.
Sheinin, R. (1966c). *Virology* **29,** 15.
Sheinin, R., and Quinn, P. A. (1965). *Virology* **26,** 73.
Stanners, C. P. (1963). *Virology* **21,** 464.
Stanners, C. P., Till, J. E., and Siminovitch, L. (1963). *Virology* **21,** 448.
Stewart, S. E. (1960). *Advan. Virus Res.* **7,** 61.
Stoker, M., and Abel, P. (1962). *Cold Spring Harbor Symp. Quant. Biol.* **27,** 375.
Sueoka, N., and Cheng, T. Y. (1962). *J. Mol. Biol.* **4,** 161.
Vogt, M., and Dulbecco, R. (1960). *Proc. Natl. Acad. Sci. U.S.* **46,** 365.
Vogt, M., Dulbecco, R., and Smith, B. (1966). *Proc. Natl. Acad. Sci. U.S.* **55,** 956.
Wanka, F., Vasil, J. K., and Stern, H. (1964). *Biochim. Biophys. Acta* **85,** 50.
Weil, R. (1962). *Cold Spring Harbor Symp. Quant. Biol.* **27,** 83.
Weil, R. (1963). *Proc. Natl. Acad. Sci. U.S.* **49,** 480.
Weil, R., Michel, M. R., and Ruschmann, G. K. (1965). *Proc Natl. Acad. Sci. U.S.* **53,** 1468.
Weissman, S. M., Smellie, R. M. S., and Paul, J. (1960). *Biochim. Biophys. Acta* **45,** 101.
Williams, J. F., and Till, J. E. (1964). *Virology* **24,** 505.
Williams, M. G., and Sheinin, R. (1961). *Virology* **13,** 368.
Winocour, E., Kaye, A. M., and Stollar, V. (1965). *Virology* **27,** 156.

Analysis of Some Viral Functions Related to Neoplastic Transformation*

V. Defendi,† F. Jensen, and G. Sauer

THE WISTAR INSTITUTE OF ANATOMY AND BIOLOGY,
PHILADELPHIA, PENNSYLVANIA

Infection by some tumor viruses may result in cell lysis or in cell proliferation, with several variations at a single cell level between the two extremes. While numerous changes may occur, one condition may be deemed, a priori, to be essential for the oncogenic process—these changes have to be compatible with cell survival and multiplication. This statement still remains a rhetorical truism, because it has been very difficult to analyze the virus-induced specific changes that are necessary for such a "permissive" condition. Studies of transformed cells are not very helpful since this population has a high genetic instability (Defendi, 1966). Analysis of the initial phases of cell-virus interaction, the phase in which the decision for lysis or proliferation presumably takes place, has been difficult because malignant transformation is a rare event if one considers the high number of virus particles necessary, and the paucity of competent cells. Thus, about 10^6 physical particles of polyoma virus represent one transforming unit for BHK-21 cells (MacPherson and Montagnier, 1964), or of SV40 for 3T3 cells (Todaro and Green, 1964). A fairly high frequency of transformation (ca. 10%) can be obtained only in cell lines like BHK-21 and 3T3, which, in some unknown way, have already progressed through some steps of spontaneous malignant transformation. Thus, in the study of the transformation phenomenon, one deals with a virus population that very probably is not homogeneous— in which the relevant fraction may be a small one indeed—and a cell

* This research was supported by Grant No. CA–04534–08 from the National Cancer Institute and by Grant No. E–89 from the American Cancer Society, Inc.

† Leukemia Society Scholar and Department of Pathology, University of Pennsylvania, Philadelphia, Pennsylvania.

population that is not homogeneous in terms of genetic properties and physiological conditions.

We attempted to test, by the use of certain markers, the concept of lack of homogeneity of the virus population, and to define the cellular events which are compatible, and possibly necessary, for neoplastic transformation. The hypothesis will be formulated that only the cells in which a defective or incomplete viral multiplication occurs are those in which the neoplastic transformation proceeds, or, stated differently, that only a portion of the viral genome is responsible and sufficient for neoplastic conversion.

In the case of SV40 and polyoma viruses, it has now been established that certain specific products are synthesized, and certain new or latent functions are activated in the infected cells. Several of these have been previously discussed and are summarized in Table I.

TABLE I

PRODUCTS AND FUNCTIONS INDUCED BY POLYOMA AND SV40

Products	Functions
CF antigen (ICFA or T antigen)	Activation of host DNA synthesis
Transplantation antigen	Loss of contact inhibition
Enzymes (specificity?)	Morphological transformation
Viral protein(s)	Chromosomal alteration
Viral DNA	Rescue from finite cell multiplication
	Viral replication

The complement-fixing antigen (ICFA* or T antigen) and the transplantation antigen (ITA) are the only two cellular products that, at the present time, can be unquestionably defined as being specifically induced by the incoming viruses. Thus their presence can be taken as evidence that an infection has taken place (Table II). These antigens are not related to structural components of the virus, and persist in transformed or tumor cells, which by all possible means of detection can be defined as virus-free. All transformed cells, including several clones, that have been tested so far in our and in other laboratories have been shown to be

* The term ICFA (induced complement-fixing antigen) is used in preference to the term T or tumor antigen (Huebner *et al.,* 1964) for the following reasons: (a) The complement-fixing antigen is demonstrable in tumor cells as well as in infected cells undergoing lysis that will not be tumor cells. All the available evidence indicates that the antigens from both systems are identical; (b) the term T for the complement-fixing antigen could create confusion with the transplantation antigen, from which it apparently differs; (c) letters of the alphabet are already used in genetic immunology to indicate genes responsible for cellular isoantigens.

positive for ICFA (Defendi *et al.,* 1966). Two exceptions to this generalization have been observed: BHK-21 cells transformed by SV40 DNA (Black and Rowe, 1965), and some hamster liver and lung lines that have been infected by SV40 (Diamandopoulos and Enders, 1965). In both cases, it could not be proved that the transformation had occurred in cells that had actually been infected.

The transplantation antigen, or at least its expression, appears to be more labile, particularly in hamster cells, than the complement-fixing antigen (Defendi and Lehman, 1965). In the case of the complement-fixing antigen, it has also been established that its synthesis occurs during

TABLE II

PROPERTIES OF THE COMPLEMENT-FIXING ANTIGEN (ICFA) AND TRANSPLANTATION ANTIGEN (ITA) INDUCED BY POLYOMA AND SV40

Properties	PV		SV40	
	ICFA	ITA	ICFA	ITA
Present in lytic infection	+	?	+	?
Precedes virus synthesis	+	?	+	?
Present in virus free transformed cells	+	+	+	+
Localization	Nucleus	Cell surface?	Nucleus	Cell surface?
Relation to structural viral antigens	None	None	None	None
Chemical nature	Protein	?	Protein	?
Molecular weight	?	?	Ca. 120,000[a]	?

[a] Approximate value obtained from rate of diffusion in gel (Porter and Defendi, unpublished observations) and from Sephadex G 100 gel filtration (Kit *et al.,* 1966).

the infectious cycle and that it proceeds without requiring DNA synthesis or synthesis of the virion (Gilden *et al.,* 1965; Rapp *et al.,* 1965). Operationally, ICFA and ITA may be considered as distinct products, although there are no contradictions to the hypothesis that they may represent expression, at different cellular levels, of the same genetic changes. The functions of ICFA and ITA are unknown, although the intimate association of the SV40 ICFA with nuclear RNA that has been previously reported (Gilden *et al.,* 1965) could indicate that they play some role in the cell's regulatory process. Notwithstanding the suggestion that the ICFA may be one of the enzymes involved in DNA synthesis whose activity is increased at about the same time after virus infection that ICFA synthesis occurs, no experimental evidence is available to support such an hypothesis, and indeed, there are several theoretical objections to it. A more subtle question that has not been properly solved as yet, is whether these two antigenic products are synthesized under the direct control of the viral genome, or whether they are produced by the cell genome that has been derepressed as a consequence of a specific interaction with the virus. Recently, Dr. Porter, in our laboratory, has been able to demonstrate, by immunodiffusion in agar gel, apparent immunological identity between the SV40-ICFA extracted from SV40 transformed human and hamster cells (Porter and Defendi, unpublished observations). This evidence, although again not conclusive since the method may not be able to detect fine differences, certainly supports the first alternative. It would appear improbable that substances with identical immunological determinants would be synthesized in infected cells from two such different species unless the virus was directly responsible for them.

Previous evidence suggesting that transcription of the whole viral genome is not necessary to obtain neoplastic transformation stems essentially from two sets of observations: (1) Only minimal viral replication occurs in BHK-21 cells (Bourgaux, 1964) or in rat cells (Gershon et al., 1965) infected with polyoma virus and does not occur at all in 3T3 cells infected with SV40 (Todaro and Barron, 1965). These are the only cell lines in which transformation occurs in a fairly predictable and quantitative fashion. (2) An adeno virus 7 strain, LLE46, in which some particles have a portion of the SV40 genome incorporated into the adeno virus capsid, produces tumors that have some properties of SV40-transformed cells, such as SV40 morphological characteristics, SV40-ICFA (Huebner et al., 1964) and ITA antigens. The portion of the SV40 genome participating in this process of symbiotic carcinogenesis is still unknown.

The evidence which we intend to bring forward arises from two approaches: a study of the effect of gamma and ultraviolet (UV) radiation on LLE46, SV40, and polyoma viruses and the effect on their respective functions, and the analysis at a single cell level of a cell population in which lytic and transforming interactions may occur.

Virus Inactivation by Gamma or UV Radiation

Polyoma, SV40, adeno-7, and LLE46 viruses were grown in the appropriate cells, and harvested at the time of maximum cytopathic effect. Most of the cellular debris was removed by low speed centrifugation, after which the supernatant was centrifuged at 30,000 rpm for 4 hours and the virus pellet was suspended in 1% tryptone solution. The final viral concentration was 10^7–10^8 TCID$_{50}$ or PFU per ml. For gamma radiation experiments, the virus suspensions were distributed in 5 ml vials and frozen at $-70°C$. The irradiations were conducted in a ^{60}Co irradiation source at Pennsylvania State University (by kindness of Dr. W. Ginoza). In order to minimize the secondary effects during irradiation, the virus was kept frozen in dry ice or liquid nitrogen. At various times, samples were removed and held at $-70°C$ until the time of testing.

For UV irradiation, the virus suspension was distributed as a thin layer in a 150-mm petri dish and exposed to a UV germicidal lamp (Westinghouse Sterile Lamp, 782L20). In these experiments, R17 bacteriophage was added to the polyoma and SV40 virus suspensions as a reference virus. Infectivity of adeno-7 and LLE46 was tested in culture tubes of human kidney, SV40 was titrated by plaque assay in green monkey kidney or CV cells and polyoma by plaque assay in secondary cultures of mouse embryo cells. SV40, polyoma, and adeno-7 ICFA were tested by an indirect immunofluorescence method as previously described (Gilden *et al.,* 1965). Viral antigens were detected by direct immunofluorescence with the respective antisera. Induction of DNA synthesis was determined by pulse or continuous labeling with ^3H-TdR followed by autoradiography, and calculated on the basis of the percentage of labeled cells. After exclusion of the background number of cells in DNA synthesis in the control culture, there is a linear relationship between the proportion of labeled cells and virus input (Sauer and Defendi, 1966).

The rate of loss of infectivity for SV40, adeno-7, and LLE46 after gamma irradiation is illustrated in Fig. 1. The inactivation curves of adeno-7 and LLE46 are very similar as would be expected from the fact that the infectivity and replication of the hybrid population of LLE46 are determined by the adeno-7 genome only. The inactivation of SV40 proceeds at a slower rate. Under the condition of "direct" inactivation used, the D_0 survival values for the three viruses are in accord with the linear relationship existing between the nucleotide content of double-stranded DNA viruses and their radiosensitivity that has been established for several other mammalian and bacterial viruses (Kaplan and Moses, 1965). For each virus we can compare the radiation sensitivity of infectivity and of the ability to induce ICFA. In the case of SV40 (Fig. 2), the ability to induce ICFA is about three times as resistant as is the infectivity. The rate of decay of ICFA induction is not strictly exponential at the lower doses of radiation. We have not yet been able to determine whether this "plateau" is the result of methodological inadequacy in this experiment or an indication that more than one function is responsible for the induction of ICFA. When SV40 is inactivated by UV (Carp and Gilden, 1965), the rate of decay for ICFA induction is exponential, as will be demonstrated later (Fig. 6). The capability of inducing DNA synthesis in CV cells also decays at a similar rate (Fig. 6).

The ability of the LLE46 virus to induce the adeno-7 and SV40 ICFA's is about five times as radio-resistant as is the infectivity (Fig. 3). The ratio of adeno-7 to SV40 ICFA positive cells was the same whether the cultures were infected with the irradiated virus samples or with the

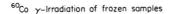

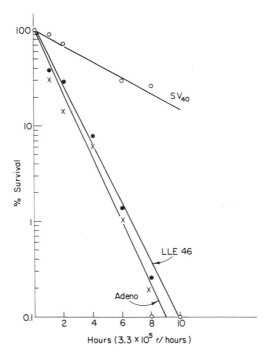

FIG. 1. Survival curves of SV40, adenovirus 7 and LLE46 (adeno 7-SV40 "hybrid") in 1% tryptone after exposure to ⁶⁰Co radiation source. Vials were maintained in dry ice during the period of irradiation.

nonirradiated control pool. Since the molecular weight of adeno-7 DNA (2.3×10^7) is much higher than that of SV40 DNA (3×10^6) and since a fixed fraction of the respective genomes is apparently responsible for the induction of ICFA, one would have expected that the induction of SV40 ICFA by the irradiated LLE46 virus would decay at a slower rate than that of adeno-7 ICFA. That this is not the case, suggests that the fractions of SV40 and adeno-7 genomes present in the hybrid population are rather intimately associated, at least for the character of ICFA induction. Rowe and Pugh (1966) have also suggested, from experiments in which the incomplete SV40 genome was transferred from one adeno virus type to another, that a linkage exists between the two viral genomes.

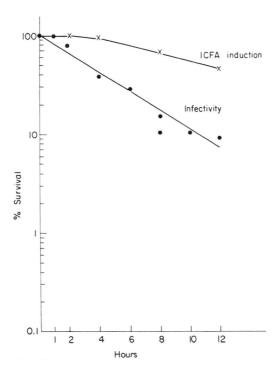

Fig. 2. Rate of inactivation by ⁶⁰Co of infectivity (plaqu⌐ forming ability) and ability to induce ICFA (proportion of ICFA positive cells) of SV40 in CV-1 cells. Samples were maintained in liquid nitrogen during irradiation. Delivery dose as measured in air at the center of the field was ca. 10^6 r/hour. No correction was made for absorption by the flask container.

In the case of polyoma virus after UV irradiation, we have also observed that the ability to induce ICFA is more resistant than the infectivity—in this instance, by a factor of 2 or 3 (Fig. 7).

These results indicate that some functions of the viral genome—the synthesis of the complement-fixing antigen, and in the case of SV40 at least, the induction of DNA synthesis—are more radiation resistant than the function of replication. If these results are interpreted on the basis of the target theory, the fraction of the genome involved in the synthesis of these products is $\frac{1}{3}$ to $\frac{1}{5}$ of the whole genome of the respective virus.

We have also been able to demonstrate that the ability to induce the SV40 transplantation antigen does not require the transcription of the whole viral genome, since LLE46 virus can protect against the induction of primary SV40 tumor (Fig. 4). Similar evidence has been obtained by Rapp *et al.* (1966) by a different method.

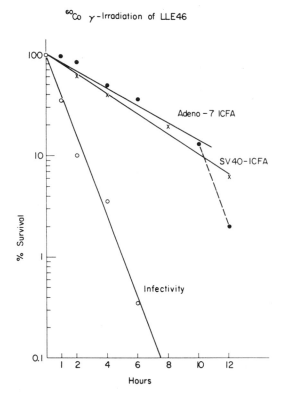

FIG. 3. Rate of inactivation of infectivity (TCID$_{50}$) and ability to induce adeno-7 ICFA and SV40 ICFA (proportion of ICFA positive cells) in human embryonic kidney cells of LLE46 virus. Experimental conditions were similar to those employed in the experiment illustrated in Fig. 2.

Inactivation studies in other laboratories have demonstrated that the ability of polyoma to transform BHK-21 cells (Benjamin, 1965; Basilico and diMayorca, 1965; Latarjet *et al.,* 1966) and to induce DNA synthesis in rat cells (Gershon *et al.,* 1965) can be dissociated from the function of viral replication.

LLE46 gamma, irradiated at a dose at which no more infectious virus could be demonstrated, rapidly transformed cultures of green monkey kidney or CV cells (Jensen and Defendi, unpublished observations).

Inoculation of gamma- or UV-irradiated LLE46, SV40, and polyoma viruses into newborn hamsters produced a further interesting phenomenon. In relation to the dosage of radiation used and to the loss of infectivity, the tumor incidence not only did not decline but actually increased. Several tumors induced by the irradiated pools were tested for presence of ICFA, and all of them were found to be positive for the respective viral-induced antigen. These results are illustrated in Figs.

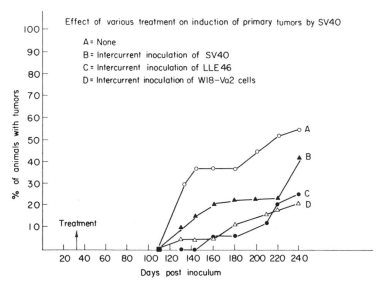

FIG. 4. Effect of various treatments on the incidence and rate of appearance of primary tumors in hamsters by SV40. Litters of newborn Lakeview hamsters were inoculated with 10^7 TCID$_{50}$ of SV40 per animal. At weaning time (32 days) animals were randomly distributed in cages and the groups treated in the following way: Group A = control; Group B = single inoculation of 10^7 TCID$_{50}$ of SV40; Group C = single injection of 10^6 TCID$_{50}$ of LLE46 virus; Group D = single inoculation of 3×10^6 cells per animal of W18 Va2 cells (human cells transformed by SV40, virus-free). Each group consisted of 20 animals.

5, 6, and 7. With all three irradiated viruses, the tumors occurred earlier and at a frequency twice that observed in animals inoculated with non-irradiated viruses. Although an explanation of this phenomenon can only be presumptive, since tumorigenesis *in vivo* is a complex phenomenon in which several factors interplay, some hypotheses can be formulated. Since there is very little virus replication in animals inoculated with these viruses, a good correlation generally exists between the amount of virus inoculated and the percentage of animals showing tumors (Stoker, 1960; Girardi *et al.*, 1963). Latarjet *et al.* (1966) have observed that Schmidt-

Ruppin Rous sarcoma virus which survives a moderate dosage of irradiation has an enhanced ability for virus synthesis. However, on the basis of results obtained by passive transfer of immunity with viral antibodies (Habel and Silverberg, 1960) or with immunocompetent cells (Defendi and Koprowski, 1959) it appears that, at least with polyoma virus, the interaction leading to the formation of tumors occurs within the first

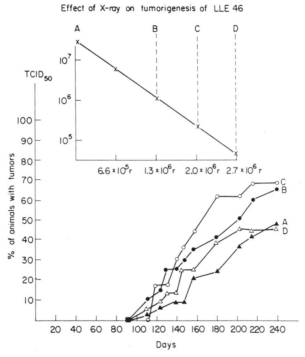

Effect of X-ray on tumorigenesis of LLE 46

FIG. 5. Effect of ⁶⁰Co irradiation on infectivity and tumorigenicity of LLE46 virus. Experimental conditions of irradiation were described in Fig. 1—twenty-five animals per group.

24 hours post-inoculation. Thus, it is highly improbable that the increased frequency of tumors with irradiated viruses can be accounted for by further replication of the virus, but rather must be due to a property of the viral population at the time of inoculation. Increased tumorigenesis of mammary tumor virus preparations after gamma irradiation has been reported by Ardashnikova and Spasskaia (1962) and Moore (1962) and a similar response has been observed with the Friend leukemia virus (Latarjet, 1962). Several interpretations of these results are possible—inactivation by radiation of an inhibitor larger in size than the virus itself; changes in the protein coat that can enhance virus absorption;

modification of the host's immunological response by the irradiated virus (for example, increase in tumor enhancing activity of the serum). The interpretation we favor at the moment and that we think is more compatible with other known facts, is that oncogenesis is a property of "defective" virus particles and that, in irradiated pools, because of inactivation of the infectious particles, the proportion of "defective" particles increases. Since the function of infectivity is repressed, more particles can express the oncogenic properties.

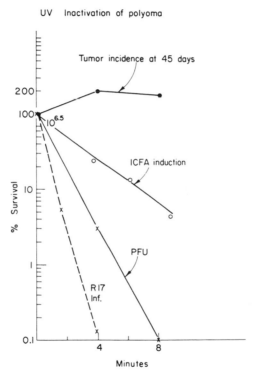

FIG. 6. Effect of UV inactiviation on infectivity (plaque forming ability), ability to induce ICFA (proportion of cells positive for ICFA) and tumorigenicity of polyoma virus (pool P-189). R17 phage was added to the polyoma pool and its infectivity tested on *E. coli* K-12 Hfr. UV irradiation source was a Westinghouse Sterile Lamp, 782 L 20, delivering 95% of energy at 2537 Å. The virus samples (5 ml in a 150 mm petri dish) were placed at a distance of 12 inches from the lamp. At this distance the energy at the surface is calculated to be 196.8 microwatts/cm². Litters of Lakeview hamsters were inoculated with the virus samples on the third day after birth; 22 animals per group. Tumor incidence with the untreated virus was 45% at 45 days. Determination of ICFA positive cells was done by the indirect immunofluorescence method with serum kindly provided by Dr. R. R. Gilden.

The fact that no decrease of oncogenic activity has been observed, even at the highest level of irradiation so far used, would further suggest the possibility that the fragment of the viral DNA responsible for oncogenesis may be more easily incorporated in the cell genome, or may more easily survive as an episome, if the integrity of the whole viral DNA is altered by radiation.

UV Inactivation of SV40

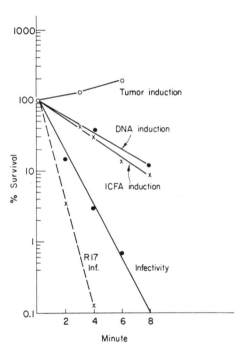

FIG. 7. Effect of UV irradiation on infectivity (plaque forming units), ability to induce ICFA and ability to induce DNA synthesis in CV cells and tumorigenicity of SV40. Conditions of irradiation were similar to those used in the experiment illustrated in Fig. 6. Newborn litters of Lakeview hamsters were inoculated with various samples of virus (3 litters per sample). In this experiment, tumor incidence in the group inoculated with untreated virus was 42%.

Host Factors

With several tumor and nontumor viruses, cellular response will vary according to the growth phase of the cells at the time of infection. Typical of such differences are the results obtained with polyoma virus during lytic interaction with cultures of mouse fibroblasts. In growing cultures there is a depression (Sheinin and Quinn, 1965), and in stationary phase cultures an activation of host cell DNA synthesis (Dulbecco *et al.,* 1965; Vogt *et al.,* 1966; Weil *et al.,* 1965; Winocour *et al.,* 1965). SV40 multiplies to a high titer in resting green monkey kidney cells but little virus is produced in the same cells if they are allowed, by frequent sub-division of the culture, to continuously multiply (Carp and Gilden, 1966). Under the latter condition, the cells may transform rather than undergo complete lysis (Carp and Gilden, 1966).

The cells in which we have chosen to investigate the relationship between cells and viral functions are those of the human diploid cell strain, WI38. In these cells there is an effective inhibition of DNA synthesis when the monolayers become confluent. Furthermore, the growth fraction of the population, i.e., the proportion of cells able to undergo DNA synthesis and mitosis, declines even in nonconfluent monolayers as a function of the numbers of passages or cell generations *in vitro.* An example of the decay of the growth fraction in this diploid strain is diagrammed in Fig. 8. The growth fraction at each serial passage was determined by exposing lightly seeded cultures (5×10^5 cells in 60-mm petri dishes) to ³H-TdR during the second day after plating (Defendi,

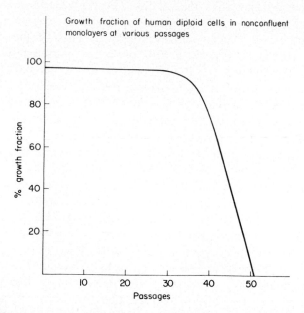

FIG. 8. Growth fraction of human diploid cells (strain WI38) in nonconfluent monolayer at various passages. Lightly seeded cultures (5×10^5 cells in 60 mm petri dish) were continuously exposed to 0.5 μCi/ml of ³H-TdR from 24 to 48 hours after plating. The fraction of cells incorporating ³H-TdR was determined by autoradiography.

unpublished observation). Thus, it was possible to study the effect of SV40 infection on DNA synthesis (1) in cells which were not synthesizing DNA, because of contact inhibition, but which still had the ability to do so; and (2) in cells which had lost the potential to synthesize DNA and to divide (permanent G_0 cells) and which, therefore, could be considered terminal cells, analogous to red cells or neurons. In the current experiments, WI38 cells, from the forty-fifth to the fiftieth passage, were seeded on coverslips. The cultures were allowed to reach confluency and were maintained in this condition for several days, at which time they were infected at a ratio of 350 $TCID_{50}$ per cell. After absorption of virus, the cultures were maintained throughout the experiment in medium containing 10% anti-SV40 calf serum and ^{3}H-TdR (.02 μCi/ml, specific activity 6 Ci/mM). At various times, coverslips were examined for the proportion of cells that had incorporated the isotopically labeled precursor, for cells that had produced ICFA and viral antigen (*VP*), and for the number of mitoses.

The results of a typical experiment are illustrated in Fig. 9. Several points are quite evident. The number of ICFA-positive cells two days post-infection is higher than that of the *VP*-positive cells, and, while the number of the former cells progressively increases, the number of the latter stabilizes at a plateau and eventually declines. If the cultures are subdivided, then the *VP*-positive cells disappear, presumably because of lysis, while the number of ICFA-positive cells continues to increase. Thus, if the presence of ICFA is considered an expression of successful infection, it is apparent that only a fraction (30 to 50%) of the infected cells proceeds to a productive interaction as detected by the synthesis of capsid antigen.

SV40 infection also induces a rapid and persistent increase in the number of cells labeled with ^{3}H-TdR. We cannot distinguish by autoradiography between viral and cellular DNA, but an indirect estimate can be made by analyzing the population of labeled cells with respect to the compartments into which they may be placed (Fig. 10). By the eleventh day, 30% of the labeled population can be considered as a cumulative background contributed by noninfected cells, since this number is observed in the corresponding noninfected cultures. About 20% of the cells which presumably have synthesized viral DNA during this period were positive for SV40 viral antigen. We are, therefore, left with a cumulative fraction of labeled cells (about 30% at the eleventh day) whose proportion is practically identical with the fraction of cells that are ICFA positive, but *VP* negative. Even though it could not be proved, except by this numerical analysis, that all cells which are ICFA positive also synthesize DNA, certainly some cells positive for ICFA do synthesize

DNA, as demonstrated by a combined analysis of immunofluorescence and autoradiography on the same cell. In fact, after infection, and throughout the period of observation, mitotic activity increases far above the control level. The majority of these mitotic cells show positive immunofluorescence for ICFA (Fig. 11). In contrast, no mitotic cells were ever found to be positive for viral protein antigen. Cultures treated similarly, except for the absence of isotope in the medium, transformed morphologically within two weeks post-infection.

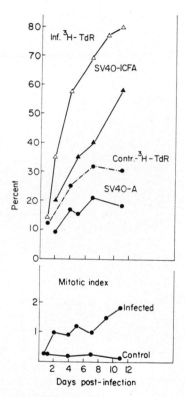

FIG. 9. Changes induced by SV40 in human diploid cells. Stationary WI38 cells at the forty-fifth passage level were infected with 350 TCID$_{50}$/cell and maintained, after the adsorption period, in the presence of 10% SV40-antiserum and 0.02 μCi ^{3}H-TdR.

In conclusion, we have observed that a fraction of human cells infected by SV40 is stimulated into a DNA synthesis that is compatible with the cell's survival and multiplication. In these cells, it seems probable that only a portion of the viral genome is transcribed, since they synthesize ICFA, but are not able to produce viral protein to any detectable amount.

It is from these cells that the permanent transformed lines must originate.

At this moment, it is not possible to decide whether the mitotically stimulated population originated from cells that had the potential to divide, but were arrested in G_1 because of contact inhibition, or from cells in G_0 that had actually been rescued by the virus infection. Experiments in which cells potentially able to divide were eliminated from the cultures prior to infection by extended exhaustion through rapid subculturing or by killing with incorporation of large amounts of isotopically labeled

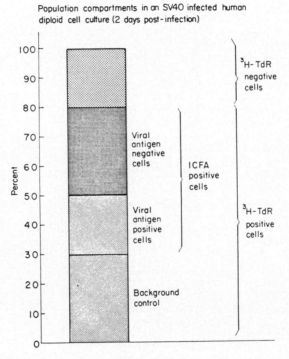

FIG. 10. Compartmentalization of diploid human cells at the eleventh day post-infection with SV40.

thymidine, have given results that seem to indicate that virus infection is unable to rescue G_0 cells. Some sort of rescue must unquestionably occur in the population since transformed cells are able to multiply indefinitely, while noninfected cells are not.

In summary, we have presented direct and indirect evidence that malignant transformation occurs in cells in which the infectious process initiated by these DNA tumor viruses is arrested at some stage prior to complete transcription of the viral genome. These experiments provide evidence that an abortive infection might result either from infection with defective or lethal mutant virus particles or from cellular processes that regulate virus transcription.

The extent of transcription necessary for neoplastic transformation is as yet a matter of speculation, but the induction of the complement-fixing antigen, whatever its exact nature may be, appears to be a necessary, minimal prerequisite. It is probably not sufficient since only some of the ICFA positive cells in our experiments will proceed to transformation.

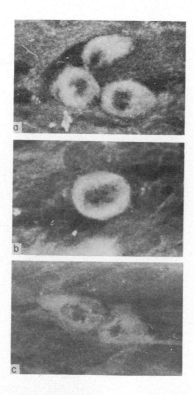

Fig. 11. Demonstration of SV40-ICFA by indirect immunofluorescence technique in WI38 cells 48 hours after SV40 infection. Mitoses *a* (tripolar telophase) and *b* (metaphase) show intensive fluorescence, while *c* is negative.

Similarly, more 3T3 cells are positive for this antigen after SV40 infection than are transformed (P. Black, 1966). Preliminary, but as yet unconfirmed results in this laboratory, suggest that the transplantation antigen is also synthesized prior to or in absence of viral replication. This antigen, which affects the cell surfaces, could well be the other viral induced product that permits transformation to occur in an infected cell.

REFERENCES

Ardashnikova, S. N., and Spasskaia, I. G. (1962). Quoted in D. H. Moore, *Ciba Found. Symp. Tumour Viruses Murine Origin* pp. 107–130.

Basilico, C., and diMayorca, G. (1965). *Proc. Natl. Acad. Sci. U.S.* **54,** 125.

Benjamin, T. L. (1965). *Proc. Natl. Acad. Sci U.S.* **54,** 121.

Black, P. H. (1966). *Virology* **28,** 760.

Black, P. H., and Rowe, W. P. (1965). *Proc. Natl. Acad. Sci. U.S.* **54,** 1126.

Bourgaux, P. (1964). *Virology* **23,** 120.

Carp, R., and Gilden, R. (1965). *Virology* **27,** 639.

Carp, R., and Gilden, R. (1966). *Virology* **28,** 150.

Defendi, V. (1966). *Progr. Exptl. Tumor Res.* **8,** 125.

Defendi, V., and Koprowski, H. (1959). *Nature* **184,** 1579.

Defendi, V., and Lehman, J. M. (1965). *J. Cellular Comp. Physiol.* **66,** 351.

Defendi, V., Carp, R. I., and Gilden, R. V. (1966). In "Viruses Inducing Cancer" (W. Burdette, ed.), p. 269. Univ. of Utah Press, Salt Lake City, Utah.

Diamandopoulos, G. T., and Enders, J. F. (1965). *Proc. Natl. Acad. Sci. U.S.* **54,** 1092.

Dulbecco, R., Hartwell, L. H., and Vogt, M. (1965). *Proc Natl. Acad. Sci. U.S.* **53,** 403.

Gershon, D., Hausen, P., Sachs, L., and Winocour, E. (1965). *Proc. Natl. Acad. Sci. U.S.* **54,** 1584.

Gilden, R. V., Carp, R. I., Taguchi, F., and Defendi, V. (1965). *Proc. Natl. Acad. Sci. U.S.* **53,** 684.

Girardi, A. J., Sweet, B. H., and Hilleman, M. R. (1963). *Proc. Soc. Exptl. Biol. Med.* **112,** 662.

Habel, K., and Silverberg, R. J. (1960). *Virology* **12,** 463.

Huebner, R. J., Chanock, R. M., Rubin, B. A., and Casey, M. J. (1964). *Proc. Natl. Acad. Sci. U.S.* **52,** 1333.

Kaplan, H. S., and Moses, L. E. (1965). *Science* **145,** 21.

Kit, S., Dubbs, D. R., Frearson, P. M., and Melnick, J. L. (1966). *Virology* **29,** 69.

Latarjet, R. (1962). *Ciba Found. Symp. Tumour Viruses Murine Origin* pp. 176–192.

Latarjet, R., Cramer, R., Golde, A., and Montagnier, L. (1966). In press.

MacPherson, I., and Montagnier, L. (1964). *Virology* **24,** 291.

Moore, D. H. (1962). *Ciba Found. Symp. Tumour Viruses Murine Origin* pp. 107–130.

Rapp, F., Butel, J. S., Feldman, L. A., Kitahara, T., and Melnick, J. L. (1965). *J. Exptl. Med.* **121,** 935.

Rapp, F., Tevethia, S. S., and Melnick, J. L. (1966). *J. Natl. Cancer Inst.* **36,** 703.

Rowe, W. P., and Pugh, W. E. (1966). *Proc. Natl. Acad. Sci. U.S.* **55,** 1126.

Sauer, G., and Defendi, V. (1966). *Proc. Natl. Acad. Sci. U.S.* **56,** 452.

Sheinin, R., and Quinn, P. A. (1965). *Virology* **26,** 73.

Stoker, M. (1960). *Brit. J. Cancer* **14,** 679.

Todaro, G. J., and Barron, S. (1965). *Proc. Natl. Acad. Sci. U.S.* **54,** 752.

Todaro, G. J., and Green, H. (1964). *Virology* **23,** 117.

Vogt, M., Dulbecco, R., and Smith, B. (1966). *Proc. Natl. Acad. Sci. U.S.* **55,** 596.

Weil, R., Michel, M. R., and Ruschmann, G. K. (1965). *Proc. Natl. Acad. Sci. U.S.* **53,** 1468.

Winocour, E., Kaye, A. M., and Stollar, V. (1965). *Virology* **27,** 156.

Discussion—Part VII

Dr. W. P. Cheevers: (Question directed to Dr. R. Sheinin.) How does the oncogenic potential of the TSP-1 variant of polyoma virus compare with that of the large plaque variant in mouse embryo cells and in the mouse?

Answer: No direct comparison has been made in the mouse system. The efficiency of morphological transformation of hamster embryo cells *in vitro* has been measured, however, and it was found to be higher with TSP1 than with the large plaque Toronto or Stewart-Eddy strain. This increased efficiency of transformation has been observed by other workers for other small plaque strains of the virus. We have not carried out comparative studies *in vivo*.

PART VIII

ONCOGENIC VIRUSES II

On the Mechanism of Transformation of Mammalian Cells by SV40[*]

Howard Green and George J. Todaro

DEPARTMENT OF PATHOLOGY, NEW YORK UNIVERSITY SCHOOL OF MEDICINE,
NEW YORK, NEW YORK

Since the introduction of culture methods for the study of the action of oncogenic viruses (Rubin and Temin, 1958; Vogt and Dulbecco, 1960) a variety of systems and cell types have been employed to analyze transformation of cellular properties by these viruses. Each cell-virus system has its own peculiarities, but the basic action of the virus may be described as the disruption of control systems. The loss of control of cell division is perhaps only the most evident, leading as it does, *in vivo,* to the production of tumors. In culture systems, polyoma virus and SV40 (which are members of the same group) have been the most extensively studied of the DNA viruses [see reviews by Dulbecco (1963), Stoker (1965), and Defendi (1966)].

The action of SV40 on human diploid fibroblasts has been studied in detail (Koprowski, 1966). Transformation occurs as part of a complicated sequence of changes in morphology, karyotype, and ability to grow, the transformants eventually developing into established lines (Girardi *et al.,* 1965). When the same virus acts on cells of line 3T3, the study of the transformation is somewhat simpler. Since all the cells remain viable, it is easier to control their state of growth, and the transformation frequency is higher. For these reasons we have concentrated on the 3T3-SV40 system for the study of the transformation process.

The Phenotype of 3T3 and its Viral Transformants

3T3 is an established cell line of mouse embryo origin. Like most established mouse lines it has an abnormal karyotype, but because of

* Aided by Grant CA 06793 and Award 1-K6-Ca 1181 (H.G.) from the United States Public Health Service.

the manner of its evolution (Todaro and Green, 1963) it has retained a very high sensitivity to contact-inhibition of cell division. In standard medium containing 10% calf serum, it grows with a doubling time of 18 hours in sparse culture. Growth stops at a saturation density of only 5×10^4 cells/cm^2 of surface, the cells remaining strictly confined to a monolayer. The cell is of fibroblastic type—in stationary cultures about 7% of the cell protein made is collagen (Green and Goldberg, 1965), and there is abundant formation of hyaluronic acid (Hamerman *et al.,* 1965).

Cells of line 3T3 are readily transformed by both polyoma virus and SV40. Cultures infected with polyoma virus support considerable viral multiplication, and there is extensive cell killing; however, transformants emerge from among the survivors. They have reduced sensitivity to contact inhibition, and their colonies have a configuration rather characteristic of polyoma transformants (Sachs and Medina, 1961). The transformants contain the characteristic virus-induced T antigen or ICFA (Todaro *et al.,* 1965a).

Though infection of 3T3 cells with SV40 does not result in any cell killing, and the virus probably does not multiply, transformants are produced which are released from contact inhibition of division to a greater extent than are the polyoma transformants. They grow to a saturation density at least twentyfold higher than do cells of the parent line. SV40 transformants may be distinguished morphologically from polyoma transformants (Todaro and Green, 1965); the cells contain the SV40 T antigen, demonstrable by complement fixation (Todaro *et al.,* 1965b), and by the fluorescent antibody technique (Black, 1966). Cells of line 3T3 may be transformed consecutively by polyoma virus and SV40, and the "double" transformants, though they may become virus-free, continue to synthesize the T antigens characteristic of both viruses. All of the viral transformants of 3T3 cells, as well as of human diploid cells, have a considerably reduced ability to synthesize collagen (Green *et al.,* 1966) and hyaluronic acid (Hamerman *et al.,* 1965). As the SV40 transformation is not accompanied by a cytocidal effect, the transformation frequency is readily measured; when high concentrations of infecting virus are used, the transformation frequency may be as high as 50% (Todaro and Green, 1966a; Black, 1966).

Though the basis for sensitivity to contact inhibition of cell division is not well understood, it appears to involve control of the overall rate of RNA synthesis (Levine *et al.,* 1965; Bloom *et al.,* 1966; Green and Todaro, 1966). All diploid cell strains show some degree of contact inhibition, but most are able to grow beyond a strict monolayer. Usually, established lines are practically insensitive to contact inhibition, but this probably results from selection in continuous culture of cell types better able to grow under crowded conditions. The sensitivity of cells to contact inhibition of division is also affected by the culture conditions. For example, it depends to a considerable extent on the serum concentration in the medium (Stanners *et al.,* 1963). This is due, at least in the case of the 3T3 cell line, to the presence in serum of a macromolecular substance which diminishes the effectiveness of contact inhibition (Todaro *et al.,* 1965b; Bloom *et al.,* 1966). However, under standard conditions (medium supplemented with 10% calf serum), the sensitivity of different cell types to contact inhibition may be easily and reproducibly compared.

Line 3T3 is more sensitive to contact inhibition than most diploid cell strains, and as shown in Table I this inhibition is reciprocal. 3T3 cells

TABLE I

COLONY-FORMING ABILITY OF 3T3 AND HUMAN DIPLOID CELLS
AND OF SV40 TRANSFORMANTS OF BOTH

Cells	Ability to form colonies on a complete monolayer of	
	3T3	Human diploid fibroblasts
3T3	—	—
SV40 Transformants of 3T3	++++	++++
Human Diploid Fibroblasts (H.D.F.)	—[a]	—
SV40 Transformants of H.D.F.	++++	++++

[a] If inoculated at a very high density on to 3T3, human diploid fibroblasts form a layer, but grow very slowly.

will not form colonies when plated on a complete layer of either 3T3 or human diploid fibroblasts, while SV40 transformants of 3T3 will form colonies readily on both. Similarly, human diploid fibroblasts will not form colonies on a monolayer of either cell type, while SV40 transformants of human fibroblasts will form colonies on both. Thus, contact inhibition operates between the two species and between euploid and heteroploid cell types. Transformants of one are freed from inhibition by both. This generalization does not cover all cell types, as examples exist of transformed cell lines which show no contact inhibition among themselves, but which are still subject to some inhibition by the parent cell type (Stoker *et al.,* 1966).

Dependence of Viral Transformation on Cell Growth

Under the proper conditions, cultures of 3T3 cells may be maintained in the nongrowing state for long periods of time, the cells being arrested in the G_1 period of the growth cycle (Nilausen and Green, 1965). If such a culture is infected with a high concentration of SV40 and maintained in the nongrowing state, no colonies of transformed cells develop. But if the cells, once infected, are transferred at a dilution that enables each cell to grow through a number of cell divisions before growth is arrested by contact inhibition, then transformed colonies will develop. Therefore, it is clear that nongrowing cells, though infected, are not transformed until they have undergone some growth (Todaro and Green, 1966a).

The amount of growth necessary to permit the development of colonies of transformed cells has been determined by replating an infected saturation-density culture at various dilutions, so that the cells completed a known number of cell divisions before confluence was again attained. For example, a 1 : 2 dilution permits an average of 1 cell doubling; a 1 : 4 dilution permits 2 doublings, etc. Experiments of this type showed that practically no transformed colonies appeared even after 2 or 3 cell generations; but further growth produced a sharp increase and by about 6 generations, the number of transformed colonies reached a maximum (Todaro and Green, 1966a).

A similar experiment was performed by infecting growing cells, allowing them to continue to grow for varying periods, and then plating them on complete monolayers of uninfected 3T3 cells. If very little growth was allowed after infection, very few colonies of transformed cells formed on the 3T3 monolayer. By the time infected cells had undergone 5–8 cell divisions, they were found to form colonies on a 3T3 monlayer with maximum efficiency.

It may be concluded that growth of the virus-infected cell is necessary for the loss of sensitivity to contact inhibition. It will be noted that the virus itself is not able to initiate cell growth in a stationary culture. The conditions permitting the growth of the cells shortly after infection are the same as those permitting the growth of uninfected cells. Further analysis has shown that growth must precede the loss of contact sensitivity in the infected cell for two reasons: First, the fixation of the transformed state as a heritable property requires one cell generation; and second, the phenotypic expression of that state requires several more.

Fixation of the Transformed State—The First Stage in Transformation

The amount of cell growth required for the virus to produce the first irreversible effect has been studied in two different ways, both of which lead to the conclusion that one cell generation is sufficient.

1. A stationary population of cells was infected and allowed to remain in the nongrowing state for varying periods of time. The cells were then transferred at high dilution so that growth could begin; two weeks later the plates were scored for colonies of transformed cells. The results of such experiments (Fig. 1, curves 1 and 2) show that the ultimate transformation frequency obtained decreases as the time during which the cells are held in the stationary phase increases, until virtually all the transformants which would have developed if the cells had been plated immediately after infection fail to develop. This decline correlates, as a first

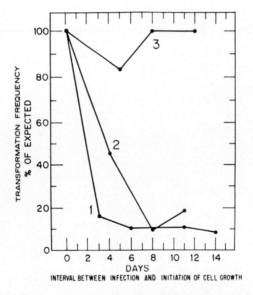

FIG. 1. Transformation frequency of virus infected cells after varying intervals in the nongrowing state. Curves 1 and 2, no cell divisions between infection and arrest of growth. Curve 3, one cell division between infection and arrest of growth.

approximation, with the destruction of intracellular virus, as determined by the number of infectious units remaining in the cell layer. The few transformants which occur, even after ten days in the stationary state, are probably due to the small number of viral particles which remain in the cells and which are able to initiate the transformation when the cells begin growth after dilution (Todaro and Green, 1966a).

On the other hand, if the experiment was modified so as to permit the infected cells to complete one doubling before they were held in the stationary state, the result was entirely different. Under these conditions no diminution in the transformation frequency was found when the cells were diluted for assay after as long as 12 days in the stationary state. (Fig. 1, curve 3).

2. Interferon is an effective inhibitor of transformation if added to cultures before or shortly after viral infection (Todaro and Baron, 1965). As the phenotype of virus-transformed cells is not affected by interferon, there must be a time after infection when the transformation process becomes interferon resistant. Figure 2 shows the results of experiments performed in order to determine when this occurs. Exponentially growing cells were infected, exposed to interferon at various times thereafter, and then replated for scoring of transformation frequency. Immediately after

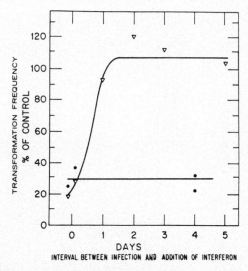

FIG. 2. Relation between transformation frequency and time of addition of interferon to infected cells. Nongrowing cells, ● ; exponentially growing cells, ▽.

infection, the addition of interferon suppressed transformation quite effectively, but its ability to do so was lost one day later, when the cells had completed one doubling. The transformation process is therefore past the interferon-sensitive stage by one generation after infection. If nongrowing cells were infected and kept in the stationary state for four days, interferon added, and the cells replated, the interferon was fully effective in preventing transformation. The transformation process therefore cannot pass the interferon-sensitive stage in nongrowing cells.

Relation of the First Stage of Transformation to Cellular DNA Synthesis

More precise analysis of the rate at which growing, infected cells pass the interferon-sensitive stage gives additional information on the mechanism of the early events. Exponentially growing populations were infected, exposed to interferon at varying times, and replated for assay of transformation frequency. From Fig. 3, it is clear that the cells begin to lose sensitivity to interferon almost immediately, and the decline is linear until all the cells become resistant at about 24 hours after infection. Such kinetics seem most easily explained by a random entry of cells into a period of the growth cycle in which they are sensitive to the action of the virus. All cells would be expected to have passed through this period within one cell generation after infection.

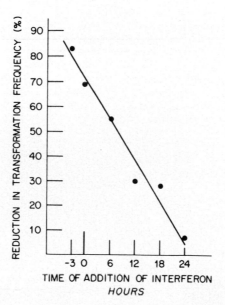

FIG. 3. Loss of ability of interferon to prevent transformation of exponentially growing cells. Exposure to interferon for three hours, beginning at indicated times. Infection at time zero.

This idea is supported by the finding that in synchronous cultures, the kinetics of the loss of sensitivity of the transformation to interferon are completely different. When a stationary culture of 3T3 cells (cells in G_1 is transferred with dilution, the cells prepare for division, synthesize DNA and divide; the first cycle is rather synchronous (Nilausen and Green, 1965). The cells increase their rate of RNA synthesis, then their rate of protein synthesis; 20 hours after plating, DNA synthesis begins. Most of the cells complete the S period by 35 hours, and mitosis by 40 hours after plating.

Cells in a stationary culture were trypsinized and transferred with dilution. After attachment, the cells were infected, either while they remained in the stage of preparation for DNA synthesis (G_1), or while DNA synthesis was occurring (Table II). After ten hours' exposure to the virus, interferon was added and the cells were plated the following day for assay of transformation frequency. In cells infected before they began to synthesize DNA interferon remained able, ten hours later, to prevent transformation with unimpaired efficiency, while cells infected during the S period became interferon resistant during a similar interval. It is known

TABLE II

EFFECT OF INTERFERON ADDITION TEN HOURS SUBSEQUENT TO INFECTION OF A SYNCHRONOUS CELL POPULATION

Time after plating		Transformation frequency		Reduction in transformation frequency (%)	Loss of interferon inhibition (%)[a]	% of nuclei labeling[b]
Infection	Interferon	− Int	+ Int			
2	12	2.4	0.5			
		7.2	0.9			
		3.4	0.5			
		4.3	0.8	82	0	0
24	34	1.8	1.0			
		1.3	1.2			
		1.3	1.0			
		1.5	1.1	27	67	78

[a] Relative to exposure to interferon prior to infection. See Fig. 3.

[b] Parallel cultures exposed for the same ten hour interval to 1 μCi/ml of thymidine-^{3}H. Values obtained from more than 2000 nuclei in each case.

that interferon begins to act almost immediately and that viral functions are very effectively suppressed within less than four hours (Lockhart and Horn, 1963). It may therefore be concluded that cells preparing for DNA synthesis by increased rates of biosynthesis of other macromolecules are not sensitive to the virus, but cells synthesizing DNA are fully sensitive to the virus. The data do not permit us to say whether sensitivity is confined to a particular time in the S period, when the replication of some particular portion of the genome is occurring, or whether many or all stages of the S period are sensitive. In addition, it must be noted that the sensitivity of the cells to virus during the G_2 period and during mitosis has not been tested by these experiments because of their short duration; but their involvement seems unlikely (Todaro and Green, 1967; also see below). It seems quite reasonable that the virus should interact with cellular DNA which is in the process of replication, especially since that is the time when the double strand of DNA is presumably opened and when a single strand could most easily interact with viral nucleic acid.

In considering how the virus acts, it is also necessary to take into account the mechanism of action of interferon. Recent experiments have shown that interferon treatment affects the translation of viral messenger RNA. According to Joklik and Merigan (1966), interferon treatment, while not interfering with transcription of DNA, prevents the formation of a polyribosome on the viral messenger, while Marcus and Salb (1966) find that though some polyribosomes may be formed, they are not able to synthesize protein. As pointed out by Joklik, these effects could be produced by combination of the interferon-induced host cell protein (Taylor, 1964) with the viral messenger. Since interferon prevents the synthesis of proteins coded for by the virus (Fig. 4), it is reasonable that it should block the synthesis of the SV40 T antigen in acutely infected cells; this was shown by Oxman and Black (1966). However, interferon also blocks the transformation (Todaro and Baron, 1965). Since interferon does not appear to affect viral DNA (Joklik and Merigan, 1966), transformation cannot result from simple interaction of viral and cellular DNA. If it is the viral DNA which interacts with cellular DNA, then a viral-directed protein must be necessary as well—possibly an enzyme necessary for insertion of the viral DNA into the host genome. However, another pos-

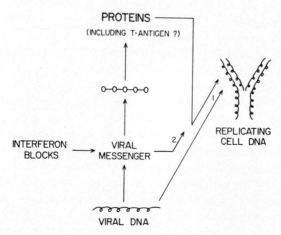

Fig. 4. Mechanisms of transformation and its prevention by interferon.

sibility is that the viral messenger itself is the transforming agent. Interferon, by combining with it, would prevent its action on the host genome as well as its translation into T antigen and other proteins (Fig. 4). This would be the simplest explanation for the observation that interferon at different concentrations suppresses transformation and T antigen synthesis (see below) with equal efficiency (Oxman and Black, 1966).

Relevance of ICFA (T Antigen) Synthesis to the Transformation Process

SV40, like other oncogenic viruses, induces in the infected cell the synthesis of an intranuclear protein antigen, whose properties have been examined in some detail (Gilden *et al.*, 1965). The relation of this antigen to transformation is of the greatest interest, since it is always found in transformed cells, where its synthesis continues indefinitely, even though the cells are free of infectious virus. However, in acutely infected cells, the antigen is induced without prior cellular DNA synthesis—for example, when the DNA synthesis is inhibited by nucleoside analogs (Gilden *et al.*, 1965; Rapp *et al.*, 1965), or when the cells are in the stationary state (Black, 1966). For this reason, an infected 3T3 cell may make the antigen acutely, fail to be transformed, and then this 3T3 cell will lose the ability to make the antigen (Black, 1966). If the infected cell does become transformed, then antigen synthesis persists.

Hypotheses regarding the nature of the transformation process must be considered in the light of these facts. Some of the formal possibilities are set forth in Table III and are considered briefly in what follows:

1. The gene for the T antigen is in the viral DNA alone. Its transcription is begun in the nongrowing cell, but as there is no viral replication in this system, the antigen cannot continue to be produced indefinitely unless the relevant DNA is incorporated into the host cell genome. When this occurs, T antigen synthesis will become a permanent feature of the cell, and the cellular growth properties will be altered. As noted above, the integration will require the assistance of a virus-directed protein.

2. The gene for T antigen is present in both viral DNA and host cell DNA. In this case there would have to be appreciable complementarity between viral and cellular DNA's. In uninfected cells, the cellular DNA coding for the antigen is not transcribed. In infected but nongrowing cells, the virus makes the mRNA for the antigen, but this mRNA cannot transform since the cellular DNA is not in replication. As in (1), if growth is not initiated in time, the viral genes are eliminated by the cellular nucleases; and the antigen is diluted out in growth. However, if DNA synthesis occurs in the freshly infected cell, the viral mRNA specifically derepresses the corresponding sequences in the replicating host

TABLE III

HYPOTHESES REGARDING THE MECHANISM OF TRANSFORMATION BY ONCOGENIC VIRUSES

Site of gene for T antigen	Origin of mRNA for T antigen in absence of transformation	Origin of mRNA for T antigen after transformation
1. Only in viral DNA	Viral DNA	Incorporated part of viral genome
2. In both viral and cellular DNA	Viral DNA	Host cell DNA, irreversibly derepressed
3. Only in cellular DNA	Host cell DNA, reversibly derepressed	Host cell DNA, irreversibly derepressed

cell DNA; and, from that time on, this region makes the messenger for the antigen. In principle, viral mRNA isolated even from a lytic infection should be capable of carrying out the transformation.

This model seemed attractive in view of the complementarity believed to exist between certain regions of mammalian DNA and polyoma virus DNA (Axelrod *et al.,* 1964; Winocour, 1965). However, the more recent experiments of Winocour (1967, this volume) show that this is probably an artifact resulting from the inclusion of some cellular DNA within the viral particles. He has concluded that there is very little complementarity between cellular mRNA synthesized *in vitro* and viral DNA. Unless there is considerable selection in the transcription process, even *in vitro,* the amount of complementarity between the two DNA's would be insufficient for the same gene to be present in both. With the same reservation, Winocour's experiments together with those of Benjamin (1966) and Fujinaga and Green (1966) would be more consistent with model (1).

3. The gene for the T antigen is present only in host cell DNA. The virus must specifically derepress the relevant DNA, but it must do so reversibly in nongrowing cells, and irreversibly in growing cells. It is conceivable that this could proceed through action on a regulator type of gene controlling both T antigen synthesis and sensitivity to contact inhibition. Very little complementarity between viral and host cell DNA would be needed to permit specificity of interaction between the two.

The Problem of Mixed versus Homogeneous Colonies

Though the probability of a single particle of SV40 or polyoma inducing a transformation is very low, the transformation frequency is proportional to the concentration of infecting virus over a very broad range (Macpherson and Montagnier, 1964; Todaro and Green, 1966b; Black, 1966). It is most likely that a single event establishes the transformation. Of the transformations occurring, most are initiated during the first generation after infection. However, there is a certain probability that the first step in the transformation may occur later. If it occurs after the final plating, the result will be a mixed or sectored colony, of which one half or less consists of transformed cells and the remainder of unchanged cells. This is known to occur in polyoma transformation of BHK21 cells (Stoker, 1963) and in SV40 transformation of 3T3 cells (Todaro and Green, 1966b). However, in both cases, most of the transformations are presumably fixed during the first S period following infection, as most of the resulting colonies are not mixed, but homogeneous. In order to achieve this, the transformation must occur by a mechanism which does not allow the development of normal segregants when the cells are plated immediately after infection. Therefore, if the transformation involves integration, the viral DNA must be inserted just prior to replication; and it must be replicated with the host DNA, so that both daughter cells will be affected. If the transformation involves derepression, then again, both strands of the cellular DNA prepared for replication must be affected just before the replication occurs. This confers DNA altered in both strands to both progeny cells at division. In neither case, it will be noted, could this occur if the interaction occurred during G_2 when the cellular DNA has already completed replication.

Concluding Remarks

In general, theories of the mechanism of action of oncogenic DNA viruses have been inclined to regard the DNA of the virus as bearing structural genes for new properties found in the transformed cells. This view has been inspired primarily by the phenomenon of phage conversion in bacteria, and by the fact that the DNA is the only self-replicating molecule in the virus which might reasonably be made a permanent part of the host cell genome.

However, the possibility that the mechanism of viral action on the host cell genome may be one of specific derepression rather than integration is still not excluded, and it may be asked how genes for the various virus-specific T antigens (polyoma, SV40, adeno etc.) could be present in a repressed form in the genome of the uninfected cell. It is well known that tumors induced by a variety of means synthesize antigens not present in the cells from which the tumors arose (Old and Boyse, 1964). In two cases, these "tumor" antigens have been identified as antigens normally present in the corresponding cell or organ type in embryonic life (Abelev *et al.,* 1963; Gold and Freedman, 1965). However, the synthesis of these antigens is normally repressed in the adult tissue from which the tumor arose. The renewed synthesis of these antigens in the tumor cells clearly results from derepression of cellular genes. There is therefore no intrinsic difficulty in supposing that genes for viral T antigens exist in normal cell DNA, where they would not be transcribed until after interaction with the virus. However, in addition to the problem of defining the viral constituent which might carry out the derepression, it would have to be explained how, following its action on replicating DNA, a change in selective transcription could be made permanent.

REFERENCES

Abelev, G. I., Perova, S. D., Khramkova, N. I., Postnikova, Z. A., and Irlin, I. S. (1963). *Biokhimiya* **28**, 625.

Axelrod, D., Habel, K., and Bolton, E. T. (1964). *Science* **146**, 1466.

Benjamin, T. L. (1966). *J. Mol. Biol.* **16**, 359.

Black, P. (1966). *Virology* **28**, 760.

Bloom, S., Todaro, G., and Green, H. (1966). *Biochem. Biophys. Res. Commun.* **24**, 412.

Defendi, V. (1966). *Progr. Exptl. Tumor Res.* **8**, 125.

Dulbecco, R. (1963). *Science* **142**, 932.

Fujinaga, K., and Green, M. (1966). *Proc. Natl. Acad. Sci. U.S.* **55**, 1567.

Gilden, R. V., Carp, R. I., Taguchi, F., and Defendi, V. (1965). *Proc. Natl. Acad. Sci. U.S.* **53**, 684.

Girardi, A. J., Jensen, F. C., and Koprowski, H. (1965). *J. Cellular Comp. Physiol.* **65**, 69.

Gold, P., and Freedman, S. O. (1965). *J. Exptl. Med.* **122**, 467.

Green, H., and Goldberg, B. (1965). *Proc. Natl. Acad. Sci. U.S.* **53**, 1360.

Green, H., and Todaro, G. J. (1966). *Symp. M.D. Anderson Hosp.* (in press).

Green, H., Todaro, G. J., and Goldberg, B. (1966). *Nature* **209**, 916.

Hamerman, D., Todaro, G. J., and Green, H. (1965). *Biochim. Biophys. Acta* **101**, 343.

Joklik, W., and Merigan, T. C. (1966). *Proc. Natl. Acad. Sci. U.S.* **56**, 558.

Koprowski, H. (1966). *Harvey Lectures* **60**, 173.

Levine, E. M., Becker, Y., Boone, C. W., and Eagle, H. (1965). *Proc. Natl. Acad. Sci. U.S.* **53**, 350.

Lockhart, R. Z., Jr., and Horn, B. (1963). *J. Bacteriol.* **85**, 996.

Macpherson, I., and Montagnier, L. (1964). *Virology* **23**, 291.

Marcus, P. I., and Salb, J. M. (1966). *Virology* **30**, 502.

Nilausen, K., and Green, H. (1965). *Exptl. Cell Res.* **40**, 166.

Old, L. J., and Boyse, E. A. (1964). *Ann. Rev. Med.* **15**, 167.

Oxman, M. N., and Black, P. H. (1966). *Proc. Natl. Acad. Sci. U.S.* **55**, 1133.

Rapp, F., Butel, J. S., Feldman, L. A., Kitahara, I., and Melnick, J. L. (1965). *J. Exptl. Med.* **121**, 935.

Rubin, H., and Temin, H. M. (1958). *Federation Proc.* **17**, 994.

Sachs, L., and Medina, D. (1961). *Nature* **189**, 457.

Stanners, C. P., Till, J. E., and Siminovitch, L. (1963). *Virology* **21**, 448.

Stoker, M. (1963). *Virology* **20**, 366.

Stoker, M. (1965). *Can. Cancer Conf.* **6**, 357.

Stoker, M., Shearer, M., and O'Neill, C. (1966). *J. Cell. Sci.* **1**, 297.

Taylor, J. (1964). *Biochem. Biophys. Res. Commun.* **14**, 447.

Todaro, G. J., and Baron, S. (1965). *Proc. Natl. Acad. Sci. U.S* **54**, 752.

Todaro, G. J., and Green, H. (1963). *J. Cell. Biol.* **17**, 299.

Todaro, G. J., and Green, H. (1965). *Science* **147**, 513.

Todaro, G. J., and Green, H. (1966a). *Proc. Natl. Acad. Sci. U.S.* **55**, 302.

Todaro, G. J., and Green, H. (1966b). *Virology* **28**, 756.

Todaro, G. J., and Green, H. (1967). *J. Virol.* **1**, 115.

Todaro, G. J., Habel, K., and Green, H. (1965a). *Virology* **27**, 179.

Todaro, G. J., Lazar, J., and Green, H. (1965b). *J. Cellular Comp. Physiol.* **66**, 325.

Vogt, M., and Dulbecco, R. (1960). *Proc. Natl. Acad. Sci. U.S.* **46**, 365.

Winocour, E. (1965). *Virology* **27**, 520.

Tumor Virus RNA and the Problem of Its Synthesis

William S. Robinson

DEPARTMENT OF MOLECULAR BIOLOGY AND VIRUS LARORATORY,
UNIVERSITY OF CALIFORNIA, BERKELEY, CALIFORNIA

Tumor Virus RNA

Certain tumor-inducing viruses possess RNA as their nucleic acid. These are the members of the avian tumor viruses, the murine leukemia viruses and the mouse mammary tumor virus. Besides their nucleic acid, they probably have other structural features in common. All contain a significant fraction of lipid in the mature virion which may represent cell material acquired at the cell surface as they mature by a mechanism similar to that described for the myxoviruses (Morgan *et al.*, 1962).

Recently the mixture of the Bryan strain of Rous sarcoma virus and its helper Rous associated virus (RSV + RAV) (Robinson *et al.*, 1965), avian myeloblastosis virus (AMV) (Robinson and Baluda, 1965), Rauscher mouse leukemia virus (MLV) (Duesberg and Robinson, 1966), and mouse mammary tumor virus (MTV) (Duesberg and Blair, 1966), have been purified and their intact RNA's isolated in our laboratory. Preliminary characterization of the RNA's indicates that all are single-stranded and very similar in sedimentation characteristics. Thus, all are probably very similar in size and much larger than the known single-stranded RNA's from other groups of viruses. The common method of virus purification and several characteristics of the RNA's are reviewed briefly in what follows.

VIRUS PURIFICATION

Precipitation with ammonium sulfate has been used as the initial step to concentrate RSV + RAV (Robinson *et al.*, 1965) and AMV (Robinson and Baluda, 1965) without loss of infectivity from tissue culture medium. However, when MLV was precipated with ammonium sulfate, two or more virus bands containing intact viral RNA were later found in density gradients, in contrast to virus purified without ammonium sulfate, where only one virus band was seen (Duesberg and Robinson, unpublished observations). This indicates that high salt concentrations may disrupt MLV. The earlier findings, of a single infectious band and multiple noninfectious bands with MLV and only one infectious band with RSV after centrifugation in concentrated salt solutions (O'Connor *et al.,* 1964), also suggest that MLV may be disrupted by high salt under conditions where RSV + RAV are stable.

A second purification step takes advantage of the high lipid content and relatively large size of the RNA tumor viruses. All may be rapidly centrifuged to density equilibrium in sucrose density gradients. Figure 1 shows the results of equilibrium centrifugation of ^{32}P-labeled RSV + RAV after layering virus over a preformed sucrose density gradient. Viral infectivity and radioactivity are seen to coincide, indicating that all ^{32}P-labeled material has the buoyant density of infectious RSV. The buoyant density of RSV + RAV in sucrose solution is about 1.16 gm/ml. The

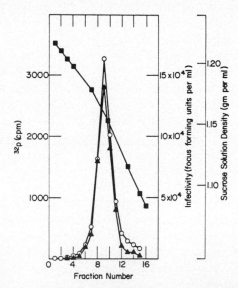

Fig. 1. Equilibrium density gradient centrifugation of ^{32}P-labeled RSV + RAV. Centrifugation was for 2 hours, 36,000 rpm, 4°C in the Spinco SW39 rotor, after layering the virus solution over a 5.0 ml preformed gradient of sucrose (15–60%) containing 0.10 *M* NaCl, 0.01 *M* Tris·HCl pH 7.4 and 0.001 *M* EDTA. RSV infectivity (▲———▲), ^{32}P (○———○) and solution density (■———■) were determined on appropriate fractions (data from Robinson *et al.,* 1965).

other RNA tumor viruses have similar buoyant densities: 1.17 gm/ml for MLV (Duesberg and Robinson, 1966) and 1.22 gm/ml for MTV (Duesberg and Blair, 1966).

Centrifugation in sucrose is a gentle method that allows recovery of all virus infectivity (Robinson *et al.,* 1965; Robinson and Baluda, 1965). More fragile viruses such as MLV can be purified by several centrifugations in sucrose solutions, thus avoiding preliminary ammonium sulfate precipitation (Duesberg and Robinson, 1966). RSV + RAV infectivity is stabilized by reagents such as EDTA and 2-mercaptoethanol (Robinson *et al.,* 1965) and EDTA is included in the buffers used during purification of all four tumor viruses. Rapid purification using gentle methods is important for recovery of intact viral RNA as well as infectivity (Robinson *et al.,* 1965).

THE VIRAL RNA's

The detergent, sodium dodecyl sulfate (SDS), effectively disrupts the lipid-containing RNA viruses (Robinson *et al.,* 1965). Its effects on ^{32}P-labeled RSV + RAV are shown in Fig. 2. Figure 2A shows the results of velocity sedimentation in a sucrose density gradient of ^{32}P-labeled

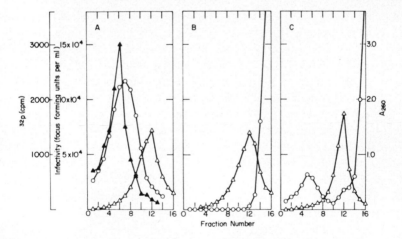

FIG. 2. Velocity sedimentation of purified ^{32}P-labeled RSV + RAV in the presence and absence of SDS. The virus solution was layered over a 5.0 ml gradient of sucrose (5 to 20%) containing the buffer described in Fig. 1. (A) Whole ^{32}P-labeled RSV + RAV with TMV (1.0 mg) was centrifuged in the Spinco SW39 rotor at 25,000 rpm, 4°C for 40 min. (B) Same as (A), except that SDS was added to virus to make a 0.1% solution before centrifugation. (C) Same as (B), except that TMV RNA was used instead of TMV and centrifugation was at 36,000 rpm for 2.25 hours. RSV infectivity (▲———▲), ^{32}P (○———○) and A$_{260}$ (△———△) were determined on each fraction (data from Robinson *et al.,* 1965).

RSV + RAV in the presence of Tobacco Mosaic Virus (TMV). The ^{32}P-labeled virus can be seen to sediment faster than TMV. Virus infectivity can be seen to follow the leading portion of the zone of ^{32}P-labeled material, indicating that all ^{32}P-labeled material is not infectious RSV. The effects of adding SDS to the virus before centrifugation are shown in Fig. 2B. It can be seen that exposure of the virus to 0.1% SDS results in marked reduction in sedimentation rate of all ^{32}P-labeled virus components without altering the sedimentation behavior of TMV. Fig. 2C shows that when ^{32}P-labeled virus is disrupted by SDS and centrifuged at a higher speed for a longer time, the fastest sedimenting ^{32}P-labeled component moves at a little more than twice the rate of TMV RNA, which was used as marker in this experiment. This ^{32}P-labeled component is the intact viral RNA. Thus, SDS disrupts RSV + RAV and releases the intact RNA.

After disruption of the virus with SDS, the RNA can be purified by phenol extraction and alcohol precipitation (Robinson *et al.,* 1965.) The RNA's of all four tumor viruses have been prepared in this way.

Figure 3 shows the results of sedimentation of radioactively labeled RNA from each of the four viruses in sucrose density gradients. Figure 3A shows sedimentation of ^{32}P-labeled RSV + RAV RNA, Fig. 3B shows ^{32}P-labeled AMV RNA sedimenting with tritium-labeled RSV + RAV RNA, Fig. 3C shows tritium-labeled MTV RNA with ^{32}P-labeled RSV + RAV RNA, and Fig. 3D shows tritium-labeled MLV RNA with ^{32}P-labeled RSV + RAV RNA. Tobacco mosaic virus (TMV) RNA has been used as carrier in each experiment and its position is shown the A_{260} tracing. Sedimentation is from right to left. It can be seen that the labeled RNA from each tumor virus consists of two major components. The labeled RNA sedimenting faster than TMV RNA in each case is thought to be the intact viral RNA, and that sedimenting more slowly is thought to be the degraded viral RNA resulting from breakdown of RNA in the virus before the time of RNA isolation (Robinson *et al.,* 1965; Bratt and Robinson, 1966). However, the possibility that the slowly sedimenting component also contains RNA from other sources has not been excluded. Small amounts of radioactively labeled cell RNA are found in many preparations of viral RNA. The tritium-labeled RNA in fractions 13 and 14 in Fig. 3B, fractions 14 to 17 in Fig. 3C and fractions 15 to 21 in Fig. 3D probably represents cellular RNA. The cell RNA probably comes from contamination of the virus preparations with cellular material of approximately the same size and buoyant density as virus. The least contamination is usually found in the RNA from RSV + RAV grown in chick embryo fibroblast cultures (Fig. 3A), suggesting that the release of cell material containing RNA into the culture medium is minimal in this kind of culture.

Similar findings have been made with AMV (J. Harel *et al.,* 1965a), RSV (L. Harel *et al.,* 1965b), and Rauscher MLV (Galibert *et al.,* 1965; Mora *et al.,* 1966).

Sedimentation constants of 71 S for both RSV + RAV RNA and AMV RNA and 74 S for MLV RNA in 0.11 *M* salt have been de-

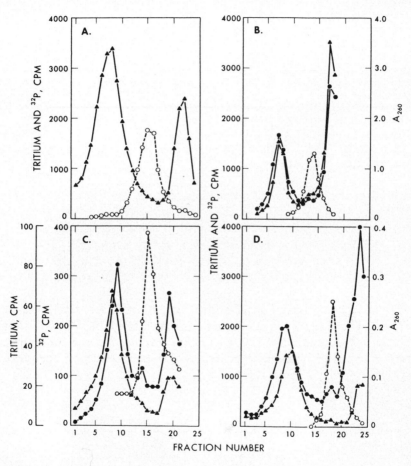

Fig. 3. Sedimentation of tritium or ^{32}P-labeled viral RNA's with carrier TMV RNA (A[260]) in sucrose density gradients (5 to 20%) containing the buffer described in Fig. 1 in Spinco SW50 rotor at 4°C. (A) ^{32}P-labeled RSV + RAV RNA and TMV RNA centrifuged at 36,000 rpm for 2.25 hours (from Robinson *et al.,* 1965). (B) ^{3}H-labeled RSV + RAV RNA, ^{32}P-labeled AMV RNA and TMV RNA centrifuged as described in (A) (from Robinson and Baluda, 1965). (C) ^{32}P-labeled RSV + RAV RNA, ^{3}H-labeled MTV RNA and TMV RNA centrifuged at 50,000 rpm for 75 min (from Duesberg and Blair, 1966). (D) ^{32}P-labeled RSV + RAV RNA, ^{3}H-labeled MLV RNA and TMV RNA centrifuged for 2 hours 10 min at 50,000 rpm (from Duesberg and Robinson, 1966). [32]P (▲————▲), tritium (●————●) and A[260] (○————○) were determined on each fraction.

TABLE I

Sedimentation Constants ($s_{20, w}$) of Tumor Virus RNA's

Tumor virus RNA's	Sedimentation constants in varying salt concentration[a]		
	0.21 M	0.11 M	0.001 M
RSV + RAV[b]	79	71	27
AMV[b]	79	71	27
MLV[c]	—	74	~26[e]
MTV[d]	—	~69[e]	~22[e]

[a] Sedimentation constants of RNA's determined in the Spinco Model E analytical ultracentrifuge.

[b] The data are from Robinson and Baluda (1965).

[c] Data are from Duesberg and Robinson (1966).

[d] Data are from Duesberg and Blair (1966).

[e] Estimated by sucrose gradient centrifugation based on analytical ultracentrifuge values for RSV + RAV RNA.

termined in the analytical ultracentrifuge (Table I). These values are in agreement with the results in Fig. 3 where RSV + RAV and AMV RNA's are seen to sediment together in 0.11 M salt (Fig. 3B) and MLV RNA sediments slightly ahead of RSV + RAV RNA (Fig. 3D). MTV RNA moves more slowly than RSV + RAV RNA (Fig. 3D) and can be estimated to have a sedimentation constant of about 69 in 0.11 M salt.

Certain characteristics suggest that the RNA's are single-stranded: (1) complete digestion to acid soluble products by RNase in 0.2 M salt; (2) marked dependence of sedimentation velocity on salt concentration (Table I); and (3) base compositions which do not suggest complementary base pairing (Table II).

TABLE II

Base Composition of ^{32}P-Labeled RNA[a]

Bases	^{32}P-Labled RNA		
	RSV = RAV	AMV	MLV
C	24.2	23.0	26.7
A	25.1	25.3	25.5
G	28.3	28.7	25.1
U	22.4	23.0	22.7

[a] Base composition of ^{32}P-labeled RNA's determined by alkaline hydrolysis and paper electrophoresis of intact RNA's from purified virus. The data for RSV + RAV RNA are from Robinson *et al.* (1965), for AMV RNA from Robinson and Baluda (1965), and for MLV RNA from Duesberg and Robinson (1966).

Using the relationship between sedimentation constant and size determined by Spirin for TMV RNA (Spirin, 1963), these RNA's would have molecular weights around 10 or 12 million. This value is only an estimate because the relationship between sedimentation constant and molecular weight may differ considerably for different single-stranded RNA's (Strauss and Sinsheimer, 1963).

The value of 10 million for the molecular weight of AMV RNA has been recently estimated by electron microscopy of viral RNA (Granboulan, *et al.,* 1966).

In summary, RSV + RAV and AMV RNA's are indistinguishable in sedimentation rate (Fig. 3B) and change in conformation with change in ionic strength (Table I), they are also very similar in base composition (Table I). This is not surprising in view of their close biological relationship (Robinson and Baluda, 1965). The RNA's of MLV and MTV are also single-stranded and very similar to the avian tumor virus RNA's in sedimentation behavior (Fig. 3C and 3D and Table I). Thus RSV + RAV, AMV, MLV, and MTV appear to contain RNA's which are similar in size and structure.

Synthesis of Viral Specific RNA

RSV + RAV Infected Cells

Viral specific RNA synthesis in cells infected with the Bryan strain of RSV and its helper virus RAV will be considered first. Eighty to 100% of the chick embryo cells in tissue culture can be infected with RSV and RAV (Trager and Rubin, 1964). The infected cells become transformed and produce large amounts of virus for many days. Using such cultures it is possible to show that actinomycin D, at concentrations above 0.1 μg/ml, rapidly and almost completely inhibits incorporation of uridine-^{3}H (ur-^{3}H) into the RNA of the mature virus recovered from the culture medium. This inhibition occurs at any time after infection. In the experiment shown in Fig. 4 actinomycin D (0.5 μg/ml) was added to the medium of a culture one week after infection. Thirty minutes later ur-^{3}H was added to the actinomycin D treated culture (B) as well as to a control culture without actinomycin D (A). Culture medium was removed at different times, virus was recovered from the medium and the amount of radioactivity in the 71 S viral RNA was determined. It can be seen that virus production in the presence of actinomycin D is reduced to around 5% of that in the absence of actinomycin D. Higher concentrations of actinomycin D inhibit virus production to an even greater extent.

This finding is in agreement with the previous studies by Temin (1963),

Bader (1964), and Vigier and Goldé (1964), who showed that production of infectious RSV is inhibited by actinomycin D.

In addition, when infected cells are incubated with ur-³H in the presence of different concentrations of actinomycin D, almost no labeled RNA can be found in the infected, actinomycin D treated cells that is not also present in uninfected control cultures. An exception to this generalization

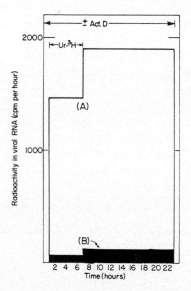

Fig. 4. Effect of actinomycin D on production of tritium-labeled virus by RSV + RAV infected cells. Two cultures of cells one week after infection with RSV + RAV were used. Actinomycin D (0.5 μg/ml) was added to one culture (B) and after 30 min ur-³H (1 μCi/ml, 1 mCi/μmole) was added to that culture and to a control culture without actinomycin D (A). After 6 hours, the medium from both cultures was removed for virus purification and fresh medium was added to both. Actinomycin D was again added to culture B. After another 16 hours the medium was again removed from both for virus recovery. Virus was purified from each sample of medium and viral RNA was extracted and the amount of radioactivity in the 71 S viral RNA determined (Robinson, 1966). The shaded area represents the radioactivity, expressed as tritium cpm per hour of incubation, in virus produced by culture B, and the unshaded area that produced by culture A.

is the irregular occurrence of very small amounts of labeled RNA sedimenting with characteristics of intact RSV + RAV RNA that has been detected at certain actinomycin D concentrations (Robinson, 1966). The amount of this labeled RNA is always less than 5% of that found in mature virus in the culture medium of infected cells labeled for a comparable time in the absence of actinomycin D. With concentrations of actinomycin D higher than about 2 μgm/ml no such labeled RNA has been observed.

From these experiments we conclude that actinomycin D under the conditions of our tissue culture inhibits the synthesis of RSV and RAV RNA.

CELLS INFECTED WITH NEWCASTLE DISEASE VIRUS (NDV)

Similar experiments have been done with two different myxoviruses, NDV (Bratt and Robinson, 1967) and influenza virus (flu) (Duesberg and Robinson, 1967) in order to compare them with RSV + RAV. NDV and flu were chosen because of several similarities with RSV + RAV. First, NDV and flu can be grown in tissue culture using the same cells and the same culture medium as are used for RSV + RAV. Second, the myxoviruses possess RNA as their nucleic acid. They are large viruses with a high lipid content, and probably have structural similarities with the RNA-containing tumor viruses. In addition neither group of viruses is found to accumulate in the infected cell as is the case of small RNA viruses such as polio. Instead, they are thought to mature by a process of budding at the cell surface.

Another feature of similarity is in the level of viral RNA synthesis by infected cells. In the case of NDV and flu, the total amount of viral RNA synthesis during one growth cycle in tissue culture (as measured by ur-^{3}H incorporation into the RNA of mature virus recovered from the culture medium at the end of the growth cycle) is less than one-half that recovered from RSV + RAV labeled in the same way for a comparable period of time. Thus the rate of viral RNA synthesis in NDV- and flu-infected cells is probably of the same order of magnitude as that in RSV + RAV-infected cells.

There is a significant difference, however, between RSV + RAV and the myxoviruses, and this difference makes it possible to study viral specific RNA synthesis in myxovirus-infected cells. As just described, production of RSV + RAV is inhibited by actinomycin D at any stage of infection. In contrast, when actinomycin D is added to cultures at 2 to 3 hours or more after infection, little inhibition of NDV or flu production is observed (Barry *et al.*, 1962; Granoff and Kingsbury, 1964).

Here I will describe, in detail, experiments with NDV-infected cells, and will mention only briefly the results of similar experiments with flu. The experiments concerned with RNA synthesis in NDV-infected cells involve the use of actinomycin D at 2 μg/ml to stop normal cell RNA synthesis as described for experiments with RSV + RAV, followed by the addition of ur-^{3}H or ^{32}P to label viral specific RNA. Figure 5 shows the results of sucrose gradient fractionation of the total RNA isolated from a culture of uninfected cells (A) and NDV-infected cells (B), following incubation with ur-^{3}H for 2 hours (6 to 8 hours after infection in the case of B) in the presence of actinomycin D. Actinomycin D almost completely inhibits incorporation of ur-^{3}H into RNA of the uninfected culture (A). In contrast, at least three tritium-labeled RNA components are seen in the NDV-infected cells (B). The fastest sedimenting labeled RNA (peak in fraction 7) is in the position (57 S) expected for intact viral RNA. A second peak of labeled RNA (peak in fractions 13 and 14) sediments just ahead of 28 S cell RNA and is designated 35 S. The third peak (fraction 19) can be seen in the position of 18 S cell RNA.

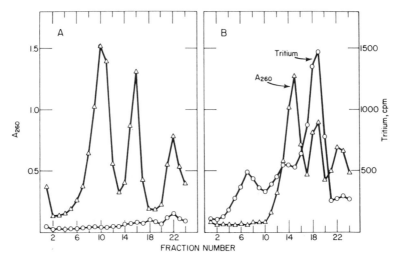

FIG. 5. RNA synthesis in normal cells (A) and NDV infected cells (B) in the presence of actinomycin D. Actinomycin D (2 μg/ml) was added to a culture of uninfected cells (A) and to cells 6 hours after infection with NDV (B). Thirty min later ur-^{3}H (5 μCi/ml, 5 mCi/μmole) was added to both. Two hours after the addition of ur-^{3}H, the cell RNA was recovered from both and fractionated by sucrose gradient centrifugation. Tritium (O———O) and A$_{260}$ ($\triangle$———$\triangle$) were determined on each fraction of the sucrose gradient (data from Bratt and Robinson, 1967).

The 18 S and 35 S labeled RNA components first appear in the cell 3 to 4 hours after infection and increase in amount up to about 10 hours, when cell death occurs. The 57 S component is usually not seen in the amount shown in Fig. 5 and frequently is not seen at all. This may be due to the rapid removal of 57 S RNA from the cell into virus.

A fourth viral-specific RNA component, designated 22 S, is always found when the RNA is sedimented for a longer time for better separation of RNA in the 22 S region (Bratt and Robinson, 1967).

The labeled RNA components were shown to be almost completely digested to acid-soluble products by pancreatic RNase in 0.2 *M* salt, indicating that they are predominantly single-stranded. However, variable but always small (0 to 15%) amounts of RNase resistance was found in the RNA of the 57 S and 35 S components in different experiments. Such RNase resistant RNA could represent base-paired RNA, analogous to that found in cells infected with other RNA viruses (Montagnier and Sanders, 1963; Baltimore *et al.,* 1964; Kelly and Sinsheimer, 1964; Weissmann *et al.,* 1964) and which is thought to be an intermediate in viral RNA synthesis.

The base compositions of the 18 and 35 S RNA components were found to be nearly complementary to that of the intact viral RNA (Bratt and Robinson, 1967). In order to test for complementarity of base sequence, tritium-labeled 18, 22, and 35 S viral specific RNA's were incubated under annealing conditions with unlabeled 57 S RNA from NDV and then tested for RNase resistance. The results are shown in Table III. The labeled RNA's were completely resistant to RNase after incubating

TABLE III
ANNEALING ³H-LABELED RNA[a]

Unlabeled RNA (0.73 µg) added to the annealing mixture	TCA insoluble cpm after annealing and digestion with RNase			
	18 S ³H-RNA from infected cells	22 S ³H-RNA from infected cells	35 S ³H-RNA from infected cells	18 S ³H-ribosomal RNA from uninfected cells
57 S NDV	7007	2200	2794	190
4 S NDV	3920	—	—	—
TMV	100	36	76	204
Uninfected cell 18 S	88	33	—	—
Uninfected cell 28 S	—	—	65	—
Total cpm added to the annealing mixture (i.e. no RNase digestion)	6736	2160	2707	3400

[a] Tritium-labeled RNA from infected cells was prepared as described in Fig. 5. Tritium labeled 18 S cell RNA was prepared from normal cells in the same way after incubation of the cells with ur ³H for 12 hours in the absence of actinomycin D. Unlabeled viral RNA was prepared from purified virus grown in eggs and was fractionated into the 57 S and 4 S components as previously described (Duesberg and Robinson, 1965). TMV was a gift of Dr. C. A. Knight and the RNA was isolated therefrom by phenol extraction. Annealing was done by placing solutions containing the appropriate nucleic acids in 0.25 *M* salt at 90°C and cooling slowly to 37°C over 6 hours. To test for annealing, each RNA sample in 0.1 *M* salt and 0.002 *M* MgCl₂ was then incubated with pancreatic RNase (10 µg/ml) at 37°C for 1 hour following which TCA precipitable radioactivity was determined. The data are from Bratt and Robinson (1967).

with 57 S viral RNA but not with TMV RNA or with ribosomal RNA from normal cells. In addition no annealing occurred when 18 S ribosomal RNA, labeled in normal cells in the absence of actinomycin D, was incubated with NDV RNA. Thus in this experiment the three RNA components from infected cells were entirely complementary to viral RNA. In all experiments the 18 S RNA component annealed completely with viral RNA. However, the 22 and 35 S RNA components did not always anneal completely with viral RNA. In different experiments from 90 to 100% of the 35 S component and from 95 to 100% of the 22 S RNA were found to be complementary to viral RNA (Bratt and Robinson, 1967). Such findings are in accord with the observation that small and variable amounts of the 22 and 35 S RNA components in different experiments are found to be in the form of base-paired RNA.

It is of interest that about 60% of the labeled 18 S viral specific RNA annealed to the 4 S RNA component from NDV (Table III). The 4 S RNA from NDV is thought to contain degraded viral RNA (Duesberg and Robinson, 1965) and the annealing, which indicates that the 4 S component has at least some of the same base sequences as 57 S viral RNA, is compatible with this idea.

The maximum fraction of the 57 S RNA component labeled in infected cells that could be annealed to 57 S RNA from virus was 30% (Bratt and Robinson, 1967). This indicates that about $\frac{1}{3}$ of the labeled 57 S RNA component found in infected cells is complementary to viral RNA and $\frac{2}{3}$ is probably identical to viral RNA.

When the 57 S RNA from NDV labeled with ^{32}P was incubated with various amounts of 18, 22 and 35 S RNA from infected cells annealing was easily observed. Figure 6 shows that the maximum fraction of ^{32}P-labeled viral RNA that could be annealed to the 35 S component was 70%, to the 18 S component 50% and to the 22 S component 40%. Only 70% could be annealed to a mixture of the 18, 22 and 35 S components (Bratt and Robinson, 1967) indicating that the three probably share base sequences in part and that their sum does not make the entire complement of the 57 S viral RNA. Annealing with labeled viral RNA it is possible to show that the complementary RNA components in the NDV-infected cell are present in approximately the same amount and with the same sedimentation distribution in cells not treated with actinomycin D as in cells treated with actinomycin D (Bratt and Robinson, 1967). This suggests that actinomycin D does not significantly influence the synthesis of these viral specific RNA components.

Fractionation of NDV-infected cells into cytoplasmic and nuclear fractions, after incubation with ur-^{3}H for about 15 min in the presence of actinomycin D, revealed that the labeled, viral-specific RNA was al-

most exclusively in the cytoplasm (Bratt and Robinson, 1967). Further
fractionation of the cytoplasmic extract showed that most of the labeled
RNA was attached to polyribosomes. Figure 7 shows the results of sedimentation of the RNA isolated from each of three parts of a sucrose
gradient after centrifugation of a cytoplasmic extract from infected cells
in such a way as to delineate the polyribosomes and single ribosomes
(Bratt and Robinson, 1967). Figure 7-1 shows the RNA from the polyribosomes. A small amount of cellular 28 and 18 S RNA (fractions 12

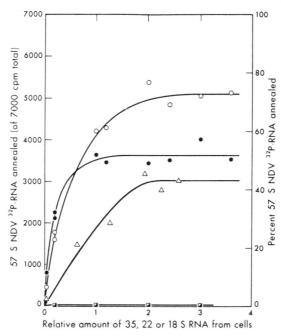

Fig. 6. Annealing ^{32}P labeled 57 S NDV RNA with varying amounts of 35 S
(O———O), 22 S (△———△) and 18 S (●———●) viral specific RNA's from
NDV infected cells and with 18 (□———□) and 28 S (■———■) RNA's from
normal cells. Annealing was carried out as described in Table III. One unit on
the abscissa represents the amount of RNA recovered from a single culture of cells
(data from Bratt and Robinson, 1967).

to 18) and a large amount of 4 S RNA (used as carrier during RNA
isolation) are indicated by the A_{260} tracing. The labeled viral specific
RNA from the polysomes consists of 57 S RNA (fractions 4 to 8), 35 S
RNA (fractions 10 to 12) and 18 S RNA (fractions 17 to 21). The
labeled 18 and 35 S RNA's annealed to viral RNA as shown in Table III.
Figure 7-2 shows the RNA from the single ribosomes. Cellular RNA and
carrier 4 S RNA are again indicated by the A_{260} tracing. Most of the

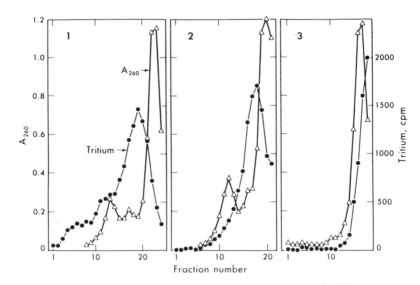

FIG. 7. Viral specific RNA recovered from polysomes (1), single ribosomes (2), and structures sedimenting more slowly than single ribosomes (3) after sucrose gradient fractionation of the cytoplasmic extraction from NDV infected cells as described by Bratt and Robinson (1967). Infected cells were incubated with ur-^{3}H in the presence of actinomycin D as described in Fig. 5B. Cells were then removed from the culture dish with trypsin and fractionated into cytoplasm and nucleus by Dounce homogenization and differential centrifugation. The cytoplasmic extract was made 0.5% with respect to deoxycholate and fractionated in a sucrose gradient in such a way as to delineate the polyribosome and single ribosomes. The regions of the sucrose gradient containing the polyribosomes, the single ribosomes and the material sedimenting more slowly than single ribosomes were recovered and the total RNA was isolated from each by phenol extraction using RNA as carrier. Each RNA preparation was then fractionated by sucrose gradient centrifugation as described in Fig. 5B and A₂₆₀ (△————△) and tritium (●————●) were determined on each fraction of each gradient (data from Robinson and Bratt, unpublished).

labeled RNA is the 18 S viral-specific RNA. Figure 7-3 shows that almost no cellular RNA and no labeled RNA sedimenting faster than 4 S RNA was recovered from the top of the polysome gradient (in structures sedimenting more slowly than single ribosomes).

SUMMARY AND DISCUSSION

In cells infected with NDV, several viral specific RNA components can be labeled with ur-^{3}H in the presence of actinomycin D. The RNA is predominantly single-stranded, although small and variable amounts of base-paired RNA are sometimes found. From 90 to 100% of three of the

RNA components is complementary to viral RNA. One-third of the RNA in the other component (57 S) is complementary to viral RNA. One-third of the RNA in the other component (57 S) is complementary to viral RNA and two-thirds is probably identical to viral RNA.

The function of the complementary RNA pieces is unknown: However, the three occurring in greatest amount (18, 22, and 35 S) have common base sequences in part and their sum does not complete the total complement of viral RNA. In addition they appear to be attached to the polyribosomes of the cell.

The finding of large amounts of single-stranded RNA complementary to viral RNA is so far unique to NDV. This is not the case for influenza virus, another myxovirus, where the predominant RNA made in the infected cell in the presence of actinomycin D is the viral strand, and less than 5% is complementary to viral RNA (Duesberg and Robinson, 1967). This appears to be true also for viral specific RNA made in cells infected with small RNA viruses such as the RNA phage (Weissmann *et al.*, 1964), and poliovirus (Darnell, 1962). In addition large amounts of RNA complementary to viral RNA are not found in cells infected with RSV + RAV (Robinson and Duesberg, unpublished observations).

A final point to be emphasized is the unique behavior of RSV + RAV with respect to viral RNA synthesis. RSV + RAV appear to be the only RNA-containing viruses studied so far where viral specific RNA is not made in the infected cell in the presence of actinomycin D. It has been known for some time that actinomycin D does not inhibit viral specific RNA synthesis in cells infected with RNA phages (Kelly and Sinsheimer, 1964) and by small RNA-containing animal viruses such as poliovirus (Darnell, 1962; Zimmerman *et al.*, 1963). The same appears to be true for arboviruses such as Western equine encephalomyelitis virus (Sreevalsan and Lockart, 1966), for Reovirus (A. Shatkin, this volume), for myxoviruses such as influenza virus (Duesberg and Robinson, 1967) and the parainfluenza virus NDV (Bratt and Robinson, 1967). Whether inhibition of viral specific RNA synthesis by actinomycin D is common for all RNA-containing tumor viruses remains to be shown.

ACKNOWLEDGMENT

I wish to thank Dr. Harry Rubin and Dr. Wendell M. Stanley for support and encouragement throughout the course of this work, and Dr. H. Robinson for helpful discussions and for aid in preparation of the manuscript. I would also like to thank Drs. P. H. Duesberg and M. A. Bratt who contributed many ideas and a great deal of work to experiments reported here.

This investigation was supported in part by U.S. Public Health Service research grants, CA 04774, CA 05619, and CA 08557 from the National Cancer Institute and AI 01267 from the National Institute of Allergy and Infectious Diseases, Public Health Service, and a grant from the Rockefeller Foundation.

REFERENCES

Bader, J. P. (1964). *Virology* **22,** 462.
Baltimore, D., Becker, Y., and Darnell, J. E. (1964). *Science* **143,** 1034.
Barry, R. D., Ines, D. R., and Cruickshank, F. G. (1962). *Nature* **194,** 1139.
Bratt, M. A., and Robinson, W. S. (1967). *J. Mol. Biol.* **23,** 1.
Darnell, J. E. (1962). *Cold Spring Harbor Symp. Quant. Biol.* **26,** 149.
Duesberg, P. H., and Blair, P. (1966). *Proc. Natl. Acad. Sci. U.S.* (in press).
Duesberg, P. H., and Robinson, W. S. (1965). *Proc. Natl. Acad. Sci. U.S.* **54,** 794.
Duesberg, P. H., and Robinson, W. S. (1966). *Proc. Natl. Acad. Sci. U.S.* **55,** 219.
Duesberg, P. H., and Robinson, W. S. (1967). *J. Mol. Biol.* (in press).
Galibert, F., Bernard, C., Chenaille, B., and Boiron, M. (1965). *Compt. Rend.* **261,** 1771.
Granboulin, N., Huppert, J., and Lacour, F. (1966). *J. Mol. Biol.* **16,** 571.
Granoff, K., and Kingsbury, D. W. (1964). *Ciba Found. Symp. Cellular Biol. Myxovirus Infections* p. 96
Harel, J., Huppert, J., Lacour, F., and Harel, L. (1965a). *Compt. Rend.* **261,** 2266.
Harel, L., Goldé, A., Harel, J., Montagnier, L., and Vigier, P. (1965b). *Compt. Rend.* **261,** 4559.
Kelly, R. B., and Sinsheimer, R. L. (1964). *J. Mol. Biol.* **8,** 602.
Montagnier, L., and Sanders, F. K. (1963). *Nature* **199,** 664.
Mora, P. T., McFarland, V. W., and Luborsky, S. W. (1966). *Proc. Natl. Acad. Sci. U.S.* **55,** 438.
Morgan, C., Riffkind, R., and Rose, H. (1962). *Cold Spring Harbor Symp. Quant. Biol.* **26,** 57.
O'Connor, T. E., Rauscher, F. J., and Zeigel, R. F. (1964). *Science* **144,** 1144.
Robinson, W. S. (1966). *In* "Viruses Inducing Cancer" (W. J. Burdette, ed.), p. 107. Univ. of Utah Press, Salt Lake City, Utah.
Robinson, W. S., and Baluda, M. A. (1965). *Proc. Natl. Acad. Sci. U.S.* **54,** 1686.
Robinson, W. S., Pitkanen, A., and Rubin, H. (1965). *Proc. Natl. Acad. Sci. U.S.* **54,** 137.
Spirin, A. S. (1963). *Progr. Nucleic Acid Res.* **1,** 301.
Sreevalsan, T., and Lockart, R. W. (1966). *Proc. Natl. Acad. Sci. U.S.* **55,** 974.
Strauss, H. J., and Sinsheimer, R. (1963). *J. Mol. Biol.* **7,** 43.
Temin, H. M. (1963). *Virology* **20,** 577.
Trager, G., and Rubin, H. (1964). *Natl. Cancer Inst. Monograph* **17,** 575.
Vigier, P., and Goldé, A. (1964). *Virology* **23,** 511.
Weissmann, C., Borst, P., Burdon, R. H., Billeter, M. A., and Ochoa, S. (1964). *Proc. Natl. Acad. Sci. U.S.* **51,** 682.
Zimmerman, E. F., Heeter, M., and Darnell, J. E. (1963). *Virology* **19,** 400.

Metabolic Requirements in Rous Sarcoma Virus Replication

John P. Bader

CHEMISTRY BRANCH, NATIONAL CANCER INSTITUTE, NATIONAL INSTITUTES OF HEALTH, BETHESDA, MARYLAND

The requirements of animal viruses for cellular metabolic processes may be different for different kinds of viruses. While evidence for the involvement of specific cellular processes is meager, it is apparent that certain processes are required for the replication of some viruses that are not required for the replication of others. Metabolic requirements for virus growth cannot yet be predicted from a knowledge of the size, structure, or chemical constitution of a virus particle. Certainly, the general classification of animal viruses into DNA and RNA types has been shown to be an inadequate criterion for the determination of cellular metabolic requirements, since certain RNA-containing viruses (e.g., avian leukosis viruses) require DNA synthesis during the early phase of the infectious cycle (Bader, 1965a). About all that can be said is that closely related viruses probably have similar metabolic requirements.

Differences Among RNA Viruses

The variation in metabolic requirements among the RNA-containing viruses is apparent from the effects of various metabolic inhibitors on the growth of a number of these agents in chick embryo cell cultures (Table I). The avian leukosis viruses (Rous sarcoma virus [RSV–RAV$_1$, or RSV], Rous-associated virus [RAV$_1$], and Schmidt-Ruppin Rous sarcoma virus [SR–RSV]), influenza virus, and vesicular stomatitis virus have been characterized as RNA-containing viruses, which are without DNA, have protein-lipid envelopes, and are completed by budding from the cell membrane. The transforming particle, RSV, has been shown to require a helper virus for reproduction in chick embryo cells (Hanafusa *et al.,* 1963). The effects of inhibitors observed in this and in some subsequent studies may be due in part to effects on the helper virus, but this does not alter the interpretation of the experimental results. For example, SR–RSV requires no helper for growth, but shares a number of biological properties in common with RSV, and is affected similarly by various antimetabolites.

Three types of metabolic interactions may be distinguished in a summary of published data (Bader, 1965a). The inhibition of the biological activity of DNA by actinomycin prevents the growth of the avian leukosis viruses and of influenza virus, but has little effect on the replication of vesicular stomatitis virus. It has further been shown that actinomycin

TABLE I

EFFECTS OF METABOLIC INHIBITORS ON GROWTH OF RNA VIRUSES[a]

Inhibitor	Virus yield (test/control)				
	RSV-RAV$_1$	RAV$_1$	SR-RSV	Influenza virus	Vesicular stomatitis virus
Actinomycin	<0.001	<0.001	—	<0.001	1
5-iododeoxy-uridine	0.008	0.053	—	1	1
Cytosine arabinoside	<0.001	<0.001	<0.002	1	1

[a] Chick embryo cells were exposed to virus, washed, and metabolic inhibitors added. Fluids were removed 14 to 24 hours later for subsequent assay.

blocks subsequent replication of RSV no matter when it is added to infected cultures (Bader, 1964), whereas White *et al.* (1965) have shown that influenza virus growth is susceptible to inhibition by actinomycin only during the early phase of the infectious cycle. Actinomycin inhibits both DNA synthesis and DNA-dependent RNA synthesis in chick embryo cells (Bader, 1964). When the effects of specific inhibitors of DNA synthesis (5-iododeoxyuridine, cytosine arabinoside) were examined, only the avian leukosis viruses were found to be affected. These results showed that DNA synthesis is a specific requirement for the growth of the avian leukosis viruses, but not for influenza virus.

Certain other experiments have shown that DNA synthesis is required for the replication of RSV only during the early phase of the infectious cycle. This suggests that an early event may be the synthesis of a stable primer which is utilized later for the synthesis of viral components. Evidence presently available does not resolve the question as to whether the DNA required for the growth of RSV is a new DNA, primed by viral

RNA, or cellular DNA, which for some reason must be synthesized in order for the infectious process to proceed.

A summary of the effects of metabolic inhibitors on the growth of RSV is shown in Fig. 1. Growth of RSV can be resolved into at least three stages: (1) initial infection (adsorption, penetration, and release of viral nucleic acid), (2) a phase requiring DNA synthesis, (3) and a phase no longer requiring DNA synthesis but still dependent upon functioning DNA. The following experiments demonstrate that cellular metabolic activities are involved in each of the three phases.

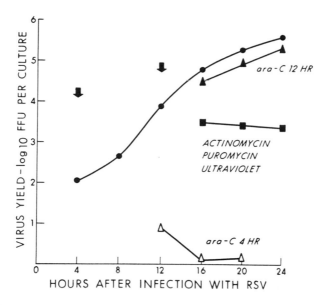

FIG. 1. Summary of the effects of metabolic inhibitors on the growth of RSV. Cytosine arabinoside ($10^{-3.5}$ M) was added at 4 ($\triangle$) or 12 ($\blacktriangle$) hours after exposure of cells to RSV. Addition of actinomycin (2 μg/ml), puromycin (10 μg/ml), or exposure to ultraviolet light (520 ergs/mm^2) was at 12 hours post-infection.

Growth of Virus during the Phase not Requiring Active DNA Synthesis

This series of experiments utilized the effects of ultraviolet light (UV) on nucleic acids (RNA and DNA) and was based on observations reported initially by Rubin and Temin (1959). These workers described an unusual, differential susceptibility of chick embryo cells to UV irradiation, with respect to their capacity to support the growth of RSV and of Newcastle disease virus. Similar results were obtained in the present experiments (Fig. 2), and, in addition, it was noted that nucleic acid syntheses are inhibited in cells exposed to UV light.

A dose of UV irradiation (520 ergs/mm²) which has little effect on free RSV, or on the capacity of the cells to support the production of vesicular stomatitis virus (VSV), will prevent subsequent growth of RSV even when administered during the late stages of virus synthesis (Fig. 3). This result suggests that the participation of some cellular component is

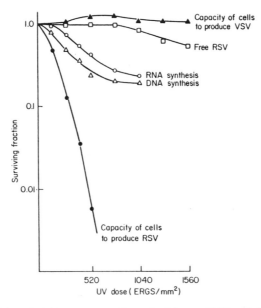

FIG. 2. Sensitivity of chick embryo cells and of intact VSV and RSV to ultraviolet light (UV). Cultures were irradiated before being infected with the virus, then exposed to virus and washed. Culture fluids were removed for subsequent assay at 14 (vesicular stomatitis virus, VSV) or 16 hours (RSV) post-infection. To study nucleic acid syntheses, uridine-³H (RNA) or thymidine-³H (DNA) was added immediately after irradiation, and 1 hour later the extent of incorporation into acid-precipitable material was measured. In all cases control cultures were treated identically except for irradiation. All values are plotted as the ratio of value for irradiated sample: control value.

required during the late stages of virus growth, but does not eliminate the possibility that the intracellular form of the virus is more susceptible than free virus to inactivation by UV light.

To test this possibility, cell cultures were exposed to graded doses of UV before or at various times after virus adsorption, and the ability of the dispersed cells to form infectious centers was measured. The rate at which the capacity of cells to support the replication of RSV was lost was found to be the same whether the cells were irradiated before exposure to virus or during the late stages of virus growth (Fig. 4). Inactivation of infectious centers is a "single-hit" process both before and after infection, and infected cells do not become more resistant to UV with increasing time after infection, as has been reported for poliovirus (Fenwick and Pelling, 1963), Newcastle disease virus (Kirvaitis and Simon, 1965), or influenza virus (White and Cheyne, 1966). A similar result

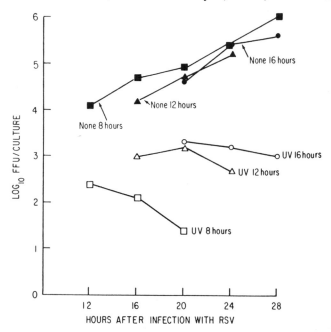

FIG. 3. Effect of UV irradiation on the growth of RSV. Cells were washed and exposed to UV irradiation at 8, 12, or 16 hours after infection with RSV. Culture fluids from both irradiated and unirradiated cultures were removed at 4-hour intervals thereafter and assayed for virus content.

was found when, instead of infectious centers, total virus yields in extracellular fluids were measured. It was found that the virus yield decreases at the same rate as does the decrease in infectious centers.

Varying the multiplicity of infection (FFU* per cell = 10, 2, or 0.4) has no effect on the rate of inactivation of infectious centers; this demonstrates that inactivation of intracellular virus is not an important factor in these results, but that the capacity of the cells to support the replication of RSV is continuously dependent on a cellular component. That this component is DNA is evident from the following considerations: (1) in-

* FFU, focus forming units.

activation is a "single-hit" phenomenon, and the only cellular molecules which can be safely assumed to be nonredundant are DNA's; (2) any cellular constituent other than DNA would probably be replenished by normal cellular synthetic processes; (3) the synthesis of both RNA and DNA is significantly depressed by doses of UV which inhibit virus growth; (4) evidence provided by other investigators (Setlow, 1960; Setlow and Setlow, 1962) indicates that the major biological activity of UV irradiation stems from its action on DNA, and (5) the protein-synthesizing machinery remains intact in UV-irradiated cells, as shown by the unimpaired capacity of irradiated cells to produce vesicular stomatitis virus.

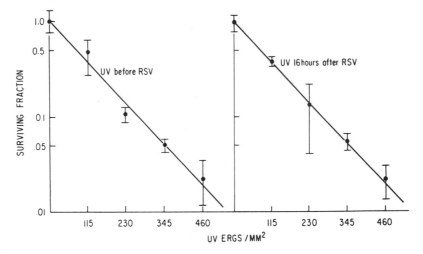

FIG. 4. The effect of UV irradiation on the survival of infectious centers. Cells were exposed to various doses of UV before, or at 16 hours after, infection with RSV. Cells were then dispersed, suspended in a small volume of nutrient agar and added to a sparse monolayer of cells. More nutrient agar was added, the cultures incubated for one week and the resulting foci counted.

A more direct experimental approach has been employed to show that the critical cellular component is indeed DNA. A number of investigators have reported that 5-bromodeoxyuridine can be incorporated into cellular DNA, and that such cells are more sensitive than cells containing normal DNA to the action of UV irradiation. In the present experiments, chick embryo cells were grown for 2 days in medium containing BUDR. This medium was then removed, the cells were infected with RSV, and the cultures incubated for 16 hours in medium containing thymidine. Under these experimental conditions cellular DNA incorporated BUDR, prior to infection with RSV, but the DNA synthesized subsequent to virus infection did not. Control cultures and BUDR-labeled cultures were then irradiated with graded doses of UV, and subsequent virus yields were measured. Cultures which had been exposed to BUDR were more sensitive with respect to the UV-inactivation of their capacity to support viral replication than were the controls (Fig. 5). Incorporation of BUDR into cellular DNA was confirmed by demonstrating (by equilibirum centrifu-

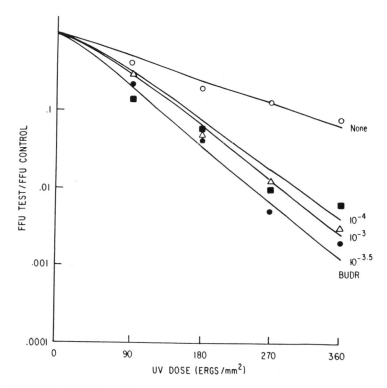

FIG. 5. Effect of UV irradiation on virus production in cells pretreated with 5-bromodeoxyuridine (BUDR). Cells were grown for 2 days in the presence of BUDR (10^{-3}, $10^{-3.5}$, or 10^{-4} M). Fluids were removed and the cells infected with RSV, after which growth medium containing thymidine, but no BUDR, was added. Sixteen hours thereafter the cultures were exposed to various doses of UV, and culture fluids were collected 12 hours later for virus assay.

gation in Cs_2SO_4 density gradients) that DNA extracted from cells grown in the presence of BUDR had a higher density than did the normal cellular DNA, and also by measuring the incorporation of BUDR-^{3}H into cellular DNA. No incorporation of BUDR-^{3}H into RNA was observed. These results provide additional evidence that functioning cellular DNA is a continuous requirement for the production of RSV.

The Phase of Viral Replication Requiring DNA Synthesis

It has already been shown that DNA synthesis is a requirement for virus growth. Other unpublished experiments from these laboratories have indicated that RSV cannot induce DNA synthesis, and conversely that DNA synthesis must be in progress for infection to be successful. Experiments designed to examine this point further were carried out using synchronized cells. Chick embryo cell cultures were synchronized by blocking DNA synthesis with excess thymidine (Xeros, 1962). At various times after release from thymidine inhibition, when cultures were in different stages of the mitotic cycle, RSV was added, and subsequent virus growth, or the formation of foci of transformed cells, was measured. It can be seen (Fig. 6) that marked differences in the number of foci produced were observed in cultures infected with virus at various stages

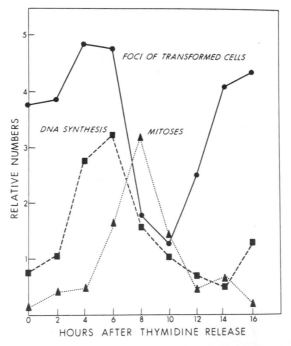

FIG. 6. The effect of position in the mitotic cycle on the susceptibility of cells to infection by RSV. DNA synthesis was blocked by exposing cells to 2 m*M* thymidine for at least 16 hours. At 2-hour intervals the medium containing excess thymidine was removed. The number of mitoses were counted in 5 microscopic fields ($\times$ 63). DNA synthesis was measured by the incorporation of deoxycytidine-^{3}H into acid-precipitable material. Cultures in various stages of the mitotic cycle were exposed to RSV for 1 hour, after which nutrient agar containing anti-RSV serum was added. Foci were counted one week later.

of the mitotic cycle. Infection is most likely to be successful if cells are infected either at the time when DNA is being actively synthesized or at a time immediately preceding the initiation of DNA synthesis. Similar results were obtained when virus yields, rather than foci formation, were measured (Fig. 7). Virus samples were taken at both 16 and 20 hours after infection with RSV to minimize the possible effect of the mitotic state of the cells on virus production (Marcus and Robbins, 1963). Since the pattern of virus growth is similar at 16 and 20 hours, it is concluded that the observed differences are due to the position of the cells in the mitotic cycle at the time of exposure to virus.

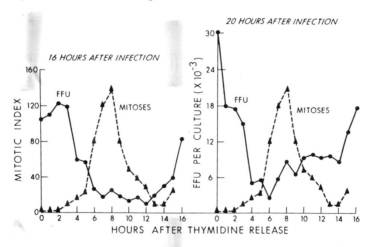

FIG. 7. The effect of position in the mitotic cycle on the susceptibility of cells to infection by RSV. Cultures were synchronized by removing excess thymidine at hourly intervals from different cultures. Cultures in various stages of the mitotic cycle were exposed to RSV for 1 hour, after which unattached virus was removed by washing and growth medium was added. Fluids were removed 16 hours later and replaced with fresh medium. These latter fluids were removed 4 hours later (20 hours post-infection) and all samples assayed for virus content.

It is possible that cells in mitosis are less likely to adsorb virus than are interphase cells. This factor would be unlikely to affect the results, however, since the cultures, while relatively synchronized, usually contained no more than 25% of the cells in mitosis at any instant. Also, similar cultures in various stages of the mitotic cycle were exposed to Newcastle disease virus, and little variation was seen in the resulting numbers of plaques that were produced (Fig. 8). While these experiments do not specifically show that infection with RSV is dependent on processes related to cellular DNA synthesis, the results are consistent with this idea.

The following experiment shows that cellular enzymes are involved in the synthesis of the DNA required by RSV. Earlier studies showed that if DNA synthesis is temporarily blocked during the early stage of infection by RSV, then virus infection is aborted, and striking decreases in virus yield or in the formation of foci of transformed cells are noted after DNA synthesis resumes (Bader, 1965b). Puromycin is an effective and reversible inhibitor of protein synthesis in chick embryo cells. It was possible, therefore, to use this drug to determine whether the formation of any new proteins are necessary for the synthesis of the DNA required for

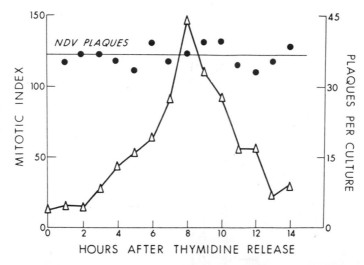

Fig. 8. The effect of position in the mitotic cycle on the susceptibility of cells to infection by Newcastle disease virus (NDV). Cultures in various stages of the mitotic cycle were exposed to NDV for 1 hour. Cultures were washed and nutrient agar added. Plaques were counted 3 days later after staining cells with neutral red.

RSV replication. The study was coupled with an investigation of the effects of cytosine arabinoside (ara C), an inhibitor of DNA synthesis, on the replication of this viral agent.

Cultures of chick embryo cells were incubated with RSV for 1 hour, after which anti-RSV serum was added to inactivate superficially adsorbed virus. Growth medium containing puromycin or ara C (or neither inhibitor) was added for a period of 8 hours, after which the medium was replaced by either normal growth medium or medium containing ara C. After an additional 8 hours of incubation, the medium was removed from all cultures, which were then overlaid with nutrient agar. Foci of transformed cells were counted one week later.

From the data summarized in Table II, it is clear that the presence of ara C during the period 0–8 hours post-infection sharply inhibits the formation of foci. On the other hand, neither the presence of this inhibitor during the period 8–16 hours post-infection, nor the presence of puromycin during the 0–8 hour period has any significant effect on subsequent foci formation. It may be concluded that the necessary DNA synthesis can take place in the presence of puromycin, at a concentration adequate to inhibit protein synthesis, and that no new proteins are required during this stage of the infection. It follows that preformed cellular enzymes—specifically those involved in DNA synthesis—are utilized.

TABLE II

THE EFFECT OF TEMPORARY INHIBITION OF PROTEIN AND/OR DNA SYNTHESIS ON THE
TRANSFORMATION OF CHICK EMBRYO CELLS BY RSV*

Hours post-infection	Inhibitor	Foci of transformed cells produced
0–8	Ara C	18
0–8	Puromycin	
then 8–16	Ara C	260
8–16	Ara C	233

* Cells were exposed to RSV for 1 hour. Puromycin (2 μg/ml) and cytosine arabinoside (ara C, $10^{-4.5}$ M) were present only during the periods indicated in the left hand column. Foci of transformed cells were counted one week later.

Evidence against Induction of an "Uncoating" Enzyme

Since the discovery of an "uncoating" enzyme (Abel, 1963), which is required for the successful infection of cells by poxviruses, there has been speculation regarding the possible role of similar enzymes in other virus systems. Results comparable to those shown in Table II were obtained in experiments in which puromycin was added to cultured cells prior to infection with RSV, and in which ara C was added for various post-infection periods. We have concluded, therefore, that no new proteins are required for the adsorption, penetration, and uncoating of RSV, and that no uncoating enzyme, comparable to that described for poxvirus systems, is required for the infection of cells by RSV.

Conclusions

The results presented here demonstrate what has long been suspected of the Rous sarcoma system, i.e., that there is an intimate association of the virus with the cell which is not as apparent with other RNA viruses. Preexisting cellular enzymes release the nucleic acid from the virion and synthesize the DNA required in the early phase of infection. Synchronization experiments show that successful infection by RSV is a function of the position of the host cell in the mitotic cycle, and experiments with UV irradiation demonstrate that functioning cellular DNA is a continous requirement for virus growth. It is only natural to speculate on the relation of these processes to the ability of RSV and related RNA-viruses to transform normal to malignant cells. With great profundity I suggest there is a relation.

REFERENCES

Abel, P. (1963). *Z. Vererbungslehre* **94,** 249.
Bader, J. P. (1964). *Virology* **22,** 462.
Bader, J. P. (1965a). *Virology* **26,** 253.
Bader, J. P. (1965b). *Science* **149,** 757.
Fenwick, M. L., and Pelling, D. (1963). *Virology* **20,** 137.
Hanafusa, H., Hanafusa, T., and Rubin, H. (1963). *Proc. Natl. Acad. Sci. U.S.* **49,** 572.
Kirvaitis, J., and Simon, E. H. (1965). *Virology* **26,** 545.
Marcus, P. I., and Robbins, E. (1963). *Proc. Natl. Acad. Sci. U.S.* **50,** 1156.
Rubin, H., and Temin, H. M. (1959). *Virology* **7,** 75.
Setlow, R. B. (1960). *Radiation Res.* Suppl. 2, 276.
Setlow, R. B., and Setlow, J. K. (1962). *Proc. Natl. Acad. Sci. U.S.* **48,** 1250.
White, D. O., and Cheyne, I. M. (1966) *Virology* **29,** 49.
White, D. O., Day, H. M., Batchelder, E. J., Cheyne, I. M., and Wainsbrough, A. J. (1965). *Virology* **25,** 289.
Xeros, N. (1962). *Science* **194,** 682.

Studies on Carcinogenesis by Avian Sarcoma Viruses* IV

Howard M. Temin

MCARDLE MEMORIAL LABORATORY FOR CANCER RESEARCH, UNIVERSITY OF WISCONSIN, MADISON, WISCONSIN

The interaction, in cell culture, between Rous sarcoma virus (RSV) and sensitive chicken embryo fibroblasts has the following characteristics which distinguish it from that observed with most other virus-cell systems, including those systems consisting of most other tumor viruses and their target cells (Temin, 1966c).

1. Under the proper conditions, essentially all cells in a culture exposed to virus are infected—as shown by plating cells for infectious centers, or by examining clones for morphological alterations.

2. No cells are killed by the infection, although cells may die due to alterations of the environment brought about by metabolic shifts in the infected cells.

3. Under the proper conditions, most cells infected by virus become altered in their properties within two to four days after infection. This alteration is called conversion (morphological or malignant).

4. Under the proper conditions, the production of infectious virus and conversion are separable—both nonconverted, virus-producing cells, and converted, nonvirus-producing cells can exist.

5. Infected cells contain one or two copies of a regularly inherited structure carrying the information for both virus production and conversion. This structure is called the "provirus."

6. The genes in the provirus are recoverable in the form of infectious virus, even from converted, nonvirus-producing cells. [See Goldé and Latarjet (1966) for a possible exception.]

* This investigation was supported in part by U.S. Public Health Service Research Grant CA-07175 from the National Cancer Institute. The author is the holder of Research Career Development Award 10K3-CA-8182 from the National Cancer Institute.

7. Under certain conditions, converted cells multiply at the same rate as do uninfected cells, while under other conditions, they may multiply at a different rate than do uninfected cells.

8. Converted cells do not form permanent cell lines.

These eight characteristics probably hold true for all the strains of avian sarcoma viruses, although they have not yet been demonstrated with all of them. There are numerous other properties of the avian sarcoma viruses, such as antigenicity, host range, type of foci produced *in vitro,* presence of leukosis virus, etc., which are not shared by all of the strains. Of the characteristics that can be compared with characteristics of the related avian leukosis viruses, and the possibly similar mouse RNA tumor viruses, the ability to cause conversion in cultured fibroblast cells is the most important characteristic not held in common.

There appear to be three stages in the conversion, by Rous sarcoma virus, of a normal chicken embryo fibroblast to a cancer cell; namely, formation of the provirus, activation of the genes in the provirus, and alteration of the host cell (Temin, 1966a).

Recent studies have shown that the synthesis of nonchromosomal DNA and cell division are both required for the formation of the provirus (Temin, 1967).

When secondary cultures of chicken embryo fibroblasts are made from primary cultures in the stationary phase, the cells divide in a partially synchronous fashion. The addition of 2.5 mM thymidine at the time of subculture sharpens this synchrony. By these means, cultures were obtained that could be infected at a known time in the cell cycle. It was found that infection did not affect mitosis, but that mitosis affected infection. If cultures were infected at the beginning of the S period, or 10 hours later, at the beginning of mitosis, a 10 hour difference in the latent period before release of progeny virus was observed. This result shows that mitosis synchronizes the infectious process.

The hypothesis was proposed that mitosis is necessary for infection. This hypothesis was tested by infecting cells and then blocking mitosis by the addition of colchicine or by the removal of serum from the medium. In neither case did virus production begin. However, after one cell division following infection with RSV at a multiplicity of infection greater than one, additional cell division is no longer needed to sustain virus production. These results support the hypothesis that mitosis is necessary for infection, and provide an explanation for the fact that the capacity of chick embryo fibroblasts to support the replication of RSV is very sensitive to inactivation by X-ray or UV irradiation when the latter are applied immediately after infection, and that this capacity becomes X-ray- and UV-resistant after one post-infection cell division (Rubin and

Temin, 1959; Temin and Rubin, 1959). Also clear, in the light of these results, is the reason for the observation that treatment of cells with actinomycin D or mitomycin C prior to infection with RSV blocks subsequent viral replication (Temin, 1963; Bader, 1964, 1965; Vigier and Goldé, 1964). The results also explain the published observations showing that the replication of RSV has a requirement for normal DNA synthesis after infection (Bader, 1964, 1965; Temin, 1964a,c; Force and Stewart, 1964).

Experiments have been carried out in an effort to determine whether or not there is a requirement, after infection, for new DNA synthesis in addition to that needed for cell division. These experiments involved infecting partially synchronized cultures in the G_2 period and then permitting the passage of sufficient time for the completion of whatever virus-specific processes were necessary before mitosis. When excess thymidine or cytosine arabinoside were added to such cultures, mitosis was essentially unaffected, but virus production was inhibited (Fig. 1). The blocking effect of cytosine arabinoside was prevented by the presence of deoxycytidine. These observations suggest that some new DNA apart from that needed for cell division, is needed to initiate virus production.

Since, after the early stages of infection of cultured chicken embryo cells by Rous sarcoma virus, DNA synthesis is not needed for virus production, it seems that the DNA synthesized early in the infectious process must make something that is required for virus production. Actinomycin D rapidly blocks the production of infectious virus and, probably, the production of viral RNA in virus-producing cells (Temin, 1963, 1964a; Bader, 1964; Vigier and Goldé, 1964). This observation suggests that a continuing DNA function is required to sustain viral synthesis. Hybridization experiments have been carried out, and the data obtained indicate that there is an increase in the homology between viral RNA and the DNA from infected cells, as compared with DNA from uninfected cells (Temin, 1964b). These results, however, are as yet unconfirmed.

These results are all consistent with the hypothesis that the viral RNA in the infected cell directs the synthesis of a DNA homologous to itself and that this DNA then acts as a template for the production of further viral RNA.

Genetic studies of virus-infected cells have shown that these cells carry one or two copies of a regularly inherited structure which carries the information for virus production, and for conversion with respect to cell morphology (Temin, 1961). These structures have been called "provirus" (Temin, 1964c). Combining these two concepts leads one to the further hypothesis that the provirus is DNA, although the virion of Rous sarcoma virus contains only RNA.

Once a cell contains the provirus, it has the potential to become converted and to become a cancer cell. However, not all infected virus-producing cells are converted to malignancy. Under the usual conditions used for assay of the virus, only 10% of the infected fibroblasts are so converted (Temin and Rubin, 1958; Rubin, 1960; Temin, 1966a). The number of converted cells is affected by the concentration of serum in the agar overlay (Table I), but the mechanism of this effect is unknown.

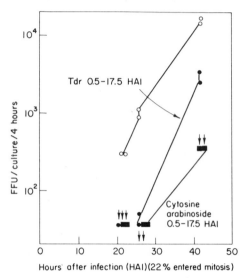

FIG. 1. Effect of blocking DNA synthesis in the G_2 period on infection of chicken cells by Rous sarcoma virus. Secondary cultures containing 8×10^5 cells were prepared in 2 ml of ETC medium containing 2.5 mM thymidine. After incubation for 19 hours, the medium was replaced with 2 ml of ETC medium containing 5×10^{-5} M deoxycytidine. One hour or $29\frac{1}{3}$ hours later, cultures were exposed to Schmidt-Ruppin Rous sarcoma virus at a multiplicity of 4 infectious units per cell. After adsorption for 30 min, the cultures were washed 2 times with ETC medium, and 2 ml of ETC medium or 2 ml of ETC medium containing either 2.5 mM thymidine or 0.5 mM cytosine arabinoside were added. Supernatants containing virus released in a four-hour period were frozen and assayed at the conclusion of the experiment. The time of mitosis was determined by adding 5×10^{-7} M colchicine to 2 cultures and examining them microscopically. Twenty-two percent of the cells had entered mitosis at the time of infection.

Another example of nonconverted, virus-producing cells is provided by liver parenchymal cells, which can be grown in primary culture, and infected. Most cells in such cultures will then plate as infectious centers, but foci of altered cells do not appear.

Conversion of cells appears to involve specific effects on normal cell functions rather than the introduction of new ones. In converted iris epithelial cells, pigment production is stopped, and production of acid mucopolysaccharides is initiated (Temin, 1965). A study of fibroblasts (Temin, 1965), which normally make acid mucopolysaccharides, indicates that conversion also increases the rate of production of this material in these cells. This increase in the production of acid mucopolysaccharides is correlated with conversion, and not with virus production, as has been shown by a study of cells infected with avian leukosis viruses.

The enzymatic basis for the increased production of acid mucopoly-saccharides has recently been elucidated (Ishimoto *et al.*, 1966). Both uninfected and converted cells contain in their membrane fraction an enzyme which makes hyaluronic acid from two nucleotide substrates.

TABLE I

EFFECT OF SERUM CONCENTRATION ON THE NUMBER OF FOCI OF TRANSFORMED CELLS PRODUCED IN RSV-INFECTED CHICK EMBRYO FIBROBLAST CULTURES[a]

	Calf serum in overlay (ml)		
	0.2	0.5	1.0
Number of foci per culture	16, 16, 18, 20	20, 25, 29, 30	67, 83, 103, 104

[a] Secondary cultures were made and infected with Schmidt-Ruppin Rous sarcoma virus. They were overlaid with agar medium containing varying amounts of calf serum.

This enzyme activity is partially destroyed if the intact cells are treated with low concentrations of trypsin, which suggests that the enzyme may be on the outer membrane of the cell. Upon infection with avian sarcoma virus, the specific enzyme activity increases markedly. The increase usually begins at 36 hours after infection, and reaches a maximum 5 to 6 days post-infection. Under the usual assay conditions, the increase is found to be five- to tenfold.

This increase in enzymatic activity is not due to the acquisition of a soluble activator, or to the loss of a soluble inhibitor. It appears probable that it is due to an increased amount of enzyme, since the kinetic properties of the enzymes from uninfected and converted cells are similar. The enzymes from both sources have the same pH optimum, and virtually identical properties with respect to metal ion activation. Heat inactivation studies at a variety of pH's failed to show any significant difference in the stability of the two enzyme preparations. The Michaelis constants for both substrates for the two enzyme preparations were similar, and both enzymes showed the same substrate activation phenomenon.

The results of this study, therefore, support the hypothesis that conversion involves specific effects on normal cell functions, rather than the introduction of new ones. A study of the altered characteristics of cell multiplication of converted cells also appears to support this hypothesis.

Cultures of either uninfected or of converted cells will continue to multiply at the same rate as long as the medium is changed frequently (Temin, 1965). It is only when the serum concentration is limiting that a difference in the rates of multiplication becomes apparent (Temin, 1966b). When secondary cultures are infected, overlaid with agar medium containing varying (and suboptimal) amounts of serum, and the saturation population density determined, it is observed that at every serum concentration the converted cultures have a saturation population density about 50% higher than that of the corresponding uninfected cultures. This effect appears to be due to the exhaustion, by the multiplying cells, of a factor in serum required for cell multiplication. The difference between the uninfected and the converted cells appears to be due to a difference in the amount of this factor required per cell division.

Since this factor in serum is bound by agar and other polyanions, these latter substances exert a differential effect on the multiplication of uninfected and converted cells (Temin, 1965, 1966b). (In addition, low concentrations of polyanions cause an increase in the saturation population density of both uninfected and converted cells, probably by binding a toxin produced by the cells. The stimulatory effect is greater with converted cells, which seem to produce more of this toxin than do uninfected cells.) The serum factor is destroyed by predigested pronase, and is nondialyzable. It is not replaced by fetuin or other serum fractions, with the exception of Fraction III (β-globulins) which has a limited ability to do so. The relation of this factor to that studied by Michl (1962) or Todaro *et al.* (1965) is not known.

The alteration in the characteristics of cell multiplication in the duck cells infected with avian sarcoma virus. In that system, too, infection leads to an alteration in cell morphology, an increase in the rate of production of acid mucopolysaccharide, and to altered growth characteristics.

The alteration in the characteristics of cell multiplication in the duck cell system appears to be controlled by the same factor responsible for the control of multiplication of chicken cells. However, the quantitative difference between uninfected and converted duck cells is greater than the quantitative difference between uninfected and converted chicken cells. Thus, a difference in the rate of cell multiplication between cultures of uninfected and converted duck cells is found even if fresh medium is added every day (Temin, 1965).

Another effect of conversion is to alter the rate of glycolysis in uninfected and converted cells. However, this difference appears to be secondary to the differences in the requirement for the serum factor for cell multiplication.

In summary, we have proposed that infection by Rous sarcoma virus leads to the production in a cell of a DNA provirus, and that genes in this provirus cause specific alterations in the metabolism of the infected cell, which in turn can convert the infected cell into a cancer cell.

There are many similarities between these ideas and those arising from studies of malignant transformation by DNA viruses discussed in the chapters by Winocour, Weil, Sheinin, Defendi, and Green. Whether these similarities are coincidental or real will remain for future work to determine.

Note Added in Proof

Insulin can replace the factor(s) in calf serum whose amount is limiting for multiplication of uninfected and converted chicken and duck cells. In serum-free, insulin-containing medium, converted cells multiply more than do uninfected ones.

REFERENCES

Bader, J. (1964). *Virology* **22**, 462.

Bader, J. (1965). *Virology* **26**, 253.

Force, E. F., and Stewart, R. C. (1964). *Proc. Soc. Exptl. Biol. Med.* **116**, 803.

Goldé, A., and Latarjet, R. (1966). *Compt. Rend.* **262**, 420.

Ishimoto, N., Temin, H. M., and Strominger, J. L. (1966). *J. Biol. Chem.* **241**, 2052.

Michl, J. (1962). *Exptl. Cell Res.* **26**, 129.

Rubin, H. (1960). *Virology* **10**, 29.

Rubin, H., and Temin, H. M. (1959). *Virology* **7**, 75.

Temin, H. M. (1961). *Virology* **13**, 159.

Temin, H. M. (1963). *Virology* **20**, 577.

Temin, H. M. (1964a). *Virology* **23**, 486.

Temin, H. M. (1964b). *Proc. Natl. Acad. Sci. U.S.* **52**, 323.

Temin, H. M. (1964c). *Natl. Cancer Inst. Monograph.* **17**, 557.

Temin, H. M. (1965). *J. Natl. Cancer Inst.* **35**, 679.

Temin, H. M. (1966a). *Cancer Res.* **26**, 212.

Temin, H. M. (1966b). *J. Natl. Cancer Inst.* **37**, 167.

Temin, H. M. (1966c). *In* "Recent Results in Cancer Research" (W. Kirsten, ed.), Vol. XX. Springer, Berlin.

Temin, H. M. (1967). *J. Cell Physiol.* **69**, 1.

Temin, H. M., and Rubin, H. (1958). *Virology* **6**, 669.

Temin, H. M., and Rubin, H. (1959). *Virology* **8**, 209.

Todaro, G. J., Lazar, G. K., and Green, H. (1965). *J. Cellular Comp. Physiol.* **66**, 325.

Vigier, P., and Goldé, A. (1964). *Virology* **23**, 511.

Discussion—Part VIII

Dr. A. K. Field: (Question directed to Dr. W. S. Robinson.) Since you have been unable to obtain infectious RNA from RSV, RAV, MTV, or the Rauscher murine leukemia virus and since you concede that your nucleic acid preparations probably contain host cell RNA, how can you be confident that the sedimentation data and the base ratio data describe the viral nucleic acids?

Virus infection certainly causes alterations in cellular metabolism as evidenced by the fact that cells are transformed. Perhaps RNA molecules found in "purified" material from infected cells represent new or altered cellular nucleic acids rather than virus nucleic acids.

Answer: The fact that the RNA's in question may represent 90% or more of the total RNA recovered from purified virus suggests that they are the viral RNA's. The amount of these RNA's increases as the amount of infectious virus increases; their purity or homogeneity increases as the purity of the infectious virus increases. Furthermore, these RNA's cannot be detected in the RNA extracted from infected whole cells or from subcellular fractions.

Dr. D. Kindig: (Question directed to Dr. H. M. Temin.) Has it been possible to isolate an enzyme from Rous virus-infected cells that makes DNA using purified viral RNA as a primer?

Answer: No. The levels of polymerase activity recovered from infected cells have been too low to find any such activity.

Subject Index

A

Acrylamide gel electrophoresis
  of proteins synthesized in *B. subtilis*
    infected with bacteriophage
    SPO1, 74
    in *E. coli* infected with bacterio-
      phage T4, 74, 76, 80
    with bacteriophage T5, and T5
      "first-step transfer" DNA, 43–
      46, 48, 50
    in poliovirus-infected HeLa cells,
      397–399

Actinomycin D
  effect of, on production of viral-
    specific RNA in Newcastle dis-
    ease virus-infected cells, 690
    on replication of herpes, vesicular
      stomatitis, and rabies viruses,
      453, 454
    of Rous sarcoma, Rous-associ-
      ated, influenza, and vesicular
      stomatitis viruses, 698
    on SV40-induced enzyme synthesis,
      503, 505
    on synthesis of viral RNA in cells
      infected with Rous sarcoma and
      Rous-associated viruses, 687–688
    on thymidine kinase activity of
      pseudorabies virus-infected cells,
      540
  toxicity of, for BHK/21 cells, 452

Adenovirus, type 5
  effect of fiber antigen from, on
    DNA-dependent RNA and DNA
    polymerases, 565
    of hexon antigen from, on DNA-
      dependent RNA and DNA poly-
      merases, 565
    of, on DNA synthesis in KB cells,
      551, 553

on protein synthesis in KB cells,
    551, 557
    on RNA synthesis in KB cells,
      551
    on synthesis of host cell en-
      zymes, 557, 559
    of host mRNA, 560
  fiber antigen from, adsorption of, to
    KB cells, 564
  association of, with viral and KB
    cell DNA, 565
  effect of, on synthesis of DNA,
    RNA and protein in uninfected
    KB cells, 563
  hexon antigen from, association of,
    with viral and KB cell DNA, 565
  inhibition of cellular DNA synthesis
    in cells infected with, 553, 554
  physical properties of, 548
  structure of, 548, 549

Adenovirus 7, survival of, after exposure to ⁶⁰Co radiation, 650

Arabinosylcytosine
  effect of, on replication of herpes and
    vesicular stomatitis viruses, 454
    on replication of Rous sarcoma,
      Rous-associated, influenza, and
      vesicular stomatitis viruses, 698–
      699
    on SV40-induced enzyme synthe-
      sis, 508
    on synthesis of DNA and RNA in
      BHK/21 cells, 455
    on transformation of chick embryo
      cells by Rous sarcoma virus, 707
  inhibition of rabies virus replication
    by, 453, 454, 456
  stimulation of thymidine kinase ac-
    tivity in monkey kidney cells by,
    511
  toxicity of, for BHK/21 cells, 452